BIOLOGICAL and MEDICAL APPLICATIONS

SPORTS APPLICATIONS

ASTRONOMY and SPACE EXPLORATION

NATURAL PHENOMENA

INQUIRY INTO PHYSICS

SECOND EDITION

Vern J. Ostdiek
Benedictine College

Donald J. Bord
University of Michigan-Dearborn

WEST PUBLISHING COMPANY
St. Paul   New York   Los Angeles   San Francisco

Artwork: Rolin Graphics; Dennis McCarthy
Composition: Carlisle Communications, Ltd.
Copyediting: June Gomez
Dummy artist: Kristin Weber
Text Design: John Edeen
Cover Design: John Edeen
Cover Image: Rainbows over, Oksfjord, Norway. TSW-Click, Chicago.

Library of Congress Cataloging-in-Publication Data

Ostdiek, Vern J.
   Inquiry into physics / Vern Ostdiek, Donald J. Bord. -- 2nd ed.
      p.   cm.

   Includes bibliographical references and index.
   ISBN 0-314-79885-4
   1. Physics.   I. Bord, Donald J.   II. Title.
QC23.088   1991   90-22499
530--dc20   CIP

Contents Photos
Chapter 1 Chris Gilbert, San Francisco. **Chapter 2** NASA. **Chapter 4** Courtesy of IBM. **Chapter 5** Doug Vargas/The Image Works. **Chapter 6** Courtesy of Dornier Medical Systems. **Chapter 7** Courtesy of IBM. **Chapter 8** Courtesy of GE Medical Systems. **Chapter 9** Doug Johnson/ SPL/Photo Researchers. **Chapter 11** Lick Observatory Photograph. **Chapter 12** Courtesy of Fermilab.

Chapter Opening Photos
Chapter 1 Courtesy of Zentrum fur angewandte Raumfahrttechnologie und Mikrogravitacion, Bremen Germany. **Chapter 2: (u)** from "Neptune" by June Kinoshita, © 1989 by *Scientific American,* Inc., all rights reserved; **(ll)** NASA; **(lr)** NASA. **Chapter 3** AP/Wide World Photos. **Chapter 4** Helena Frost. **Chapter 5: (u)** Chris Gilbert; **(inset)** Courtesy of Southern California Edison Corporation. **Chapter 6** Courtesy of Dornier Medical Systems. **Chapter 7** Courtesy of IBM. **Chapter 8** Courtesy of GE Medical Services. **Chapter 9** Jim Grace/SPL/Photo Researchers. **Chapter 10** Courtesy of IBM. **Chapter 11** Religious News Service Photo. **Chapter 12** Courtesy of Fermilab.

Chapter 1 Photos
Page 5(u) NASA. **Page 10** Lauros-Giraudon/Art Resource. **Page 11** National Institute of Standards and Technology. **Page 17** National Institute of Standards and Technology. **Page 21** The Bettmann Archive. **Page 22** TSW-Click, Chicago. **Page 28** Chris Gilbert. **Page 37** *PSSC Physics,* 2nd edition, 1965; D.C. Health & Company and Educational Development Center. **Page 40** Historical Pictures Service. **Page 41** *PSSC Physics,* 2nd edition, 1965; D.C. Health & Company and Educational Development Center. **Page 42** The Granger Collection, New York. **Page 43** Art Resource.

(credits continued following index)

Dedicated to my mother,
for all she's done

Vern Ostdiek

To CLS, Toby, and Gherkin(?),
for all the joy you've
brought to my life

Don Bord

CONTENTS

CHAPTER 3 _____

ENERGY AND CONSERVATION LAWS   94

CHAPTER 4 _____

PHYSICS OF MATTER   148

CHAPTER 5 ⎯⎯⎯⎯⎯⎯⎯⎯⎯⎯⎯⎯⎯⎯⎯⎯⎯⎯⎯

TEMPERATURE AND HEAT 196

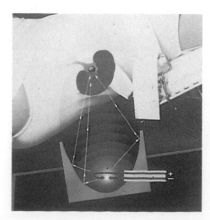

CHAPTER 6 ⎯⎯⎯⎯⎯⎯⎯⎯⎯⎯⎯⎯⎯⎯⎯⎯⎯⎯⎯

WAVES AND SOUND 248

CHAPTER 7 ——————————————————————————————

ELECTRICITY   294

CHAPTER 8 ——————————————————————————————

ELECTROMAGNETISM AND EM WAVES   332

CHAPTER 9

OPTICS   380

CHAPTER 10 _____
ATOMIC PHYSICS 434

CHAPTER 11 _____
NUCLEAR PHYSICS 478

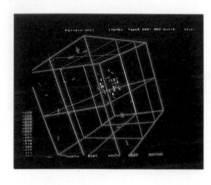

CHAPTER 12

SPECIAL RELATIVITY AND ELEMENTARY PARTICLES   518

APPENDIXES

PREFACE

Since the first edition of *Inquiry Into Physics* was published at the beginning of 1987, physics has been in the public eye a great deal. In 1987 alone, three of the cover stories in the weekly magazine *Time* dealt with physics. Two books principally about physics—*A Brief History of Time* by Stephen Hawking and *The Making of the Atomic Bomb* by Richard Rhodes—spent months on the *New York Times* best-seller list. The dramatic stories about new high-temperature superconductors and the possibility of cold fusion were front page news for weeks. Other normally esoteric physics terms such as supernova, spherical aberration, superconducting supercollider, quarks, magnetic resonance imaging, and electron tunneling were used frequently on the nightly news and in newspapers and news magazines. In ways that we could not have foretold in 1987, the events of the past four years have shown that the ancient field of study known as physics still is capable of yielding exciting discoveries that can capture the world's attention.

In revising *Inquiry Into Physics* we have tried to highlight what is new in the world of physics and to show how physics-related technologies that were new or unheard of a decade ago are now routinely used. The basic tenets of the first edition—that knowing only a few basic physical laws facilitates the understanding of a myriad of applications, and that the things and events around us often present the best examples of physics in action—are maintained. The second edition is still designed for use in a one semester course; it uses *simple* mathematics to show the important quantitative side of science, and it incorporates *Historical Notes* that describe the process of scientific discovery and *Physics Potpourris* that present closer looks at specific applications of physical principles to astronomy and to other areas of science.

The major changes in the second edition are:

New in the Second Edition

- A chapter on elementary particles (Chapter 12). Einstein's special theory of relativity has been moved from Chapter 11 into this chapter.

- *Prologues* at the beginnings of the chapters. These are designed to motivate the reader by describing something current in physics that makes use of principles that are investigated in the succeeding chapter.

- New material on magnetic resonance imaging, digital sound, curved mirrors, superconductivity, moment of inertia, photocopying, tunneling microscopes, and much more.

- Full-color illustrations in four of the chapters. This addition facilitates the use of better contrast in complex illustrations and the use of color photographs where color is clearly important (e.g., for rainbows).

- Over 150 new photographs and drawings, plus improved versions of dozens more.

- Over 90% more end-of-chapter Questions, plus dozens of new Problems.

- Quick-reference lists of applications and other items of interest (on the inside covers).

- New information on metric prefixes, metric units, and the Greek alphabet (in the Appendixes).
- The use of more conventional notation, as suggested by users of the first edition.

The book is divided into 12 relatively long chapters to minimize the number of breaks within each broad area of physics. The sequencing of topics is traditional, with each chapter incorporating the following features:

Chapter Features

- *Prologues* motivate the reader by highlighting current applications of physics principles. The topics include Fallturm Bremen, magnetic resonance imaging, and Supernova 1987A.
- *Examples* present worked problems based on the role of physics in everyday situations.
- *Do-It-Yourself Physics* experiments give students the chance to investigate physics themselves, without the need of special equipment.
- *Learning Checks* midway through each chapter test the student's comprehension of the basic material.
- *Physics Potpourris* provide interesting digressions on selected topics including meteors, telescope mirrors, holograms, electricity and life, and many others.
- *Historical Notes,* the final section of each chapter, present a brief look at the human side of physics by considering the lives and work of the men and women who devoted their energy to the development of the principles and laws discussed in the chapter.
- A *Summary* gives a review of the important points in the chapter.
- A *Summary of Important Equations* is included in most of the chapters.
- *Questions* are designed to check basic understanding of the material as well as the ability to extend that understanding to new or hypothetical situations.
- *Problems* provide additional realistic applications of physics using simple mathematics.
- *Challenges,* a collection of more advanced questions and problems, test the reader's understanding of physical concepts at a deeper level. Many are good starting points for class discussions.
- *Suggested Readings* describe nontechnical articles and books that supply more information about selected topics in the chapter.

At the back of the book you will find a table of *Conversion Factors and Metric Prefixes,* a *Periodic Table of the Elements,* a list of *Winners of the Nobel Prize in Physics,* a *Math Review, Answers* to Odd-Numbered Problems and Challenges (as well as to all of the Learning Checks), a list of the *Greek Alphabet,* and a *Glossary.*

There is more than enough material in the text for a typical one-semester course. However, about 30 of the more specialized sections can be omitted without affecting the material that follows. Those sections are the Historical Notes, and

Sections 2.8, 2.10, 3.6, 3.8, 4.6, 4.7, 5.2, 5.5, 5.6, 5.7, 6.4, 6.5, 6.6, 8.4, 8.6, 8.7, 9.2, 9.4, 9.6, 10.7, and 10.8.

Ancillaries

The *Instructor's Manual* accompanying *Inquiry Into Physics* contains: suggestions for selecting and ordering topics, relevant class demonstrations, discussion topics, and laboratory experiments; answers to Questions and even-numbered Problems and Challenges; and samples of examination questions. Also available is a computerized test bank, a set of transparency masters, a set of slides, a selection of videos for lectures, and the new booklet *Great Ideas For Teaching Physics*. Contact your local West sales representative for further information.

Acknowledgments

As with the first edition of the book, we received help from many talented individuals during the revision process. Their assistance was greatly appreciated. At Benedictine College, Scott Baird and Doug Brothers shared their expertise in physics and astronomy, as well as their libraries. Denny McCarthy breathed more life into the book with new drawings. Roger Siau supplied needed translations of correspondence. Mary Asher, Pat Asher, Beverly Cook, Bonnie Faudere, and Dave McGarry made the nuts-and-bolts part of the process much easier. At the University of Michigan-Dearborn, Paul Zitzewitz once again provided helpful advice about avoiding common pitfalls when writing physics for students and about new developments in particle physics. Many thanks to all of you.

We wish to thank the people at West Educational Publishing for managing the revision. Jerry Westby and Denis Ralling organized the extensive reviewing process and supplied many helpful suggestions. Tom Hilt and Nancy Roth handled the production process with skill and dedication that is greatly appreciated. Kristi Shuey and others in the Marketing group did an excellent job on the promotional materials. To them, and to the many other professionals behind the scenes at West, we say thank you.

Last, but certainly not least, we wish to thank the following reviewers for their careful analyses, criticisms, and suggestions. The perspectives of those who used the first edition were particularly useful in helping us decide how to improve the text.

Edward Borchardt
Mankato State University

Howard Bryant
University of New Mexico

Art Cary
California Polytechnic State University

Rory Coker
University of Texas at Austin

Robert Cole
University of Southern California

Nelson Duller
Texas A & M University

John Flinner
Mankato State University

Robert Gowdy
Virginia Commonwealth University

Kenneth Hahn
Northeast Missouri State University

Ghazi Hassoun
North Dakota State University

S.B. Joshi
York University

John Hubisz
Texas College of the Mainland

Gabor Kemeny
Michigan State University

Joon Lee
University of Southern Mississippi

Robert Packard
Baylor University

William Ploughe
Ohio State University

Grayson Rayborn
University of Southern Mississippi

Stan Shepherd
Pennsylvania State University

David B. Slavsky
Loyola University-Chicago

Paul Varlashkin
East Carolina University

Robert Whitaker
Southwest Missouri State University

Clarence Wolff
Western Kentucky University

OUTLINE

THE STUDY OF MOTION

PROLOGUE: FALLING FOR SCIENCE

One of the earliest subjects of speculation and investigation in the area of physics was the motion of a body as it falls. Aristotle, Galileo, Newton, and many other famous persons in physics studied this motion. When idealized by assuming that a falling body is not affected by the air through which it moves, the motion is relatively simple and can be described with basic high-school level mathematics. It is referred to as *free fall* because the body is assumed to be free of all influences except that of gravity. Virtually all beginning physics students learn about free fall. In real life things are affected by the air as they fall, so the simple model of free fall is limited in its application. Recently a unique research tool was completed that uses true free fall in a big way.

The *Fallturm "Bremen" (Drop Tower "Bremen")* houses a vertical tube some 110 meters tall (about the height of a 35-story building) and 3.5 meters (11.5 feet) in diameter. Almost all of the air is pumped out of the tube to create a near vacuum. Anything dropped from the top of the tower undergoes almost perfect free fall for about five seconds before reaching the bottom. Built and operated by the *Centre for Applied Space Technology and Microgravity* at the University of Bremen, Germany, the Drop Tower is used to conduct experiments in a *microgravity* (weightless, or zero g) environment. A large test chamber is hoisted to the top of the tower and then released. As it undergoes free fall, conditions inside the chamber are nearly the same as if gravity were somehow completely switched off. The chamber and everything in it fall together and consequently seem to be weightless, just as a basketball held in your hand would suddenly seem to have no weight if you walked off a diving board and fell toward the water below.

Of course the Drop Tower is not the only way to achieve a microgravity environment for research purposes. A spacecraft in orbit provides one, as does an airplane put through a maneuver much like a car driven too fast over the crest of a hill. (In each of these cases the body is continuously falling toward the earth with no supporting force.) So why build the Drop Tower when all you can conduct are experiments lasting less than 5 seconds? The entire cost of the installation is only a fraction of the cost to launch a manned spacecraft. Also, some experiments involving open flames and the combustion of gases—hazards to have in a spacecraft or aircraft—can be conducted in the Drop Tower chamber. Five seconds is long enough to study such processes and many others, including the formation and interactions of fluid droplets, the movement of liquids in capillary tubes, the way melted materials resolidify, and many more. Eventually the experiment time may be doubled to over 9 seconds. A catapult is to be installed at the bottom of the tower that will fling the chamber upward at just the right speed for it to rise to the top and then fall back down, like a ball thrown straight up into the air.

The physical principle exploited by the Drop Tower is simple, but the apparatus is high technology. The test chamber has its own on-board computer for operating experiments and storing data. Telemetry can be sent to the control room during a fall, and high-speed cameras can record what happens.

Perhaps you are curious about what happens at the end of a "drop." How do you stop a 300-kilogram (660-pound) chamber going about 45 meters/second (100 miles/hour)? A pit at the bottom is filled with polystyrene beads to a depth of about 8 meters. These beads are much like the beans in a bean-bag chair.

In this first chapter we establish the groundwork for the study of physics. We begin by introducing some of the key concepts that are used in the scientific analysis of our physcial world. Most of this chapter is an introduction to the branch of physics called mechanics—the study of motion and its causes. The principal topic of this chapter is motion and how it is described using the concepts of speed, velocity, and acceleration. Both examples and graphs are used to help illustrate and define these three concepts. An important example treated in detail is the motion of a body undergoing free fall. We conclude with a brief account of the work of two men who made important contributions to the development of modern mechanics.

1.1   INTRODUCTION

Why study physics? The answer is easy for those majoring in physics, engineering, or other sciences: physics is useful. The technology that our modern society relies on comes from applying the discoveries of physics and other sciences. From designing safe, efficient passenger jets to producing sophisticated, inexpensive personal computers, engineers apply physics everyday. Landing astronauts on the moon and returning them safely to earth, one of the greatest feats of this century, is a nice example of physics applied on many levels (see Figure 1.1). The machines involved—from powerful rockets to on-board computers—were designed, developed, and tested by people who knew a lot about physics. The planning of the orbits and the timing of the rocket firings to change orbits involved scientists with a keen understanding of basic physics, like gravity and the "laws of motion."

Often the payoff is not so tangible or immediate as a successful moon landing. Behind great technological advances are years or even decades of basic research into the properties of matter. For instance, take a simple solar-powered calculator (see Figure 1.2). Solar cells convert light into electrical energy, integrated circuit chips inside use this energy to perform computations, and a liquid crystal display presents the answer to the user. If you could take this marvelous device back in time, say, 50 years, you would totally astound the greatest physicists and electrical engineers of the day. But even at that time, scientists were studying the

FIGURE 1.1
It took a lot of physics to get to the moon.

properties of semiconductors (the raw material for solar cells and integrated circuit chips) and liquid crystals.

For you and others like you, taking perhaps just one physics course in your life, the usefulness of physics is probably not a big reason for studying it. We will see that with even one course, you can use physics to, for example, determine how large a raft has to be to support you or to determine whether using a toaster and a hair dryer at the same time will blow a fuse. But you are not going to make a living with your understanding of physics, nor will you be using it every day. So why should you study physics? There is enjoyment and satisfaction in having a better understanding of the physical world around you. The fact that you are continuing your education indicates that you are curious about things, that you want to learn more. We hope that all who study physics—those who will use it every day and those who take only one course—will take some delight in what they learn. Knowing how sunlight and raindrops combine to make a rainbow deepens one's appreciation of its beauty. Knowing about centripetal force helps one understand why ice on a curved road is dangerous. Knowing the basics of nuclear physics helps one understand the danger of radon gas and the promise of nuclear fusion. Knowing the principles behind stereo speakers, aircraft altimeters, refrigerators, lasers, microwave ovens, guitars, and Polaroid sunglasses gives you a better appreciation of what they do. Simply put, there is something rewarding about learning nature's ways and about learning how devices work.

We cannot promise you a great revelation on each page of this book. As with any field, the process of learning physics involves studying the fundamentals. But even this part of it shows you the order that exists in nature and how that order can be expressed in the form of basic laws.

FIGURE 1.2
One product of decades of research in physics.

Since physics is one of the basic sciences, it is important to have some idea of just what *science* is. In the most fundamental sense, science is the *process* of seeking and applying knowledge about our universe. Science also refers to the *body of knowledge* about the universe that has been amassed by the human race. Pursuing knowledge for its own sake is pure or basic science; developing ways to use this knowledge is applied science. Astronomy is mostly a pure science, while engineering fields are applied science. The material we present in this book is a combination of fundamental concepts that we believe are important to know for their own sake and examples of the many ways that some of these concepts are applied in the world around us.

So how does one "do" science? How did the human race come by this mountain of scientific knowledge that has been amassed over the ages? There exists a blueprint for scientific investigation that makes an interesting starting point for answering these questions. It is at best an oversimplification of how scientists operate. Perhaps we should regard it as a game plan that is quickly modified when the action starts. It is called the *scientific method.* One version of it goes something like this. Careful *observation* of a phenomenon induces an investigator to question its cause. A *hypothesis* is formed that purports to explain the observation. The scientist devises an *experiment* that will test this hypothesis, hoping to show that it is correct—at least in one case—or that it is incorrect. The outcome of the experiment often raises more questions that lead to a *modification* of the hypothesis and subsequent, further experimentation. Eventually an accepted hypothesis that has been verified by different experiments can be elevated to a *theory,* or a *law.* (The term that is used—theory or law—is not particularly important in physics. Newton's Second *Law* of Motion and Einstein's Special *Theory* of Relativity are held in roughly the same regard by physicists in terms of their validity and their importance).

One nice thing about the scientific method is that it is a logical procedure that is practiced by nearly everyone from time to time. Let's say that you get into your car and find that it won't start (observation). You speculate that maybe the battery is dead (hypothesis). To see if this is true you turn on the radio or the lights to see if they work (experiment). If they don't you may look for someone to give you a jump start. If they do, you probably guess something else, such as the starter, is causing the problem.

Clearly a good mechanic must be proficient at this. A doctor making a medical diagnosis uses similar procedures. But does the outline of the scientific method appear in some "how-to" manual for scientific discovery, or are scientists required to take an oath to follow it faithfully everyday at work? Hardly. But the individual elements of the methodology are essential tools of the scientist.

The scientific method is important, but it is not the whole story. Even a brief look at the history of physics reveals that there is no simple formula that scientists have followed to lead them to discovery. Throughout this book the "Historical Notes" sections give you some idea of the variety of ways that discoveries have been made.

It is appropriate to talk about the scientific method in this chapter because one of the architects of the way we do science was Galileo Galilei. He thought a great deal about how science should be done and applied his ideas to his study of motion. Galileo believed that science had to have a strong logical basis that included precise definitions of terms and a mathematical structure with which to express relationships. He introduced the use of controlled experiments, which he used with great success in his studies of how objects fall. By ingenious experimental design he overcame the limitations of the crude timing devices that existed and measured the acceleration of falling bodies. Galileo was a critical

FIGURE 1.3
Examples of the systems we will be examining.

observer of natural phenomena, from the swinging of a pendulum to the moons orbiting Jupiter. He was able to draw logical conclusions from what he saw and thereby remove much of the "magic" that seemed to govern the universe.

The goal of learning physics and any other science is to gain a better understanding of the universe and the things in it. To be practical, we generally consider only a small segment of the universe at one time, a segment we call a *system*. The size of a system is chosen so that its structure and the interactions within it are manageable. Some of the systems we will talk about are the nucleus of an atom, the atom itself, a collection of atoms inside a laser, a sound source in a room, a rock moving near the earth's surface, and the earth with satellites in orbit about it (see Figure 1.3).

The kinds of things a person might want to know about a system include (1) its structure or configuration, (2) what is going on in it and why, and (3) what will be happening in it in the future. The first step is easy—identifying the objects in the system. Protons, electrons, chromium atoms, a flute, a rock, and the moon are some of the things in the systems mentioned above. Often the items in a physical system are already familiar to us. A host of other intangible things in a system must also be identified and labeled before real physics can begin. We must define things like the *speed* of the rock, the *pressure* variations in air, the *energy* of the atoms in the laser tube, and the *angular momentum* of a satellite to understand what is going on in a system and where it is headed. These things and others like them we will call *physical quantities.* Most will be unfamiliar to you unless you have studied physics before. Together with the named objects, they form what could be called the vocabulary of physics. There are hundreds of physical quantities in regular use in the various fields of physics, but we will need only a fraction of these.

Physics seeks to discover the basic ways in which things interact. *Laws* and *principles* express quantitative relationships that exist between physical quantities. For example, the law of fluid pressure expresses how the pressure at some location in a fluid depends on the weight of the fluid above. This can be used to find the water pressure on a submerged submarine or the air pressure on a person's chest. These "rules" are used to understand the interactions in a system and to predict what will happen in the system in the future. The laws and principles themselves were formed after repeated, careful observations of countless systems by scientists throughout history. They have withstood the test of time and repeated experiment before being elevated to this status. You might regard physics as the continued search for and the application of basic rules that govern the interactions in the universe.

The process of learning physics has two main thrusts: (1) identifying or defining the physical quantities needed (establishing a vocabulary); and (2) learning the laws and principles that express the relationships among these physical quantities.

As parts of these tasks are completed in the various chapters, several examples will show you how your knowledge can be applied.

1.2   PHYSICAL QUANTITIES: THE VOCABULARY OF PHYSICS

Now let us take a closer look at these physical quantities. To be useful in physics, they must satisfy some conditions. A physical quantity must be unambiguous, its meaning clear and universally accepted. Understanding the meaning of a term involves more than just memorizing the words in its definition. Words can only describe something; to understand, you must go beyond words. For example, the simple definition "speed equals distance divided by time" does not really convey the conceptual understanding of speed that we have. Many physical quantities (speed, pressure, power, density, and others) can be defined by an equation. Mathematical statements tend to be more precise than literary ones, making the meanings of these terms clearer.

Observation yields *qualitative* information about a system. Measurement yields *quantitative* information, which is also needed in any science that strives for exactness. Consequently, physical quantities must be *measurable,* directly or indirectly. One must be able to assign a numerical value that represents the amount of this quantity that is present. It is easy to visualize a measurement of distance, area, or even speed, but other things are a bit more abstract: pressure, voltage, and power. However, each of these quantities can be measured in prescribed ways; they would be useless if this were not so.

The basic act of measuring is one of comparison. To measure the height of a person, for instance, one would compare the distance from the floor to the top of the person's head with some chosen standard length, such as a foot or a meter (Figure 1.4). The height of the person is the number of units 1-foot long (including fractions) that have to be put together to equal that distance. The *unit of measure* is the standard used in the measurement—the foot or meter in this case. A complete measurement of a physical quantity, then, consists of a number and a unit of measure. For example, a person's height might be expressed as

$$\text{height} = 5.75 \text{ feet}$$

or

$$h = 5.75 \text{ ft}$$

Here *h* represents the quantity (height) and *ft* the unit of measure (feet). The same height in meters is:

$$h = 1.75 \text{ m}$$

So, when we introduce a physical quantity into our physics vocabulary—another "tool," so to speak—we must specify more than just a *definition*. We should also give a mathematical definition (if possible), relate it to other familiar physical quantities, and include the units of measure that are used.

In the world today there are two separate systems of measure. The United States uses the *English system,* and the rest of the world, for the most part, uses the *metric system.* An attempt has been made in the United States to switch completely to the metric system, but so far it has not succeeded.

The metric system has been used by scientists for quite some time, and we will use it a great deal in this book. It is a convenient system to use because the different units for each physical quantity are related by powers of 10. For example, a kilometer equals 1,000 meters, and a millimeter equals 0.001 meter. The prefix itself designates the power of 10. *Kilo-* means 1,000, *centi-* means 0.01 or $\frac{1}{100}$, and *milli-* means 0.001 or $\frac{1}{1,000}$. A kilometer, then, is 1,000 meters. Table 1.1 illustrates the common metric prefixes. You may not know what an *ampere* is, but

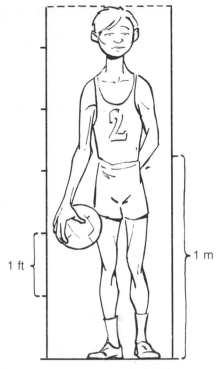

FIGURE 1.4
Measurement is an act of comparison. A person's height is measured by comparison with the length of a chosen standard. In this case, height equals 5 1-foot lengths plus a segment 0.75-feet long. The same person's height is also equal to 1 meter plus 0.75 meter.

1 ft ⊣   ⊢ 1 m

TABLE 1.1. COMMON METRIC PREFIXES AND THEIR EQUIVALENTS	
1 *centi*meter = 0.01 meter	1 meter = 100 centimeters
1 *milli*meter = 0.001 meter	1 meter = 1,000 millimeters
1 *kilo*meter = 1,000 meters	1 meter = 0.001 kilometer
EXAMPLES	
189 centimeters = 1.89 meters	72.39 meters = 7,239 centimeters
25 millimeters = 0.025 meter	0.24 meter = 240 millimeters
7.68 kilometers = 7,680 meters	23.4 meters = 0.0234 kilometer

The French Revolution, beginning with the storming of the Bastille on 14 July 1789, gave birth not only to a new republic but to a new system of weights and measures (see Figure 1.5). Eighteenth-century France's system of weights and measures had fallen into a chaotic state, with unit names that were confusing or superfluous and standards that differed from one region to another. Seizing upon the opportunity presented to them by the political and social turmoil accompanying the Revolution, scientists and merchants, under the leadership of Charles-Maurice Talleyrand, Bishop of Autun, presented a plan to the French National Assembly in 1790 to unify the system. The plan proposed two changes: (1) the establishment of a decimal system of measurement, and (2) the adoption of a "natural" scale of length. Neither of these two notions was new to scholars of this period. The first had been discussed as early as 1585 by Simon Stevin, a dike inspector in Holland, in a pamphlet called *La Disme* (ie, *The Tenth Part*). The second notion was introduced in 1670 by Abbé Gabriel Mouton, a Lyons vicar, who proposed that a standard of length be defined in terms of a fraction of the length of the meridian arc extending from the North Pole to the equator.

The plan was finally adopted into law on 7 April 1795. The new legislation defined the meter as the

FIGURE 1.5
The storming of the Bastille.

measure of a length equal to one 10 millionth of the meridian arc passing through Paris from the North Pole to the equator and the gram as the absolute weight of pure water contained in a cube one hundredth of a meter on a side at the temperature of melting ice. It also made this system obligatory in France.

The tasks of actually determining the sizes of these newly defined units were assigned to Messrs. Delambre and Mechain, who were to survey the length of the meridian arc through Paris, and to Messrs. Léfèvre-Gineau and Fabbroni, who were to determine the specific gravity (see Chapter 4) of water. As it turned out, these measurements were not made without difficulty and, sometimes, danger. For example, during the period between 1792 and 1798, Delambre and Mechain made measurements along the meridian between Dunkirk, France and Barcelona, Spain, amid the riot and turmoil still present in many parts of Europe. They were frequently arrested as spies, often had their equipment confiscated, and were generally harassed at every turn. Finally in 1798, with the job done and the length of the centimeter accurately known, Léfèvre-Gineau and Fabbroni set about their work. They, too, encountered difficulties, largely having to do with reaching and maintaining the required measurement temperature, but they completed their task in 1 year.

Beginning in 1798, an international committee including representatives of nine nations undertook to carry out the calculations required to produce the standards needed to define the new system of weights and measures. It submitted its report to the French legislature for ratification on 27 June 1799, and the bill passed on 10 December 1799. That document is the first official text in which the metric system is mentioned. According to this law, the definitive standards of length and mass to be used in commercial and scientific interactions throughout France were "the meter and kilogram of platinum deposited with the legislative body" (see Figure 1.6).

Since then the definitions of the standards of length and mass have undergone several revisions, and other units of measure have been incorporated into the scheme. Nevertheless the basic tenets of the metric system—simplicity and convenience stemming from its use of a decimal system of measure and uniformity and reproducibility deriving from its basis on a set of standards—survive. Since its adoption in Europe, first in France, then in Holland (1816) and Greece (1836), its use has so spread that no civilized nation is without knowledge of it. Without a doubt, the metric system has become, in the motto adopted by its founders, a system "for all people, for all time."

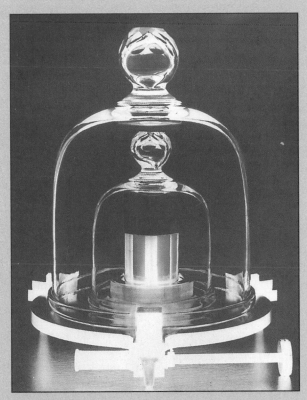

FIGURE 1.6
A standard kilogram, in its protective bell jar, is kept in a vault at the National Institute of Standards and Technology (formerly the National Bureau of Standards).

you should see immediately that a *milliampere* is 1 one–thousandth of an ampere.

Having to use two systems of units is like living near the border between two countries and having to deal with two systems of currency. Most of us have a better feel for the size of English system units (like feet, miles per hour, and pounds) than for metric system units (like meters, kilometers per hour, and newtons). Often the examples in this book will use units from both systems, so that you can compare them and develop a feel for the sizes of the metric units. A table relating the units in the two systems is included in Appendix A. Fortunately we won't have to deal with two systems of units after we reach electricity (Chapter 7).

Now we can get down to business. In physics there are three basic things that we must first talk about and quantify in various ways: *space, time,* and *matter.* All of the physical quantities used in this textbook will involve measurements (or combinations of measurements) of space, time, and the properties of matter. The units of measure of all of these quantities can be traced back to the units of measure of distance, time, and two properties of matter called mass and charge. We will not deal with charge until Chapter 7.

Distance, time, and mass are such basic concepts that it is difficult to define them, particularly time. Distance represents a measure of space in one dimension. Length, width, and height are examples of distance measurements. The following table lists the common distance units of measure and their abbreviations. This same format will be used for all physical quantities that have several common units.

Physical Quantity	Metric Units	English Units
Distance d (or l, w, h)	meter (m)	foot (ft)
	millimeter (mm)	inch (in)
	kilometer (km)	mile (mi)
	centimeter (cm)	

—19 mm—

—0.000019 km—

FIGURE 1.7
The two measurements represent the same distance, but the one in millimeters is more convenient to use and to visualize.

Why are there so many different units in each system? Generally it is easier to use a unit that fits the scale of the system being considered. The meter is good for measuring the size of a house, the millimeter for measuring the size of a coin, and the kilometer for measuring the distance between cities. Eighty kilometers is the same as 80,000 meters, but the former measurement is a more "manageable" number to most people. Similarly it might be correct to say that a coin is 0.000019 kilometers in diameter, but 19 millimeters is a more convenient measure (see Figure 1.7).

The sizes of all of the distance units, including the English units, are defined in relation to the meter. In this book we will use meters most often in our examples. You should try to get used to distance measurements expressed in meters and have an idea, for example, of how long 25 meters or 0.2 meters are. Table 1.2 shows some representative distances expressed in metric and in English units.[1]

---

[1] We will use scientific notation occasionally. See Appendix D for a review.

It is actually a rather simple matter to convert a distance expressed in, say, meters to a distance expressed in another unit. For example, you might solve a problem and find that the answer is "The sailboat travels 23 meters in 10 seconds." Just how far is 23 meters? Think of any numerical measurement as a number multiplied by a unit of measure. In this case, 23 meters equals 23 *times* 1 meter. Then the 1 meter can be replaced by the corresponding number of feet—a conversion factor found in Appendix A.

$$23 \text{ meters} = 23 \times 1 \text{ meter}$$

But, 1 meter = 3.28 feet:

$$23 \text{ meters} = 23 \times 1 \text{ meter} = 23 \times 3.28 \text{ feet}$$

$$\textit{23 meters} = \textit{75.4 feet}$$

Two other physical quantities that are closely related to distance are *area* and *volume*. Area commonly refers to the size of a surface, such as the floor in a room or the outer skin of a basketball. The concept of area can apply to surfaces that are not flat and to "empty," two-dimensional spaces such as holes and open windows (see Figure 1.8). Area is a much more general idea than "length times width," an equation you may have learned in another course that applies only to rectangles. The area of something is just the number of squares 1 inch by 1 inch (or 1 meter by 1 meter, or 1 mile by 1 mile, etc) that would have to be put together to cover it.

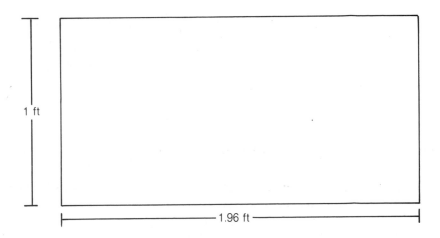

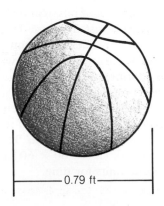

FIGURE 1.8
Every surface, whether it is flat or curved, has an area. The area of this rectangle and the area of the surface of the basketball are equal.

TABLE 1.2 SOME REPRESENTATIVE SIZES AND DISTANCES		
Size/Distance	Metric	English
Size of a nucleus	1×10^{-14} meters	4×10^{-13} inches
Size of an atom	1×10^{-10} m	4×10^{-9} in
Size of a red blood cell	8×10^{-6} m	3×10^{-4} in
Height of a person	1.75 m	5.75 feet
Tallest building	443 m	1,454 ft
Diameter of Earth	1.27×10^{7} m	7,920 miles
Earth-Sun distance	1.5×10^{11} m	9.3×10^{7} mi
Size of our galaxy	9×10^{20} m	6×10^{17} mi

In a similar manner, the volume of a solid is the number of cubes 1 inch or 1 centimeter on a side needed to fill the space it occupies. For simple geometric shapes such as a rectangular box or a sphere, there are equations to compute the volume. We use volume measures nearly every time we buy food in a supermarket.

	Metric	English
Area (A)	square meter (m^2)	square foot (ft^2)
	square centimeter (cm^2)	square inch (in^2)
	square kilometer (km^2)	square mile (mi^2)
	hectare	acre
Volume (V)	cubic meter (m^3)	cubic foot (ft^3)
	cubic centimeter (cm^3 or cc)	cubic inch (in^3)
	liter (L)	quart, pint, cup
	milliliter (mL)	teaspoon, tablespoon

Area and volume are examples of physical quantities that are based on other physical quantities, in this case just distance. Their units of measure are called *derived units* because they are derived from more basic units (1 square meter = 1 meter $\times$ 1 meter = 1 meter^2). Until we reach Chapter 7, on electricity, all of the physical quantities will have units that are derived from units of distance, time, mass, or a combination of these.

The measure of *time* is based on periodic phenomena—processes that repeat over and over at a regular rate. The rotation of the earth was originally used to establish the universal unit of time, the *second*. The time it takes for one rotation was set equal to 86,400 seconds (24 $\times$ 60 $\times$ 60). Both the metric system and the English system use the same units for time.

	Metric	English
Time (t)	second (s)	second (s)
	minute (min)	minute (min)
	hour (h)	hour (h)

Clocks measure time by using some process that is repetitive. Many mechanical clocks use a swinging pendulum. The time it takes to swing back and forth is always the same. This is used to control the speed of a mechanism that turns the hands on the clock face (see Figure 1.9). Mechanical wristwatches and stopwatches use an oscillating balance wheel for the same purpose. Quartz electric clocks and digital watches use regular vibrations of an electrically stimulated crystal made out of quartz.

The first step in designing a clock is to determine exactly how much time it takes for one cycle of the oscillation. If it were two seconds for a pendulum, for example, the clock would then be designed so that the second hand would rotate once during 30 oscillations. The *period* of oscillation in that case is 2 seconds.

FIGURE 1.9
The regular swinging of the pendulum—by its repetitive motion—controls the speed of this clock.

> **PERIOD**   The time for one complete cycle of a periodic process. It is abbreviated "T," and the units are seconds, minutes, etc.

There is another way to look at this. The clock designer must determine how many cycles must take place before 1 second elapses (or 1 minute or 1 hour). In the example, one half of a cycle takes place during each second. This is called the *frequency* of the oscillation.

> **FREQUENCY**   The number of cycles of a periodic process that occur per unit time. It is abbreviated "f."

The standard unit of frequency is the hertz (Hz), which equals one cycle per second.

$$1\,\text{Hz} = 1/\text{s} = 1\,\text{s}^{-1}$$

The frequency of AM radio stations is expressed in kilohertz (kHz) and those of FM stations in megahertz (MHz). "Mega" is the metric prefix signifying 1 million. So, 102.5 MHz equals 102,500,000 Hz.

The relationship between the period of a cyclic phenomenon and the frequency is simple: the period equals 1 divided by the frequency, and vice versa.

$$\text{period} = \frac{1}{\text{frequency}}$$

$$T = \frac{1}{f}$$

While the notion of time may be hard to define, the process of timekeeping is relatively straightforward, requiring only a system for accurately counting the cycles of regularly occurring events. The operative word here is "regularly": to keep time or to measure a time interval, we need something that cycles or swings or oscillates at a constant rate. Up until quite recently, the fundamental clock used to tell time was the earth-sun system. It has the desirable characteristics of being generally available to all persons interested in keeping time, of being highly reliable and not likely to stop, as many conventional timepieces, and of being reasonably stable and not prone to significant (at least by most human standards) variations in its rate of cycling, that is, its rotation. The fact that the rate of rotation of the earth is not *strictly* constant is of importance to many scientists and engineers, and so it has been necessary within the last three decades to establish a better, more stable unit of time than that based on the spin of the earth.

Up until 1956, the fundamental unit of time, the second, was defined as $\frac{1}{86,400}$ of a *mean solar day*. A mean solar day is the average amount of time between successive crossings of your local meridian by the sun. This is to be distinguished from an *apparent solar day,* which is the actual interval of time separating any two successive meridian passages by the sun. Apparent solar time is the time kept by sundials, and the apparent solar day is not at all uniform, differing in length by up to 16 minutes from the mean solar day.

Despite being more uniform than the apparent solar day, a mean solar day also varies in length over long time scales, because of a gradual slow-down in the earth's rotation rate brought about by friction between its oceans and its solid surfaces. In addition, there are short-term variations in the rate of spin of the earth, some of which are seasonal in nature, which alter the length of a mean solar day. For these reasons, in the late 1950s, a still more uniform cycle within the earth-sun system upon which to base the unit of time was sought. Consequently, the second was redefined as 1/31,556,925.9747 of the length of the year beginning in January, 1900. This so-called "ephemeris second" is based on the motion of the earth around the sun, which is governed (as we shall see in Chapter 2) by Newton's laws of motion. As a result, its evaluation requires astronomical observation and calculation, and is not easily obtained with high accuracy except after many years of careful work. Thus, although ephemeris time appears to be extremely uniform, it suffers from not being able to be found quickly and accurately.

In 1967 a new *atomic second* was defined. In this case, 1 second equals the interval of time containing 9,192,631,770 oscillations of light waves given off by undisturbed cesium atoms (see Chapter 10 for a discussion of atoms and their properties). With modern cesium atomic clocks (see Figure 1.10), it is possible to establish the length of a second in less than 1 minute, with an accuracy of a few billionths of a second.

Time measurement is so accurate that the length of the standard meter is set by how long it takes light to travel that distance—3.33564095 billionths of a second.

Also:

$$f = \frac{1}{T}$$

EXAMPLE 1.1   A mechanical stopwatch uses a balance wheel that rotates back and forth 10 times in 2 seconds. What is the frequency of the balance wheel?

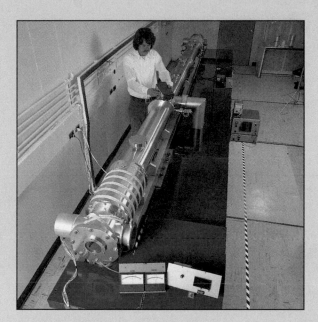

FIGURE 1.10
This cesium atomic clock, at the National Institute of Standards and Technology in Boulder, Colorado, is accurate to within about 3 millionths of a second per year.

The introduction of atomic time leaves us with a problem. Our daily lives are generally governed by day-night cycles, that is, by the earth-sun clock, not by phenomena associated with cesium atoms. As the earth continues to spin down, the length of a mean solar day (and hence the mean solar second) will continue to grow larger relative to the atomic

second. As time goes on, earth-sun clocks will gradually fall further and further behind atomic clocks. This is clearly not a desirable situation.

In the early 1970s, the French Bureau International de l'Heure, the world's official timekeepers, introduced Coordinated Universal Time (UTC). Under this system the length or duration of the second is dictated by atomic time, but the time commonly reported by time services is required to remain within 0.9 seconds of mean solar time (corrected for some slight variations caused by changes in the location of the earth's poles). In this system a "leap second" is added or subtracted as needed to compensate for changes in the rate of the earth's rotation relative to atomic time. Generally leap seconds are added during the last minute of the last day in June or December, and leap seconds are subtracted during the last minute of the last day in March or September.

If all this discussion seems a bit complicated to you, that's understandable. The topic of time and its determination is one with which physicists and philosophers have struggled for centuries. The point to be emphasized is that as more sophisticated experiments have been devised to probe nature's secrets, the need for ever more reliable and stable clocks has grown. The development of atomic clocks is the latest attempt to meet the requirements of scientists and engineers in this regard, and their use has now spilled over into our everyday lives as a result of the adoption of UTC. One thing remains, however. Regardless of how it is measured, *tempus fugit*, "time flies."

$$\text{frequency} = \text{Number of cycles per time}$$

$$f = \frac{10 \text{ cycles}}{2 \, s}$$

$$f = 5 \, Hz$$

What is the period of the balance wheel?

$$\text{period} = \text{Time for one cycle}$$

period = 1 divided by the frequency

$$T = \frac{1}{5\,Hz}$$

$$T = 0.2\,s$$

The balance wheel oscillates 300 times each minute.

The third basic physical quantity is mass. The *mass* of an object is basically a measure of how much matter is in it. (This statement illustrates the sort of circular definition that arises when one tries to define fundamental concepts.) We know intuitively that a large body, such as a locomotive, has a large mass because it is composed of a great deal of material. Mass is also a measure of what we sometimes refer to in everyday speech as inertia. The larger the mass of an object, the greater its inertia and the more difficult it is to speed up or slow down.

	Metric	English
Mass (m)	kilogram (kg)	slug
	gram (g)	

Mass is not in common use in the English system; note the unfamiliar unit, the slug. Weight, a quantity that is related to mass but is *not* the same thing, is used instead. We might contrast the two ideas as follows: when you try to lift a shopping cart, you are experiencing its weight. When you try to speed it up or slow it down, you are experiencing its mass (see Figure 1.11). We will take another look at mass and weight in Chapter 2.

FIGURE 1.11
(a) When lifting something, you must overcome its weight. (b) When changing its speed, you experience its mass.

1.3   SPEED AND VELOCITY

The key concept to use when quantifying motion is, of course, *speed.*

> **SPEED**   Rate of movement. Rate of change of distance from a reference point. The distance something travels divided by the time elapsed.

	Metric	English
Speed (v)	meter per second (m/s)	foot per second (ft/s)
	kilometer per hour (km/h)	miles per hour (mph)

A couple of aspects of speed are quite important in physics and need to be mentioned. First, speed is *relative.* A person running on the deck of a ship cruising at 20 miles/hour might have a speed of 8 miles/hour relative to the ship, but the speed relative to the water would be 28 miles/hour (if headed toward the front of the ship). If we use the ship as the reference point, the speed is 8 miles/hour. With the water or another ship at rest in the water as the reference point, the speed is 28 miles/hour (Figure 1.12). If you are traveling at 55 miles/hour on a highway and a car passes you going 60 miles/hour, its speed is 5 miles/hour relative to your car. Most of the time speed is measured relative to the surface of the earth; this will be the case in this book, unless stated otherwise.

Second, we must distinguish between *instantaneous speed* and *average speed.* A car's speedometer registers the instantaneous speed, the speed at that instant in time. You might measure the instantaneous speed of an object by measuring how much time it takes for it to travel some short distance. Then:

$$\text{speed} = \frac{\text{short distance}}{\text{short time}}$$

$$v = \frac{\Delta d}{\Delta t} \quad \text{(instantaneous speed)}$$

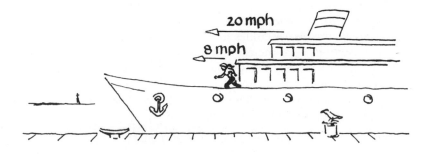

FIGURE 1.12
Speed is relative. The speed of a person running on a ship is 8 miles/hour relative to the deck. If the ship's speed is 20 miles/hour, the person's speed relative to the water is either 28 miles/hour (if headed toward the front of the ship) or 12 miles/hour (if headed toward the rear).

(The symbol Δ is the Greek letter delta and signifies a "small amount of" or a "small change in." We will use it often.) The distance must be small enough that the speed does not change significantly during the measurement. The concept of instantaneous speed is what is important more so than how it might be measured.

The *average speed* of an object is the total distance it travels divided by the time it takes to complete the trip.

$$\text{speed} = \frac{\text{total distance}}{\text{total time}}$$

$$v = \frac{d}{t}$$

The speed of a car being driven around in a city changes quite often. The speed is zero at stop signs and red lights, and negative when the car backs up (see Figure 1.13). During a trip across town, the instantaneous speed of a car may vary from − 10 miles/hour to 45 miles/hour. The average speed is the total distance the car travels divided by the total time, maybe 20 miles/hour.

When the speed of an object is constant, the average speed and the instantaneous speed are the same. In this case we can express the relationship between the distance traveled and the time that has elapsed as follows:

$$d = vt \quad \text{(constant speed)}$$

This is an example of what is called a proportionality: we say that *d is proportional to t* (abbreviated $d \propto t$). If the time is doubled, the distance is doubled. The constant speed v is called the *constant of proportionality*. We will encounter many examples in which one physical quantity is proportional to another.

DO-IT-YOURSELF PHYSICS

Look up the record times for Olympic racing events in running, swimming, ice skating, or bicycling. (These can be found in an encyclopedia, a world almanac, or a book of world records.) For several events, calculate the average speed and notice how it is lower for longer races. Why is this so? For comparison, compute the average speeds for the same distance in two different events, such as 1,500 meters running and swimming.

Most Americans are accustomed to measuring speed in miles per hour. This is most convenient when talking about travel times for distances greater than a few miles. Often it is more enlightening to use feet per second. For example, a car going 65 miles/hour travels 130 miles in 2 hours. But for a potential accident, it is relevant to consider how far the car will travel in a matter of seconds. Sixty-five miles per hour equals 95.6 feet per second. If a driver takes 2 seconds to decide how to avoid an accident, the car will have traveled 191 feet.

Try to develop a feel for the different speed units, particularly meters per second. You might keep in mind this comparison:

$$65 \text{ mph} = 95.6 \text{ ft/s} = 29.1 \text{ m/s}$$

FIGURE 1.13
In traffic, a car's instantaneous speed changes often.

This conversion works the same as before:

$$65 \text{ mph} = 65 \times 1 \text{ mph}$$

1 mph = 0.447 m/s (from Appendix A), so

$$65 \text{ mph} = 65 \times 1 \text{ mph} = 65 \times 0.447 \text{ m/s}$$

$$65 \text{ } mph = 29.1 \text{ } m/s$$

Apparently the universe was created with an absolute speed limit: the speed of light in empty space. This speed is represented by the letter *c*. Nothing has ever been observed traveling faster than *c*. The value of *c* and some other speeds are included in Table 1.3.

DO-IT-YOURSELF PHYSICS

When lightning strikes (see Figure 1.14), the flash of light reaches us in a fraction of a second, but the sound (thunder) is delayed. That is because the speed of sound is much slower than the speed of light. You can use the time delay to estimate the distance to the lightning. The next time you see a lightning flash, time how long it takes before you hear the thunder. The distance to the strike point is 340 meters for each second of time. Put another way, the distance is about one fifth of a mile for each second. (If you are more than a mile away, you may not be able to hear the thunder.)

TABLE 1.3 SOME SPEEDS OF INTEREST		
Description	Metric	English
Speed of light, c (in vacuum)	3×10^8 m/s	186,000 miles/second
Speed of sound (air, room temperature)	344 m/s	771 mph
Highest instantaneous speeds:		
Running (cheetah)	28 m/s	63 mph
Swimming (sailfish)	30.4 m/s	68 mph
Flying—level (merganser)	36 m/s	80 mph
Flying—dive (peregrine falcon)	97 m/s	217 mph
Humans:		
Swimming	2.24 m/s	5.0 mph
Running	11.8 m/s	26.3 mph
Ice skating	13.8 m/s	30.9 mph
Bicycling (recumbent bike)	29 m/s	65 mph

An important aspect of motion is *direction*. We will see that changing the direction of motion of a moving body is in a way equivalent to changing the speed itself. *Velocity* is a physical quantity that incorporates both ideas.

VELOCITY   Speed in a particular direction (same units as speed)

The speed of a ship might be 10 meters/second, while its velocity might be 10 meters/second east. Whenever a moving body changes direction, such as a car going around a curve or someone walking around a corner, the velocity changes

FIGURE 1.14
A distant lightning strike is a good demonstration of the difference between the speed of sound and the speed of light.

even if the speed does not. Being told that an airplane flies for 2 hours from a certain place and averages 100 mph does not tell you enough to determine where it is. Knowing that it travels in a straight line due north would allow you to pinpoint its location.

Velocity is an example of a physical quantity that is called a *vector.* Vectors have both a numerical size (magnitude) and a direction associated with them. Quantities that do not have a direction are called *scalars.* Speed by itself is a scalar; only when the direction of motion is included do we have the vector, velocity. In a similar way, we can define the vector *displacement* as distance in a specific direction. For the airplane referred to earlier, the *distance* it travels in 2 hours is 200 miles. Its actual location can be determined only from its *displacement*—200 miles due north, for example.

We can classify all physical quantities as scalars or vectors. Time, mass, area, and volume are all scalars because there is no direction associated with them.

Vectors are represented by arrows in drawings, the length of the arrow being proportional to the size of the vector (see Figure 1.15). Often it will be obvious in a discussion of some physical system that the motion is taking place in a straight line and the direction is not important. An example would be the motion of an object after it is dropped. In such cases, the words speed and velocity are often used interchangeably.

FIGURE 1.15
Vector quantities can be represented by arrows. Arrow length indicates the vector's size, and arrow direction shows the vector's direction. In both figures velocity is represented with an arrow. The car's velocity (and speed) is much greater than the pedestrian's, so the arrow representing the car's velocity is longer.

Vector Addition

Sometimes a moving body has two velocities at the same time. The runner on the deck of the ship in Figure 1.12 has a velocity relative to the ship and a velocity because the ship itself is moving. A bird flying on a windy day has a velocity relative to the air and a velocity because the air is moving relative to the ground. The velocity of the runner relative to the water or that of the bird relative to the ground is found by adding the two velocities together to give the *net,* or *resultant, velocity.* Let us consider how two velocities (or two vectors of any kind) are combined in the process known as *vector addition.*

When adding two velocities, you represent each as an arrow with its length proportional to the magnitude of the velocity—the speed. For the runner on the ship, the arrow representing the ship's velocity is $2\frac{1}{2}$ times longer than the arrow representing the runner's velocity because the two speeds are 20 miles/hour and 8 miles/hour ($8 \times 2\frac{1}{2} = 20$). Each arrow can be moved around for convenience provided *its length and its direction are not altered.* (Any such change would make it a different vector). The procedure for adding two vectors is:

> Two vectors are added by representing them as arrows and then positioning one arrow so its tip is at the tail of the other. An arrow drawn from the *tail of the first arrow* to the *tip of the second* is the arrow representing the *resultant vector*—the sum of the two vectors.

Figure 1.16 shows this for the runner on the deck of the ship. In (a) the runner is running forward so the two arrows are parallel. When the arrows are positioned "tip to tail," the resultant velocity vector is parallel to the others and its magnitude—the speed—is 28 miles/hour (8 miles/hour + 20 miles/hour). In (b) the runner is running toward the rear of the ship, so the arrows are in opposite directions. The resultant velocity is parallel to the ship's velocity, and its magnitude is 12 miles/hour (20 miles/hour − 8 miles/hour).

FIGURE 1.16
(a) The resultant velocity of a runner on the deck of a ship is found by adding the runner's velocity and the ship's velocity. The result is 28 miles/hour forward. (b) Using the same procedure when the runner is headed toward the rear of the ship, the resultant velocity is 12 miles/hour forward.

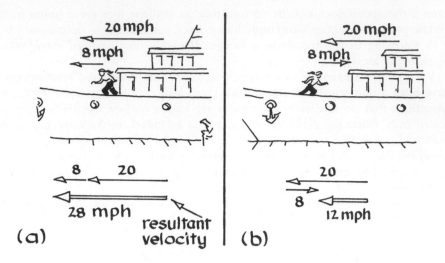

Vector addition is done the same way when the two vectors are not along the same line. Figure 1.17 shows a bird with velocity 8 meters/second north in the air while the air itself has velocity 6 meters/second east. The bird's velocity observed by someone on the ground, (b), is the sum of these two velocities. This we determine by placing the two arrows representing the velocities tip to tail as before and drawing an arrow from the tail of the first to the tip of the second (Figure 1.17c and d). The direction of the resultant velocity is toward the northeast. (Watch for this when you see a bird flying on a windy day: often the direction the bird is moving is not the same as the direction its body is pointed.)

What about the magnitude of the resultant velocity? It is not simply 8 + 6 or 8 − 6 because the two velocities are not parallel. With the numbers chosen for this example, the magnitude of the resultant velocity—the bird's speed—is 10 meters/second. If you draw the two original arrows with correct relative lengths and then measure the length of the resultant arrow, it will be 5/4 times the length of the arrow representing the 8 meters/second vector. Then 8 times 5/4 equals 10 meters/second.[2]

Vector addition is performed in the same manner, no matter what the directions of the vectors. Figure 1.18 shows two other examples of a bird flying with different wind directions. The magnitudes of the resultants are best determined by measuring the lengths of the arrows.

---

[2]The Pythagorean theorem can be used to calculate the magnitude of the resultant vector. The arrows in Figure 1.17 c form a right triangle. For any right triangle:

$$c^2 = a^2 + b^2$$

$$c^2 = a^2 + b^2$$

In this case, $a = 8$ and $b = 6$. Therefore: $c^2 = a^2 + b^2 = 8^2 + 6^2$
$$c^2 = 64 + 36 = 100$$
$$c = \sqrt{100} = 10$$

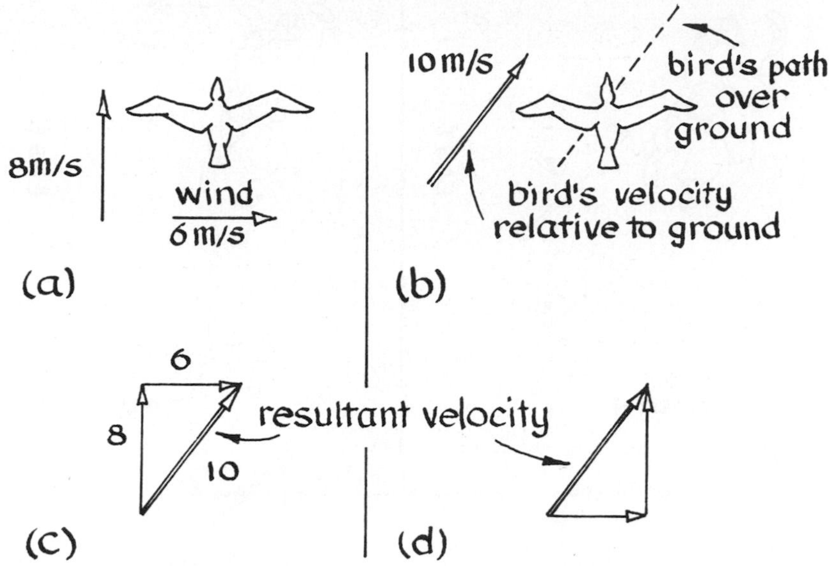

8m/s **wind** **6m/s**

(a)

10 m/s **bird's path over ground**

bird's velocity relative to ground

(b)

6 **8** **10** resultant velocity

(c)

(d)

FIGURE 1.17
The velocity of a bird relative to the ground (b) is the vector sum of its velocity relative to the air and the velocity of the air (wind). (c) and (d) show that the vectors can be added two different ways, but the resultant is the same vector.

There are many other situations in which a body's velocity is the sum of two (or more) velocities (eg, a swimmer or boat crossing a river).

The process of vector addition can be "turned around." Any vector can be thought of as the sum of two other vectors, called *components* of the vector. When we observe the bird's one velocity in Figure 1.17b, we would likely realize that the bird has two velocities that have been added. Even when a moving body only has one "true" velocity it may be convenient to think of it as two velocities that have been added together. For example, a soccer player running southeast across a field can be thought of as going south with one velocity and east with another velocity at the same time (Figure 1.19). A car going down a long hill has one velocity component that is horizontal and another that is vertical (downward).

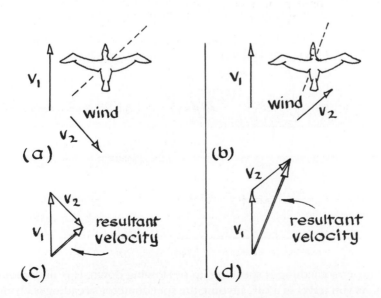

FIGURE 1.18
Other examples of vector addition. The bird has the same speed and direction in the air, but the wind direction is different.

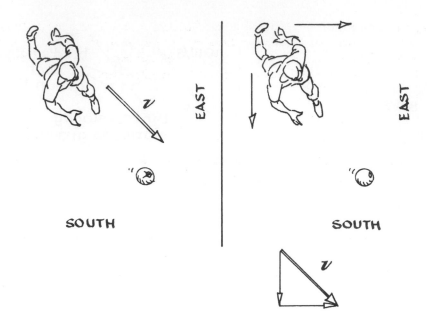

FIGURE 1.19
A soccer player running southeast can be thought of as having a velocity toward the east and a velocity toward the south at the same time. When these two velocities, called components, *are added together, the resultant is the original velocity.*

1.4   ACCELERATION

The physical world around us is filled with motion. But think about this for a moment: it is rare that something in motion doesn't change that motion. Cars, bicycles, pedestrians, airplanes, trains, and other vehicles all change their speed or direction often. They start, stop, speed up, slow down, and make turns. The velocity of the wind usually changes from day to day, or even from hour to hour. Even the earth as it moves around the sun is constantly changing its direction of motion and its speed, though not by much. The main thrust of Chapter 2 is to show how the change in velocity of an object is related to the force acting on it. For these reasons, a very important concept in physics is *acceleration*.

ACCELERATION   Rate of change of velocity. The change in velocity divided by the time elapsed.

$$a = \frac{\Delta v}{\Delta t}$$

	Metric	English
Acceleration (a)	meter per second^2 (m/s^2)	foot per second^2 (ft/s^2)
		mph per second (mph/s)

Whenever something is speeding up or slowing down, it is undergoing acceleration. As you travel in a car, anytime the speedometer's reading is changing, the car is accelerating. Acceleration is a vector quantity; it has both magnitude and direction. Note that the relationship between acceleration and velocity is the same as the relationship between velocity and displacement. Acceleration indicates how rapidly velocity is changing, while velocity indicates how rapidly displacement is changing.

A car accelerates from 20 to 25 meters/second in 4 seconds as it passes a truck (Figure 1.20). What is its acceleration?   EXAMPLE 1.2

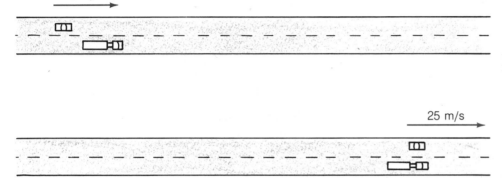

FIGURE 1.20
While passing a truck, a car increases its speed from 20 to 25 meters/second.

Since the direction of motion is constant, the change in velocity is just the change in speed—the later speed minus the earlier speed.

$$a = \frac{\Delta v}{\Delta t} = \frac{\text{final speed} - \text{initial speed}}{\Delta t}$$

$$a = \frac{\Delta v}{\Delta t} = \frac{25 \text{ m/s} \quad 20 \text{ m/s}}{4 \text{ s}}$$

$$a = \frac{5 \text{ m/s}}{4 \text{ s}}$$

$$a = 1.25 \text{ m/s}^2$$

This means that the car's speed increases 1.25 meters/second during each second.

When something slows down (when its speed decreases), we say that it *decelerates*. Deceleration is the same as negative acceleration.

After a race a runner may take 5 seconds to come to a stop from a speed of 9   EXAMPLE 1.3
meters/second. The (negative) acceleration, or deceleration, is

$$a = \frac{\Delta v}{\Delta t} = \frac{0 \text{ m/s} - 9 \text{ m/s}}{5 \text{ s}}$$

$$a = \frac{-9 \text{ m/s}}{5 \text{ s}}$$

$$a = -1.8 \text{ m/s}^2$$

Perhaps the most important example of accelerated motion is that of an object falling freely near the earth's surface. By an object falling freely, we mean that only the force of gravity is acting on it, and we can ignore things like air resistance. A rock falling a few meters would satisfy this condition, but a feather would not (see Figure 1.21).

Freely falling bodies move with a constant downward acceleration. This acceleration is represented by the letter *g*.

$$g = 9.8 \text{ m/s}^2 \qquad (\text{acceleration of gravity})$$

$$= 32 \text{ ft/s}^2 = 22 \text{ mph/s}$$

The downward speed of a falling rock increases 22 miles/hour each second that it falls. The letter *g* is often used as a unit of measure of acceleration. An acceleration of 19.6 m/s² equals 2 *g*. (Some representative accelerations are given in Table 1.4.)

FIGURE 1.21
A freely falling body has constant acceleration.

DO-IT-YOURSELF PHYSICS

It's easy to compute the acceleration of a car (in *g*'s) as it speeds up or slows down. Simply time how long it takes for the speed to change 22 miles/hour. (Do this when you are a passenger and can see the speedometer.) The acceleration in *g*'s is 1 divided by the number of seconds. For example, if it takes four seconds to go from 20 mph to 42 miles/hour, the acceleration is one-fourth *g*.

TABLE 1.4 SOME ACCELERATIONS OF INTEREST		
Description	Acceleration	
Freely falling body (on the moon)	1.6 m/s²	0.16 *g*
Freely falling body (on the earth)	9.8 m/s²	1 *g*
Space shuttle (maximum)	29 m/s²	3 *g*
Drag racing—car (average for ¼ mile)	32 m/s²	3.3 *g*
Highest (sustained) survived by human	245 m/s²	25 *g*
clothes—spin cycle in a washing machine	400 m/s²	41 *g*
Tread of a car tire at 65 mph	2,800 m/s²	285 *g*
Click beetle jumping	3,920 m/s²	400 *g*
Bullet in a high-powered rifle	2,000,000 m/s²	200,000 *g*
Projectile in an electromagnetic "rail gun"	1.96×10^9 m/s²	2×10^8 *g*

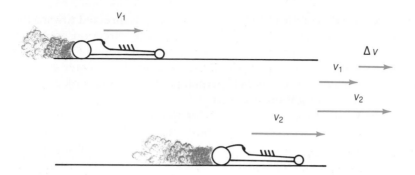

FIGURE 1.22
As a car accelerates in a straight line, the length of the arrow representing its velocity increases. The arrow marked Δv represents the change in velocity from v_1 to v_2 and indicates the direction of the acceleration—forward.

Another way to think of acceleration is in terms of the changes in the arrows that can be drawn to represent velocity. When a race car accelerates from one speed to a higher speed, the *length* of the arrow used to represent the velocity *increases* (see Figure 1.22). If we place the two arrows side by side, the change in velocity can be represented by a third arrow, Δv, drawn from the tip of the first arrow to the tip of the second. The later velocity equals the initial velocity plus the change in velocity. The original arrow plus the arrow representing Δv equals the later arrow. The acceleration vector is this change in velocity divided by the time. Note that the acceleration vector points forward; increasing the velocity is a forward acceleration.

Centripetal Acceleration

The concept of acceleration includes changes in *direction* of motion as well as changes in speed. A car going around a curve and a billiard ball bouncing off a cushion are accelerated, *even* if their speeds do *not* change.

We can once again use arrows to indicate the change in velocity. Figure 1.23 shows a car at two different times as it goes around a curve. If we place the two arrows representing the car's velocity next to each other, we see that the direction of the arrow has changed. The change in velocity, Δv, is represented by the arrow drawn from the tip of v_1 to the tip of v_2. In other words, the original arrow plus the arrow representing Δv equals the later arrow, just like with straight-line acceleration (Figure 1.22).

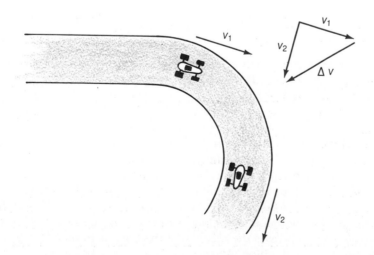

FIGURE 1.23
As a car, traveling at constant speed, rounds a curve, it undergoes centripetal acceleration. Here the arrow representing the car's velocity changes direction. Since the change in velocity, Δv, is directed towards the center of the curve, so is the acceleration.

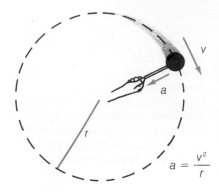

FIGURE 1.24
An object moving along a circular path is accelerated because its direction of motion, and therefore its velocity, are changing. Its centripetal acceleration equals the square of its speed, divided by the radius of its path. Doubling its speed would quadruple its acceleration. Doubling the radius of its path would halve the acceleration.

Note the direction of the change in velocity Δv: it is directed toward the *center* of the curve. Since the acceleration is in the same direction as Δv, it also is directed toward the center. For this reason the acceleration of an object moving in a circular path is called *centripetal acceleration* (for "center-seeking"). The centripetal acceleration is always perpendicular to the object's velocity— directed either to its right or to its left.

So we know the direction of the acceleration of a body moving in a circular path, but what about its magnitude? The faster the body is moving, the more rapidly the direction of its velocity is changing. Consequently the magnitude of the centripetal acceleration depends on the speed, v. It also depends on the radius, r, of the curve (Figure 1.24). A larger radius means the path is not as sharply curved, so the velocity changes more slowly and the acceleration is smaller. The actual equation for the size of the acceleration is:

$$a = \frac{v^2}{r} \qquad \text{(centripetal acceleration)}$$

We can also describe the relationship between a, v, and r by saying that the *acceleration is proportional to the square of the speed:*

$$a \propto v^2$$

and the *acceleration is inversely proportional to the radius r:*

$$a \propto \frac{1}{r}$$

This means that when the speed is doubled, the acceleration becomes four times as large. If the radius is doubled, the acceleration becomes one-half as large. We will encounter these two relationships again.

EXAMPLE 1.4   Let us estimate the acceleration of a car as it goes around a curve. The radius of a segment of a typical cloverleaf is 60 meters, and a car might take the curve with a speed of 20 meters/second (about 45 miles/hour; Figure 1.25).

Since the motion is circular,

$$a = \frac{v^2}{r} = \frac{(20 \text{ m/s})^2}{60 \text{ m}}$$

$$a = \frac{400 \text{ m}^2/\text{s}^2}{60 \text{ m}} = 6.67 \text{ m/s}^2$$

If the car could go 40 meters/second and stay on the road (it could not), its acceleration would be four times as large—26.67 m/s^2 or 2.7 g.

Many people have difficulty accepting the idea of centripetal acceleration, or they don't see how changing the direction of motion of a body is the same kind of thing as changing its speed. But some common experiences show that it is. Let us say that you are riding in a bus and that a book is resting on the slick seat next to you. The book will slide over the seat in response to the bus's acceleration. As the bus speeds up, the book slides backwards. As the bus decelerates, the book slides forward. In both cases the reaction of the book is to move in the *direction opposite* the bus's acceleration. What happens when the bus goes around a curve?

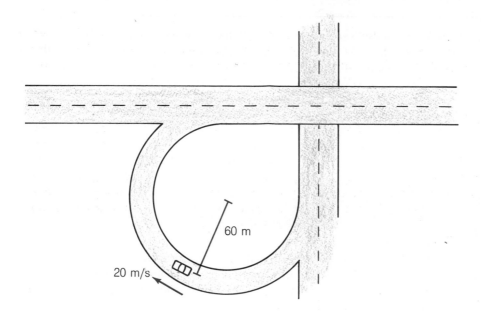

FIGURE 1.25
*Car on a cloverleaf with a
60-meter radius.*

The book slides towards the *outside* of the curve, indicating that the bus is accelerating toward the *inside* of the curve—a *centripetal acceleration.*

The cornering ability of a car is often measured by the maximum centripetal acceleration it can have when it rounds a curve. Automotive magazines often give the "cornering acceleration" or "lateral acceleration" of a car they are evaluating. A typical value for a sports car is 0.85 *g*, which is equal to 8.33 meters/second².

LEARNING CHECK

(Note: Simple quizzes like this are included midway through each chapter. They are not designed to be a thorough examination of the material; they are more of a check to see if you have been "awake." If you have any trouble answering the questions, it means you probably have not been reading carefully enough. Answers can be found at the back of the book in Appendix E.)

1. The two main systems of measure are the _____ and the _____ .

2. Any physical quantity that has a direction associated with it is called a _____ .

3. For objects moving with constant speed,
 a) the instantaneous speed is always zero.
 b) the average speed is always zero.
 c) the instantaneous speed and average speed are always equal.
 d) the acceleration is always greater than zero.

4. When an object's velocity is changing, we say that it is _____ .

5. A freely falling body experiences
 a) constant velocity.
 b) centripetal acceleration.
 c) constant acceleration.
 d) steadily increasing acceleration.

continued on next page

6. When something moves in a circle with constant speed,
   a) its acceleration is perpendicular to its velocity.
   b) Its acceleration is zero.
   c) its velocity is constant.
   d) its acceleration is parallel to its velocity.

1.5   SIMPLE TYPES OF MOTION

Now let's take a look at some simple types of motion. In each example we'll consider a single body moving in a particular way. Our goal is to show how distance and speed depend on time. Take note of the different ways that the relationships can be shown or expressed.

Zero Velocity

The simplest situation is one in which *no motion* occurs: a single body sitting at a fixed position in space. We characterize this system by saying that the distance of the object, from whatever reference point we choose, is constant. Nothing is changing. This means that the object's velocity and acceleration are both zero.

Constant Velocity

The next simplest case is *uniform motion.* Here the body moves with a constant velocity—a constant speed in a fixed direction. An automobile traveling on a straight, flat highway at a constant speed is a good example. A hockey puck sliding over smooth ice almost fits into this category, because it slows down only slightly due to friction. Note that the *acceleration equals zero.* That much should be obvious. The interesting relationship here is between distance and time. We can express that relationship in four different ways: with words, mathematics, tables, or graphs.

Let us take the example of a runner traveling at a steady pace of 7 m/s. If you are standing on the side of the road, how does the distance from you to the runner change with time, after the runner has gone past you? In words, *the distance increases 7 meters each second.* Two seconds after the runner passes by, the distance would be 14 meters (see Figure 1.26). Stated mathematically, *the distance in meters equals the time multiplied by the speed, 7 meters/second.* The time is measured in seconds, starting just as the runner passes you. The shorthand way of writing this is the mathematical equation

$$d = 7t$$

This is just an example of the general equation that we saw earlier with $v = 7$ meters/second:

$$d = vt$$

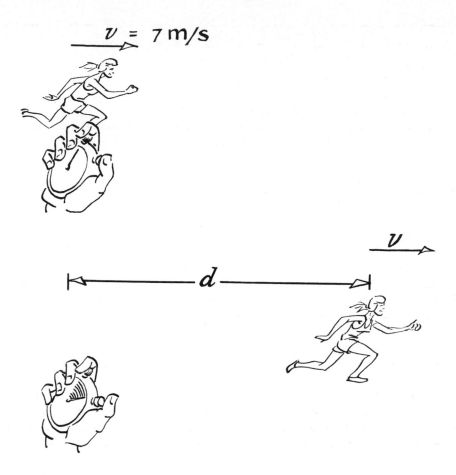

FIGURE 1.26
The speed of a runner is 7 meters/second. This means that the runner's distance from a fixed point (you standing behind the runner) increases 7 meters each second.

The same information can also be put in a table of values of time and distance (see the margin table).

These values all satisfy the equation $d = 7t$. The table just shows the values of distance (d) at certain times (t). You could make the table longer or shorter by using different time increments.

The fourth way to show the relationship between distance and time is to graph the values in the table (see Figure 1.27a). The usual practice is to graph distance versus time (distance on the vertical axis). Note that the data points lie on a straight line. Remember this simple rule: when the speed is constant, the graph of distance vs time is a straight line. (In general, when one quantity is proportional to another, the graph of the two quantities is a straight line.)

An important feature of this graph is its *slope*. The slope of a graph is a measure of its steepness. In particular, the slope is equal to the rise between two points on the line divided by the run between the points. This is illustrated in Figure 1.27b. The rise is a distance, Δd, and the run is a time interval, Δt. So the slope equals Δd divided by Δt, which is also the object's speed. The *slope of a distance versus time graph equals the speed.*

The graph for a faster-moving body, a racehorse for instance, would be steeper—it would have a larger slope. The graph of d versus t for a slower object (a person walking) would have a smaller slope (see Figure 1.28). When an object is standing still (when it has no motion), the graph of distance versus time is a flat line that coincides with the horizontal axis. The slope is zero because the speed is zero.

Time (s)	Distance (m)
0	0
1	7
2	14
3	21
4	28

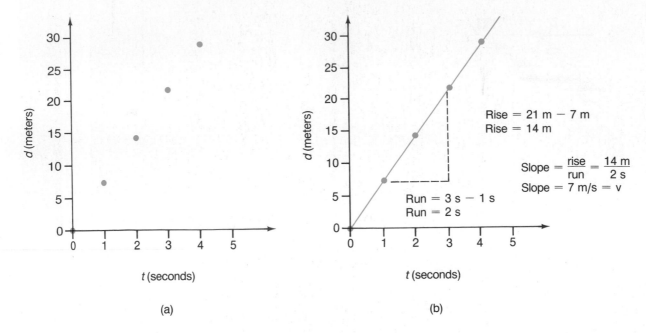

(a)   (b)

FIGURE 1.27
(a) Graph of distance versus time when speed is a constant 7 meters/second. (b) Same graph with the slope indicated.

Even when the speed is not constant, the slope of a *d* versus *t* graph is still equal to the speed. In this case the graph is not a straight line because as the slope changes (a result of the changing speed), the graph curves, or bends. The graph in Figure 1.29 represents the motion of a car that starts from a stop sign, drives down a street, then stops and backs into a parking place. When the car is stopped, the graph is flat; the distance is not changing, and the speed is zero. When the car is backing up, the graph is slanting downward; the distance is decreasing and the speed is negative.

Constant Acceleration

The next simple motion example is *constant acceleration in a straight line.* This means that the object's speed is changing at a fixed rate. A freely falling body is the best example. A ball rolling down a straight, inclined plane is another. Often

FIGURE 1.28
The slope of a distance versus time graph equals the speed. A body with a higher speed (the horse) has a larger slope, that is, a steeper graph. When the speed is very low, the graph is nearly a horizontal line.

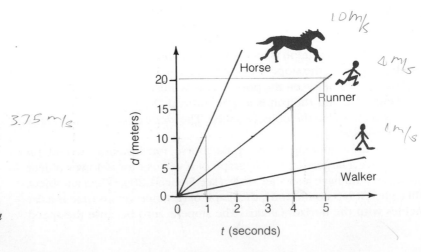

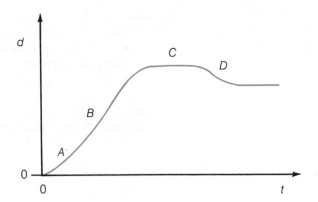

FIGURE 1.29
Graph of distance versus time for a car with varying speed. At point A *on the graph, the slope starts to increase as the car accelerates. At* B *its speed is constant. The slope decreases to zero at point* C *when the car is stopped. At* D *the car is backing up, so its speed is negative.*

cars, runners, bicycles, trains, and aircraft have nearly constant acceleration when they are speeding up or slowing down.

Let us use free fall as our example. Assume that a heavy rock is dropped from the top of a building and that we can measure the instantaneous speed of the rock and the distance that it has fallen at any time we choose (see Figure 1.30). The rock falls with an acceleration equal to g. This is 9.8 m/s^2, which is the same as 22 miles/hour/second. First, consider how the rock's speed changes. *The speed increases 9.8 meters/second, or 22 miles/hour each second.* This means that *the rock's speed equals the time* (in seconds) *multiplied by 9.8* (for meters/second) or *22* (for miles/hour). Mathematically,

$$v = 9.8t \quad \text{(in meters/second)} \qquad v = 22t \quad \text{(in mph)}$$

The general form of the equation that applies when the acceleration equals some constant value a is

$$v = at \qquad \text{(constant acceleration)}$$

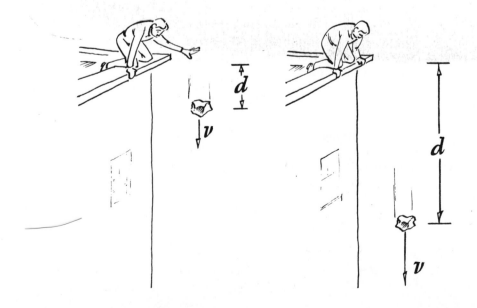

FIGURE 1.30
A rock falls freely after it is dropped from the top of a building. The distance d *is measured from the top of the building. The rock's speed increases at a steady rate—9.8 meters/second each second.*

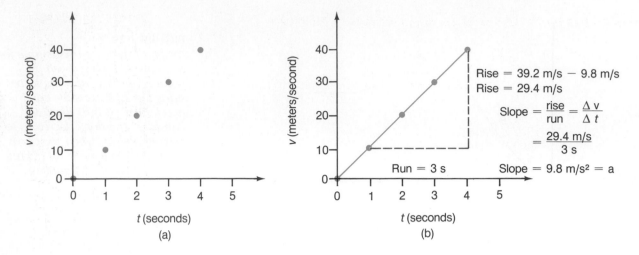

FIGURE 1.31
(a) Graph of speed versus time for a freely falling body. (b) The same graph with the slope indicated.

	Speed	
Time (s)	(m/s)	(mph)
0	0	0
1	9.8	22
2	19.6	44
3	29.4	66
4	39.2	88

So in constant acceleration, the *speed is proportional to the time.* The proportionality constant is the acceleration *a.* A table of values for this example is shown in the margin.

The graph of speed versus time is just a straight line (see Figure 1.31). *The slope of a graph of speed versus time equals the acceleration.* This is because the rise is a change in speed, Δv, and the run is a change in time, Δt. So:

$$\text{Slope} = \frac{\Delta v}{\Delta t} = a$$

The corresponding graph for a body with a smaller acceleration, say a ball rolling down a ramp, would have a smaller slope.

The similarity between the graphs in Figure 1.31 and the graphs in Figure 1.27 for uniform motion is obvious. But keep in mind that it is the graph of *distance* versus time that is a straight line for uniform motion. Here it is the graph of *speed* versus time.

What is the relationship between distance and time when the acceleration is constant? It is a bit more complicated, as expected. Figure 1.32 shows that the distance a falling body travels during each successive time period grows larger as it falls. Since the speed is continually changing, the distance equals the *average speed* times the time. What is the average speed? The object starts with speed equal to zero, and after accelerating for a time *t*, its speed is *at.* Its average speed is just:

$$\text{average speed} = \frac{0 + at}{2} = \frac{1}{2}at$$

(If you take two quizzes and get 0 on the first and 8 on the second, your average grade is 4—one half of 0 plus 8.)

The distance traveled is this average speed times the time.

$$d = \text{average speed} \times t = \frac{1}{2}at \times t$$

$$d = \frac{1}{2}at^2 \quad \text{(constant acceleration)}$$

In the case of a falling body, the acceleration is 9.8 m/s². Therefore:

$$d = \frac{1}{2}at^2 = \frac{1}{2} \times 9.8 \times t^2$$

$$d = 4.9t^2 \qquad (d \text{ in meters, } t \text{ in seconds})$$

Time (s)	Distance (m)
0	0
1	4.9
2	19.6
3	44.1
4	78.4

So in the case of constant acceleration, the *distance is proportional to the square of the time.* The constant of proportionality is one half the acceleration.

A table of distance values for the falling rock is shown above.

This distance increases rapidly. The graph of distance versus time curves upward (see Figure 1.33). This is simply because the speed of the rock is increasing with time, and the slope of this graph equals the speed. Table 1.5 summarizes these three simple types of motion.

Rarely does the acceleration of an object stay constant for long. As a falling body picks up speed, air resistance causes its acceleration to decrease (see Section 2.6). When a car is accelerated from a stopped position, its acceleration usually decreases, particularly when the transmission is shifted into a higher gear. Figure 1.34 shows the speed of a car as it accelerates from 0 to 80 miles/hour. Note that acceleration decreases (the slope gets smaller). During the short time the transmission is shifted, the acceleration is zero.

At this level of physics, graphs may be the best way to show the relationships between physical quantities. Mathematics is the most precise way, and it is not limited to two (or sometimes three) quantities, as are graphs. However, math is inherently abstract and also somewhat like a language. To persons familiar with it, math is very useful, efficient, and precise. But to those who do not use it regularly or who don't feel comfortable with it, mathematics can be confusing and intimidating. If you wanted to study one of Aristotle's works, it would be best to read it in the original Greek. But unless your reading comprehension of the language is very good, you would be better off to read a translation.

In this book, we will "translate" most of the mathematics we use into statements and will also express many of the relationships graphically. But we will also retain some of the "original" language, mathematics, and show that even an understanding of high school math allows you to solve physics problems. Here is where mathematics is indispensable. The application of physics in our society by engineers and others *requires* mathematics. The Golden Gate Bridge could not have been designed and built through the use of words only.

When you see a graph, the first thing you should do is take careful note of which quantities are graphed. You should notice that some of the graphs for the examples of motion are just straight lines. When the graph shows *distance* versus time, a straight line means the *speed* is constant. But when it shows *speed* versus

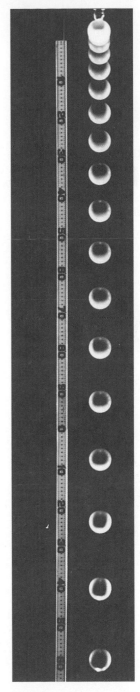

FIGURE 1.32
A falling ball photographed with a flashing strobe light. Each image shows where the ball was at the instant when the light flashed. The images are close together near the top, because the ball is moving more slowly at first. As it falls, it picks up speed, and thus travels farther between flashes.

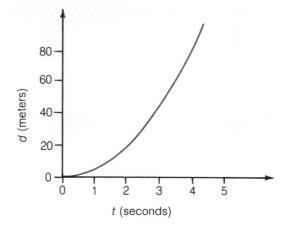

FIGURE 1.33
Graph of distance versus time for a freely falling body. The slope increases, indicating that the speed increases.

time, a straight line means that the *acceleration* is constant. These two situations are quite different. To a business executive, a rising graph showing profits will inspire joy, and a rising graph showing expenses will cause concern.

The most important thing to look for in the shape of a graph is trends. Is the slope always positive or always negative (is the graph continually going up or going down)? If the slope changes, consider what that signifies. If it is a graph of speed versus time, for example, choose a particular time, and note whether the speed is increasing, decreasing, or remaining the same. A graph may be worth a thousand words, but you can't grasp its full meaning by simply glancing at it.

Let's look at an application of the kinds of graphs we've been using. During a karate demonstration, a concrete block is broken by a person's fist. The upper graph in Figure 1.35 shows the distance of the hand above the block vs time, measured using high-speed photography. The fist travels downward until it con-

TABLE 1.5 SUMMARY OF EXAMPLES OF MOTION		
Type of Motion	Behavior of Physical Quantities	Equation
Stationary Object	Distance constant	$d =$ constant
	Speed zero	$v = 0$
	Acceleration zero	$a = 0$
Uniform motion*	Distance increasing	$d = vt$
	Speed constant	$v =$ constant
	Acceleration zero	$a = 0$
Uniform acceleration* (from rest)	Distance increasing	$d = \frac{1}{2}at^2$
	Speed increasing	$v = at$
	Acceleration constant	$a =$ constant

*Distance measured from object's initial location.

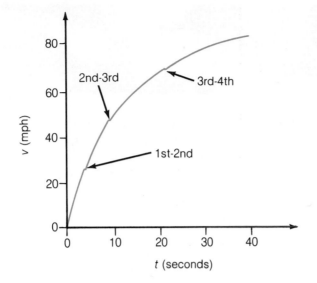

FIGURE 1.34
Realistic graph of a car's speed versus time as the car is accelerated. The notches in the curve occur at transmission shifts when the engine is momentarily disconnected from the drivetrain. Between the shifts, the acceleration (slope) decreases as the speed increases.

tracts the block, at about 6 milliseconds. This causes a rapid deceleration of the fist (accompanied by some pain, one would imagine). The lower graph shows the speed of the fist, just the slope of the distance graph at each time. Contact with the concrete is indicated by the steep part of the graph as the speed goes to zero. If we take the slope of this segment of the speed graph, we find that the acceleration of the fist at that moment was about *3,500 m/s²*, or *360 g* (ouch!). What happened at about 25 milliseconds?

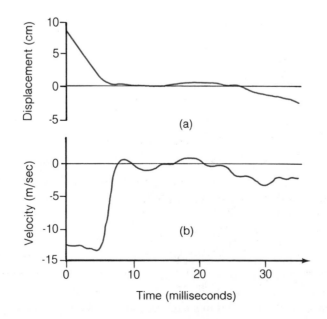

FIGURE 1.35
(a) Graph of the position of a fist versus time during a karate blow. Contact is made at about 6 milliseconds. (b) Graph of the speed of the fist versus time. At contact, the fist decelerates rapidly. (From Wilke, S. R; McNair, R. E.; and Feld, M S. The physics of karate. American Journal of Physics 51 (September 1983): 783–90.

We finish this chapter with a brief look at how our method of analyzing motion evolved. The concepts of distance, time, and even speed seem to be basic to human perception of the natural world. Daily life has always been synchronized with the periodic motion of the earth's rotation. A day's journey incorporates both distance and speed.

The development of written language provided a great boost to science, as it did to all scholarly endeavors. Some might even argue that science is impossible without literacy. But there are too many examples of primitive societies accomplishing remarkable feats, particularly in architecture and astronomy, for this view to go unquestioned. However, written languages did make it possible for people to learn from their predecessors without relying solely on word-of-mouth.

The first great strides toward the development of exact sciences were made by the ancient Greeks. Pythagoras, in the sixth century B.C., showed that numbers, often regarded merely as mental abstractions, are related to the natural world and even to human perception. His most famous discoveries were in geometry and acoustics. Other ancients, before the year 200 B.C., accomplished accurate measurements of the diameter of the earth, the moon, and the moon's orbit.

Aristotle

The natural philosopher Aristotle (384−322 B.C.; Figure 1.36) is regarded as the first person to attempt physics and to actually give *physics* its name. Born during the height of ancient Greek civilization, Aristotle became one of the central figures in the explosion of intellectual development that took place in that era. Starting at the age of seventeen, Aristotle was a student of the great philosopher Plato for twenty years. He became the tutor of Alexander the Great and later spent years studying the cultures, flora, and fauna of the exotic lands conquered by Alexander. In 335 B.C., Aristotle opened a school at the Lyceum, near Athens, where he taught until his death. Most of what we know about Aristotle's thoughts and teachings are based on the lectures he gave at the Lyceum.

Aristotle was a master of virtually all of the academic disciplines that existed at the time and actually invented some of them; psychology is one example. His ideas about physics were influenced a great deal by the methodology that he used in logic, biology, and other areas. Aristotle took the important step of realizing that there are a number of complicating factors in the physical world that often mask hidden order. His analysis of motion, the first mechanics, rested on the distinction between *natural motion,* like that of a falling body, and *unnatural motion,* like that of a cart being pulled down a road.

According to Aristotle, heavy objects fall because they are seeking their natural place. In his model, the speed of a falling body is constant and depends on its weight and the medium through which it falls. Heavy objects fall faster than lighter ones. Also, a rock drops faster through air than it does through water (Figure 1.37). This analysis is only partly correct: it applies only after an object has been falling for a period of time and friction has become important. Speed first increases at a fixed rate—9.8 m/s each second, as we have seen. A marble and a

FIGURE 1.36
Aristotle: philosopher, physicist, biologist, teacher.

FIGURE 1.37
(a) In Aristotle's model of falling bodies, an object falls with a constant speed that depends on its weight and the medium through which it falls. A heavy object (right) falls faster than a light object (left). (b) An object falls faster through air than through water.

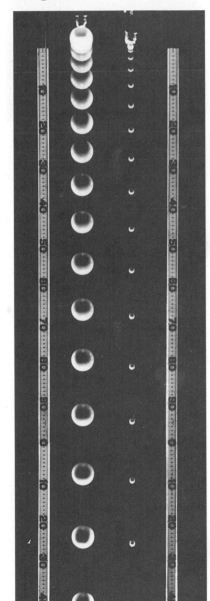

rock dropped at the same time will build up speed together and hit the ground at the same time (Figure 1.38). Only after a rock has fallen a much greater distance will air resistance cause its speed to level off at some constant value. This final speed, called the terminal speed, depends as much on the shape and size of the object as it does on the weight. The same factors affect the terminal speed of bodies falling through water, but such speeds are much slower (as Aristotle pointed out) and are reached much more quickly. We will take a closer look at this in Section 2.6.

Aristotle's description of the motion of falling bodies fits only after their motions are dominated by air resistance. In a similar way, his analysis of "unnatural motion" indirectly includes friction as the dominating influence. A rock's "natural" tendency is to remain at rest on the ground or to fall toward the earth's center if dropped. Making the rock move horizontally by pushing it is unnatural and requires some external agent or force. Aristotle's view that a force is always required to maintain horizontal motion fits well whenever there is a great deal of friction but not so well in cases like those of a rock thrown horizontally or of a smooth, heavy ball rolling over a flat, had surface. In these situations, motion can continue for quite some time with no force acting to maintain it. Because of Aristotle's overwhelming reputation as a scholar, zealous supporters of his ideas allowed his theories to dominate physics, almost without question, for about 2,000 years.

Galileo

Galileo Galilei (1564–1642; Figure 1.39) lived during one of the most fruitful periods of human civilization—The Renaissance. His birth came in the same year as the death of Michelangelo and the birth of Shakespeare. Galileo made many important discoveries in mechanics and astronomy, some of which were disputed bitterly because they contradicted accepted views passed down from Aristotle's time. Galileo's strong support of the heliocentric (sun-centered) model of the

FIGURE 1.38
Strobe photograph of two falling bodies. Even though one is much heavier than the other, they have the same acceleration.

FIGURE 1.39
Galileo Galilei

solar system and other factors led to his being placed on trial by leaders of the Inquisition. But this was the age when superstition was giving way to rational thought, when evidence and logical proof were beginning to win out over blind acceptance of doctrines. Though censured and forced to publicly recant his support of the heliocentric theory, Galileo's work gained wide acceptance, and he is now considered one of the founders of modern physical science.

Galileo was one of the first to rely on observation and experimentation. In 1583 he noticed that a lamp hanging from the ceiling of the cathedral at Pisa would swing back and forth with a constant period, even though the length of the arc of its motion decreased. His discovery, that the frequency of a pendulum depends only on its length, is the basis for the pendulum clock.

Another important contribution by Galileo was his insistence that scientific terms, statements, and analyses be logically consistent. He sought to establish a "scientific method" and believed that, as a first step, the language of science should be unambiguous. To Galileo, mathematics was necessary to help accomplish this task.

Although it was his astronomical discoveries that brought Galileo most of his fame (and controversy), it is his conclusions about motion that are of interest here. Like Aristotle, Galileo recognized that the simple rules that govern motion can be hidden by other phenomena. He realized that the two things that have the largest effect on objects moving on the earth are gravity and friction. More importantly, he reasoned that friction is often complicated and unpredictable, and that it is best to first focus on systems in which friction doesn't dominate completely.

The physical system that Galileo used to analyze uniform motion and uniformly accelerated motion is a smooth, heavy ball rolling on a smooth, straight surface that can be tilted. If the surface is tilted (an inclined plane), the ball's speed will increase as it rolls down. The opposite occurs when the ball is initially rolling up the plane—its speed decreases. But what if the surface is level? Logically, the ball's speed should neither increase nor decrease but should stay the same (Figure 1.40). So, Galileo reasoned correctly that if there is no friction, an object moving on a level surface will proceed with constant speed.

Galileo discovered the law of falling bodies, also using a ball rolling on an inclined plane (see Figure 1.41). He realized that the speed of a dropped object increases as it falls. The biggest problem he faced was measuring the speed. The only clocks available at that time were water clocks, based on how long it took water to drain out of (or to fill) a container. Galileo realized that a ball rolling down an inclined plane is also accelerated because of gravity. He reasoned that tilting the surface more and more steeply would make the ball's motion become closer to that of freefall (see Figure 1.42). Using the inclined plane and his crude clocks, Galileo discovered the correct relationship between distance and time for uniformly accelerated motion.

FIGURE 1.40
(a) As a ball rolls down an inclined plane, its speed increases. (b) As a ball rolls up an inclined plane, its speed decreases. (c) When a ball rolls on a level surface, it should therefore have a constant speed.

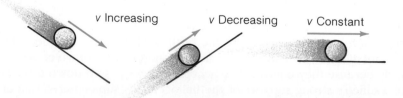

v Increasing   v Decreasing   v Constant

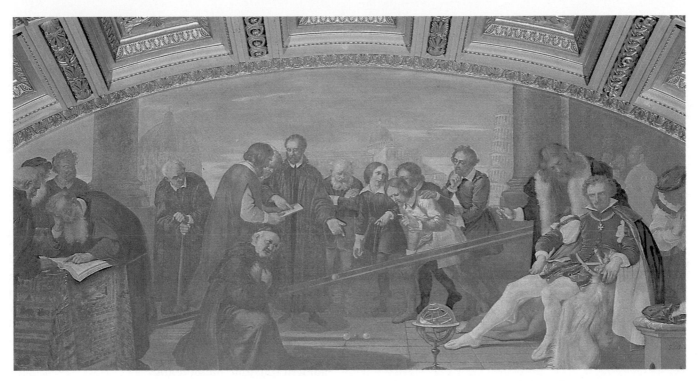

FIGURE 1.41
Galileo using an inclined plane to experiment with acceleration.

Legend has it that Galileo once dropped two different-sized objects from the top of the Leaning Tower of Pisa to show that they would hit the ground at nearly the same time. Whether he actually did this or not is unknown. His conclusion about the motion of falling bodies is correct whenever air resistance is negligible. This was illustrated rather dramatically by astronaut Dave Scott while performing experiments on the moon. He simultaneously dropped a hammer and a feather; both fell at the same rate and hit the lunar surface at the same time because there is no air on the moon to slow the feather.

Galileo used an approach that is now standard procedure in physics: if you wish to understand a real physical process, first consider an idealized system in which complicating factors (like friction) are absent. He promoted the idea of imagining how bodies would move in a perfect vacuum, devoid of air resistance and other forms of friction. Only by understanding a simplified model of reality can one hope to comprehend its complexities and subtleties.

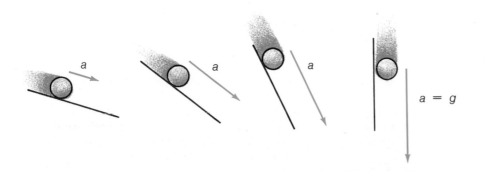

FIGURE 1.42
A ball is accelerated as it rolls down an inclined plane. Making the plane steeper increases the ball's acceleration. When the plane is vertical, the acceleration equals g.

Aristotle and other ancient Greeks initiated the science of physics. Galileo corrected Aristotle's mechanics and established the importance of mathematics and experimentation in physics. But the greatest name in the development of mechanics is that of Isaac Newton, who was born on Christmas Day in the year of Galileo's death.

SUMMARY

Physics is a science that relies on well-defined physical quantities, laws and principles, and mathematics to describe and make sense of the physical world. Distance, time, and mass are the three fundamental physical quantities used in mechanics, the branch of physics dealing with the motion of objects. Speed, velocity, and acceleration are derived quantities (based on distance and time) that are used to specify how an object moves. Acceleration, which is the rate of change of velocity, is zero for bodies that are moving with a constant speed in a straight line. On the earth, freely falling objects have a constant downward acceleration of 9.8 m/s^2 or 32 ft/s^2.

Velocity and acceleration are vectors, meaning that they have direction as well as magnitude. A body moving along a circular path undergoes centripetal acceleration because the direction of its velocity is changing.

Aristotle was the first to attempt a scientific analysis of motion. His conclusions about "natural motion" and "unnatural motion," while flawed, were still on the right track. Galileo made several important discoveries in physics and correctly described the motion of freely falling bodies. He was one of the first to use logical observation, experimentation, and mathematical analysis in his work. These are all important components of modern scientific methodology.

SUMMARY OF IMPORTANT EQUATIONS

EQUATION	COMMENTS	EQUATION	COMMENTS
FUNDAMENTAL EQUATIONS		*SPECIAL-CASE EQUATIONS*	
$T = \dfrac{1}{f}$	Relates frequency and period	$d = vt$	Distance from starting point when speed is constant
$f = \dfrac{1}{T}$	Relates frequency and period	$d = \dfrac{1}{2}at^2$	Distance when acceleration is constant and object starts from rest
$v = \dfrac{d}{t}$	Definition of average speed	$v = at$	Speed when acceleration is constant and object starts from rest
$v = \dfrac{\Delta d}{\Delta t}$	Instantaneous speed	$a = g$	Acceleration of freely falling body
$a = \dfrac{\Delta v}{\Delta t}$	Definition of acceleration	$a = \dfrac{v^2}{r}$	Centripetal acceleration (object moving in a circular path)

QUESTIONS

1. Suppose you leave a note for a friend asking him or her to meet you at a certain time. If the friend does not show up, describe the steps you might take to determine why, based on the model of the scientific method.

2. A pendulum clock is taken to a repair shop. Its pendulum is replaced by a shorter one that oscillates with a smaller period than the original. What effect, if any, does this have on how the clock runs?

3. What are the fundamental physical quantities? Why are they called that?

4. Many countries that formerly use the English system of measure have converted to the metric system. Why is the metric system simpler to use, once you are familiar with it?

5. To prepare an aircraft carrier to launch airplanes it is steered "into the wind" (so it is sailing against the wind). Why does this aid in launching the aircraft?

6. Scenes in films or television programs sometimes show people jumping off moving trains and having unpleasant encounters with the ground. If someone is on a moving flat-bed train car and wishes to jump off, how could the person use the concept of relative speed to make a safer dismount?

7. List the physical quantities identified in the chapter. From which of the fundamental physical quantities is each derived? Which of them are vectors, and which are scalars?

8. What is the distinction between speed and velocity? Describe a situation in which an object's speed is constant but its velocity is not.

9. What is "vector addition," and how is it done?

10. Can the resultant of two velocities have zero magnitude? If so, give an example.

11. A swimmer heads for the opposite bank of a river. Make a sketch showing the swimmer's two velocities and the resultant velocity.

12. A basketball player shoots a free throw. Make a sketch showing the basketball's velocity just after the ball leaves the player's hands. Draw in two components of this velocity, one horizontal and one vertical. Repeat the sketch for the instant just before the ball reaches the basket. What is different?

13. What is the relationship between velocity and acceleration?

14. How does the speed of a freely falling body change with time? How does the distance it has fallen change? How about the acceleration?

15. If caught in an elevator falling out of control, one suggested way to improve the chances of survival is to jump upward just before impact. Explain why this would not help very much.

16. What is centripetal acceleration? What is the direction of the centripetal acceleration of a car going around a curve?

17. As a car goes around a curve, the driver increases its speed. This means the car has two accelerations. What are the directions of these two accelerations?

18. If a ball is thrown straight up into the air, what is its acceleration as it moves upward? What is its acceleration when it reaches its highest point and is stopped for an instant?

19. What does the slope of a distance versus time graph represent physically?

20. Sketch a graph of speed versus time for the motion illustrated in Figure 1.29. Indicate what the car's acceleration is at different times.

21. To what extent was Aristotle's model of falling bodies correct? How was it wrong?

PROBLEMS

1. A yacht is 20 m long. Express this length in feet. 65.6 Feet

2. Express your height (**a**) in meters and (**b**) in centimeters. 177.8 cm   1.778 m = 1.8 m

3. A convenient time unit for short time intervals is the *millisecond*. Express 0.0452 s in milliseconds. 45.2 ms

4. One mile is equal to 1,609 m. Express this distance in kilometers and in centimeters. 1.609 km   160900 cm

5. A hypnotist's watch hanging from a chain swings back and forth every 0.8 s. What is the frequency of its oscillation?

1.25 Hz

6. The quartz crystal used in an electric watch vibrates with frequency 32,768 Hz. What is the period of the crystal's motion? 30.5 microseconds

7. A passenger jet flies from one airport to another 1,200 mi away in 2.5 h. Find its average speed. 480 m/ph

8. In 1990, the world record in the 200-meter dash was 19.72 s. What was the average speed in m/s? In mph?

9. In Figure 1.18, assume that $v_1 = 8$ m/s and $v_2 = 6$ m/s. Use a ruler to estimate the magnitudes of the resultant velocities in (c) and (d).

10. On a day when the wind is blowing toward the south at 3 m/s, a runner jogs west at 4 m/s. What is the velocity (speed and direction) of the air relative to the runner?

11. How far does a car going 25 m/s travel in 5 s? How far would a jet going 250 m/s travel in 5 s?

12. A long-distance runner has an average speed of 4 m/s during a race. How far does the runner travel in 20 min?

13. Draw an accurate graph showing the distance versus time for the car in problem 11. What is the slope?

14. The graph below shows the distance versus time for an elevator as it moves up and down in a building. Compute the elevator's speed at the times marked a, b, and c.

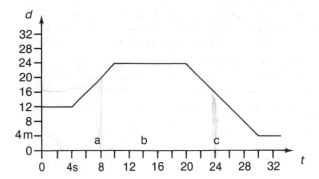

15. A Ferrari GTO can go from 0 to 100 mph (44.7 m/s) in 11 s.
    a) What is its average acceleration?
    b) The same car can come to a complete stop from 40 m/s in about 5 s. What is its average acceleration (deceleration)?

16. As a baseball is being thrown, it goes from 0 m/s to 40 m/s in 0.15 s.
    a) What is the acceleration of the baseball?
    b) What is the acceleration in g's?

17. A child attaches a rubber ball to a string and whirls it around in a circle overhead. If the string is 0.5 m long, and the ball's speed is 10 m/s, what is the ball's centripetal acceleration?

18. A beetle sits on the edge of a spinning record that has a radius of 0.15 m. The beetle's speed is about 0.5 m/s when the record is turning at 33⅓ rpm. What is the beetle's acceleration?

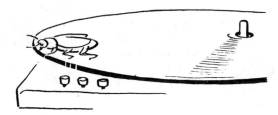

19. A rocket accelerates at a rate of 60 m/s².
    a) What is its speed after it accelerates for 40 s?
    b) How long does it take to reach a speed of 7,500 m/s?

20. A train, initially stationary, has a constant acceleration of 0.5 m/s².
    a) What is its speed after 15 s?
    b) What would be the total time it would take to reach a speed of 25 m/s?

21. a) Draw an accurate graph of the speed versus time for the train in problem 20.
    b) Draw an accurate graph of the distance versus time for the train in problem 20.

22. Draw an accurate graph of the speed versus time for the elevator in problem 14.

23. A skydiver jumps out of a helicopter and falls freely for 3 seconds before opening the parachute.
    a) What is the skydiver's downward speed when the parachute opens?
    b) How far below the helicopter is the skydiver when the parachute opens?

24. A rock is dropped off the side of a bridge and hits the water below 2 s later.
    a) What was the rock's speed when it hit the water?
    b) What was the rock's average speed as it fell?
    c) What is the height of the bridge above the water?

25. A roller coaster starts at the top of a straight track that is inclined 30° with the horizontal. This causes it to accelerate at a rate of 4.9 m/s² (½ g).
    a) What is the roller coaster's speed after 3 s?
    b) How far does it travel during that time?

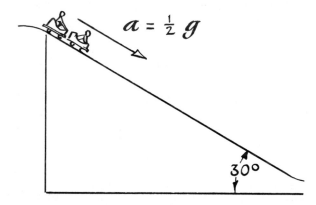

26. During takeoff, an airplane goes from 0 to 50 m/s in 8 s.
    a) What is its acceleration?
    b) How fast is it going after 5 s?
    c) How far has it traveled by the time it reaches 50 m/s?

27. The graph below shows the speed versus time for a bullet as it is fired from a gun, travels a short distance, and enters a block of wood. Compute the acceleration at the times marked a, b, and c.

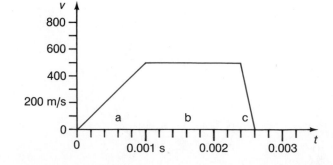

CHALLENGES

1. Communication satellites are stationed 22,000 mi above the earth's surface. How long does it take a radio signal traveling at the speed of light to travel from the earth to the satellite and back?

2. The maximum deceleration of most cars is much larger than the maximum acceleration. (See problem 15.) What factors contribute to cause this?

3. The moon's mass is 7.35×10^{22} kg, and it moves in a nearly circular orbit with radius 3.84×10^8 m. The period of its motion is 27.3 days. Use this information to determine the moon's (a) speed and (b) acceleration.

4. A sports car is advertised to have a maximum cornering acceleration of 0.85 g.
   a) What is the maximum speed that the car can go around a curve with a 100-m radius?
   b) What is its maximum speed for a 50-m radius curve?
   c) If wet pavement reduces its maximum cornering acceleration to 0.6 g, what do the answers to (a) and (b) become?

5. A spacecraft lands on a newly discovered planet orbiting the star Antares. To measure the acceleration of gravity on the planet, an astronaut drops a rock from a height of 2 m. A precision timer indicates that it takes the rock 0.71 s to fall to the ground. What is the acceleration of gravity on this planet?

6. When an object is thrown straight upward, gravity causes it to decelerate at a rate of 1 g—its speed decreases 9.8 m/s or 22 mph each second.
   a) Explain how we can use the equations, tables, and graphs for a freely falling body in this case.
   b) A baseball is thrown vertically at a speed of 39.2 m/s (88 mph). How much time elapses before it reaches its highest point, and how high above the ground does it get?

7. For an object starting at rest with a constant acceleration, derive the equation that relates its speed to the distance it has traveled. In other words, eliminate time in the two equations relating d to t and v to t. Test the equation by showing that an object reaches a final speed of 9.8 m/s if it is dropped from a height of 4.9 m.

SUGGESTED READINGS

Abers, Ernest S., and Kennel, Charles F. *Matter in Motion: The Spirit and Evolution of Physics*. Boston: Allyn and Bacon, Inc., 1977. An introductory physics text emphasizing the history of physics.

Boslough, John. "The Enigma of Time." *National Geographic* 177, no. 3 (March 1990): 109–132. A well-illustrated look at the history of timekeeping and attempts to understand what time is.

Brancazio, Peter J. *Sport Science*. New York: Simon and Schuster, 1984. A good source for applications of physics to sports. Chapter 1 has discussions on speed and acceleration during racing, including graphs of speed versus time for sprinters.

Cohen, Morris R., and Drabkin, I. E. *A Source Book in Greek Science*. New York: McGraw-Hill, 1948. Aristotle's mechanics is presented, beginning on page 200.

"Great Guns." *Scientific American*. 253, no. 4 (October 1985): 80–82. This article describes the high accelerations—up to 200,000,000 g's—accomplished by 'electromagnetic accelerators.'

Hawking, Stephen W. *A Brief History of Time*. New York: Bantam Books, 1988. Best-selling book by a famous physicist on the "role of time in physics." (Only one equation included.)

Segrè, Emilio. *From Falling Bodies To Radio Waves*. New York: W. H. Freeman and Co., 1984. History of physics up to a century ago by a Nobel Prize winner. It begins with "Whimsical Prelude," a visit to Galileo's home as it was in 1610.

Wilke, S. R.; McNair, R. E.; and Feld, M. S. "The Physics of Karate." *American Journal of Physics* 51, no. 9 (September 1983): 783–790. Ronald E. McNair was one of the seven astronauts killed in the disaster of the space shuttle Challenger. For an obituary of this remarkable man, see *Physics Today*, 39(4): 72–73 (April 1986). Part IV of the article by Wilke et al. (page 786) explains how the graphs in Figure 1.32 were produced.

OUTLINE

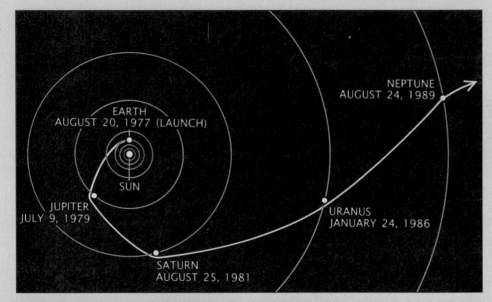

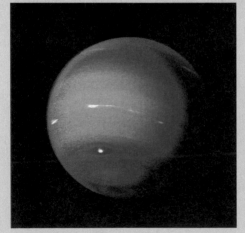
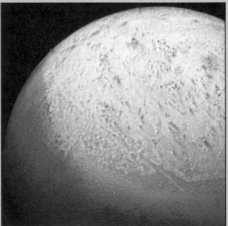

Top: Path of Voyager II. Bottom left: Neptune photographed through filters. Right: Triton, Neptune's largest moon.

NEWTON'S LAWS

PROLOGUE: VOYAGE OF DISCOVERY

The Voyager 2 spacecraft is on its way out of the solar system after completing one of the greatest journeys of scientific discovery of all time. Launched in 1977, it took advantage of a once-in-176-years alignment to complete a "grand tour" of the planets Jupiter, Saturn, Uranus, and Neptune. In addition to stunning color photographs of these planets and their moons that appeared on television newscasts and magazine covers, Voyager 2 sent back a wealth of scientific data and made several major discoveries, including rings around Jupiter and Neptune, volcanoes on Io (one of Jupiter's moons), several moons never before seen, and much more. Nearly everything now known about Uranus and Neptune came from Voyager 2, the first spacecraft to visit these distant planets. The mission completed, in the words of Voyager Chief Project Scientist Edward Stone, "nothing less than the encyclopedia of the planets."

Voyager 2's journey was a true odyssey. The decision to send it on to Uranus and Neptune was made 3½ years *after* launch, although the project designers had the foresight to incorporate that possibility. The spacecraft experienced several technical problems on its way. A computer chip failed, and the computer lost part of its memory. The main radio receiver failed, and the backup receiver, because of a faulty component, could receive only one radio-wave frequency. Ever since, controllers have had to adjust precisely the frequency of signals sent to Voyager to take into account the relative motion between the transmitter on earth and the spacecraft. After passing Saturn, its scan platform, which allows cameras and other instruments to be pointed in different directions, jammed. This problem was corrected before Voyager reached Uranus. It spite of these difficulties, most of the spacecraft's subsystems had actually been *improved* during its journey. For example, computer programs that process images were altered so Voyager 2 could send back photographs more rapidly.

Voyager 2's mission also represents a triumph of the basic physics of motion presented in the first part of this book. After traveling for 12 years, it arrived at Neptune, 4.5 billion kilometers (2.8 billion miles) from earth, on schedule and a mere 34 kilometers (21 miles) off its intended path. The rocket that launched Voyager 2 had to accelerate it to just the right speed in just the right direction to send it on its journey. On the way, the gravitational pull of each planet it visited was carefully exploited to increase Voyager 2's speed and to alter its path so it would reach the next planet. (Small rockets on board were used to make fine adjustments to its velocity.) In planning the mission, scientists had to predict the exact positions of the planets years in advance. In the case of Neptune, this meant determining where it would be over a decade in the future, even though it had not finished one complete orbit around the sun since its discovery in 1846.

All of this was possible largely because of the work of a seventeenth century physicist named *Isaac Newton*. He formulated three *laws of motion,* which are essential for determining how to accelerate things such as a spacecraft, and the *law of universal gravitation,* which governs the paths of the planets and spacecraft. He even invented a branch of mathematics, calculus, necessary for applying these laws. Voyager 2's success attests to the accuracy of the science of mechanics, which is based on these laws, as well as to the expertise of the scientists who applied it.

In this chapter we present Newton's three laws of motion and his law of universal gravitation. Formulated more than 300 years ago, these laws form the basis of modern mechanics. Force, an essential concept in physics, is introduced, and several common examples are given. Each law is used to extend your understanding of motion, of how forces affect motion, and of gravity. The process of accelerating an object, the principle behind rockets, and the nature of orbits are some of the applications of these laws that are described. The chapter concludes with a close look at the life of Sir Isaac Newton.

2.1   FORCE

Sir Isaac Newton (1642–1727) was an English scholar who made many fundamental discoveries in both physics and mathematics. He was the coinventor of calculus, a branch of mathematics that is now essential in physics. Newton is often regarded as the father of modern physical science, and for this reason it is difficult to overestimate his importance in the development of today's civilization. The dominant scientific, industrial, and technological advancements of the last two centuries were triggered in part by Newton's work.

In formulating his mechanics, Newton began with Galileo's ideas about motion and then sought systematic rules that govern motion and, more importantly, *changes in motion.* The key concept in Newtonian mechanics is *force.*

FORCE   A push or a pull acting on a body. Force usually causes some distortion of the body, a change in its velocity, or both. Force is a vector.

	Metric	English
Force (F)	newton (N)	pound (lb)
	dyne	ounce (oz)
	metric ton	ton

The conversion factors between the primary units of force in the two systems are:

$$1\,N = 0.225\,lb \qquad 1\,lb = 4.45\,N$$

This means that a force of 150 pounds is equal to 668 newtons.

The distortion caused by a force is often obvious, such as the compression of a sofa cushion when you sit on it. Sometimes it cannot be observed without help. For example, high-speed photography reveals that a supposedly rigid golf ball is flattened as a club hits it (Figure 2.1). Spring scales, like the produce scales in supermarkets, use distortion to measure forces. The greater the force acting to stretch or compress a spring, the greater the distortion (Figure 2.2).

Force is a bit difficult to define and is sometimes regarded as a fundamental quantity like time and distance. The English system units should indicate to you that the idea of force is quite common. The words "push," "shove," "lift," "pull," and "yank" are some everyday synonyms that we use to describe force. The concepts of force and energy are probably the two most ubiquitous and useful in all of physics. Newton's laws of motion are simple, direct statements about forces in general and their relationship to motion.

Before formally stating and illustrating Newton's laws, let's consider some examples of forces and try to show just how versatile the concept is. We exert forces on objects in dozens of everyday situations, such as in pushing or pulling a door open, lifting a box, pulling down a window shade, and throwing a ball

FIGURE 2.1
Force on the golf ball distorts it and accelerates it.

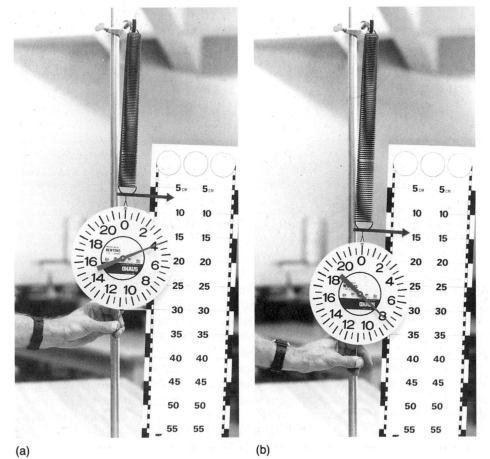

(a)   (b)

FIGURE 2.2
(a) A force of 4 newtons stretches the spring 7 centimeters.

(b) Doubling the force on the spring doubles the distance it is stretched. The scale itself has a spring inside and makes use of this principle.

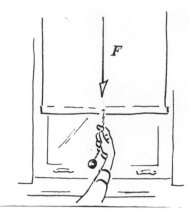

FIGURE 2.3
We exert forces every day.

(Figure 2.3). Many of the machines we use are designed to exert forces. Cranes, hoists, jacks, and vises are all obvious examples of such machines. Automobiles and other propelled vehicles function by causing forces to act on them. If this last concept seems a bit odd to you, come back to it after you have read about Newton's third law of motion in Section 2.7.

The most common force in our lives is *weight.* Most of us measure the weight of our bodies regularly. Many things that we buy, such as flour, cat food, and nails, are sold by weight.

> **WEIGHT**   The force of gravity acting on a body, abbreviated *W.*

The direction of the force of gravity is what determines the directions of up and down. We are so accustomed to living with this force that pulls everything towards the earth's surface that we often forget that it is a force (Figure 2.4). A child's simple observation that things naturally "fall down" leads one to Aristotle's idea of motion toward a natural place. But it was Newton who made the important observation that objects are pulled toward the earth by a force in much the same way that a sled is pulled by a person. The *law of universal gravitation,* the topic of Section 2.9, generalizes the concept of gravitational force.

The weight of any given object depends on two things: the amount of matter comprising the object (its mass) and whatever it is outside the object that is causing the gravitational force. This last point means that a body's weight depends on where it is. The weight of an object on the moon is about one sixth the weight it would have on earth because the gravitational pull of the moon is less than the earth's. Even on the earth itself, the weight of an object varies slightly with location. A 190-pound person would weigh about 1 pound less at the equator than at the North or South Pole (even excluding the difference the weight of extra clothes would make!).

Another important example of force is *friction.*

> **FRICTION**   A force of resistance to relative motion between two bodies or substances in physical contact.

Friction is at work when a chair slides across a floor, when brakes keep a car from rolling down a hill, when air resistance slows down a baseball, and when a boat glides over the water. Notice that friction occurs at the surfaces or boundaries of the materials involved. We can distinguish two types of friction: static and kinetic. When there is no relative motion between two objects, the friction that acts is *static friction.* If you try to push a refrigerator across a floor, it is the force of static friction between it and the floor that you must first overcome. A person who is walking or running relies on the static friction between his or her shoes and the ground. Even though the person's body is moving relative to the ground, there is no relative motion between the shoes and the ground when they are in contact. It is difficult to walk on ice because the force of static friction is limited. Even a moving car relies on the static friction between its tire surfaces and the pavement. As long as the tires are not skidding or spinning, there is no relative motion between them and the pavement where they are in contact with one another.

FIGURE 2.4
Weight (W) is the downward force of gravity. This force acts on objects whether they are stationary, moving horizontally, or moving vertically.

A block of wood resting on an inclined plane is a simple system that relies on static friction (Figure 2.5). If the plane is horizontal, there is no friction. When the plane is tilted a small amount, friction acts to keep the block from sliding down. Here the force of friction opposes the component of the force of gravity—weight—which acts parallel to the plane. The steeper the plane is, the greater the force of static friction must be to keep the block from sliding down. When the angle of the plane reaches a certain value, the block will begin to slide down; the component of the weight parallel to the plane will have exceeded the maximum force of static friction. Notice that the force of static friction between two surfaces can have any value between zero and some maximum.

Refer back to "vector addition" in Section 1.3.

Kinetic friction acts when there is relative motion between two substances in contact, such as an aircraft when moving through the air, a fish swimming underwater, or a tire skidding on pavement. In Figure 2.5, kinetic friction acts on the block of wood when it is sliding down the inclined plane. The force of kinetic friction that acts between two solids is usually less than the maximum static friction that can act. This means that a car can be stopped more quickly when its tires are not skidding.

F_g = Component of weight
F_f = Force due to friction

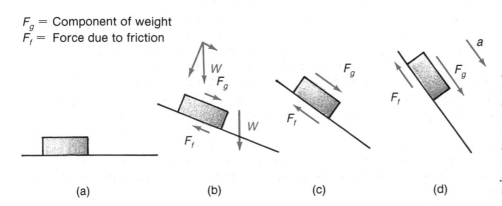

(a)   (b)   (c)   (d)

FIGURE 2.5
Static friction opposes gravity, which acts to pull the block down the incline. As the plane is tilted, both forces increase until the maximum possible force of static friction is reached (c). Then the force of gravity overcomes the frictional force, and the block slides down.

FIGURE 2.6
*Bicycle brakes use kinetic friction
as the pads rub against the rim
of the wheel.*

Place a pencil, an eraser, a small book, or some other object on the cover of this book, and gradually tilt the book until the object slides off. Notice that the angle at which the different objects begin to slide varies. Why is this so? Tilt the book until a particular object slides off. Decrease the angle slightly, and put the object back on it. It stays put unless you give it a slight push, then it slides off. This shows that the maximum static friction is greater than the kinetic friction.

Kinetic friction is often undesirable. A car that is traveling on a flat road at a constant speed consumes fuel mainly because it must act against the forces of kinetic friction—air resistance acting on the car's exterior and friction between various moving parts in the axles, transmission, and engine. This also applies to aircraft and ships. Brakes represent a useful application of kinetic friction. On most bicycles, the brakes consist of pads that rub against the wheel rims (Figure 2.6). When the brakes are applied, the force of kinetic friction between the pads and the rim slows the bicycle.

Often it is difficult to include the effect of friction in mathematical models of physical systems. Consequently we will find it helpful to consider situations in which frictional forces are small enough to be ignored. It is easier to analyze the motion of a falling rock than that of a falling feather. We need to understand these simple systems before we can incorporate the complexities of friction.

2.2   NEWTON'S FIRST LAW OF MOTION

As we have just seen, there are many forces that act in everyday situations. Newton's laws make no reference to specific types of forces. They are perfectly general statements that apply whether a force is caused by gravity, friction, or a direct push or pull.

NEWTON'S FIRST LAW OF MOTION   An object will remain at rest or in motion with constant velocity unless acted on by a net external force.

The last phrase needs some explanation. An *external* force is one that is caused by some outside agent. Weight is an external force because it is caused by something outside an object (eg, the earth). If your car stalls, you cannot move it by sitting in the driver's seat and pushing on the windshield. This would be an internal force. You must get out of the car and push from the outside. The *net* force is the vector sum of the forces acting on the body. If one person pushes forward on your car while another pushes backward with equal effort, the net force is zero. If two forces act in the same direction, the net force is simply the sum of the two (Figure 2.7). If two forces act in different directions, the net force is found by adding the two vectors together (Figure 2.8).

*Refer back to "vector addition"
in Section 1.3.*

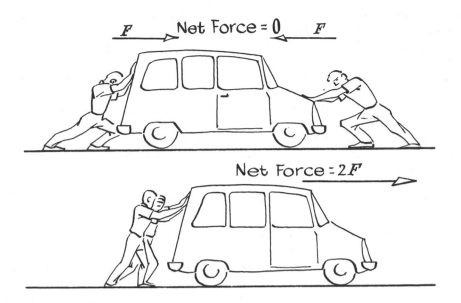

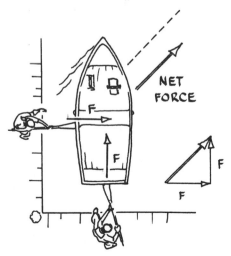

FIGURE 2.7
(a) Equal forces in opposite directions produce a net force equal to zero. (b) Equal forces in the same direction add together.

At first glance, Newton's first law does not seem to be very profound. Obviously an object will remain stationary unless a net force causes it to move. But the law also states that anything that is already moving will not speed up, slow down, or change direction unless a net force acts on it. This means that the states of *no motion* and of *uniform motion* are *equivalent* as far as forces are concerned. Aristotle's concept of unnatural motion (pp. 40–41) implies that a force is required to maintain motion. Newton's first law implies that a force is required only to *change* the state of motion. This may seem to run counter to your intuition, but that is because you rarely see a moving object with no forces acting on it. A car traveling at a constant velocity actually has zero net force acting on it: the various forces (including air resistance and gravity) cancel each other.

As you throw a ball, your hand exerts a net force that is in the same direction as the ball's velocity. The ball's speed increases because the force acts in the same direction as its motion. When you catch the ball, the force is opposite to the direction of the ball's velocity. Here the force slows down, or decelerates, the ball.

Another important point implied by the first law is that a force is required to *change the direction* of motion of an object. Velocity includes direction; constant velocity implies constant direction as well as constant speed. To make a moving object change its direction of motion, a net external force must act on it. To deflect a moving soccer ball, a player must exert a sideways force on it with the foot or head (Figure 2.9). Without such a force, the ball will not change direction.

The implication of Newton's first law is that a net external force must act on an object to speed it up, slow it down, or change its direction of motion. Since each of these is an acceleration of the object, the first law tells us that *a force is required to produce any acceleration.* The centripetal acceleration inherent in circular motion requires a force; it is called the *centripetal force.* Like centripetal acceleration, the centripetal force on an object is directed toward the center of the object's path. As a car goes around a curve, the centripetal force that acts on it is a sideways frictional force between the tires and the road. Without this friction, as is nearly the case on ice, there is no centripetal force, and the car will go in a straight line—off the curve.

FIGURE 2.8
Two sailors push on a boat in different directions. The net force is the vector sum of the two forces.

FIGURE 2.9
Force is needed to change the direction of motion.

Refer back to "acceleration" in Section 1.4.

Imagine a rubber ball tied to a string and whirled around in a circle overhead. The ball "wants" to move in a straight line (by the first law) but is prevented from doing so by the centripetal force acting through the string. If the string breaks, the path of the ball will be a straight line in the direction the ball was moving *at the instant the string broke* (Figure 2.10). You might be tempted to think that the ball would move in a direct line away from the center of its previous circular motion. But that is not the case. If you ever see a hammer thrower, a discus thrower, or someone using a sling (like the one David used in killing Goliath), notice the path of the projectile after it is released. It moves along a line tangent to its original circular path.

There are other examples of this effect. Children (or even physics professors) riding on a spinning merry-go-round must hold on because their bodies "want" to travel in a straight line, not in a circle. Similarly, clothes are partially dried during the spin cycle in an automatic washer because the water droplets tend to travel in straight lines and move through the holes in the side of the tub. In essence, the clothes are "pulled away" from the water as they spin. The same effect could be used to create an "artificial gravity" on a space station by making it spin (Figure 2.11). (The classic movie *2001: A Space Odyssey* showed this nicely.) You your-

FIGURE 2.10
Centripetal force must act on any object to keep it moving on a circular course. If the force is removed, the object will move at the same speed in a straight line.

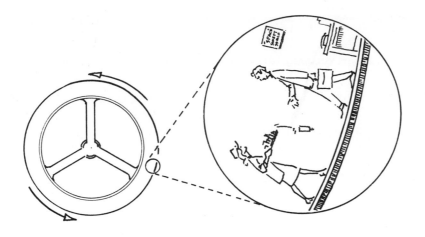

FIGURE 2.11
Model of a rotating space station. The circular paths of the occupants would make them experience artificial gravity.

self may have tried out a "spinning room" at an amusement park; it uses the same principle.

2.3 MASS

Newton's second law of motion states the exact relationship between the net force acting on an object and its acceleration. But before stating the second law, we will take another look at *mass*. Imagine the effect of a small net force acting on a car (see Figure 2.12). The resulting acceleration would be quite small. The same net force would cause a much larger acceleration if it acted on a shopping cart. Obviously, the effect of a force on an object depends on some measure of the object's size. The acceleration of a car would be small because the car is big. The mass of an object determines its acceleration when it experiences a net force.

Bearing that in mind, consider this definition of mass:

> **MASS** A measure of an object's resistance to acceleration. A measure of the quantity of matter in an object.

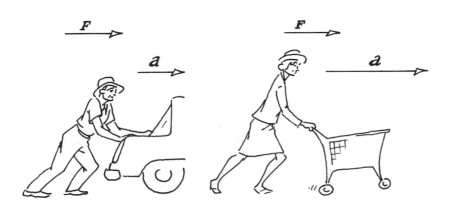

FIGURE 2.12
The effect of a net force on an object depends on the object's mass. If the mass is very large, the acceleration of the object will be very small. Conversely, if the mass is very small, the acceleration will be very large.

The equality of all persons under the law is one of the fundamental tenets of our judicial system. The equality of inertial and gravitational mass is a fundamental assumption in Einstein's theory of general relativity (or gravity). This assumption takes form in what is often called the *principle of equivalence*. It is the equivalence between inertial and gravitational mass which ensures that all objects, regardless of their composition or structure, fall at the same rate. This result was, of course, demonstrated by Galileo. But the first really accurate experiments to show that all bodies fall with the same acceleration were done by Baron Lorand (Roland) von Eötvös (pronounced *"út vush"*).

Eötvös (1848–1919) was a professor of physics in Budapest whose interests included capillary pressure, paleomagnetism, and the shape of the earth. Beginning in about 1886, he concentrated his research on gravity and devised a *null experiment* to demonstrate the equivalence of inertial and gravitational mass. Specifically he sought to balance the acceleration of one object against the acceleration of another and to look for small departures from equilibrium. If the accelerations were the same, no differences would be found, and he would have arrived at a null (zero) result.

Eötvös performed his experiment by hanging two weights—one made of wood, the other of platinum—from the ends of a 40 centimeter beam suspended by a fine wire (Figure 2.13). Any inequality in the ratios of the inertial to gravitational masses of the two bodies would have resulted in a twisting of the wire from which the balance was hung. The wire did not twist, however, and Eötvös concluded that the earth imparts the same acceleration to wood as it does to platinum to an accuracy of five parts in 10^9.

By 1922 this result had been extended to such materials as copper, asbestos, water, and tallow. In 1964 researchers led by Robert H. Dicke at Princeton University found that the *sun* accelerates cylinders of gold and aluminum at the same rate to within one part in 10^{11}! With experiments of such high accuracy to rely upon, acceptance of the "uniqueness of free fall," that is, the principle that any and all objects fall at the same rate, can be made with confidence.

Or can it? In 1986, investigators at Purdue University carefully reanalyzed the data from Eötvös' experiment and reported evidence that common objects of high density fall *more slowly* than those of low density! If this were true, then ignoring air resistance, a feather would reach the ground slightly ahead of an iron ball if both were released simultaneously from the same height. The Purdue scientists account for their findings by proposing the existence of a weak repulsive force, effective only over distances of a few hundred meters or so,

	Metric	English
Mass (m)	kilogram (kg) gram (g)	slug

The concept of mass is not in everyday use in the English system. For example, flour is purchased by weight (pounds), not by mass (slugs). In the metric system, the situation is the reverse. Mass is commonly used instead of weight. Flour is purchased by the kilogram, not by the newton.[1]

---

[1]To help you get a feel for the size of the kilogram, we offer the following "mixed" conversion:

1 kilogram *weighs* 2.2 pounds (on earth)

One kilogram *does not equal* 2.2 pounds! On earth, anything with that mass has a *weight* of 9.8 newtons—which happens to equal 2.2 pounds.

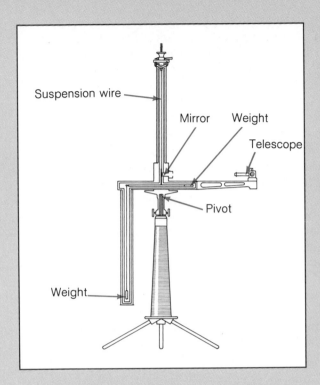

FIGURE 2.13
Torsion balance used by Eötvös to demonstrate the equivalence of inertial mass and gravitational mass.

which is composition-dependent: the force increases in strength in direct proportion to the number of neutrons and protons (or "baryons"; see Section 12.3) in the nucleus. This is why the iron ball falls more slowly: it feels a greater repulsive force from the earth than does the feather, which slightly reduces its downward gravitational acceleration relative to the feather.

Recent attempts to confirm the existence of this "fifth force of Nature" (joining the gravitational, the electromagnetic, and the strong and weak nuclear forces [see Section 12.2]) have produced conflicting data. Torsion balance experiments conducted by groups at Princeton and Seattle found no evidence for such a force, while measurements using sophisticated accelerometers produced mixed results, some supporting, some refuting the need for a composition-dependent fifth force. Gravity measurements along a 600-meter high television tower in North Carolina, a 2000-meter deep borehole in Greenland, and a 1000-meter long mine in Australia also seem to show deviations from the predictions of strict Newtonian gravitation, but it is not clear at present whether such observations require the postulated fifth force or not for their interpretation. Until this issue is resolved we must remain content in the recognition that, for now, with regard to the laws of free fall, just as with the laws of the land, everything is "created equal."

The actual mass of a given object depends on its size (volume) and its composition. A rock has more mass than a pebble because of its size; it has more mass than an otherwise identical piece of Styrofoam because of its composition. The rock is harder to accelerate in the sense that it requires a larger force. Fundamentally the mass of a body is determined mainly by the total numbers of subatomic particles that comprise it (protons, neutrons, and electrons).

Mass is *not* the same thing as weight. The distinction is often a source of confusion for beginning physics students. Mass and weight are related to each other in that the weight of an object is proportional to its mass. But *weight* is a *force* that is an effect of gravity. Mass is an intrinsic property of matter that does not depend on any external phenomenon. One major cause of the confusion is that users of the English system and users of the metric system often incorrectly lump together the two concepts in everyday use. Weight is often incorrectly used in place of mass in countries using the English system, and mass is often incorrectly used in place of weight in countries using the metric system.

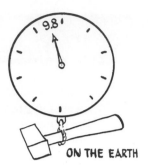

FIGURE 2.14
The hammer's weight depends on where it is, but its mass is always the same—1 kilogram.

If a hammer has a mass of 1 kilogram, that will not change if it is taken into space aboard a spacecraft and then to the moon's surface. Its mass is 1 kilogram wherever it is. The hammer's weight, however, varies with the location, because weight depends on gravity (see Figure 2.14). Its weight on earth is 9.8 newtons, its weight would appear to be zero in a spacecraft traveling to the moon, and its weight would be only 1.6 newtons on the moon's surface. Though "weightless" in the spacecraft, the hammer is *not massless* and would still resist acceleration as it does on earth.

Another way of approaching mass and weight is to think of them as two different characteristics of matter. We might call these aspects "inertial" and "gravitational." Mass is a measure of the inertial property of matter—how difficult it is to change its velocity. Weight illustrates the gravitational aspect of matter: any object experiences a pull by the earth, the moon, or any other body near it.

2.4   NEWTON'S SECOND LAW OF MOTION

We are now ready to state Newton's second law of motion. This law is our most important tool for applying mechanics in the real world.

> NEWTON'S SECOND LAW OF MOTION   An object is accelerated whenever a net external force acts on it. The net force equals the object's mass times its acceleration.[2]
>
> $$F = ma$$

This law expresses the exact relationship between force and acceleration. For a given body, a larger force will cause a proportionally larger acceleration (see Figure 2.15). Force and acceleration are both vectors. The direction of the acceleration of an object is the same as the direction of the net force acting on it.

---

[2]The unit of measure of force, the newton, is established by this law. One newton is the force required to give a 1-kilogram mass an acceleration of 1 m/s². In other words:

$$1 \text{ newton} = 1 \text{ kilogram–meter/second}^2$$

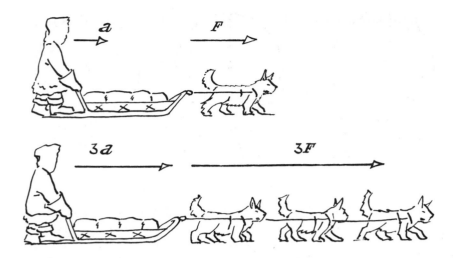

FIGURE 2.15
The acceleration of an object is proportional to the net force. Tripling the force on an object triples its acceleration.

An airplane with a mass of 2,000 kilograms is observed accelerating at a rate of 4 meter/second². What is the net force acting on it?

$$F = ma = 2{,}000 \text{ kg} \times 4 \text{ m/s}^2$$

$$F = 8{,}000 \text{ N}$$

EXAMPLE 2.1

Another way of stating Newton's second law is that an object's acceleration is equal to the net force acting on the object divided by its mass.

$$a = \frac{F}{m}$$

This means that a large force acting on a large mass can result in the same acceleration as a small force acting on a small mass. A force of 8,000 newtons acting on a 2,000-kilogram airplane causes the same acceleration as a force of 4 newtons acting on a 1-kilogram toy.

The following example illustrates what can be done with the mechanics that we have learned so far.

An automobile manufacturer decides to build a car that can accelerate uniformly from 0 mph to 60 mph in 10 s (see Figure 2.16). In metric units this is 0 m/s to 27 m/s. The car's mass is to be about 1,000 kg. What is the force required?

EXAMPLE 2.2

$v = 0$

60 mph = 27 m/s

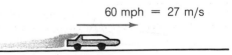

FIGURE 2.16
Car accelerating from rest to 60 miles/hour.

First, we must determine the acceleration and then use Newton's second law to find the force.

$$a = \frac{\Delta v}{\Delta t} = \frac{27 \text{ m/s} - 0 \text{ m/s}}{10 \text{ s}}$$

$$a = 2.7 \text{ m/s}^2$$

So the force needed to cause this acceleration is

$$F = ma = 1{,}000 \text{ kg} \times 2.7 \text{ m/s}^2$$

$$F = 2{,}700 \text{ N}$$

(This is equal to 2,700 × 0.225 pounds = 607.5 lb) In the next chapter we will use this information to determine the size (power output) of the car's engine.

It is Newton's second law that establishes the relationship between mass and weight. Recall that any freely falling body has an acceleration equal to g (9.8 m/s²). By the second law, the size of the gravitational force needed to cause this acceleration is:

$$F = ma = mg$$

We call this force the object's weight. So in this case the expression "force is equal to mass times acceleration" translates to "weight is equal to mass times acceleration of gravity."

$$F = ma \quad \rightarrow \quad W = mg$$

On earth, where the acceleration of gravity is 9.8 meter/second², the weight of an object is:

$$W = m \times 9.8 \quad \text{(on earth, } m \text{ in kg, } W \text{ in N)}$$

The weight of a 2-kilogram brick is:

$$2 \times 9.8 = 19.6 \text{ N}$$

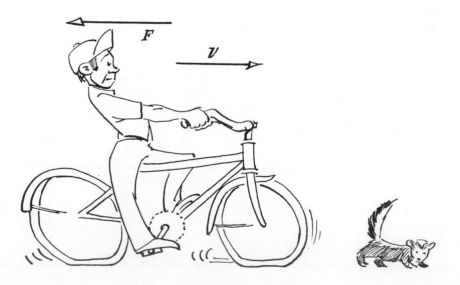

FIGURE 2.17
The bicycle slows down because the net force on it (F) is opposite its velocity (v).

On the moon the acceleration of gravity is 1.6 meter/second². Therefore:

$$W = m \times 1.6 \qquad \text{(on the moon, } m \text{ in kg, W in N)}$$

Forces that cause deceleration or centripetal acceleration also follow the second law. To slow down a moving object, a force must act in the direction opposite the object's velocity (see Figure 2.17). A force acting sideways on a moving body causes a centripetal acceleration; the object moves along a circular path as long as the net force remains perpendicular to the velocity. The size of the centripetal force that is required is

Refer back to "centripetal acceleration" in Section 1.4.

$$F = ma, \qquad \text{and since} \qquad a = \frac{v^2}{r}$$

$$F = \frac{mv^2}{r} \qquad \text{(centripetal force)}$$

In Example 1.4, we computed the centripetal acceleration of a car going 20 m/s around a curve with a radius of 60 m. If the car's mass is 1,000 kg, what is the centripetal force that acts on it?

EXAMPLE 2.3

$$F = \frac{mv^2}{r} = \frac{1,000 \text{ kg} \times (20 \text{ m/s})^2}{60 \text{ m}}$$

$$F = 6,670 \text{ N} = 1,500 \text{ lb}$$

Since we had already computed the acceleration ($a = 6.67$ m/s²), we could have used $F = ma$ directly.

$$F = ma = 1,000 \text{ kg} \times 6.67 \text{ m/s}^2$$

$$F = 6,670 \text{ N}$$

The centripetal force acting on a car going around a curve is supplied by the friction between the tires and the road. Note that the faster the car goes, the greater the force required to keep it moving in a circle. If two identical cars go around the same curve, and one is going *two times* as fast as the other, the faster car needs *four times* the centripetal force (see Figure 2.18). The required centripetal force is inversely proportional to the radius of the curve. A tighter curve

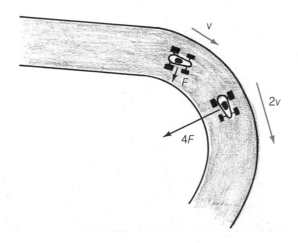

FIGURE 2.18
The centripetal force necessary to keep a car on a curved road is supplied by static friction between the tires and the road. The size of the force is proportional to the square of the speed. One car going twice as fast as the other would need four times the centripetal force.

(one with a smaller radius) requires a larger force or a smaller speed. If a car goes around a curve too fast, the force of friction will be two small to maintain the centripetal force, and the car will go off the outside of the curve.

This form of Newton's second law, $F = ma$, is not the original version, but it is the most useful one for our purposes. It applies only when the mass of the object doesn't change. This is almost always the case, but there are important exceptions. For example, the mass of a rocket decreases rapidly as it consumes its fuel. Mathematics more advanced than we will use is needed to deal with such cases.

2.5   THE SI SYSTEM OF UNITS

The metric unit of force—the newton—is the force required to cause a 1-kilogram mass to accelerate 1 m/s^2. So when a mass in kilograms is multiplied by an acceleration in meters per second squared, the result is a force in newtons. If grams or cm/s^2 were used instead, the force unit would not be newtons.

At this point, you may be getting confused by the different units of measure, particularly when different physical quantities are combined in an equation. To help alleviate this problem, a separate system of units within the metric system was established. It is called the *SI system* after the French phrase "Système International d'Unités." The SI system consists of only one unit of measure for each physical quantity (Table 2.1). Each unit is chosen so that the system is internally consistent; that is, when two or more physical quantities are combined in an equation, the result will be in SI units if the original physical quantities are in SI units. For example, the SI units of distance and time are the meter and the second, respectively. Consequently, the SI unit of speed is the meter per second. Table 2.1 gives the SI units of the physical quantities that we have used so far.

You may notice that, for the most part, we have been using SI units in our examples. From now on, we will use SI units when working in the metric system.

TABLE 2.1 SI UNITS (PARTIAL LIST)	
Physical Quantity	**SI Unit**
Distance (d)	Meter (m)
Area (A)	Square meter (m^2)
Volume (V)	Cubic meter (m^3)
Time (t)	Second (s)
Frequency (f)	Hertz (Hz)
Speed and velocity (v)	Meter per second (m/s)
Acceleration (a)	Meter per second squared (m/s^2)
Force (F) and weight (W)	Newton (N)
Mass (m)	Kilogram (kg)

2.6   EXAMPLES: DIFFERENT FORCES, DIFFERENT MOTIONS

In Chapter 1, we considered three simple states of motion: zero velocity, constant velocity, and constant acceleration. Now, let us see how force is involved in these cases and then look at some other types of motion. Here the particular *cause* of the force is not important. A constant net force acting on a moving object has the same effect whether it is due to gravity, friction, or someone's pushing the object. What interests us now is the relationship of the force—both its magnitude and direction—to the motion. Our concerns are the direction of the force compared to the object's velocity, whether the force is constant, and the way in which the force varies if it is not constant.

The acceleration of a stationary object or of one moving with uniform motion (constant velocity) is zero. By the second law, this means that the net force is also zero (see Figure 2.19). It is just that simple. When the net force on a body is zero, we say that it is in *equilibrium,* whether it is stationary or moving with constant velocity.

In uniform acceleration, the second law implies that a constant force acts to cause the constant acceleration. A steady net force acting in a fixed direction will make an object move with a constant acceleration. A freely falling body is a good example: the force (weight) is constant and always acts in a downward direction.

A net force that acts opposite to the direction of motion of the object will cause it to slow down. If the force continues to act, the object will come to a stop for an instant and then accelerate in the direction of the force, opposite its original direction of motion. This is what happens when you throw a ball straight up into the air (see Figure 2.20). It is accelerating downward the whole time. While on the way up, its speed decreases until it reaches its highest point. There it has zero speed for an instant and then it falls with increasing speed. Its acceleration is always g, even at the instant that it stops.

What happens when an object is thrown upward at an angle to the earth's surface? This is one example of a classic mechanics problem—*projectile motion.* The motion is a composite of horizontal and vertical motions. The path that a projectile takes has a characteristic arc shape that is nicely illustrated by the stream of water from a water fountain (see Figure 2.21). This shape, an important one in mathematics, is called a *parabola.*

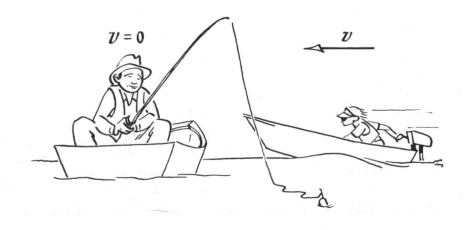

FIGURE 2.19
Whether moving with constant velocity or sitting at rest, the net force on both boats is zero.

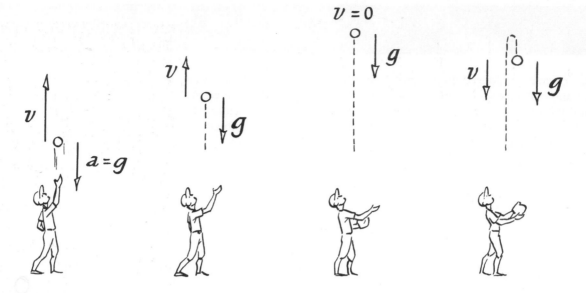

FIGURE 2.20
When a ball is thrown straight up, its acceleration is 1 g downward. This causes it to decelerate on the way up, stop for an instant, then fall downward.

The key to understanding projectile motion is to realize that the vertical force of gravity has no effect on the horizontal motion. An object that is initially moving horizontally has the same downward acceleration, g, as an object that is simply dropped. So, ignoring air resistance, a *projectile moves horizontally with constant speed and vertically with constant downward acceleration* (Figure 2.22). As an object moves along its path, the vertical component of its velocity decreases to zero (at the high point of the arc) and then increases downward. The horizontal component of its velocity stays constant. Over level ground, a projectile will travel farthest if it starts at an angle of 45° to the ground. When the force of air resistance is large enough to affect the motion, such as when a softball is thrown very far, the maximum range occurs at smaller angles.

DO-IT-YOURSELF PHYSICS

Investigate projectile motion with a garden hose on a day with no wind. Shoot the water up at an angle to the ground, and note how the distance it travels varies with the angle of the noz- zle. What angle gives the maximum range? For any range shorter than this, there are two different angles that will give the same range. Demonstrate this.

FIGURE 2.21
The stream of water from a water fountain shows the shape of the path of projectiles—a parabola.

Simple Harmonic Motion

Here is a simple situation with a force that is not constant. Imagine an object, say a block, attached to the end of a spring (Figure 2.23). When it is not moving, the block is in equilibrium, and the net force on it is zero. If you lift the block up a

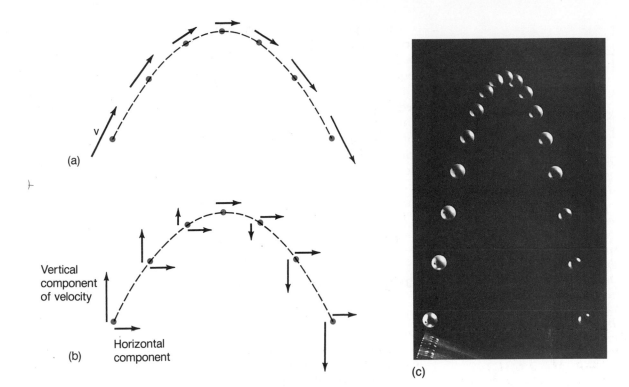

FIGURE 2.22
(a) The path of a projectile is shown with its velocity vector at selected moments. At the very top its velocity is horizontal. (b) The same projectile with the vertical and horizontal components of the velocity shown separately. The horizontal velocity component stays constant, while the vertical component decreases, then increases downward, because of the constant downward acceleration of gravity. (c) A strobe view of a projectile.

bit and then release it, it will experience a net force downward, toward its original (rest) position. The reverse occurs if you pull it down a bit below its rest position and release it. Then the net force on it will be up, again toward its original position. Such a force is called a *restoring force* because it acts to restore the system to the original configuration. In this case, the net force is proportional to the distance from the rest position. The farther the block is displaced, the greater the force acting to move it back. (Figure 2.2 also shows this.)

What kind of motion does this force cause? If an object is pulled down and then released, the upward force will cause it to accelerate. It will pick up speed as it moves upward, but the force and acceleration decrease as it nears the rest position. When it reaches that point, the force is zero, and it stops accelerating but continues its upward motion (Newton's first law). Once it has moved past the rest position, the force is downward. The object will slow down, stop for an instant, and then gain speed downward. This process is repeated over and over: the object oscillates up and down.

This type of motion, which is very important in physics, is called *simple harmonic motion*. It occurs in many other systems: a pendulum swinging through a small angle, a cork bobbing up and down in the water, a car with very bad shock absorbers, and air molecules vibrating with the sound from a tuning fork. In fact, simple harmonic motion is involved in all kinds of waves, as we shall see.

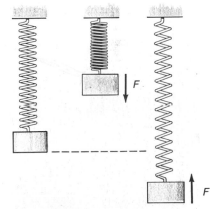

FIGURE 2.23
The net force that acts on a block hanging from a spring depends on the displacement of the block from its rest (equilibrium) position. If raised and then released, the block experiences a net force downward. If the block is pulled downward and then released, the net force is upward.

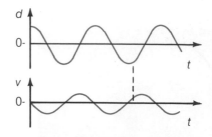

FIGURE 2.24
Graphs of distance versus time and speed versus time for simple harmonic motion. Notice that the speed is zero when the distance is largest, and vice versa. This agrees with what we see: at the high and low points of the motion, the mass is not moving for an instant. As it passes through the rest position, it is at its highest speed.

The graphs of distance versus time and speed versus time show the characteristic oscillation (Figure 2.24). Both graphs have what is known as a *sinusoidal shape*. This shape arises often in physics, and we will see it again when we talk about waves (Chapter 6).

Simple harmonic motion is a cyclic motion with a constant frequency. The frequency of oscillation depends only on the mass of the object and the strength of the spring.[3] The same mass on a stronger spring will have a higher frequency. For a given spring, a larger mass will have a smaller (lower) frequency. This principle is employed by an *inertial balance,* a device that uses simple harmonic motion to measure the mass of an object. To use it one measures the frequency of oscillation of the object and then compares this frequency to those of known masses listed in a table or plotted on a graph. In an orbiting spacecraft, this sort of scale must be used because of the weightless environment (see Figure 2.25).

Falling Body with Air Resistance

Without air resistance the constant force on a falling body (its weight) gives it a constant acceleration, *g.* Its speed increases steadily until it hits the ground. But as any body falls through the air, the force of air resistance grows as the speed increases and eventually affects the motion (Figure 2.26). This increasing force acts opposite to the downward force of gravity, so the *net force decreases*. This continues as the body gains speed until the force of air resistance acting upwards equals the weight acting downwards. At this point, the net force is zero, so the speed stays constant from then on. This speed is called the *terminal speed* of the body.

Rocks and other dense objects have large terminal speeds and may fall for many seconds before air resistance affects their motion appreciably. Feathers, dandelion seeds, and balloons take less than a second to reach their terminal speeds, which are quite low. If a skydiver (see Figure 2.27) jumps out of a hovering helicopter or balloon, he or she will fall for about 2 or 3 seconds before the force of air resistance starts to have a major effect. The terminal speed depends on the skydiver's size and orientation when falling, but it is typically around 120 mph (about 54 m/s).[4] Then, when the parachute opens, the increased air resistance decelerates the skydiver to a much lower terminal speed—maybe 10 mph.

In the spring and block example in Figure 2.23, the net force depends on the object's *distance* from its equilibrium position. When air resistance affects the motion of a falling body, the net force depends on the body's *speed*. These

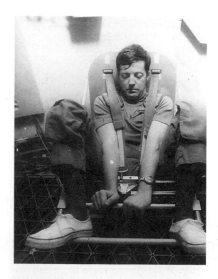

FIGURE 2.25
An astronaut tries out an inertial balance while training for a Skylab mission. The chair and astronaut oscillate back and forth while an electronic timer and calculator measure the frequency and compute the astronaut's mass.

---

[3]For the mathematically inclined: The strength of a spring is found by measuring the force *F* needed to stretch it a distance *d.* Then one computes:

$$k = \frac{F}{d}$$

"k" is called the spring constant. A strong (stiff) spring has a large *k,* and a weak spring has a small *k.* A mass *m* attached to the spring would oscillate with a frequency given by:

$$f = \frac{1}{2\pi} \times \sqrt{\frac{k}{m}} = 0.159 \times \sqrt{\frac{k}{m}}$$

[4]Terminal speed also depends on how thin the air is. In 1960 Captain Joseph Kittinger jumped from a balloon at 100,000 feet altitude and reached a speed of 825 miles/hour—faster than the speed of sound! As he descended into denser air he decelerated—even before he opened his parachute.

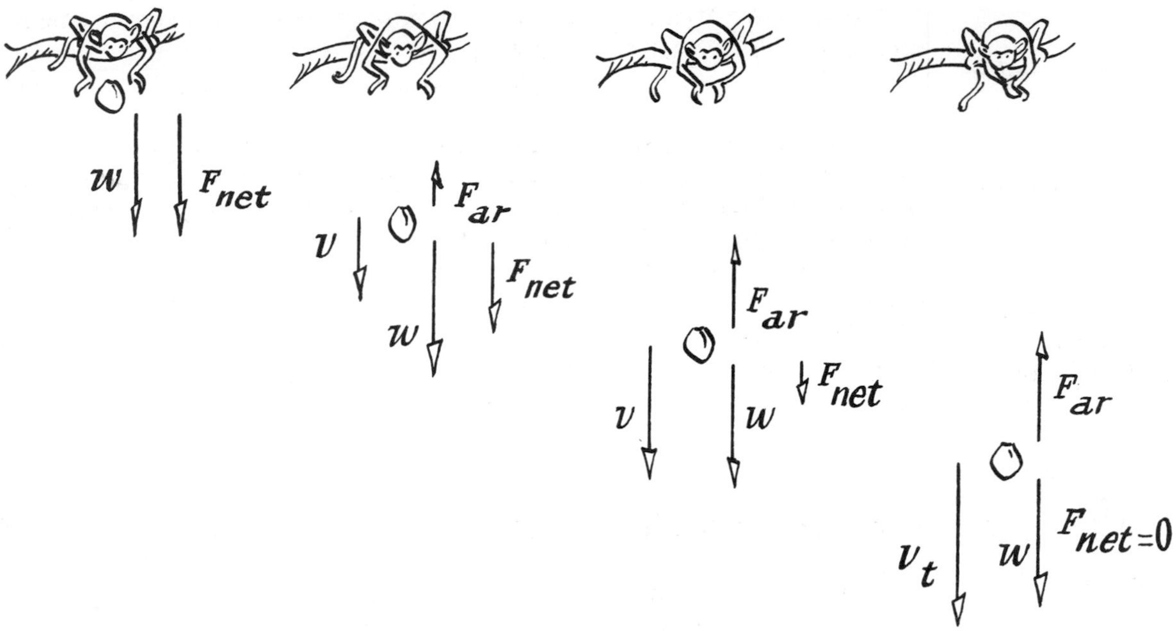

w F_{net}

v F_{ar} F_{net} w

F_{ar} v w F_{net}

F_{ar} v_t w $F_{net} = 0$

FIGURE 2.26
Successive views of a falling body affected by air resistance (ar). The upward force of air resistance increases as the object's speed increases. When this force is large enough to offset the downward weight (W), the net force (F_{net}) is zero (right). The object's speed is constant and is called the terminal speed (v_t).

examples illustrate how interconnected the different physical quantities are in real physical systems. Things can get quite complicated: imagine a spring and mass suspended under water. Here the net force would depend on both the block's position and its speed. See if you can describe how the block would move. Table 2.2 summarizes the types of forces that we have considered thus far.

TABLE 2.2 SUMMARY OF EXAMPLES OF FORCES	
Nature of Net Force	Description of Motion
Zero net force	Constant velocity: stationary or motion in straight line with constant speed
Constant net force	Constant acceleration
Force parallel to velocity	Motion in a straight line with increasing speed
Force opposite to velocity	Motion in a straight line with decreasing speed
Force perpendicular to velocity	Motion in a circle: radius depends on speed and force
Restoring force proportional to displacement	Simple harmonic motion (oscillation)
Net force decreases as speed increases	Acceleration decreases: velocity reaches a constant value

FIGURE 2.27
A skydiver uses air resistance to fall at a steady, slow speed.

✍ | **LEARNING CHECK**

1. The force of gravity acting on an object is called _____ .

2. The air exerts a force on a person standing in a strong wind. This type of force is one example of _____ friction.

3. For an object to move with constant velocity,
   a) the net force on it must be zero.
   b) a constant force must act on it.
   c) it must move in a circle.
   d) there must be no friction acting on it.

4. When a given net force acts on an object, the object's acceleration depends on its _____ .

5. The weight of a body equals its mass times _____ .

6. In which of the following situations is the net force constant?
   a) Simple harmonic motion
   b) Freely falling body
   c) A dropped object affected by air resistance
   d) All of the above

2.7   NEWTON'S THIRD LAW OF MOTION

Newton's third law of motion is a basic statement about the nature of forces in general. It is a simple law that adds an important perspective to the understanding of forces.

> **NEWTON'S THIRD LAW OF MOTION**   Forces always come in pairs: when one object exerts a force on a second object, the second exerts an equal and opposite force on the first.

If object A causes a force on object B, then object B exerts an equal force in the opposite direction on A.

$$F_{B\,on\,A} = {}^{-}F_{A\,on\,B}$$

If you push against a wall with your hand, the wall exerts an equal force back on your hand (see Figure 2.28). A book resting on a table exerts a downward force on the table equal to the weight of the book. The table exerts an upward force on the book also equal to the book's weight. The earth pulls down on you with a force that is called your weight. Consequently *you exert an equal but upward force on the earth.* When you dive off a diving board, the earth's gravitational force accelerates you downward. At the same time your equal and opposite pull upward on the earth accelerates it toward you. But since the earth's mass is about

FIGURE 2.28
If one body exerts a force on a second body, the second exerts an equal and opposite force on the first.

one hundred thousand billion billion (10^{23}) times your mass, its acceleration is negligible.

The third law gives a new insight into what actually happens in many physical systems. If you are on roller skates and push hard against a wall, you accelerate backwards (see Figure 2.29). Think about this for a moment: a *forward* force by your hands makes you accelerate in the *opposite direction*. In reality it is the equal and opposite force of the wall on your hands that causes the acceleration. If you stand in the middle of the floor, you can't use your hands to push yourself because you need to push *against* something. Similarly, when a car speeds up, the engine causes the tires to push backwards on the road. It is the road's equal and opposite force on the tires that causes the car to accelerate forward. The same thing happens when the brakes slow the vehicle. The recoil of a gun is caused by the third law: the large force accelerating the bullet produces an equal and opposite force on the gun—the "kick."

Figure 2.30 illustrates Newton's third law. A wooden cart is fitted with a spring plunger that can be pushed into the cart and retained. When the spring is released, the plunger pushes out. If there is nothing for the plunger to push against (left), the cart is not moving afterwards. The plunger can exert no force, and so there is no equal and opposite force to accelerate the cart. If the cart is next to

FIGURE 2.29
When a roller skater pushes against a wall, the wall's equal and opposite force on the skater causes the skater's acceleration backwards.

FIGURE 2.30
A wooden cart is fitted with a spring-loaded plunger and trigger. The cart is accelerated only if the plunger can exert a force on something else; then the equal and opposite force on the cart causes the acceleration.

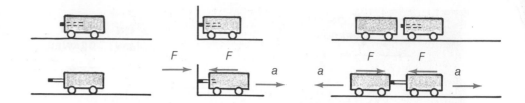

a wall when the spring is released (center), the force on the wall results in an opposite force on the cart, and it accelerates away. If a second cart is next to the plunger (right), both carts feel the same force in opposite directions, and they accelerate away from each other. In Chapter 3, we will show that the ratio of the speeds of the carts depends on the ratio of their masses. If one has twice the mass of the other, it will have one half the speed.

Rockets and jet aircraft are propelled by ejecting combustion gases at a high speed (Figure 2.31). The engine exerts a force that ejects the gases, and the gases exert an equal and opposite force on the rocket or jet. Hence, unlike cars, they do not need to push against anything to be propelled.

Birds, airplanes, and gliders can fly because of the upward force exerted on their wings as they move through the air (Figure 2.32a). As the air flows under and over the wing, it is deflected (forced) downward. This results in an equal and opposite upward force on the wing, called *lift* (see Figure 2.32b). Propellers on airplanes, helicopters, boats, and ships employ the same basic idea. They pull or push the air or water in one direction, resulting in a force on the propeller in the opposite direction. Propellers are useless in a vacuum.

DO-IT-YOURSELF PHYSICS

When riding in a car on a wide highway, put your hand a short distance out of the window, palm down, and use it as a wing to illustrate lift. The vertical force is the lift; the force acting backwards is known as *drag*. Vary the angle of your hand, and notice how the lift varies. At a certain large angle, the lift actually goes to zero: this is known as a *stall*.

Whenever an object is accelerating, it exerts an equal and opposite force on whatever is accelerating it. You have probably noticed this when riding in a car, bus, or airplane as it accelerates: the seat exerts a forward force on you that causes you to accelerate. Your body pushes back on the seat with an equal and opposite force. From your point of view, it seems that there is some force "pulling" you back against the seat. This is not a real force, just a reaction of your mass to acceleration. The same effect is observed when an object is undergoing a centripetal acceleration. As a car or bus goes around a curve, you seem to be "pulled" to the side. Again, it is just a reaction to a net force causing an acceleration.

Like Newton's first law, the third law is primarily conceptual rather than mathematical. It emphasizes that forces arise only during interactions between two or more things. By thinking in terms of pairs of forces, we can more easily distinguish the causes and effects in such interactions.

FIGURE 2.31
Rockets use Newton's third law.

(a)

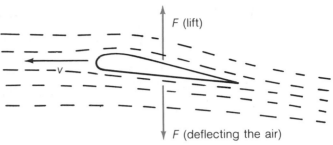

F (lift)

v

F (deflecting the air)

FIGURE 2.32
(a) Author Vern Ostdiek having fun with Newton's third law. (b) A wing on a flying aircraft deflects the air downward. This downward force on the air causes an equal and opposite upward force on the wing.

2.8 THE SPACE SHUTTLE: NEWTON'S LAWS ON A GRAND SCALE

Let us apply your new understanding of mechanics to the motion of the space shuttle as it is launched into orbit (see Figure 2.33). The winged orbiter is attached to an external fuel tank, as are two solid rocket boosters. (It was a leak in one of these boosters that caused the tragic explosion of the Challenger on 28 January 1986.) The external fuel tank actually has two main compartments, one containing liquid hydrogen and the other liquid oxygen. The hydrogen and oxygen are fed through 12-inch-diameter pipes to each of the three main engines on the orbiter. The 775 tons of hydrogen and oxygen are burned by the engines in 8½ minutes. The exhaust is water vapor. The two solid rocket boosters consume their 1,100 tons of solid fuel in 2 minutes and then separate from the external tank.

It takes approximately 10 minutes for the shuttle to go from the launch pad to its orbit, about 170 miles above the earth's surface. The shuttle starts out moving straight up but gradually slants until it is traveling parallel to the earth's surface. This means that the force of gravity acts opposite the direction of motion at the beginning of the launch sequence and perpendicular to the motion at the end. During this time the mass of the shuttle changes constantly as the fuel is consumed and the solid rocket boosters and then the external tank are jettisoned.

The upshot of all this is that the acceleration of the shuttle varies considerably during the launch sequence. To make matters simpler for the moment, let us consider the acceleration of the shuttle caused by the five engines alone and ignore the effect of gravity. At launch, the shuttle's engines are consuming more than 10 tons of fuel per second and ejecting that amount of combustion gases downward. The resulting upward force on the shuttle gives it an initial acceleration of about 1.5 *g* (the net acceleration is 0.5 *g*). In metric units, the lift-off mass is 2,000,000 kilograms, the total force (thrust) is about 30,000,000 newtons, so the acceleration is:

The German mathematician Johannes Kepler wrote the following words to Galileo in April 1610:
"There will certainly be no lack of human pioneers when we have mastered the art of flight. . . . Let us create vessels and sails adjusted to the heavenly ether, and there will be plenty of people unafraid of the empty wastes. In the meantime, we shall prepare, for the brave sky-travelers, maps of the celestial bodies—I shall do it for the Moon, you, Galileo, for Jupiter."

FIGURE 2.33
The space shuttle Challenger at lift-off on its first mission.

$$a = \frac{F}{m} = \frac{30{,}000{,}000 \text{ N}}{2{,}000{,}000 \text{ kg}} = 15 \text{ m/s}^2$$

The graph in Figure 2.34 shows how the acceleration changes during the launch sequence. Note that it is a graph of acceleration—not speed—versus time. At first the graph slants upward; the acceleration increases because the mass of the shuttle is decreasing. (By Newton's second law, the acceleration is inversely proportional to the mass.) At about the 55-second mark, the engines are throttled back because the shuttle is exceeding the speed of sound, and it encounters the greatest stress of the ascent. Shortly thereafter, full power is resumed, and the acceleration continues to increase. (It was during this part of the launch that the Challenger exploded.) The solid rocket boosters run out of fuel at the 2-minute mark and are jettisoned. This causes the acceleration to drop to about 1 *g* because only the three orbiter main engines are left, and the huge external tank is still about 75% full. During the next 5½ minutes, the acceleration increases steadily to 3 *g* as the fuel is consumed. At that point the engines are again throttled back so that the acceleration does not go beyond 3 *g*. This is done for the comfort of the crew members and to reduce the stress on any sensitive equipment on board. (In some of the earlier space missions, when the astronauts were professional test pilots, the accelerations reached around 9 *g*.)

At the 8½-minute mark, the orbiter main engines are shut off, and the external tank is jettisoned. The final boost into orbit is supplied by two other rocket engines with their own fuel supplies on board the orbiter. These same engines, part of the orbital maneuvering system, are used to propel the orbiter for any needed changes in its orbit and to slow it down to begin the reentry sequence. The rest of the time the orbiter is coasting in its orbit, with no force acting on it

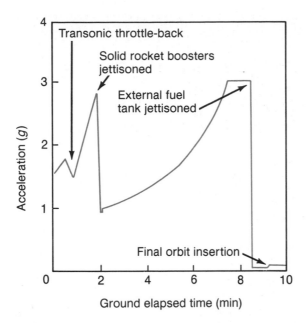

FIGURE 2.34
Graph of the acceleration of the space shuttle versus time during the launch sequence. (From NASA)

except the earth's gravity. It is in a state of constant free fall toward the earth, but its large horizontal speed makes its path nearly circular instead of parabolic.

The orbiter's speed in its normal orbit is about 17,600 miles/hour (7,850 meters/second). Its average net acceleration during the 10-minute (600-second) launch sequence is:

$$a = \frac{\Delta v}{\Delta t} = \frac{7{,}850 \text{ m/s}}{600 \text{ s}}$$

$$a = 13.1 \text{ m/s}^2 = 1.34\,g$$

(This calculation ignores the initial speed, about 400 meters/second, that the shuttle has while on the launch pad because of the rotation of the earth.)

Space shuttle astronauts return the shuttle to earth by first pointing the tail in the direction of motion and then firing the orbital maneuvering system engines. This slows the shuttle and puts it on a new trajectory that takes it back into the earth's atmosphere. The rest of the deceleration, from about 17,000 miles/hour to about 200 miles/hour when it touches down, is caused by the force of air resistance acting on the shuttle's outer surface. The heat that is generated by this friction would melt the aluminum spacecraft if it weren't protected by insulating tiles. Some places on the nose, wings, and tail reach 3,000°F.

2.9 THE LAW OF UNIVERSAL GRAVITATION

Newton's fourth major contribution to the study of mechanics is not a law of motion but a law relating to gravity. Newton made an important intellectual leap: he realized that the force that pulls objects towards the earth's surface also holds the moon in its orbit. Moreover, he claimed that *every* object exerts an attractive

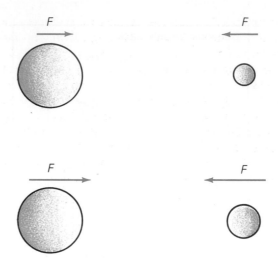

FIGURE 2.35
The equal and opposite gravitational forces that act on any pair of objects depend on the masses of both objects. If either object is replaced by another with a larger mass, the two forces are proportionally larger.

force on every other object. This concept is called *universal gravitation:* gravity acts everywhere and on all things. The force that the earth exerts on objects near its surface—weight—is just one example of universal gravitation.

What determines the size of the gravitational force that acts between two bodies? Newton used his deep understanding of mathematics and mechanics along with information about the orbits of the moon and the planets to reason what the force law must be. The third law of motion states that when two bodies exert forces on one another, the forces are equal in size. Since the earth's gravitational force depends on an object's mass, Newton reasoned, the force on each object in a pair is proportional to each object's mass (see Figure 2.35). If the mass of *either* object is doubled, the sizes of the forces on *both* are doubled.

The size of the gravitational force between two bodies must also depend on the distance between them. The sun is much more massive than the earth, but its force on you is much less than the earth's, because you are much closer to the earth. For geometric reasons, Newton felt that the size of the gravitational force

FIGURE 2.36
The gravitational force that the earth exerts on other objects is inversely proportional to the square of the distance from the earth's center to the object's center. That is why the moon's centripetal acceleration is much less than g. (Not drawn to scale.)

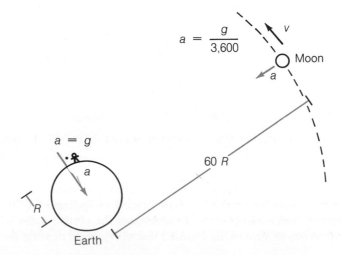

is inversely proportional to the square of the distance between the bodies. But he needed proof—proof that he got from examining the moon's orbit.

Long before Newton's time, astronomers had measured the radius of the moon's orbit and, knowing this and the period of its motion, had determined its orbital speed. So Newton could calculate the moon's (centripetal) acceleration, v^2/r, which turned out to be about $g/3{,}600$. In other words, the moon's acceleration is about 3,600 times smaller than that of an object falling freely near the earth. Newton then realized that the moon is about 60 times farther from the earth's center than an object on the earth's surface. The acceleration of the moon is 60 squared, or 3,600 times, smaller because it is 60 times as far away (see Figure 2.36). Since acceleration is proportional to force, Newton had his proof that the gravitational force is inversely proportional to the square of the distance between the centers of the two objects.

NEWTON'S LAW OF UNIVERSAL GRAVITATION   Every object exerts a gravitational pull on every other object. The force is proportional to the masses of both objects and inversely proportional to the square of the distance between their centers.

$$F \propto \frac{m_1 m_2}{d^2}$$

(m_1 and m_2 are the masses of the two objects, and d is the distance between their centers.)

This force acts between the earth and the moon, the earth and you, two rocks in a desert—any pair of objects. Your weight depends on the distance between you and the earth's center. If you were twice as far from the earth's center (8,000 miles instead of 4,000 miles), you would weigh one fourth as much (see Figure 2.37). If you were three times as far, you would weigh one ninth as much.

In 1798 the English physicist Henry Cavendish performed precise measurements of the actual gravitational forces acting between masses. Like Eötvös did years later, Cavendish used a delicate torsion balance (Figure 2.38). Two masses were balanced on a beam that was suspended from a thin wire attached to the beam's midpoint. Two large masses were placed next to the smaller ones on the beam. The gravitational forces on the smaller masses were strong enough to rotate the beam slightly and consequently twist the wire. Cavendish used the amount of twist to measure the size of the gravitational force. His results showed that the force between two 1-kilogram masses 1 meter apart would be 6.67×10^{-11} N. With this result, Newton's law of universal gravitation can be expressed as an equation:

$$F = \frac{6.67 \times 10^{-11} \, m_1 m_2}{d^2}$$

The constant of proportionality in the equation is called the gravitational constant, G.

$$G = 6.67 \times 10^{-11} \, \text{N-m}^2/\text{kg}^2$$

Since G is such a small number, the gravitational force between objects is usually quite small. For example, the force between two persons, each with mass equal to 70 kilogram, when they are 1 meter apart is only 0.00000033 newtons,

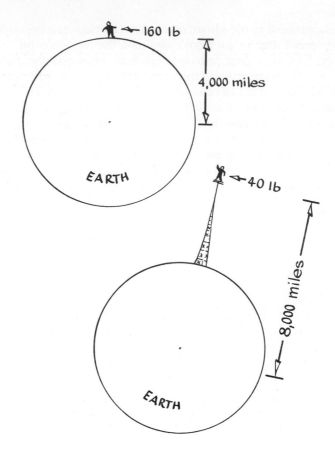

FIGURE 2.37
On a tower 4,000 miles high, you would be twice as far from the earth's center as you are when standing on its surface. On the tower, you would weigh one fourth as much.

or 0.0000012 ounces. The earth's gravitational force on us (and our force on the earth) is so large because the earth's mass is huge.

In fact, we can use the law of universal gravitation to compute the mass of the entire earth. First compute the gravitational force exerted on a body by the earth—the body's weight—using the equation $W = mg$. This force can also be calculated using:

$$F = \frac{GmM}{R^2}$$

Here M represents the mass of the earth, and R is the radius of the earth, the distance the body is from the earth's center. The value of R, first measured by the Greeks over 2,000 years ago, is about 6.4×10^6 meters. Since these two forces are equal to each other:

$$F = W$$

$$mg = \frac{GmM}{R^2}$$

Canceling the m:

$$g = \frac{GM}{R^2}$$

The values of g, G, and R can be inserted and the resulting equation solved for M, the mass of the earth. (The result: $M = 6 \times 10^{24}$ kg.)

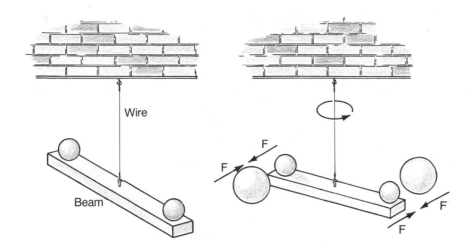

FIGURE 2.38
Simplified sketch of the torsion balance used by Cavendish to measure the gravitational force between masses. The two large objects exert forces on the two small ones, causing the suspending wire to twist.

The acceleration of gravity on the moon is not the same as it is on the earth. Likewise, each of the other planets has its own *"g."* This variation exists simply because the masses and the radii are all different. The actual values of the acceleration of gravity have been computed for the moon, the sun, and the planets, using the preceding equation that relates g to their mass, M, and their radius R (see Table 2.3).

Newton used his law of universal gravitation to explain a variety of phenomena that had been mysteries before. These include the cause of tides, the motion of comets, and the theoretical basis for orbital motion.

Orbits

Newton used an elegant "thought experiment" to illustrate that *orbital motion about the earth is actually an extension of projectile motion.* Imagine that a cannon is placed at the top of a very high mountain and that it can shoot a cannonball at any desired speed (see Figure 2.39). If the cannonball just rolls out of the barrel, it will fall in a straight line to the earth. Given some small initial speed, it will follow a parabolic trajectory to earth (path D in Figure 2.39). But if

TABLE 2.3 ACCELERATION OF GRAVITY IN OUR SOLAR SYSTEM	
Location	Acceleration of Gravity
Earth	1.0 g
Sun	27.9 g
Moon	0.16 g
Mercury	0.38 g
Venus	0.88 g
Mars	0.39 g
Jupiter	2.65 g

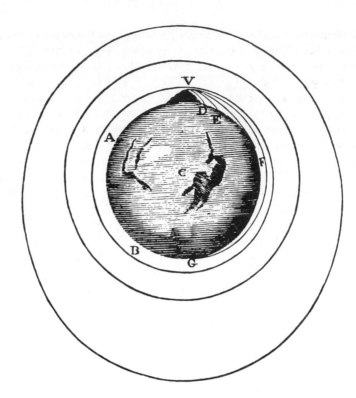

FIGURE 2.39
The original drawing of Newton's cannon "thought experiment." The cannon is placed on a very high mountaintop (V). The path of the cannonball depends on its initial speed.

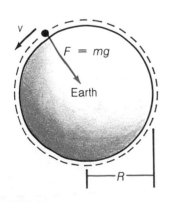

FIGURE 2.40
Satellite in a low, circular orbit about the earth. The gravitational (centripetal) force on the satellite is about equal to the satellite's weight when on the earth, mg, and the radius of the orbit is about equal to the earth's radius, R.

its speed is increased more, the ball travels farther and farther around the earth before it hits the ground (paths E, F, and G). If it were possible to shoot a cannonball with a high enough speed, it would actually travel in a full circle around the earth and hit the cannon in the rear. It would be in orbit about the earth, just as the moon is. An object in orbit is continually "falling" toward the earth. On an even higher mountain, one could place the cannonball in an orbit with a larger radius.

Realistically this could not be done because of the force of air resistance. To orbit the earth, an object must be above (outside) the earth's atmosphere. But the idea is a nice way of showing the connection between a terrestrial phenomenon—projectile motion—and the motion of the moon about the earth. It also shows what Newton predicted—that we can put satellites into orbit.

It is quite easy to estimate how fast something has to move to stay in orbit. When an object is moving in a circle about the earth, the centripetal force on it is the earth's gravitational pull. For a satellite orbiting just above the earth's surface, the gravitational force is approximately equal to the satellite's weight when it is on the earth's surface—mg. Also, the radius of the satellite's orbit is approximately equal to the radius of the earth, R (see Figure 2.40). The required speed is found by equating the centripetal force to the gravitational force. The centripetal force is mv^2/R, where R is the radius of the earth. So

$$\frac{mv^2}{R} = mg$$

$$v^2 = gR = 9.8 \text{ m/s}^2 \times (6.4 \times 10^6 \text{ m})$$

$$v^2 = 63{,}000{,}000 \text{ m}^2/\text{s}^2$$

$$v = 7{,}900 \text{ m/s}$$

The space shuttle's speed in its normal orbit is about 7,850 meters/second. Our estimate is good, but is a bit too high because we did not include the fact that the shuttle is 170 miles above the earth's surface. There the gravitational force is slightly less than *mg,* and the radius of its orbit is slightly more than *R.*

Newton's mathematical analysis of orbital motion is not restricted to earth orbits. It also applies to the motions of the planets about the sun and to the moons in orbit about the other planets. His results allowed astronomers to calculate the orbits of celestial objects with higher accuracy and with less observational data than before. He showed that comets, such as Halley's Comet, are actually moving in orbits about the sun. The orbits of most of them are flattened ellipses with the sun near one end, at a point called the "focus" of the ellipse (Figure 2.41). This is why Halley's Comet is only near enough to the earth to be seen every 76 years. It spends most of its time far from the sun and the earth.

The orbits of all of the planets, earth included, are actually ellipses, although they are much closer to being circular than the orbit of Halley's Comet. The sun is at one focus of these orbits. Note in Figure 2.41 that the elliptical nature of Pluto's orbit causes it to spend part of the time inside the orbit of Neptune.

DO-IT-YOURSELF PHYSICS

You can draw the correct shape of an ellipse using two tacks or pins, some string, a ruler, a piece of paper, and a surface into which you can push the tacks (a bulletin board works). Place the tacks into the paper 10 centimeters apart (or use 4 inches). Wrap the string about the tacks, and tie it to form a tight

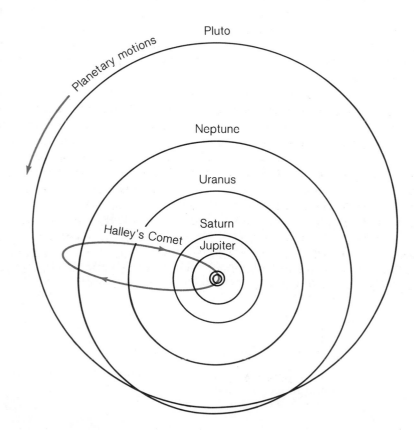

FIGURE 2.41
Orbit of Halley's Comet around the sun. The comet is seen from earth only when it is in that part of its orbit near the sun. The earth's orbit is the smallest circle.

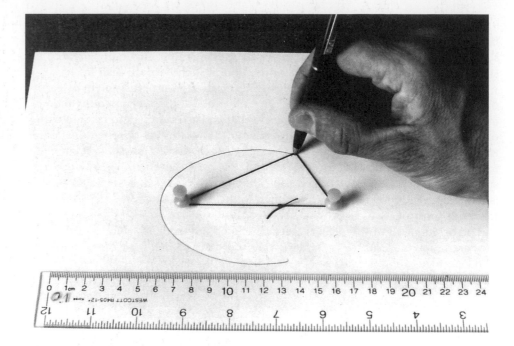

FIGURE 2.42

loop around them. Move the tacks closer together a bit. Insert your pen into the loop, move it outward until the string is tight, and draw the complete path around the tacks while holding the pen against the string. The resulting figure is an ellipse (Figure 2.42) with each tack at one focus of the ellipse. Place the tacks 4 centimeters apart (or use 1.6 inches). The ellipse you draw has the shape of Pluto's orbit. Note that it is hard to distinguish it from a circle. Place the tacks 8.6 centimeters apart (or use 3.44 inches) to draw the shape of the orbit of Nereid, a moon of Neptune that has a highly elliptical orbit. A spacing of 9.8 centimeters (or use 3.92 inches), if you can do the drawing, gives you the shape of the orbit of Halley's Comet.

Gravitational Field

Gravitation is an example of "action at a distance." Objects exert forces on each other, even though they may be far apart and there is no matter between them to transmit the forces. The other forces we have talked about involves direct contact between things; gravitation does not.

How is force possible without contact? One way to get better insight into this situation is to use the concept of a *field*. Imagine an object situated in space. The matter in the object causes an effect or a disturbance in the space around it. We called this a *gravitational field*. This field extends out in all directions but becomes weaker at greater distances from the object. In this model it is the field itself that causes a force to act on any other object. It plays the role of an invisible agent for the gravitational force. Whenever a second body is in the field of the first, it experiences a gravitational force. But the field is present even when there is no other object around to feel its effect.

We might call this a "force field" because it causes forces on other bodies. (But one must not imagine it to be the kind of "invisible wall" one sees in science fiction movies.) One way to represent the shape of the gravitational field around an object is by drawing arrows at different points in space. They show the magnitude and the direction of the force that *would* act on a particular object placed at each point (Figure 2.43a). These arrows are long near the object and short farther away because the gravitational force decreases as the distance increases. Another way to represent the gravitational field is to connect the arrows, making "field lines." The direction of the field line at any point in space again shows the direction of the force that would act on an object placed there. The strength of the gravitational field is represented by the spacing of the lines: the lines are farther apart where the field is weaker (Figure 2.43b).

All of the forces in nature can be traced to four fundamental forces.[5] The gravitational force is one of them. The others are the electromagnetic force, responsible for electric and magnetic effects; the strong nuclear force; and the weak nuclear force. These will be discussed in later chapters. But what about friction and other forces involving direct contact? Most of these can be traced back to the electromagnetic force. This force determines the sizes and shapes of atoms and molecules. For example, a stretched spring pulls back because of the electrical forces between the atoms in it.

2.10   TIDES

We have heard that "time and tide wait for no man." The truth of this statement is evident to anyone who has longed for the sun to pause, even briefly, on the horizon to permit another moment's enjoyment of a sunset's beauty or to any coastal dweller whose boat has been left high and dry by the receding ocean while he or she has been occupied elsewhere. In Chapter 1 we talked a little about time. Let us now consider some aspects of tides.

---

[5]See the last two paragraphs of the *Physics Potpourri* "That All Men are Created Equal" on pp. 58–59.

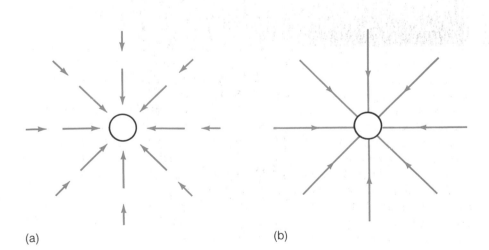

(a) (b)

FIGURE 2.43
Two ways of showing the gravitational field around an object: (a) with arrows and (b) with field lines.

FIGURE 2.44
Gravitational forces exerted by the moon on material at three different locations on earth. (Not drawn to scale.)

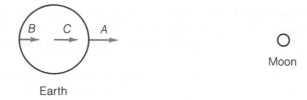

For the most part tides are the result of gravitational forces exerted on the earth by the moon. To understand this, consider Figure 2.44. Points *A, B,* and *C* all lie on a line connecting the center of the earth to the center of the moon. According to Newton's law of universal gravitation, earth material at *A* experiences a stronger attractive force toward the moon than does material at *C* because it is closer to the moon. Similarly, material at *C* experiences a greater attractive force toward the moon than does material at *B*. Thus, applying Newton's second law of motion, material at *A* accelerates ("falls") toward the moon at a greater rate than does material at *C*, while material at *C* accelerates in the same direction but at a higher rate than that at *B*. The net effect of these accelerations is to separate these three points, *A* moving out ahead of *C* in a given period of time by an amount equal to that by which *C* moves out ahead of *B* (assuming a spherical earth). Thus the shape of the earth is elongated or stretched by the gravitational pull of the moon along a line connecting the two bodies.

This view is one that might be seen by an observer outside the earth-moon system looking in. One might ask what an observer at point *C* sees. Looking toward *A*, such an observer would notice the separation between them growing larger as *A* moves away because of its higher acceleration. Alternatively, looking toward *B* from *C*, one would see the separation increasing as *B* lags farther and farther behind because of its smaller acceleration. From *C*'s point of view, *B* would appear to undergo a deceleration (a negative acceleration). Again applying Newton's second law, we can redraw the first figure (Figure 2.45), now indicating how an earth-bound observer at *C* views the forces exerted by the moon at points *A* and *B*. Notice that the same stretching or elongation of the earth along the direction of the moon occurs as before, but now the symmetry of the situation as seen from *C* is manifested.

But what has this to do with tides? Suppose we were to perform this same kind of analysis for points *D, E, F, G, H,* and *J*. We would find that, relative to *C*, forces

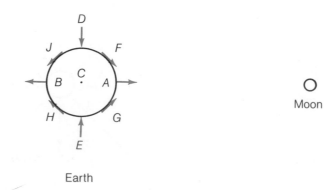

FIGURE 2.45
Gravitational forces exerted by the moon on material at different locations on earth, relative to point C, the earth's center.

along the directions of the arrows shown in Figure 2.44 would exist because of the gravitational presence of the moon. Now imagine the earth to be covered with an initially uniform depth of water. How would this fluid move in response to these forces?

Again, applying the laws of mechanics, we are led to the following, perhaps somewhat startling, results. (1) Water at points D and E would weigh slightly more than water elsewhere because in addition to the earth's own gravitational pull toward C, there is an additional small acceleration toward C produced by the moon. (2) Conversely, water at points A and B would weigh slightly less than water elsewhere because the moon exerts small forces in directions opposite to the earth's own inward gravitational attraction toward C. These forces work to counteract gravity and, hence, to reduce the weight of the water. (3) Water at points F, G, H, and J would feel a force parallel to the earth's surface and would begin to flow in the directions of the arrows at each point, much as water flows down any inclined surface in response to gravity. This flow causes the water to pile up at opposite sides of the earth along the line joining the earth to the moon. These "piles" of water are called the "tidal bulges," and as the earth rotates, points on its surface move into and then out of the bulges, resulting in the high and low tides observed at intervals of roughly 6 hours (Figure 2.46).

Needless to say, many complications to this simple picture cause the behavior of real tides to deviate somewhat from that predicted by this model (see Figure 2.47). In particular, we have ignored such things as tidal effects caused by the sun, the influence of the earth's rotation on the motion of the tidal bulges, and the effects on the local heights of tides caused by the earth's rugged, uneven surface. The first of these complications leads to such phenomena as the lower-than-average *neap tides*, which occur when the moon is in either the first or third quarter phase (and hence at right angles to the sun in the sky), and the higher-than-average *spring tides*, which happen when the moon is in either the new or the full phase (and hence in line with the sun in the sky). Despite having ignored these problems, we are still able to understand the basic characteristics of tides in terms of this simple Newtonian model. It is critically important to realize that they result principally from the gravitational interaction between the earth and its nearest celestial neighbor, the moon. So while the tides may not wait for us, their behavior is sufficiently predictable that we can at least be aware of the time of their arrivals and departures.

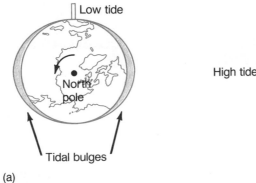

(a)

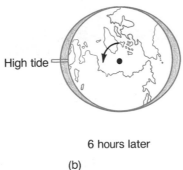

(b)

FIGURE 2.46
(a) The "tidal bulges" in the oceans caused by the gravitational pull of the moon. (b) As the earth rotates, places on its surface move from low tide to high tide, back to low tide, and so on. (Not drawn to scale.)

FIGURE 2.47
High and low tides at Friendship, Maine.

HISTORICAL NOTES

Sir Isaac Newton

Isaac Newton was born in rural Woolsthorpe, England on Christmas Day in 1642 (Figure 2.48). He seems to have had a normal, though inquisitive, childhood and entered Trinity College, Cambridge in 1661. There he studied mathematics and physics until the university was closed in the summer of 1665 because of the Great Plague. Newton returned to Woolsthorpe and during the next two years made his great intellectual discoveries in mechanics—the laws of motion and gravitation. As if this were not enough for someone in his early twenties, he also made important contributions to mathematical analysis, independently invented calculus, and explained how prisms produce the spectrum of colors from sunlight. Any of these accomplishments alone would have ensured him a place in the history of science and mathematics. The fact that they were performed in only two years by an "amateur" in isolation from the great intellectual centers of the world makes Newton's feat one of the greatest in the history of human thought.

Modern physical science and mathematics would be impossible without calculus. (Newton invented this entirely new branch of mathematics at about the same time as did the German philosopher and mathematician Baron Gottfried Wilhelm Leibnitz.) Calculus extended the reach of mathematics to the realm of continuously changing physical processes. With his calculus and his laws of motion, Newton was able to analyze physical systems that have nonconstant forces and, consequently, varying accelerations. A mass on a spring and the resulting simple harmonic motion is a good example (see Section 2.6). Many of the equations in this book were originally derived using calculus. After Newton, mechanics was no longer limited to the ideal systems of Galileo. Perhaps the best indication of the importance of calculus is that virtually all students in mathematics or physical science take courses in calculus at the beginning of their college education.

After Cambridge reopened in 1667, Newton returned to pursue his studies, but he did not publish his discoveries. He recognized their importance but was apparently reluctant to reveal them before they were completely developed. One of the details that he eventually proved, using his calculus, was that the gravitational force between spherical bodies depends on the distance between their centers. Depending on how you look at it, this may or may not seem logical, but Newton wanted proof. If this were not true, Newton's law of universal gravitation would not tie together the earth's force on the moon and the earth's force on objects on the earth. Newton's instructors were aware of his accomplishments, and one of his mathematics professors turned over his position to Newton in 1669.

During his twenty-year tenure as a professor of mathematics at Cambridge, Newton concentrated more on the study of light than on mechanics. He invented the reflecting telescope, thereby giving astronomers their most valuable tool (Figure 2.49). Nearly all of the great telescopes in use today, including the Hubble Space Telescope, are reflectors. Newton's work with lenses and the prismatic analysis of light were important contributions to optics. He did publish some of these discoveries, but they were met with contentious criticism, and Newton resolved to withhold his other works. His major critic argued that he had made some interesting findings about how light behaved but hadn't dealt with the fundamental question of what exactly light is. Newton realized that one must first know as much as possible about the properties and behavior of light before one can hope to explain what it is.

The great astronomer Edmond Halley, after whom Halley's Comet was named, recognized the importance of Newton's mechanics to physics in general and to astronomy in particular. Halley succeeded in convincing Newton to publish a treatise on mechanics. After eighteen months of prodigious concentration and labor, Newton completed his greatest work, the *Principia Mathematica Philosophiae Naturalis* (*Mathematical Principles of Natural Philosophy;* Figure 2.50). Published in 1687, the *Principia* was soon recognized as one of the greatest books ever written. In it Newton presented his basic principles of mechanics and

FIGURE 2.48
Isaac Newton (1642–1727).

FIGURE 2.49
Newton's reflecting telescope, which he presented to the Royal Society in London.

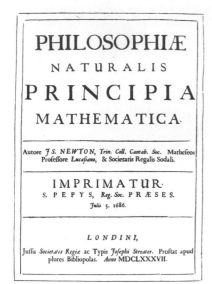

PHILOSOPHIÆ

NATURALIS

PRINCIPIA

MATHEMATICA·

Autore *JS. NEWTON,* Trin. Coll. Cantab. Soc. Matheseos
Professore *Lucasiano,* & Societatis Regalis Sodali.

IMPRIMATUR·
S. P E P Y S, *Reg. Soc.* P R Æ S E S.
Julii 5. 1686.

L O N D I N I,

Jussu *Societatis Regiæ* ac Typis *Josephi Streater.* Prostat apud
plures Bibliopolas. *Anno* MDCLXXXVII.

FIGURE 2.50
Title page of Newton's Principia
Mathematica.

gravitation and used them to explain a number of physical phenomena that had puzzled scientists for centuries. He also calculated the masses of the earth, the moon, the sun, and some of the planets. Most of the material in this chapter is based on the *Principia.*

In 1696 Newton left Cambridge and the academic world and became Warden of the Mint in London. Three years later he was appointed Master of the Mint and served in this capacity until his death. Newton put England on the gold standard and introduced milling on the edges of coins to discourage people from shaving off the precious metals. In 1704 he published another important book, *Opticks,* which summarized his work with light and color that spanned several decades. In 1705 he was knighted by Queen Anne. Sir Isaac Newton was buried with full honors at Westminster Abbey in 1727.

Newton was not simply a scientist and mathematician. He spent much of his "spare" time throughout his life analyzing biblical text. He was a religious man and considered his great discoveries in physical science evidence of the hand of God.

It is a bit difficult to fully appreciate Newton's work at this point in history because we are on the inside looking out, so to speak. But the very manner in which we "do" science is based, to a large extent, on Newton's discoveries. He showed that nature has special built-in rules (laws) that govern diverse phenomena. Individual types of motion need not be considered in isolation; they are all subject to three basic laws. He fully united the terrestrial world and the cosmos by showing that these rules apply in both realms. Newton revealed the immense usefulness and power gained by incorporating advanced mathematics into physics. He broke from the abstract, speculative approach to science and showed a new way with a combination of ingenious experimentation and theoretical analysis. One can argue that Newton's intellectual accomplishments are unsurpassed in their influence on history.

SUMMARY

Force is the most important physical quantity in Newtonian mechanics. Forces are all around us; weight and friction play dominant roles in our physical environment. The three laws of motion are direct, universal statements about forces in general and how they affect motion. Newton's *first law* states that *an object's velocity is constant unless a net force acts on it.* The *second law* states that the net force on an object equals its mass times its acceleration. This law is the key to the application of mechanics. By knowing the size of the force acting on a body, one can determine its acceleration. Then its future velocity and position can be predicted using the concepts in Chapter 1. Newton's third law of motion states that *whenever one body exerts a force on a second one, the second exerts an equal and opposite force on the first.* This deceptively simple statement gives us new insight into the nature of forces. The law of universal gravitation states that equal and opposite forces of attraction act between every pair of objects. The size of each force is proportional to the masses of the objects and inversely proportional to the square of the distance between their centers.

Isaac Newton is the founder of modern physical science. His laws of motion, law of gravitation, and calculus form a nearly complete system for solving problems in mechanics. He also made important discoveries in optics. Newton's combination of logical experimentation and mathematical analysis shaped the way science has been done ever since.

SUMMARY OF IMPORTANT EQUATIONS

EQUATION

COMMENTS

FUNDAMENTAL EQUATIONS

$F = ma$

Newton's second law of motion

$a = \dfrac{F}{m}$

Newton's second law of motion

$W = mg$

Relates weight and mass

$F = \dfrac{Gm_1m_2}{d^2}$

Law of universal gravitation

SPECIAL CASE EQUATIONS

$F = \dfrac{mv^2}{r}$

Centripetal force (object moving in a circular path)

QUESTIONS

1. What is force? Identify several of the forces that are acting on or around you.

2. What is weight? Under what circumstances might something be weightless?

3. A person places a book on the roof of a car and drives off without remembering to remove it. As the book and the car move down a street at a steady speed, there are two horizontal forces acting on the book. What are they?

4. What are the two types of friction? Can both types act on the same object at the same time?

5. What do we mean by "external" force? In light of Newton's third law of motion, why can't an internal force cause a net force?

6. How does an object move when it is subject to a steady centripetal force? How does it move if that force suddenly disappears?

7. Discuss the distinction between mass and weight.

8. Newton's second law of motion establishes a definite relationship between the net force on an object and its acceleration. Is there a similar relationship between the net force and the object's velocity?

9. As a rocket ascends, its acceleration increases even though the net force on it stays constant. Why?

10. What is the SI system?

11. An archer aims an arrow exactly horizontal over a flat field and shoots it. At the same instant the archer's watchband breaks and the watch falls to the ground. Does the watch hit the ground before, at the same time as, or after the arrow hits the ground? Defend your answer.

12. Describe the variation of the net force on, and the acceleration of, a mass on a spring as its executes simple harmonic motion.

13. Explain how the change in the force of air resistance on a falling body causes it to eventually reach a terminal speed.

14. The terminal speed of a ping pong ball is about 20 mph. From the top of a building a ping pong ball is thrown *downward* with an initial speed of 50 mph. Describe what happens to the ball's speed as in moves downward from the moment it is thrown to the moment when it hits the ground.

15. At least two forces are acting on you right now. What are these forces? On what is the equal and opposite force of each of these acting?

16. How is Newton's third law of motion involved when you jump straight upward?

17. Jane and John are both on roller skates and are facing each other. First Jane pushes on John with her hands and they move apart. Later they get together and John pushes equally hard on Jane with his hands and they move apart. Do they move any differently in the two cases? Why or why not?

18. What would be the gravitational force on you if you were at a point in space a distance R (the earth's radius) above the earth's surface?

19. Describe how the magnitude of the gravitational force between objects was first measured.

20. If suddenly the value of G, the gravitational constant, increased to a billion times its actual value, what sorts of things would happen?

21. The first "Lunar Olympics" is to be held on the moon inside a huge dome. Of the usual Olympic events—track and field, swimming, gymnastics, and so on—which would be drastically affected by the moon's gravity? In which events would earth-based records be broken? In which events would the performances be no better—or perhaps worse—than on the earth?

22. In the broadest terms, what causes tides?

23. What mathematical "tool" did Isaac Newton develop that has been essential to advanced physics since then?

PROBLEMS

1. Express your weight in newtons. From this determine your mass in kilograms.

2. A child weighs 300 N. What is the child's mass? *30.6 kg*

3. A certain airline allows a maximum of 30 kg for each suitcase a passenger brings along.
   a) What is the weight (in newtons) of a 30-kg suitcase? *294 N*
   b) What is the weight in pounds? *66.15 lb*

4. The mass of a certain elephant is 1,130 kg.

   a) Find the elephant's weight in newtons.
   b) Find the weight in pounds.

5. A 400-kg sailboat accelerates at a rate of 1.5 m/s². What is the net force acting on it?

6. A motorcycle and rider have a total mass equal to 300 kg. The rider applies the brakes, causing the motorcycle to decelerate at a rate of 5 m/s². What is the net force on the motorcycle?

7. As a 2-kg ball rolls down a ramp, the net force on it is 10 N. What is the acceleration?

8. In an experiment performed in a space station, a force of 60 N causes an object to have an acceleration equal to 4 m/s². What is the object's mass?

9. The engines in a supertanker carrying crude oil produce a net force of 20,000,000 N on the ship. If the resulting acceleration is 0.1 m/s², what is the ship's mass?

10. A person stands on a scale inside an elevator at rest. The scale reads 800 N.

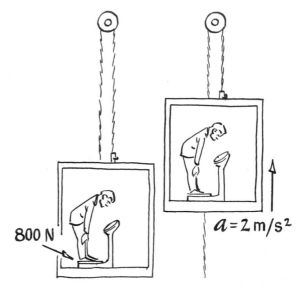

   a) What is the person's mass?
   b) The elevator accelerates upward momentarily at the rate of 2 m/s². What does the scale read then?
   c) The elevator then moves with a steady speed of 5 m/s. What does the scale read?

11. A jet aircraft with a mass of 4,500 kg has an engine that exerts a force (thrust) equal to 60,000 N.
   a) What is the jet's acceleration when it takes off?
   b) What is the jet's speed after it accelerates for 8 s?
   c) How far does the jet travel during the 8 s?

12. At the end of Section 1.5, we mentioned that the maximum acceleration of a fist during a particular karate blow was measured to be about 3,500 m/s² (Figure 1.32).
   a) If the mass of the fist was approximately 0.7 kg, what was the maximum force?
   b) What was the maximum force on the concrete block?

13. A sprinter with a mass of 80 kg accelerates from 0 m/s to 9 m/s in 3 s.
   a) What is the runner's acceleration?
   b) What is the net force on the runner?
   c) How far does the sprinter go during the 3 s?

14. As a baseball is being caught, its speed goes from 30 m/s to 0 m/s in about 0.005 s. Its mass is 0.145 kg.
a) What is the baseball's acceleration in m/s^2 and in g's?
b) What is the size of the force acting on it?

15. An airplane is built to withstand a maximum acceleration of 6 g. If its mass is 1,200 kg, what size force would cause this acceleration?

16. Under certain conditions the human body can safely withstand an acceleration of 10 g.
a) What net force would have to act on someone with mass of 50 kg to cause this acceleration?
b) Find the weight of such a person in pounds, then convert the answer to (a) to pounds.

17. A race car rounds a curve at 60 m/s. The radius of the curve is 400 m, and the car's mass is 600 kg.
a) What is the car's (centripetal) acceleration? What is it in g's?
b) What is the centripetal force acting on the car?

18. A hang glider and its pilot have a total mass equal to 120 kg. While executing a 360° turn, the glider moves in a circle with an 8-m radius. The glider's speed is 10 m/s.
a) What is the net force on the hang glider?
b) What is the acceleration?

19. A 0.1-kg ball is attached to a string and whirled around in a circle overhead. The string breaks if the force on it exceeds 60 N. What is the maximum speed the ball can have when the radius of the circle is 1 m?

20. On a highway curve with radius 50 m, the maximum force of static friction (centripetal force) that can act on a 1,000-kg car going around the curve is 8,000 N. What speed limit should be posted for the curve so that cars can negotiate it safely?

21. A space probe is launched from the earth, headed for deep space. At a distance of 10,000 miles from the earth's center the gravitational force on it is 600 lb. What is the size of the force when it is at each of the following distances from the earth's center:
a) 20,000 miles
b) 30,000 miles
c) 100,000 miles

CHALLENGES

1. The force on a baseball as it is being hit with a bat can be over 1,000 lbs. No human can push on a bat with that much force. What is happening in this instance?

2. Why does banking a curve on a highway allow a vehicle to go faster on it?

3. As a horse and carriage are accelerating from rest, the horse exerts a force of 400 N on the carriage. Illustrating Newton's third law, the carriage exerts an equal and opposite force of 400 N. Since the two forces are in opposite directions, why don't they cancel each other and produce zero acceleration?

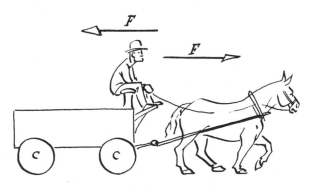

4. Complete the calculation of the force acting between two 70-kg people standing 1 meter apart as given on p. 77.

5. Perhaps you've noticed that the rockets used to put satellites and spacecraft into orbit are usually launched from pads near the equator. Why is this so? Is the fact that rockets are usually launched to the east also important? Why?

6. The acceleration of a freely falling body is not exactly the same everywhere on earth. For example, in the Galapagos Islands at the equator, the acceleration of a freely falling body is 9.780 m/s^2, while at the latitude of Oslo, Norway, it is 9.831 m/s^2. Why does the acceleration differ?

7. A 200-kg communications satellite is placed into a circular orbit around the earth with a radius of 4.23×10^7 m (26,300 miles).

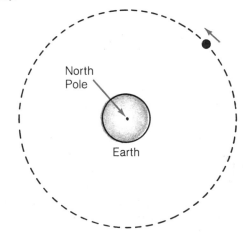

a) Find the gravitational force on the satellite. (There is some useful information in Section 2.9.)
b) Use the equation for centripetal force to compute the speed of the satellite.
c) Show that the period of the satellite—the time it takes to complete one orbit—is 1 day. (The distance it travels during one orbit is 2π, or 6.28, times the radius.) This is

a *geosynchronous orbit:* the satellite stays above a fixed point on the earth's equator.

8. Complete the calculation of the mass of the earth as outlined in Section 2.9.

SUGGESTED READINGS

Boslough, John. "Searching for the Secrets of Gravity." *National Geographic* 175, no. 5 (May 1989):563–583. Describes current research on gravity, including search for the "fifth" force.

Brancazio, Peter J. *Sport Science.* New York: Simon and Schuster, 1984. Chapter 3 deals with forces in sports; Chapter 7 deals with gravity in sports; and Chapters 8 and 9 discuss how thrown baseballs and such move through the air.

Kyle, Chester R. "Athletic Clothing." *Scientific American* 254, no. 3 (March 1986):104–110. Includes a segment on reducing air resistance on bicycles and runners.

Laeser, Richard P.; McLaughlin, William I.; and Wolff, Donna M. "Engineering Voyager 2's Encounter with Uranus." *Scientific American* 255, no. 5 (November 1986):36–45. Describes the technical problems of the Voyager 2 spacecraft and how they were solved.

O'Neill, Gerard K. *The High Frontier.* New York: Morrow, 1976. A physicist's description of how space colonies could be built. They would be designed so that rotation produced an artificial gravity inside.

Sir Isaac Newton's Mathematical Principles of Natural Philosophy and His System of the World. Translated by A. Motte (1729); edited by F. Cajori. Berkeley, Calif: University of California Press, 1934. English translation of Newton's *Principia.*

Smith, Bradford A. "Voyage of the Century." *National Geographic* 178, no. 2 (August 1990): 49–65. A summary of Voyager 2's mission written by a project scientist. It includes many stunning color photographs.

The following books all have discussions of Newton's life and work.

Abers, Ernest S., and Kennel, Charles F. *Matter in Motion: The Spirit and Evolution of Physics.* Boston: Allyn and Bacon, Inc., 1977. (Also describes how the radius of the moon's orbit was measured over 2,000 years ago.)

Cohen, Morris R., and Drabkin, I. E. *A Source Book in Greek Science.* New York: McGraw-Hill, 1948.

Segrè, Emilio. *From Falling Bodies to Radio Waves.* New York: W. H. Freeman and Co., 1984.

CHAPTER

3

OUTLINE

Greg LeMond during his dramatic come-from-behind victory in the last stage of the 1989 Tour de France. The helmet, clothing, special handlebars, disc rear wheel, and other features helped LeMond average 34.5 miles per hour for 15.5 miles.

ENERGY AND CONSERVATION LAWS

PROLOGUE: PHYSICS AND SPORTS

Although human physiology is probably the most important science for analyzing sports activities, physics is also quite useful. The basic concepts of mechanics presented in the first two chapters apply in many situations. Changing the *velocity* of a moving ball, *accelerating* it, is a principal part of tennis, baseball, soccer, and many other sports. Weight lifting and high jumping involve overcoming the *force* of *gravity.* Throwing a softball, shot (used in the shot-put), javelin, or football is an application of *projectile motion.* The force of *air resistance* affects these projectiles to differing degrees, and it is a major consideration for runners, bicyclists, ice skaters, skiers, long jumpers, and many other competitors.

Athletes learn an intuitive, applied knowledge of physics in the process of practice and training. By throwing a discus several hundred times, a thrower develops a "feel" for what angle to throw it so it will go farthest. Drills and game experience help soccer players develop a keen understanding about what force is necessary to alter the velocity of a soccer ball and to guide it into the goal. It is not practical, nor is it necessary, to calculate the speed and direction a basketball must be given when shooting. The subconscious mind and muscles work automatically to judge the proper velocity based on countless previous shots.

Physics can be used to design sports equipment that improves performance. For example, the world record in pole vaulting climbed dramatically after the fiberglass pole was introduced (around 1960). Bicycles and the clothing worn by skaters, skiers, racers, swimmers, and others can be designed to reduce the force of air (or water) resistance and thereby improve performance. Greg LeMond used the latest bicycling technology in winning the 1989 *"Tour de France."* The *Tour* consisted of 21 days of racing over a distance of 3,250 kilometers, like riding from St. Paul to San Francisco. After riding a total of over 87 hours, LeMond won the race by *8 seconds.* Even a slight reduction in air resistance could have made the difference in that race.

Sometimes it is instructive to analyze certain sports activities to better understand them and to estimate the limits that exist on human performance. The long jump is a good example. The long jumper's body must become a projectile and travel as far as possible. This is done by running as fast as possible—most world class long jumpers are also world class sprinters—and then jumping upwards so the body stays in the air a short time while its large horizontal speed carries it a long distance. In the 1968 Olympics, Bob Beamon broke the world record in the long jump by about *60 centimeters*. Since records generally increase a few centimeters at a time and since no one jumped that far during the following 20 years, this one jump has attracted a great deal of interest. Because Mexico City is 2,240 meters above sea level—Denver is 1,600 meters—it is logical to assume that the thinner air resulted in lower air resistance and that the greater distance from the earth's center allowed Beamon to remain in the air longer because of the reduced force of gravity. But careful analysis of the jump by several physicists indicates that the altitude benefit amounted to 10 centimeters at most. This result, coupled with the fact that no other long jumper has reproduced Beamon's feat, even when trying in Mexico City, suggests that it was a one-time-only, perfectly executed jump.

Forces and Newton's laws of motion are essential for analyzing sports activities, but there are other valuable "tools" as well. For example, a bicyclist has to fight two main forces when racing: the force of air resistance and the downhill component of the force of gravity when riding up a hill. But simply using the maximum force at all times to propel the bicycle will most likely result in an exhausted rider falling behind. The cyclist must understand the ideas of *energy* and *power* as well as *force* to compete successfully. In a similar way the basics of *linear momentum* offer insight into how to hit a golfball or baseball farther. The motion of a spinning figure skater or diver is most easily analyzed using *angular momentum* and *moment of inertia*.

Most of this chapter deals with how these important concepts and the associated *conservation laws* apply to a wide range of processes, including several related to sports. These conservation laws are yet another approach to mechanics. As we discuss these laws the physical quantities are defined: linear momentum, work, energy, power, and angular momentum. Conservation laws are stated for linear momentum, energy, and angular momentum. The concept of energy is extremely important, and its usefulness extends beyond mechanics alone.

3.1   CONSERVATION LAWS

Newton's laws of motion, in particular the second law, govern the instantaneous behavior of a system. They relate the forces that are acting at any instant in time to the resulting changes in motion. *Conservation laws* involve a different approach to mechanics, more of a "before-and-after" look at systems. A conservation law simply states that the total amount of a certain physical quantity present in a system stays constant (is conserved). For example, we might state the following conservation law.

The total mass in an isolated system is constant.[1]

---

[1] As we will see in Chapter 11, Einstein showed that energy must be included in this law because energy can be converted into matter and vice versa. We will not need this refinement until then.

An "isolated system" in this case means that no matter enters or leaves the system. The law states that the total mass of all the objects in a system doesn't change, regardless of the kinds of interactions that go on in it.

A simple example of this rather obvious conservation law is given by aerial refueling. One aircraft, called a tanker, pumps fuel into a second aircraft while both are in flight (Figure 3.1). If we ignore the small amount of fuel that both aircraft consume during the refueling, this can be regarded as an isolated system. The conservation of mass tells us that the total mass of both aircraft remains constant. So if the lower aircraft gains 2,000 kilograms of fuel, we automatically know that the tanker loses 2,000 kilograms of fuel. If one tanker refuels several aircraft in a formation, we include all of the aircraft in the system. The total mass of fuel dispensed by the tanker equals the total mass of fuel gained by the other aircraft. This fact may be useful. Let's say that the fuel gauge on one of the receiving aircraft is faulty. To determine how much fuel that aircraft was given, the aircrews use the conservation of mass: the total mass of fuel unloaded from the tanker minus the total mass of fuel given to the other aircraft in the formation equals the mass of fuel given to the aircraft in question.

The preceding example illustrates in a simple way how we can use a conservation law. Without knowing the details about an interaction (eg, the actual rate at which fuel is transferred between the aircrafts), we can still extract quantitative information by simply comparing the total amount of mass before and after. Three more conservation laws are presented in this chapter. Although they are a bit less intuitive and are a bit more complicated to use than the law of conservation of mass, they are applied in the same way.

FIGURE 3.1
Aerial refueling represents a simple example of the conservation of mass.

3.2   LINEAR MOMENTUM

The conservation law for linear momentum follows nicely from Newton's laws of motion, and so we will consider it first.

> **LINEAR MOMENTUM**   The mass of an object times its velocity. Linear momentum is a vector.
>
> $$Linear\ momentum = mv$$

Linear momentum is often referred to simply as "momentum." It is a vector quantity (since velocity is a vector), and its SI unit of measure is the kilogram-meter/second (kg-m/s).

Linear momentum incorporates both mass and motion. Anything that is stationary has zero momentum. The faster a body moves, the larger its momentum. A heavy object moving with a certain velocity has more momentum than a light object moving with the same velocity (Figure 3.2). For example, the momentum of a bicycle and rider with a total mass of 80 kilograms and a speed of 10 meters/second is:

$$mv = 80\ kg \times 10\ m/s = 800\ kg\text{-}m/s$$

The linear momentum of a 1,200 kilogram car with the same speed is 12,000 kg-m/s. The bicycle and rider would have to be going 150 meters/second (not likely) to have this same momentum.

Newton originally stated his second law of motion using linear momentum. In particular, we can restate this law as follows.

> **NEWTON'S SECOND LAW OF MOTION (alternate form)**   The net external force acting on an object equals the rate of change of its momentum.
>
> $$Force\ =\ \frac{change\ in\ momentum}{change\ in\ time}$$
>
> $$F\ =\ \frac{\Delta\,(mv)}{\Delta\,t}$$

To change an object's linear momentum, a net force must act. The larger the force, the faster the momentum will change. If the mass of the object stays constant, which is true in most cases, this equation is equivalent to the first form (Section 2.4) because

$$\frac{\Delta(mv)}{\Delta t} = m\,\frac{\Delta v}{\Delta t} = ma$$

Either form of the second law could be used for cars, airplanes, baseballs, and so on. For rockets and similar things with changing mass, only the alternate form can be used.

We can get yet another useful form of the second law by multiplying both sides by Δt to get

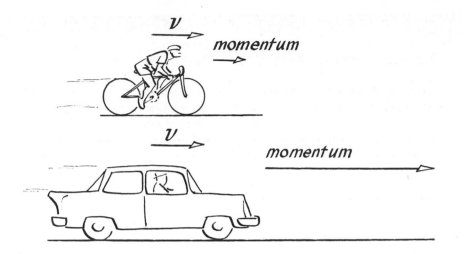

FIGURE 3.2
Linear momentum depends on the mass and the velocity. If a car and a bicycle have the same velocity, the car has a larger momentum because it has a larger mass.

$$\Delta(mv) = F\,\Delta t$$

The quantity on the right side is called the *impulse*. The same change in momentum can result from a small force acting for a long time or a large force acting for short time. When you throw a tennis ball, a small force acts on it for a relatively long period of time. When the ball is served at the same speed with a raquet, a large force acts for a short period of time.

This equation is useful for analyzing what goes on during impacts in sports that use balls and clubs. When a tennis racquet hits a tennis ball, a large force, F, acts for a short time, Δt. The result is a change in momentum of the ball, $\Delta(mv)$. One reason to have good follow through on a shot is to prolong the time of contact, Δt. This leads to a greater change in momentum, so the ball will leave the racquet with a higher speed.

Let's estimate the average force on a tennis ball as it is served (Figure 3.3). The ball's mass is 0.06 kilograms, and it leaves the racquet with a speed of, say, 40 m/s (90 mph). High-speed photographs indicate that the contact time is about 5 milliseconds (0.005 s).

EXAMPLE 3.1

Since the ball starts with zero speed, its change in momentum is

$$\Delta(mv) = \text{momentum afterwards} = 0.06\ \text{kg} \times 40\ \text{m/s}$$

$$\Delta(mv) = 2.4\ \text{kg-m/s}$$

So the average force is

$$F = \frac{\Delta(mv)}{\Delta t} = \frac{2.4\ \text{kg-m/s}}{0.005\ \text{s}}$$

$$F = 480\ \text{N} = 108\ \text{lb}$$

Our main application of the idea of linear momentum is based on the following conservation law.

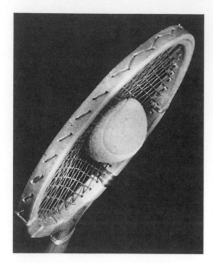

FIGURE 3.3
The tennis racquet exerts a large force on the tennis ball for a short time.

> LAW OF CONSERVATION OF LINEAR MOMENTUM   The total linear momentum of an isolated system is constant.

For this law an isolated system means that there are *no outside forces* causing changes in the linear momenta of the objects inside the system. The momentum of an object can change only because of interaction with other objects in the system. For example, once the cue ball on a pool table has been shot (ie, given some momentum), the pool table and the balls form an isolated system. If the cue ball collides with another ball, its momentum is changed (decreased; Figure 3.4). This change occurs because of an interaction with another object in the system. The momentum of the ball with which it collides is increased. If someone put a hand on the table and stopped the cue ball, the system would no longer be isolated.

The most important use of the linear momentum conservation law is in the analysis of *collisions*. Two pool balls colliding, a traffic accident, and two skaters running into each other are some familiar examples of collisions. We will limit our analysis to collisions involving only two objects that are moving in one dimension (along a line).

During any collision the objects exert equal and opposite forces on each other that cause them to accelerate (in opposite directions). These forces are usually quite large and often are due to direct contact between the bodies, as is the case with billiard balls and automobiles. They can also be "action-at-a-distance" forces such as the gravitational pull between a spacecraft and a planet. The key to applying the law of conservation of linear momentum to a collision is the following statement:

> The total linear momentum of the objects in the system before the collision is the same as the total linear momentum after the collision.
>
> total mv before = total mv after

EXAMPLE 3.2   We can use the law of conservation of linear momentum to analyze a simple automobile collision. A 1,000-kilogram automobile (car 1) runs into the rear of a stopped car (car 2) that has a mass of 1,500 kilograms. Immediately after the

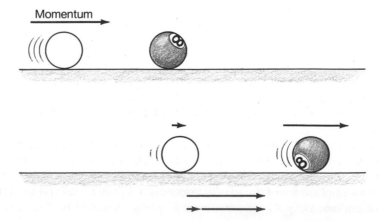

FIGURE 3.4
During this collision the cue ball loses momentum and the eight ball gains momentum. The total momentum of the two balls is the same before and after. The momentum vector before the collision equals the sum of the two momentum vectors after.

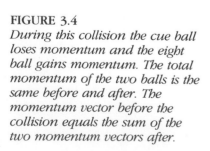

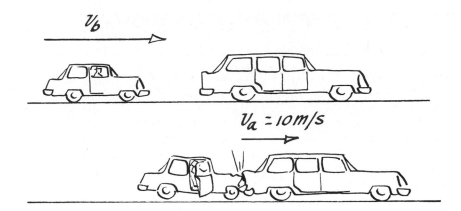

FIGURE 3.5
A 1,000-kilogram car collides with a 1,500-kilogram car that is stationary. Afterwards, the two cars are hooked together and move with a speed of 10 meters/second. The conservation of linear momentum allows us to determine that the speed of the first car was 25 meters/second.

collision the cars are hooked together, and their speed is estimated to have been 10 m/s (Figure 3.5). What was the speed of car 1 just before the collision?

Our conservation law tells us that the total linear momentum of the system will be constant.

$$\text{total } (mv)_{\text{before}} = \text{total } (mv)_{\text{after}}$$

Before the collision, only car 1 is moving. The total linear momentum before is:

$$(mv)_{\text{before}} = m_1 \times v_b = (1{,}000 \text{ kg}) \times v_b$$

After the collision, both cars are moving as one body. The total momentum after is:

$$(mv)_{\text{after}} = (m_1 + m_2) \times v_a = (1{,}000 \text{ kg} + 1{,}500 \text{ kg}) \times 10 \text{ m/s}$$

$$(mv)_{\text{after}} = 2{,}500 \text{ kg} \times 10 \text{ m/s} = 25{,}000 \text{ kg-m/s}$$

These two linear momenta are equal. Therefore:

$$(1{,}000 \text{ kg}) \times v_b = 25{,}000 \text{ kg-m/s}$$

$$v_b = \frac{25{,}000 \text{ kg-m/s}}{1{,}000 \text{ kg}}$$

$$v_b = 25 \text{ m/s}$$

This type of analysis is routinely used to reconstruct traffic accidents. For example, it can be used to determine whether a vehicle was exceeding the speed limit. In this problem the car was going 25 meters/second, or about 56 mph. If the accident happened in a 35- or 45-mph speed zone, the driver of car 1 would have been speeding. If it was a 55-mph zone, one could not immediately conclude that the driver was speeding because the values used for the masses and the final speed are usually uncertain.

The speed of a bullet or a thrown object can be measured in a similar way. A bullet is fired into and becomes embedded in a block of wood that is hanging from a string (Figure 3.6). If one measures the masses of the bullet and the wood block and the speed of the block immediately afterwards, the initial speed of the bullet can be found by using the conservation of linear momentum. You can measure

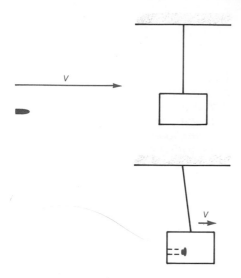

FIGURE 3.6

A bullet becomes embedded in a block of wood. If the speed of the block and the masses of the block and the bullet are measured, the initial speed of the bullet can be computed using the conservation of linear momentum.

the speed of your "fast ball" by throwing a lump of sticky clay at a block and proceeding in the same way. The key in both cases is measuring the speed of the block of wood after the collision. One can do this most easily by using the law of conservation of energy. We will show how in Section 3.5.

In Section 2.7, we described a simple experiment in which we used two wooden carts (Figure 2.30, right side). One cart has a spring-loaded plunger that pushes on the other cart, causing both carts to be accelerated. By the law of conservation of linear momentum,

$$(mv)_{before} = (mv)_{after}$$

Before the spring is released, neither cart is moving, so the momentum is zero:

$$(mv)_{before} = 0$$

Therefore the total linear momentum afterwards is zero. But since both carts are moving, this momentum equals:

$$(mv)_{after} = 0 = (mv)_1 + (mv)_2$$

Therefore:

$$(mv)_1 = - (mv)_2$$

The momentum of one of the carts is negative. This has to be the case, since they are moving in opposite directions. (Remember: linear momentum is a vector.) Since $(mv)_1 = m_1 \times v_1$:

$$m_1 \times v_1 = - m_2 \times v_2$$

$$v_1 = - \frac{m_2}{m_1} \times v_2$$

$$\frac{v_1}{v_2} = - \frac{m_2}{m_1}$$

So we have derived the statement made in Section 2.7: the ratio of the speeds of the two carts is the inverse of the ratio of their masses. If the mass of #2 is 3 kilograms and the mass of #1 is 1 kilogram, the ratio is 3 to 1. Cart 1 moves away with three times the speed of cart 2 (Figure 3.7). Note that this only gives us the ratio of the speeds, not the actual speed of each cart. These would depend on the strength of the spring. With a weak spring the speeds might be 1 m/s and 3 m/s; with a stronger spring, the speeds might be 2.5 meters/second and 7.5 meters/second.

FIGURE 3.7

The mass of the cart on the right, cart 2, is three times the mass of the cart on the left, cart 1. After the spring is released, the speed of the lighter cart is three times the speed of the more massive cart.

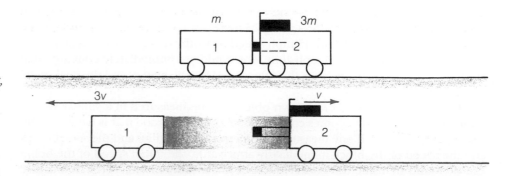

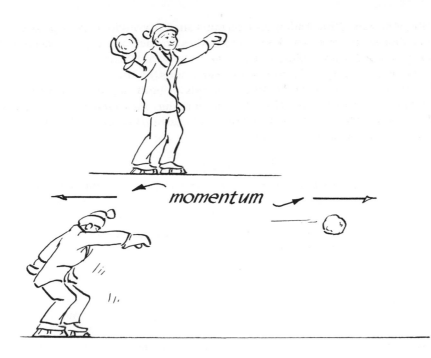

momentum

FIGURE 3.8
The skater and the snowball have the same amount of momentum but in opposite directions.

DO-IT-YOURSELF PHYSICS

The next time you go skateboarding, ice skating, or roller skating with a friend, experiment with the conservation of linear momentum in various kinds of collisions. Try to reproduce the collisions illustrated in Figures 3.5, 3.7, and 3.34. You can also experiment with Newton's third law of motion. See Figure 2.30 for examples.

We can use linear momentum conservation to get a different view of some of the situations described in Section 2.7. There we used forces and Newton's third law of motion. When a gun is fired, the bullet acquires momentum in one direction, and the gun gains equal momentum in the opposite direction—the kick. If an ice skater throws an object forward, he or she will move backwards with equal momentum (Figure 3.8). Rockets and jets give momentum to the ejected exhaust gas. They in turn gain momentum in the forward direction. For a rocket with no external forces on it, the increase in momentum during each second will depend on how much gas is ejected, which equals the mass of fuel burned, and on the speed of the gas (Figure 3.9).

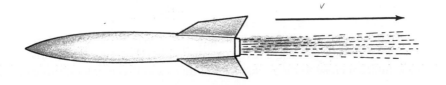

FIGURE 3.9
As a rocket gives momentum to the exhaust gases, it gains momentum in the opposite direction.

Why is linear momentum conserved? Because of Newton's second and third laws of motion. When two objects exert forces on each other by colliding or via a spring plunger, the forces are *equal and opposite*. They push on each other with the same size force but in opposite directions. By the second law (alternate form), these equal forces cause the momenta of both objects to change at the same rate. As long as the objects are interacting, *they change one another's momentum by the same amount but in opposite directions*. The momentum gained (or lost) by one object is exactly offset by the momentum lost (or gained) by the other. The total linear momentum is not changed. In Example 3.2, the 1,000-kilogram car is slowed from 25 meters/second before the collision to 10 meters/second after the collision (Figure 3.5). Its momentum is *decreased by 15,000 kilogram-meters/second* (1,000 kilograms × 25 meters/second − 1,000 kilograms × 10 meters/second). The 1,500-kilogram car goes from 0 m/s before to 10 meters/second after. So its momentum is *increased by 15,000 kilogram-meters/second* (1,500 kilograms × 10 meters/second).

These examples illustrate the usefulness of conservation laws. The approach is different from that used with Newton's second law in Chapter 2, where it was necessary to know the size of the force that acts on the object *at each moment* to determine its velocity. With the law of conservation of linear momentum, we do not have to know the details of the interactions—how large the forces are and how long they act. All we need is some information about the system *before and after* the interaction. Note that in Example 3.2 we used information from after the collision to determine the speed of the car before the collision. In the example with the wooden carts, we determined the ratio of the speeds after the interaction.

3.3   WORK: THE KEY TO ENERGY

The law of conservation of energy is the most important of the conservation laws. Not only is it useful for solving problems, it is a powerful theoretical statement that can be used to understand widely diverse phenomena and to show what hypothetical processes are or are not possible. As we mentioned earlier, the concept of energy is one of the most important in physics. This is because energy takes many forms and is involved in all physical processes. One might say that every interaction in our universe involves a transfer of energy or a transformation of energy from one form to another.

One might compare the concept of energy to that of financial assets, which can take the form of cash, real estate, material goods, or investments, among other things. The study of economics is in part a study of these forms of financial assets and how they are transferred and transformed. Much of physics deals with the forms of energy and the transformations that occur during interactions.

When first encountered, the concept of energy is a bit difficult to understand because there is no simple way to define it. As an aid we will first introduce *work*, a physical quantity that is quite basic but gives us a nice foundation for understanding energy.

The idea of work in physics arises naturally when one considers *simple machines* like the *lever* and the *inclined plane* when used in situations with negligible friction. Let's say that you use a lever to raise a heavy rock (Figure 3.10). If you place the fulcrum close to the rock, you find that a small, downward force on your

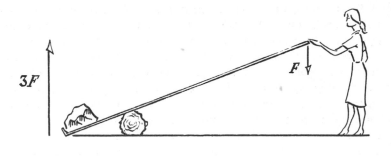

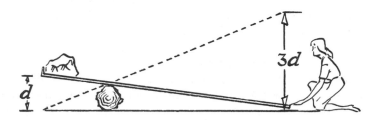

FIGURE 3.10
When a lever is used to raise a rock, a small force on the right end results in a larger force on the left end. But the right end moves a greater distance than the left end. The force multiplied by the distance moved is the same *for both ends.*

end results in a larger, upward force on the rock. However, the distance your end moves as you push it down is correspondingly larger than the distance the rock is raised. By measuring the forces and distances, we find that dividing the larger force by the smaller force gives the same number (or ratio) as dividing the larger distance by the smaller distance. In particular:

$$\frac{F \text{ on left end}}{F \text{ on right end}} = \frac{d \text{ right end moves}}{d \text{ left end moves}}$$

We can multiply both sides by *F on right* and *d left moves* and get the following result:

$$(F \text{ on left}) \times (d \text{ left moves}) = (F \text{ on right}) \times (d \text{ right moves})$$

$$F_{\text{left}} d_{\text{left}} = F_{\text{right}} d_{\text{right}}$$

In other words, even though the two forces and the two distances are different, the quantity *force times distance* has the same value for both ends. We might say that raising the rock is a fixed task. One can perform the task by lifting the rock directly or by using a lever. In the former case the force is large, equal to the rock's weight, but the distance moved is small. When using the lever, the force is smaller, but the distance is larger. But the quantity *Fd* is the same, regardless of which way the task is accomplished.

We reach the same conclusion when considering an inclined plane. Let us say that a barrel must be placed on a loading dock (Figure 3.11). Lifting the barrel directly requires a large force acting through a small distance, the height of the dock. If the barrel is rolled up a ramp, a smaller force is needed, but the barrel moves a greater distance. Again the product of the force and the distance moved is the same for the two methods.

$$(F \text{ lifting}) \times \text{height} = (F \text{ rolling}) \times (\text{ramp length})$$

$$F_{\text{lifting}} d_{\text{lifting}} = F_{\text{rolling}} d_{\text{rolling}}$$

The quantity *force times distance* is obviously a useful way of measuring the "size" of a task. It is called *work.*

FIGURE 3.11
Two identical barrels need to be placed on the loading dock. One is lifted directly, requiring a large force. The other is rolled up the ramp, an inclined plane. A smaller force is needed, but it must act on the barrel over a longer distance.

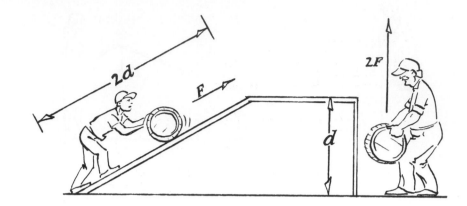

> **WORK**   The force that acts times the distance moved in the direction of the force:
>
> $$\text{Work} = Fd$$

	Metric	English
Work	joule (J)	foot-pound (ft-lb)
	erg	British thermal unit (Btu)
	calorie (cal)	
	kilowatt hour (kW-h)	

Since force is a vector, *work equals the distance moved times the component of the force parallel to the motion.* Work itself is not a vector. There is no direction associated with work.

Whenever an object moves and there is a force acting on the object *in the same or opposite direction* that it moves, work is done.

EXAMPLE 3.3   Because of friction, a constant force of 100 newtons is needed to slide a box across a room (Figure 3.12). If the box is moved 3 meters, how much work is done?

$$\text{Work} = Fd$$
$$= 100\,\text{N} \times 3\,\text{m}$$
$$\text{Work} = 300\,\text{N-m}$$

The unit in the answer is the newton-meter, N-m. This is called the *joule.*

$$1\,\text{joule} = 1\,\text{newton-meter} = 1\,\text{newton} \times 1\,\text{meter}.$$

$$1\,\text{J} = 1\,\text{N-m}$$

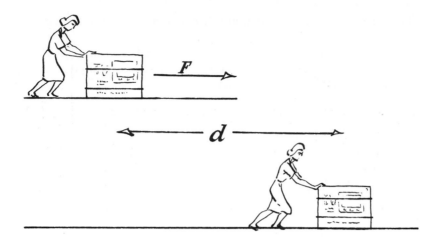

FIGURE 3.12
Pushing on a box with a force of 100 newtons causes it to slide over the floor. If the box moves 3 meters, then you have done 300 joules of work.

The joule is a derived unit of measure, and it is the SI unit of work and energy. Just remember that if you use only SI units for force, distance, and so on, your unit for work and energy will always be the joule (J).

Let's say that the barrel in Figure 3.11 has a mass of 30 kilogram and that the height of the dock is 1.2 meters. How much work do you do when lifting the barrel?

EXAMPLE 3.4

$$\text{Work} = Fd$$

The force is just the weight of the barrel, *mg*.

$$F = W = mg = 30 \text{ kg} \times 9.8 \text{ m/s}^2 = 294 \text{ N}$$

Hence:

$$\text{Work} = Fd = Wd$$
$$= 294 \text{ N} \times 1.2 \text{ m}$$
$$\text{Work} = 353 \text{ J}$$

The work you do when rolling the barrel up the ramp would be the same. The force would be smaller, but the distance would be larger.

It is important to note that work is *not done* if the force is *perpendicular* to the motion. When you simply carry a box across a room, your force on the box is *vertical,* while the motion of the box is *horizontal.* Hence, you do no work in this case (Figure 3.13).

Uniform circular motion is another situation in which a force acts on a moving body but no work is done. Recall that a centripetal force must act on anything to keep it moving along a circular path (Figure 3.14). This force is always toward the center of the circle and perpendicular to the object's velocity at each instant. Therefore the force does not do work on the object.

Work can be done in a circular motion by a force that is not a centripetal force. When you turn the crank on a pencil sharpener, for instance, you exert a force on the handle that is in the same direction as the handle's motion. Hence, you do work on the handle.

Work is done on an object when it is accelerated in a straight line. The following example shows how the amount of work can be computed. (In the next section, we will see that there is an easier way to do it.)

EXAMPLE 3.5

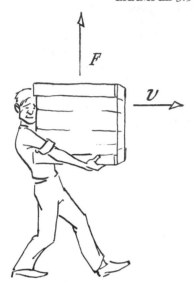

FIGURE 3.13
When you carry a box across a room, the force on the box is perpendicular to the direction the box moves. No work is done on the box.

In Example 2.2 we used Newton's second law to compute the force needed to accelerate a 1,000-kilogram car from 0 to 27 meters/second in 10 seconds. Our answer was F = 2,700 newtons. How much work is done?

$$\text{Work} = Fd$$

To find the distance that the car travels, we use the fact that the car accelerates at 2.7 meters/second^2 for 10 s. Using the equation from Section 1.5:

$$d = \frac{1}{2}at^2 = \frac{1}{2} \times 2.7 \text{ m/s}^2 \times (10 \text{ s})^2$$

$$d = 1.35 \text{ m/s}^2 \times 100 \text{ s}^2 = 135 \text{ m}$$

So the work done is:

$$\text{Work} = Fd$$
$$= 2,700 \text{ N} \times 135 \text{ m}$$
$$\text{Work} = 364,500 \text{ J}$$

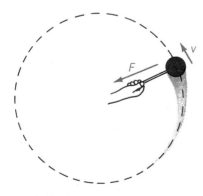

FIGURE 3.14
The string exerts a force on the ball that is perpendicular to its motion. No work is done by this force.

We have seen that work is done (1) when a force acts to oppose the force of gravity (Figures 3.10 and 3.11), (2) when a force acts against friction (Figure 3.12), and (3) when a force accelerates an object. There are many other possibilities. When a force distorts something, work is done. For example, to compress or stretch a spring, a force must act on it. This force acts through a distance in the same direction as the force, so work is done.

Work is also done when a force causes something to decelerate. When you catch a ball, your hand exerts a force on the ball. As the ball slows down, it pushes your hand back with an equal and opposite force (see Figure 3.15). In this case the *ball does work on your hand.* The amount of work that the ball does on your hand is equal to the amount of work that was originally done on the ball to accelerate it (ignoring the effect of air resistance). If you allow your hand to move back as you catch the ball, the force of the ball on you will be less than if you try to keep your hand stationary. The work that the ball will do on your hand is the same either way. Since work = force × distance, the force on your hand will be smaller if the distance that the ball moves while you catch it is larger.

One last example: When an object falls freely, the force of gravity does work on it. As in the previous example with the car, this work goes to accelerate the object. In particular, if a body falls a distance *d,* the work done on it by the force of gravity is:

$$\text{Work} = Fd$$
$$F = W = mg$$

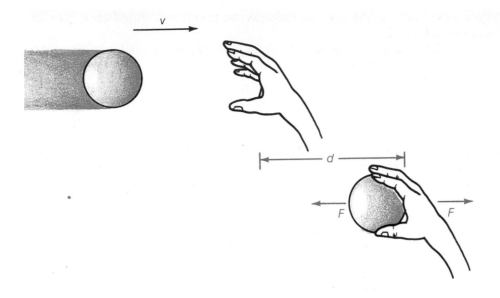

FIGURE 3.15
As you catch a ball, your hand exerts a force on the ball. By Newton's third law, the ball exerts an equal and opposite force on your hand. This force does work on you as you decelerate the ball. The work done is equal to the work that was done to accelerate the ball in the first place.

FIGURE 3.15
As you catch a ball, your hand exerts a force on the ball. By Newton's third law, the ball exerts an equal and opposite force on your hand. This force does work on you as you decelerate the ball. The work done is equal to the work that was done to accelerate the ball in the first place.

So:

$$\text{Work} = mgd$$

The work that the force of gravity does on an object as it falls is equal to the work that was done to lift the object the same distance (Figure 3.16). When something is lifted, we say that work is done *against* the force of gravity. The movement is in the opposite direction of the force. When something falls, work is done *by* the force of gravity. The movement is in the same direction as the force.

In summary, work is done by a force whenever the point of application moves in the direction of the force. Work is done against the force whenever the point of application moves opposite the direction of the force. Forces always come in pairs that are equal and opposite (Newton's third law of motion). Consequently when work is done *by* a force, this work is being done *against* the other force.

FIGURE 3.16
The work done lifting an object equals the work done by gravity as the object falls.

3.4   ENERGY

In Section 3.3 we saw that work can often be "recovered." The work that is done when a ball is thrown is equal to the work done by the ball when it is caught. The work done when lifting an object against the force of gravity is equal to the work done by the force of gravity when the object falls. When work is done on something, it gains some *energy*. This energy can then be used to do work. A thrown ball is given energy. This energy is given up to do work when the ball is caught.

> ENERGY   The measure of a system's capacity to do work. That which is transferred when work is done. Abbreviated E.

The units of energy are the same as the units of work. Like work, energy is a scalar.

The more work that is done on something, the more energy it gains, and the more work it can do in return. We might say that energy is "stored work." To be able to do work, an object must have energy. When you throw a ball, you are transferring some energy from you to the ball. The ball can then do work. In Figures 3.13 and 3.14, no energy is transferred because no work is done.

There are many forms of energy corresponding to the many ways in which work can be done. Some of the more common forms of energy are chemical, electrical, nuclear, and gravitational, as well as heat, sound, light, and other forms of radiation. Anything possessing any of these forms of energy is capable of doing work (see Figure 3.17).

In mechanics there are two main forms of energy, which we can classify under the single heading of *mechanical energy*. Anything that has energy because of its motion or because of its position or configuration has mechanical energy. We refer to the former as *kinetic energy* and to the latter as *potential energy*.

> KINETIC ENERGY   Energy due to motion. Energy that an object has because it is moving. Abbreviated KE.

Anything that is moving has kinetic energy. The simplest example is an object moving in a straight line. The amount of kinetic energy an object has depends on its mass and speed. In particular:

$$KE = \frac{1}{2}mv^2 \qquad \text{(kinetic energy)}^2$$

FIGURE 3.17
The crane can do work because it has energy—chemical energy in its fuel.

---

[2](For the mathematically inclined.) The kinetic energy that a body has equals the work done to accelerate it from rest. In the case of constant acceleration, we can compute the work:

$$\text{Work} = Fd$$

Now we use what we learned in Chapters 1 and 2. The force needed is: $F = ma$, and the distance traveled is: $d = 1/2\, at^2$. So:

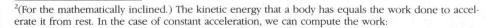
$$\text{Work} = Fd = ma \times \tfrac{1}{2}at^2 = \tfrac{1}{2}m \times a^2 \times t^2 = \tfrac{1}{2}m \times (at)^2$$

But at equals the speed v that the body has. So:

$$\text{Work} = \tfrac{1}{2}mv^2 = KE$$

The kinetic energy that an object has is equal to the work done when accelerating the object from rest. So another way to determine the amount of work done when accelerating an object is to compute its kinetic energy. This also shows that the work done depends only on the object's final speed and not on how rapidly or slowly it was accelerated.

In Example 3.5 we computed the work that is done on a 1,000-kilogram car as it accelerates from 0 to 27 meters/second. The car's kinetic energy when it is traveling 27 meters/second is:

EXAMPLE 3.6

$$KE = \frac{1}{2}mv^2 = \frac{1}{2} \times 1,000\,\text{kg} \times (27\,\text{m/s})^2$$

$$KE = 500\,\text{kg} \times 729\,\text{m}^2/\text{s}^2$$

$$KE = 364,500\,\text{J}$$

This is the same as the work done as the car accelerates. The car can do 364,500 joules of work because of its motion.

The kinetic energy of a moving body is proportional to the square of its speed. If one car is going twice as fast as a second, identical car, the faster one has four times the kinetic energy. It takes four times as much work to stop the faster car. Note that the kinetic energy of an object can never be negative.

Since speed is relative, kinetic energy is also relative. A runner on a moving ship has KE relative to the ship and a different KE relative to something at rest in the water.

Another way that an object can have kinetic energy is by rotating. To make something spin, work must be done on it. A dancer or skater performing a pirouette has kinetic energy (see Figure 3.18). A spinning top has kinetic energy,

FIGURE 3.18
A spinning dancer (far left) has kinetic energy, even though she stays in one place.

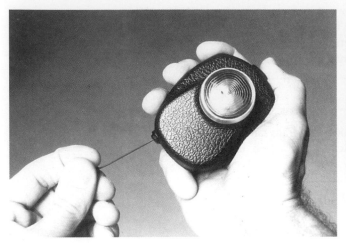

FIGURE 3.19
(a) This shaver uses rotational kenetic energy given to a flywheel inside by pulling on the cord.
(b) The shaver with its back removed showing the large flywheel.

as do the earth, moon, sun, and other astronomical objects as they spin about their axes. The amount of kinetic energy that a spinning object has depends on its mass, its rotation rate, and the way its mass is distributed. Some devices and toys use rotational kinetic energy as a way to "store" energy. Figure 3.19 shows a shaver that operates this way.

> **POTENTIAL ENERGY**   Energy due to an object's position. Energy that a system has because of its configuration. Abbreviated *PE*.

The amount of potential energy that a system has is equal to the work done to put it in that configuration. When an object is lifted, it is given potential energy. It can use this energy to do work. For example, when the weights on a cuckoo clock or any other gravity-powered clock are raised, they are given potential energy (Figure 3.20). As they slowly fall, they do work in operating the clock.

In Section 3.3 we computed the work done when an object is lifted. This is equal to the potential energy that it is given. Since the work is done against the force of gravity, it is called *gravitational potential energy.*

$$PE = \text{Work done} = \text{weight} \times \text{height}$$

$$= Wd = mgd$$

$$PE = mgd \quad \text{(gravitational potential energy)}$$

Gravitational potential energy is the most common type of potential energy, and it is often referred to simply as *potential energy.*

Potential energy is a relative quantity because height can be measured relative to different levels.

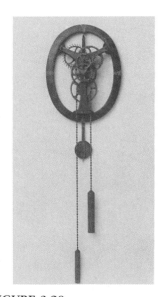

FIGURE 3.20
The source of energy to run this clock is the potential energy of the weights.

EXAMPLE 3.7

A 3-kilogram brick is lifted to a height of 0.5 meters above a table (Figure 3.21). Its potential energy relative to the table is

$$PE = mgd$$

$$= 3\,\text{kg} \times 9.8\,\text{m/s}^2 \times 0.5\,\text{m}$$

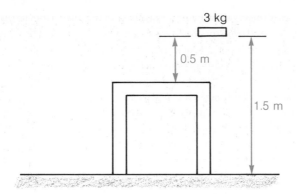

FIGURE 3.21
The potential energy (PE) *of an object depends on how its height* (d) *is measured. The brick has 14.7 joules of potential energy relative to the table and 44.1 joules of potential energy relative to the floor. In both cases, the potential energy equals the work done to raise the brick* from that level.

$$PE = 14.7\,J \quad \text{(relative to the table)}$$

But the tabletop itself may be 1 meter above the floor. The brick's height above the floor is 1.5 meters, and so its potential energy relative to the floor is

$$PE = mgd$$

$$= 3\,kg \times 9.8\,m/s^2 \times 1.5\,m$$

$$PE = 44.1\,J \quad \text{(relative to the floor)}$$

A person sitting in a chair has potential energy relative to the floor, to the basement of the building, and to the level of the oceans. Usually some convenient reference level is chosen for determining potential energies. In a room it is logical to use the floor as the reference level for measuring heights and, consequently, potential energies.

An object's potential energy is negative when it is *below* the chosen reference level. Often the reference level is chosen so that negative potential energies signify that an object cannot "leave." For example, on flat ground outdoors it is logical to measure potential energy relative to the ground level. Anything on the ground has zero potential energy, and anything above the ground has positive potential energy. If there is a hole in the ground, any object in the hole will have negative potential energy (Figure 3.22). Any object that has zero or positive

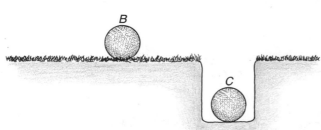

FIGURE 3.22
The potential energy of golf ball A *is positive relative to the ground. The potential energy of* B *is zero and that of* C *is negative because it is below ground level.* A *and* B *can move horizontally.* C *is restricted to the hole.*

FIGURE 3.23
As Angie pulls the plunger on the dart gun, she does work compressing the spring inside. This gives the spring elastic potential energy.

potential energy can move about horizontally if it has any kinetic energy. Objects with negative potential energy are confined to the hole. They must be given enough energy to get out of the hole before they can move horizontally.

Springs and rubber bands can possess another type of potential energy: *elastic potential energy.* Work must be done on a spring to stretch or compress it. This gives the spring potential energy (see Figure 3.23). This "stored energy" can then be used to do work. The actual amount of potential energy a spring has depends on two things: how much it was stretched or compressed and how strong it is. In Figure 3.7, the combined kinetic energies of the carts after the spring is released equals the original elastic potential energy of the spring. A stronger spring would possess more potential energy and would give the carts more KE, so they would go faster.

A number of devices use elastic potential energy. Toy dart guns have a spring inside that is compressed by the shooter. When the trigger is pulled, the spring is released and does work on the dart to accelerate it. The potential energy of the spring is converted to kinetic energy of the dart. The bow and arrow operate this same way, with the bow acting as a spring. Windup devices such as watches, toys, and music boxes use energy stored in springs to operate the mechanism. Usually the spring is in a spiral shape, but the principle is the same. Rubber bands provide lightweight energy storage in some toy airplanes and birds (Figure 3.24).

Heat energy[3] is another form of energy that is important in mechanical systems. Heat energy arises whenever there is kinetic friction. In Figure 3.12, the work that is done on the box is converted into heat energy because of the friction

---

[3]In Chapter 5 we define the more precise quantities *heat* and *internal energy.*

FIGURE 3.24
This mechanical bird uses energy stored in a rubber band inside. Patricia is holding the crank used to wind it up.

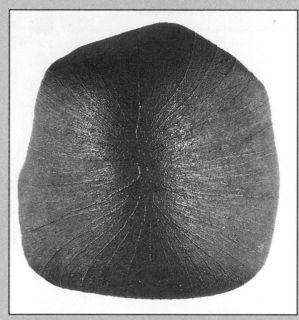

FIGURE 3.26
The Lost City meteorite showing ridges where melted rock flowed as it traveled through the atmosphere.

Persons wishing to observe meteors will have more success before sunrise than during the late evening hours. The reason for this is that after midnight the observer is located on that side of the earth facing in the direction in which the earth is moving in its orbit about the sun (see Figure 3.27). Every meteoroid entering our atmosphere at this time undergoes energetic "head-on" collisions with the air molecules at high relative speeds (equal to their own speeds *plus* that of the earth) to produce bright meteors. Conversely, during the evening hours the only meteoroids passing into the atmo-

sphere are those moving fast enough to catch up with the earth. Since the earth moves with an orbital speed of about 30 kilometers per second, this means that only relatively rare, high-velocity objects with speeds in excess of 40 kilometers per second will be capable of overtaking the earth and interacting with its atmosphere with sufficient residual energy to produce noticeable meteors. This situation is analogous to driving a car down a highway during a rainstorm. Many raindrops vigorously spatter the front windshield, while relatively few droplets strike the rear window. On a clear night in a dark location, one might expect to see about 15 meteors per hour just before dawn as compared to only about 5 or 6 per hour before midnight.

The preceding discussion applies to what are called *sporadic meteors,* that is, meteors resulting from meteoroids that reach the earth with no preferred direction in space or time of appearance. Their observation is largely a random or chance event. They are to be distinguished from *shower meteors,* which result from the earth's crossing the orbit of an often defunct comet and encountering its remains left circling the sun. In such cases, for several nights while the earth is plowing through this cometary debris, one may see up to several hundred meteors per hour emanating from a particular point in the sky called the *shower radiant,* which marks the location of the swarm of incoming particles. Meteor showers are recurrent phenomena, happening each time the earth passes through the path once followed by a comet. Meteor showers are usually named after the constellations in which their radiants are located. The table at the right gives the names and dates for some of the meteor showers having the highest sighting rates.

and melt. Gravitational potential energy can also be converted into heat energy. A box slowly sliding down a ramp has its gravitational potential energy converted into heat energy by friction. If you climb a rope and then slide down it, you can burn your hands severely as some of your potential energy is converted into heat energy because of the friction between your hands and the rope.

On its annual circuit around the sun, the earth routinely encounters many small, solid particles called *meteoroids*. They enter the atmosphere with velocities between about 12 and 75 kilometers per second and collide with air molecules. This results in a deceleration of the meteoroid and a corresponding acceleration of the atmospheric particles.

The meteoroid loses mass as well as speed as it penetrates the atmosphere. This loss of mass, called *ablation,* happens as a result of energy exchanged between the meteoroid and the air molecules during collisions. In particular, some of the kinetic energy of the meteoroid is converted to heat energy, which causes the meteoroid to vaporize, melt, and/or fragment. When ablation occurs, atoms and molecules are lost from the body of the meteoroid and collide with atmospheric molecules with sufficient energy to ionize and to excite them. It is from such interactions that light is produced. (Chapter 10 discusses how collisionally ionized and excited atoms radiate.) The resultant streak of light is called a *meteor,* or, if it is very bright and spectacular, a fireball (Figure 3.25).

Most meteoroids completely dissolve into vapor and dust grains, the latter of which gradually settle to the earth's surface and help to increase its mass by some 3,000 tons daily. Occasionally a portion of a larger meteoroid survives its passage through the atmosphere to reach the ground, whereupon we refer to it as a *meteorite.* Meteorites typically display fusion crusts, relics of their ablative pasts (Figure 3.26). Moreover, evidence of thermal damage often exists a few millimeters beneath this crust. Aside from these effects, the interiors of meteorites are generally left unscathed by their travels through the atmosphere and so provide geochemists with unaltered samples of interplanetary material. Studies of such material are important because they provide clues to the origin of the meteoroids and to the conditions prevailing in the early solar system.

FIGURE 3.25
A *brilliant fireball seen over Lost City, Oklahoma on 3 January, 1970.*

between the box and the floor. As a car or a bicycle brakes to a stop, its kinetic energy is converted into heat energy in the brakes. Automobile disc brakes can become red hot under extreme braking. Meteors (shooting stars) are a spectacular example of kinetic energy being converted into heat energy. As they enter the atmosphere at very high speed, the air resistance heats them enough to glow

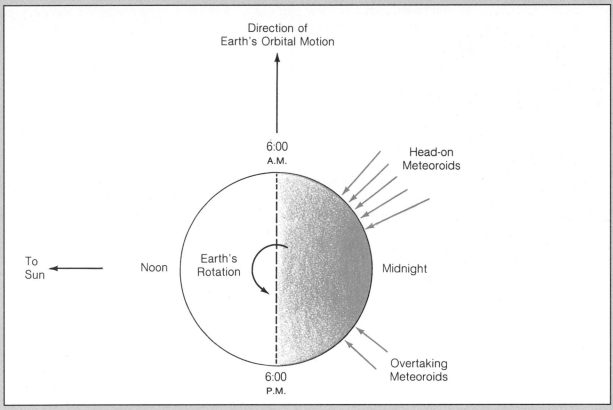

FIGURE 3.27
Diagram showing the factors influencing the rates of meteor sightings by nighttime observers on earth.

SOME CHARACTERISTICS OF METEOR SHOWERS

Shower Name	Date of Maximum Display	Associated Comet	Shower Name	Date of Maximum Display	Associated Comet
Lyrid	April 21	1861 I	Orionid	October 20	Halley
Eta Aquarid	May 4	Halley	Taurid	October 31	Encke
Perseid	August 11	1862 III	Leonid	November 16	1866 I
Draconid	October 9	Giacobini-Zinner	Geminid	December 13	

Heat energy can also be produced by internal friction when something is distorted. The work you do when stretching a rubber band, pulling taffee, or crushing an aluminum can generates heat energy. When you drop something and it doesn't bounce, like a book, most of the energy the object had is converted into heat energy on impact.

DO-IT-YOURSELF PHYSICS

Take a paper clip and straighten one of the ends. Bend the clip back and forth several times, then touch the part that was bending to your cheek. The work you do in bending the clip is converted into heat energy. You should be able to feel that the clip is warmed. You can get the same result by stretching a rubber band.

Heat energy arising from friction, unlike kinetic energy and potential energy, usually cannot be recovered. Work done to lift a box and give it potential energy can be recovered as work or some other form of energy. Work done to slide a box across a floor becomes heat energy that is "lost" (made unavailable). In every mechanical process, some energy is converted into heat energy. We will postpone until Chapter 5 a formal discussion of heat.

In summary, work always results in a transfer of energy from one thing to another, in a transformation of energy from one form to another, or both. In our earlier analogy in which we compared energy to financial assets, work plays the role of a transaction, such as buying, selling, earning, or trading. These transactions can be used to increase or decrease the net worth of an individual or to convert one form of asset into another. Work done *on* a system increases its energy. Work done *by* the system decreases the energy of the system. Work done *within* the system results in changes from one form of energy into another.

LEARNING CHECK

1. In a collision, the total _____ is the same before and after.

2. Work is done on an object when
   a) it moves in a circle with constant speed.
   b) it is accelerated in a straight line.
   c) it is carried horizontally.
   d) all of the above.

3. To be able to do work, a system must have _____ .

4. An object has kinetic energy when it is _____ .

5. As a skier gains speed while gliding down a slope _____ energy is being converted into _____ energy.

3.5   THE CONSERVATION OF ENERGY

In the preceding section we described several situations in which one form of energy was converted into another. These included a dart gun (potential energy in a spring converted to kinetic energy of the dart), a car braking (kinetic energy

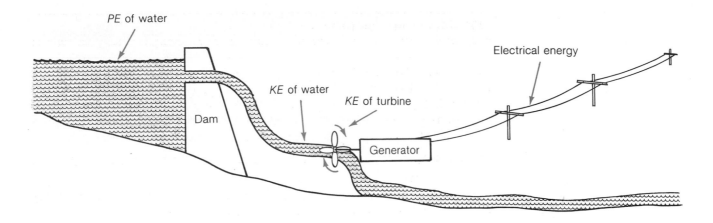

converted into heat energy), and a box sliding down an inclined plane (gravitational potential energy becoming heat energy). There are countless situations involving many other forms of energy.

Many devices in common use are simply energy converters. Some examples are listed in Table 3.1. Some of these actually involve more than one conversion. In a hydroelectric dam (see Figure 3.28), the potential energy of the water behind the dam is converted into kinetic energy of the water. The moving water then hits a turbine (propeller), which gives it kinetic energy of rotation. The rotating turbine turns a generator, which converts the kinetic energy into electrical energy. Heat energy is an intermediate form of energy in both the car engine and the nuclear power plant. In a car engine the chemical energy in fuel is first converted into heat energy as the fuel burns (explodes). The heated gases expand and push against pistons (or rotors in rotary engines). These in turn make a crankshaft rotate. In a nuclear power plant, nuclear energy is converted into heat energy in the reactor core. This heat energy is used to boil water into steam. The steam is used to turn a turbine, which turns a generator that produces electrical energy.

Heat energy is also a "wasted" by-product in all of the devices in Table 3.1. Any device that has moving parts has some kinetic friction. Some of the input energy is converted into heat energy by this friction. The generator, electric motor, car engine, and the electrical power plants all have unavoidable friction (Figure 3.29). In some of the devices, heat energy is produced because of the basic nature of the process. Over 95% of the electrical energy used by an incandescent light bulb is converted to heat energy, not usable light. Over 60% of the available energy in coal-fired and nuclear power plant goes to unused heat energy. In later chapters we will investigate this further.

Even though there are so many different forms of energy and countless devices that involve energy conversions, the following law always holds.

FIGURE 3.28
A hydroelectric power station uses several energy conversions.

> LAW OF CONSERVATION OF ENERGY   Energy cannot be created or destroyed, only converted from one form to another. The total energy in an isolated system is constant.

To be an isolated system, energy cannot leave or enter the system. For a mechanical system, work cannot be done on the system by an outside force, nor can the system do work on anything outside of it.

TABLE 3.1 SOME ENERGY CONVERTERS	
Device	Energy Conversion
Light bulb	Electrical energy to light
Car engine	Chemical energy (in fuel) to kinetic energy
Battery	Chemical energy to electrical energy
Elevator	Electrical energy to gravitational potential energy
Generator	Kinetic energy to electrical energy
Electric motor	Electrical energy to kinetic energy
Solar cell	Light to electrical energy
Flute	Kinetic energy (of air) to sound wave
Nuclear power plant	Nuclear energy to electrical energy
Hydroelectrical dam	Gravitational potential energy (of water) to electrical energy

This law means that energy is a commodity that cannot be produced from nothing or disappear into nothing. If work is being done or a form of energy is "appearing," then energy is being used or converted somewhere. Unlike money, which you can counterfeit or burn, you cannot manufacture or eliminate energy.

The law of conservation of energy is both a practical tool and a theoretical tool. It can be used to solve problems, notably in mechanics, and it is a necessary condition for proposed models to satisfy. As an example of the latter, a theoretical astrophysicist may develop a model that explains how stars convert nuclear energy into heat and radiation. A first test of the validity of the model is whether or not energy is conserved.

FIGURE 3.29
Wind generators convert kinetic energy of the wind into electrical energy plus some heat energy because of friction.

TABLE 3.2 SPEED VS DISTANCE FOR A FREELY FALLING BODY			
A. SI Units		**B. English Units**	
Distance *(d)*	Speed *(v)*	Distance *(d)*	Speed *(v)*
0 m	0 m/s	0 ft	0 mph
1	4.4	1	5.5
2	6.3	2	7.7
3	7.7	3	9.4
4	8.9	4	11
5	9.9	5	12
10	14	10	17
20	20	20	25
100	44	100	55

Note: A and B are independent. The distances and speeds in A and B are not equivalent.

Now let's consider a similar problem. A roller coaster starts from rest at a height *(d)* above the ground. It rolls without friction or air resistance down a hill (see Figure 3.31). What is its speed when it reaches the bottom?

Again, the only forms of energy are gravitational potential energy and kinetic energy, since we assume there is no friction. The total energy, which is kinetic energy + potential energy, is constant. The kinetic energy of the roller coaster at the bottom of the hill must equal its *PE* at the top.

$$KE \text{ (at bottom)} = PE \text{ (at top)}$$

$$\frac{1}{2}mv^2 = mgd$$

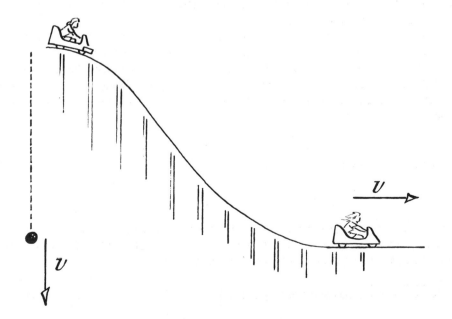

FIGURE 3.31
As a roller coaster travels down a hill, its potential energy is converted into kinetic energy. If there is no friction, its kinetic energy at the bottom equals its potential energy at the top. Its speed at the bottom is the same as that of an object dropped from the same height.

This is the same result obtained for a freely falling body. The speed at the bottom is consequently given by the same equation:

$$v = \sqrt{2gd}$$

It would have been impossible to solve this problem using only the tools from Chapter 2. To use Newton's second law, $F = ma$, one needs to know the net force that acts at every instant. The net force pulling the roller coaster along its path varies as the slope of the hill changes, and so it is a very complicated problem. The principle of energy conservation allows us to solve easily a problem that we could not have before. In the process we also come up with the following general result: the law of conservation of energy tells us that for an object affected by gravity but not friction, *the speed that it has at a distance* (d) *below its starting point is given by the preceding equation regardless of the path it takes.* The speed of a roller coaster that rolls down a hill is the same as that of an object that falls vertically the same height. The roller coaster does take more *time* to build up that speed (its acceleration is smaller), and therefore the falling body reaches the ground sooner.

The motion of a pendulum involves the continuous conversion of gravitational potential energy into kinetic energy and back again. Let's say that a child on a swing is pulled back (and up; Figure 3.32). The child then has gravitational potential energy because of his or her height above the rest position of the swing. When released the child swings downward, and the potential energy is converted into kinetic energy. At the lowest point in the arc, the child has only kinetic energy, which equals the original potential energy. The child then swings upward and converts the kinetic energy back into potential energy. This continues until

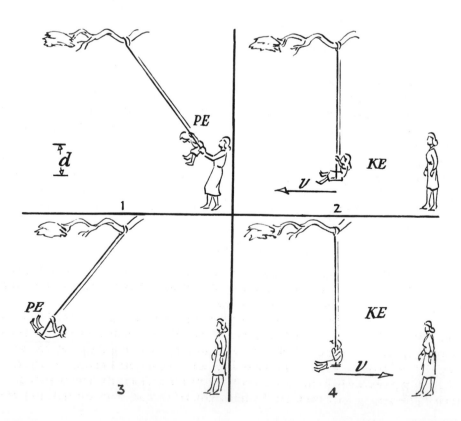

FIGURE 3.32
As a child swings back and forth, gravitational potential energy is continually converted into kinetic energy and back again. The potential energy at the highest (turning) points equals the kinetic energy at the lowest point.

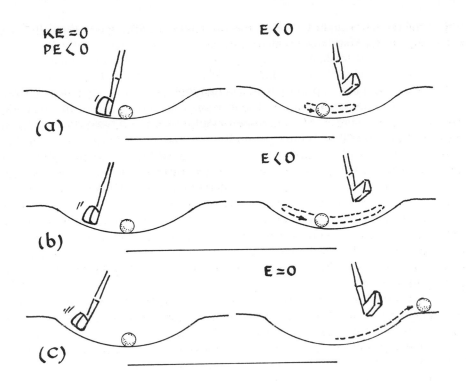

FIGURE 3.33
A golf ball at rest in the small valley has negative potential energy. Hitting the golfball gives it kinetic energy but it oscillates inside the valley if its total energy is negative (a) and (b). If the golfball is given enough kinetic energy to make its total energy zero it rolls out of the valley and stops (c).

the swing stops at a point nearly level with the starting point. This process is repeated over and over as the child swings. Air resistance takes away some of the kinetic energy. If the child is pushed each time, the work done puts energy into the system and counteracts the effect of the air resistance. Without air resistance or any other friction, the child would not have to be pushed each time.

The maximum height that a pendulum reaches (at the turning points) depends on its total energy. The more energy a pendulum has, the higher the turning points. In Section 3.2 we described a way to measure the speed of a bullet or a thrown object (Figure 3.6). The law of conservation of linear momentum is used to relate the speed of the bullet before the collision to the speed of the block (and bullet) afterwards. If the wood block is hanging from a string, the kinetic energy it gets from the impact causes it to swing up like a pendulum. The more energy it gets, the higher it will swing. We can determine the speed of the block after impact by measuring how high the block swings. The potential energy of the block (and bullet) at the high point of the swing equals the kinetic energy of the block (and bullet) right after impact. This results in the same equation relating the speed at the low point to the height reached.

$$v = \sqrt{2gd}$$

In Section 2.6 we discussed the motion of an object hanging from a spring (Figure 2.23). The motion also consists of a continual conversion of potential energy into kinetic energy and back again.

Figure 3.33 shows someone putting a ball at a miniature golf course. Since the ball rests in a small valley below ground level, its potential energy is negative relative to the level ground. When the ball is not moving, its total energy is negative because its kinetic energy is zero and its potential energy is negative. The golfer gives the ball kinetic energy by hitting it with a club. A weak putt gives it enough energy to roll back and forth but not to "escape" from the hole (a). The ball's total energy is larger but still negative. In (b) the golfer gives the ball more energy by hitting it harder, but since its total energy is still negative, the ball again oscillates back and forth, although reaching a higher point on each side before turning around. In (c) the golfer hits the ball just hard enough for the ball to roll out of the valley and stop once it is out. The ball is given just enough kinetic energy to make its total energy equal to zero, and it "escapes" from the little valley. If the ball were hit even harder, it would escape and have excess kinetic energy—it would still be rolling on the level ground.

This is the principle behind rocking a car when it is stuck. If a tire is in a hole, it is best to make the car oscillate back and forth. By giving it some energy during each cycle, by pushing or by using the engine, one can often give the car enough energy to leave the hole.

There are many analogous systems in physics in which an object is bound unless its total energy is equal to or greater than some value. A satellite in orbit about the earth is an important example. The satellite's motion from one side of the earth to the other and back is similar to the motion of the golf ball in the valley. If it is given enough energy, the satellite will escape from the earth and move away, much like the golf ball. The minimum speed that will give a satellite enough energy to leave the earth is called the *escape velocity*. Its value is approximately 11,200 meters/second or 25,000 miles/hour.

When water boils the individual water molecules are given sufficient energy to break free from the liquid (Chapter 5). Sparks and lightning occur only after electrons are given enough energy to break free from their atoms (Chapter 7). The transition of a system from a bound state to a free state is quite common in physics and not limited to mechanical systems.

These examples illustrate the qualitative and quantitative usefulness of the law of conservation of energy. It allows us to treat some systems that we couldn't before, and it provides another way of looking at other, familiar systems.

3.6   COLLISIONS: AN ENERGY POINT OF VIEW

Earlier in this chapter we pointed out that the main "tool" for studying *all* collisions is the law of conservation of linear momentum (Section 3.2). In this section we look at collisions from an energy standpoint. In some collisions the only form of energy involved, before and after, is kinetic energy. In other collisions forms of energy like potential energy and heat energy play a role. Collisions can be classified as follows.

> An **elastic collision** is one in which the total kinetic energy of the colliding bodies after the collision *equals* the total kinetic energy before the collision.
>
> An **inelastic collision** is one in which the total kinetic energy of the colliding bodies after the collision *is not equal to* the total kinetic energy before. The total kinetic energy after can be greater than or less than the total kinetic energy before.

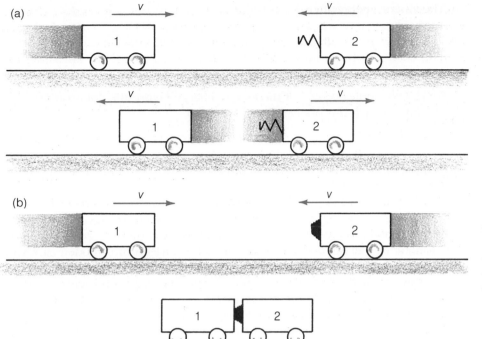

FIGURE 3.34
(a) Two carts with the same mass m *and speed* v *collide head on and bounce apart. The total kinetic energy of the carts is the same before and after the collision. (b) This time the two carts stick together after the collision. In this case the kinetic energy after the collision is zero.*

In an *elastic collision kinetic energy is conserved*. The total energy is *always* conserved in both types of collisions, but in elastic collisions no energy conversions take place that make the total kinetic energy after different than the total kinetic energy before.

Figure 3.34 illustrates examples of these two types of collisions. Two equal-mass carts traveling with the same speed but in opposite directions collide. In both collisions the *total linear momentum before the collision equals the total after.* (This total is equal to zero. Why?) In (a) the carts bounce apart because of a spring attached to one of them. After the collision each cart has the same speed it had before but it is going in the opposite direction. Consequently the total kinetic energy of the two carts is the same after the collision as it was before. This is an elastic collision.

Figure 3.34(b) is an example of an inelastic collision. This time the two carts stick together (because of putty on one of them) and stop. The total kinetic energy after the collision is zero in this case. The automobile collision analyzed in Example 3.2 (Figure 3.5) is also an inelastic collision. To see this, simply use the information to calculate the total kinetic energy before and then after the collision.

Recall the automobile collision analyzed in Example 3.2 (Figure 3.5). Compare the amounts of kinetic energy in the system before and after the collision.

EXAMPLE 3.9

The kinetic energy before the collision was

$$KE \text{ before} = \frac{1}{2} \times 1{,}000 \text{ kg} \times (25 \text{ m/s})^2$$

$$KE \text{ before} = 312{,}500 \text{ J}$$

The kinetic energy after the collision was:

$$KE \text{ after} = \frac{1}{2} \times 2{,}500 \text{ kg} \times (10 \text{ m/s})^2$$

$$KE \text{ after} = 125{,}000 \text{ J}$$

So 187,500 joules (60%) of the kinetic energy before the collision was converted into other forms of energy.

In these two examples of inelastic collisions, part of the original kinetic energy of the colliding bodies is converted into other forms of energy, mostly heat energy, but also some sound (the "crash" that one would hear). In Figure 3.34(b) *all* of the kinetic energy is converted into other forms of energy.

In some collisions the total kinetic energy after the collision is *greater than* the total kinetic energy before the collision. If our old friend the wooden cart, with its plunger in ("loaded"), is struck by a second cart, the plunger will be released by the shock (Figure 3.35). The plunger's potential energy is transferred to both carts as kinetic energy. Therefore the total kinetic energy after the collision is greater than the total kinetic energy before the collision; it is an *inelastic collision*. The stored energy is released by the collision.

The use of collisions is an invaluable tool in studying the structure and properties of atoms and nuclei. Much of the information in Chapters 10, 11 and 12 was gleaned from the careful analysis of countless collisions. Linear accelerators, cyclotrons, betatrons, and other devices produce very high-speed collisions between atoms, nuclei, and subatomic particles. The collisions are recorded and analyzed using the law of conservation of linear momentum and other principles. If a collision is inelastic, the amount of kinetic energy "lost" or "gained" in the collision is useful for determining the properties of the colliding particles.

Collisions are also responsible for other phenomena such as gas pressure and the conduction of heat. The collisions of air molecules with the inner surface of a balloon is what keeps it inflated. When you touch a piece of ice, the molecules in your finger collide with, and lose energy to, the molecules of the ice.

Elastic collisions are also used in space exploration. The paths of the Voyager 1 and Voyager 2 space probes were chosen so that as they passed by Jupiter and Saturn they gained kinetic energy from these planets. This is called the "slingshot effect." (The result is similar to a beach ball bouncing off the front of a moving

FIGURE 3.35
A wooden cart, cart 2, has energy stored in its spring-loaded plunger. When this cart is struck by cart 1, this potential energy is converted into kinetic energy, which is then shared by both carts. The total kinetic energy after the collision is greater than the total kinetic energy before the collision.

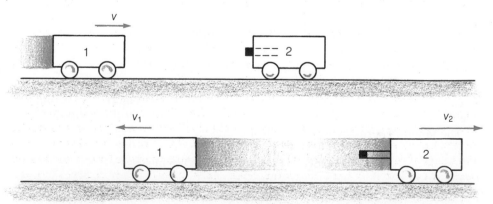

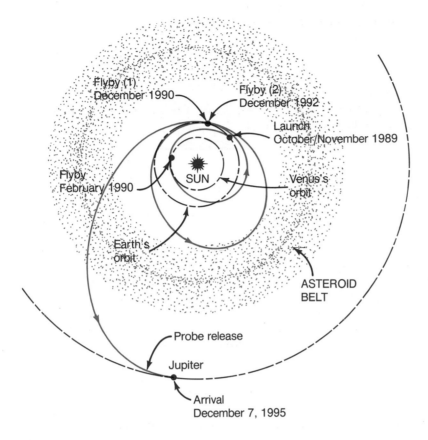

FIGURE 3.36
The path of the Galileo space probe on its journey to Jupiter.

truck; the beach ball gains kinetic energy.) On its way to a 1995 arrival at Jupiter, the Galileo space probe is following a path that uses two encounters with earth and one with Venus to gain energy (Figure 3.36). Such "collisions" don't involve physical contact; energy is gained because of the force of gravity that acts between the masses.

3.7 POWER

We have seen many examples of work being done and energy being transformed into other forms. The amount of *time* involved in the processes has not entered into the discussion until now. Let us say that a ton of bricks needs to be loaded from the ground onto a truck (see Figure 3.37). This might be done two ways. First, a person could lift the bricks one at a time and place them on the truck. This might take the person 1 hour to do. Second, a forklift could be used to load the bricks all at once. This might take only 10 seconds. In both cases the same amount of work is done. The force on each brick (its weight) times the distance it is moved (the height of the truck bed) is the same whether the bricks are loaded one at a time or all at once. The work done, *Fd,* is the same in both cases, but the *power* is different.

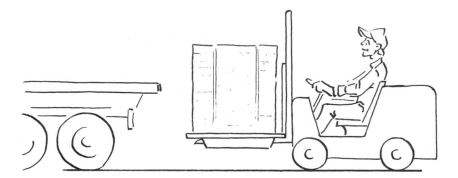

FIGURE 3.37
A ton of bricks is loaded onto a truck in two different ways. The work done is the same, but since the forklift does the job much faster, the power is much greater.

> **POWER**   The rate of doing work. The rate that energy is transferred or transformed. Work done divided by the time. Energy transferred divided by the time.
>
> $$P = \frac{\text{Work}}{t} \quad P = \frac{E}{t}$$

	Metric	English
Power *(P)*	watt (W)	foot-pound/second (ft-lb/s)
		horsepower (hp)

In SI units the weight of a ton of bricks is 8,900 newtons. If the height of the truck bed is 1.2 meters, then the work done is:

$$\text{Work} = Fd = 8{,}900 \text{ N} \times 1.2 \text{ m} = 10{,}680 \text{ J}$$

If the forklift does this work in 10 seconds, the power is:

$$P = \frac{\text{Work}}{t} = \frac{10{,}680 \text{ J}}{10 \text{ s}}$$

$$P = 1{,}068 \text{ J/s} = 1{,}068 \text{ W}$$

The unit *joule per second,* J/s, is defined to be the *watt W.*

$$1 \text{ watt} = \frac{1 \text{ joule}}{1 \text{ second}}$$

$$1\text{W} = 1\text{ J/s}$$

The watt should be quite familiar to you because it is commonly used to measure the power consumption of electrical devices. A 60-watt light bulb uses electrical energy at the rate of 60 joules each second. A 1,500-watt hair dryer uses 1,500 joules of energy each second (see Figure 3.38).

The horsepower is the most commonly used unit of power in the English system of units. Automobile engines, lawn mowers, and many other motorized devices are rated in horsepower. The basic power unit, foot-pound per second, is the unit of work, foot-pound, divided by the unit of time, second. The conversion factors are

$$1\text{ hp} = 550\text{ ft-lb/s} = 746\text{ W}$$

A device that could raise 550 pounds a distance of 1 foot in 1 second would develop 1 hp. Raising 110 pounds a distance of 5 feet in 1 second would also require 1 hp.

The relationship between power and work (or energy) is the same as that between speed and distance. In each case the former is the rate of change of the latter.

$$P = \frac{\text{Work}}{t} \qquad v = \frac{d}{t}$$

A runner and a car can both travel a distance of 20 miles, but the car can do it much faster because it is capable of much higher speeds. A person and a forklift can both do 10,680 joules of work raising the bricks, but the forklift can do it much faster because it has more power.

In Examples 2.2 and 3.5 we computed the acceleration, force, and work for a 1,000-kilogram car that goes from 0 to 27 meters/second in 10 seconds. We can

EXAMPLE 3.10

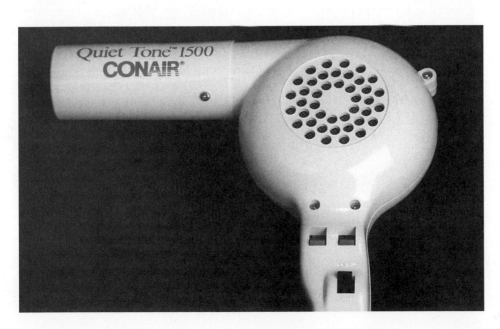

FIGURE 3.38
The number indicates how much power this hair dryer requires.

now determine the required power output of the engine. The work, 364,500 joules, is done in 10 seconds. Hence the power is

$$P = \frac{\text{Work}}{t} = \frac{364,500 \, \text{J}}{10 \, \text{s}}$$

$$P = 36,450 \, \text{W} = 48.9 \, \text{hp}$$

The car's kinetic energy when going 27 meters/second is also 364,500 joules (see Example 3.6). It takes the engine 10 seconds to give the car this much kinetic energy, so we get the same result using energy divided by time.

Given enough food or fuel, there is usually no limit to how much work a device or a person can do. But there is a *limit on how fast the work can be done*; the power is limited. In other words, only so much work can be done each second. The power output can be anything from zero (no work) to some maximum. For example, a 100-horsepower automobile engine can put out from 0 to 100 horsepower. When accelerating as fast as possible, the engine is putting out its maximum power. While cruising down a flat highway, the engine may only be putting out 10 to 20 horsepower, enough to counteract the effect of air resistance and other friction.

The human body has a maximum power output that varies greatly from person to person. In the act of jumping, an outstanding athlete can develop more than 8,000 watts, but only for a fraction of a second. The same person would have a maximum of less than 800 watts if the power level had to be maintained for an hour. The average person can produce 800 watts or more for a few seconds and perhaps 100-200 watts for an hour or more (see Figure 3.39). When running, the body uses energy to overcome friction and air resistance. In short races the best

FIGURE 3.39
Fourteen-time Greek bicycling champion Kanellos Kanellopoulos pedaled the human-powered Daedulus on a 72-mile flight between the islands of Crete and Santorin on 23 April, 1988. The aircraft required a light-weight pilot who could maintain a high power output for the nearly 4-hour flight.

runners can maintain a speed of 10 meters/second for about 20 seconds. In longer races the speeds are slower because the power level has to be maintained longer. For a race lasting about 30 minutes, the best average speed is about 6 meters/second.

DO-IT-YOURSELF PHYSICS

Compute your own power output when walking or running up a flight of stairs (Figure 3.40). First you need to compute the work you do by measuring the vertical height of the stairs and then multiplying this number by your weight. (You may want to use SI units—meters and newtons.) The power is this work divided by the time it takes to climb the stairs.

If you go hiking in the mountains, you can estimate your steady power output by performing the same calculation. The vertical distance can be found using a topographic map. (Do not forget to include the weight of your backpack.)

FIGURE 3.40
To measure your power output when going up a flight of stairs, multiply the height of the stairs by your weight, and divide this by the time it takes you to climb the stairs.

3.8   ROTATION AND ANGULAR MOMENTUM

Our final conservation law applies to rotational motion. A spinning ice skater or a satellite moving in a circular path about the earth are examples. You might say that this law is the rotational analog or counterpart of the law of conservation of linear momentum.

> **LAW OF CONSERVATION OF ANGULAR MOMENTUM**   The total angular momentum of an isolated system is constant.

There are two types of angular momentum, one for the case of an object moving in a circular path and the other for something spinning about an axis. In the first case we use the concept of *orbital angular momentum* of the object; in the second case we employ *spin angular momentum*. Once the angular momentum is determined, however, the law of conservation of angular momentum applies to both types of motion.

> The **orbital angular momentum** of an object moving in a circle equals the product of its mass, its speed, and the radius of its path.
>
> Orbital angular momentum $= mvr$

Notice that the orbital angular momentum is also equal to the linear momentum of the mass multiplied by the radius of the circular path.

In the case of orbital angular momentum, the system is isolated if the only external force acting on the mass is directed toward or away from the center of its motion.

Once again imagine tying a rubber ball to the end of a string and whirling it about in a circle over your head (Figure 3.41a). (We assume there is no friction or air resistance.) The faster the ball goes, the greater its orbital angular momentum. Using a longer string for the same speed would also make its angular momentum larger. To illustrate the law of conservation of angular momentum, imagine suddenly shortening the string by pulling downward on the end with your free hand, letting the string slide through the hand. This makes the ball move in a circle with a smaller radius but with a higher speed (Figure 3.41 b). Since the force exerted on the ball is directed toward the center of its motion, angular momentum is conserved—it has the same value before and after the change. Because the *radius* of the ball's path is now *smaller,* its *speed must be higher.* If you let the string out so the radius is *larger,* the ball will *slow down,* keeping the angular momentum constant.

With caution we can use this definition of orbital angular momentum—*mvr*— for an object moving in a path other than a circle. Figure 3.42 shows the elliptical orbit of a satellite moving around the earth. At the two points labeled A and B, the satellite's velocity will be perpendicular to a line from it to the earth's center. At those points the satellite's path is like a short segment of a circle. Consequently its angular momentum is mvr. At point B, *r is smaller* than at point A. Because the angular momentum is the same at A and at B, the satellite's *speed is greater* at B. For example, if the satellite is 13,000 kilometers (about 8,000 miles, twice the earth's radius) from the earth's center at point B and 26,000 kilometers from the earth's center at point A, its speed at B will be two times its speed at A. The actual

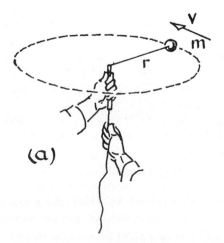
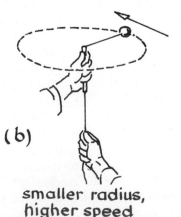

FIGURE 3.41
(a) The angular momentum of an object moving in a circle equals m *times* v *times* r. *(b) If the radius is decreased, the object speeds up so that the angular momentum stays the same.*

(a)   (b)

smaller radius, higher speed

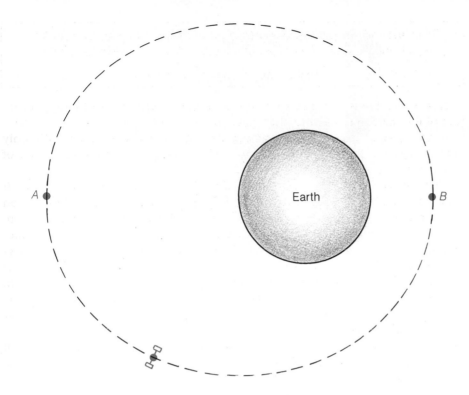

FIGURE 3.42
A satellite in orbit about the earth is twice as far from the earth's center at point A as it is at point B. Conservation of angular momentum then tells us that its speed at A is one half its speed at B.

values for the speeds are about 6,400 meters/second (about 14,000 miles/hour) when it is closest to the earth and about 3,200 meters/second when it is farthest away.

To handle the rotational motion of an object that is spinning, such as a dancer, we use *spin angular momentum.* Spin angular momentum depends on how fast the object is rotating. The faster it spins about the axis the greater its spin angular momentum. The *angular speed* of a spinning object indicates its rate of rotation. It can be measured in "rpm" (revolutions per minute—the unit of measure used by automobile tachometers), revolutions per second, degrees per second (360 degrees in one revolution), or similar units of measure. Any object with angular speed 500 rpm has twice the angular momentum of an identical object spinning at 250 rpm.

Spin angular momentum also depends on the mass of a spinning body and on how the mass is distributed relative to the axis. To incorporate both of these factors, we introduce the quantity *moment of inertia* (sometimes called rotational inertia). To determine the moment of inertia of an object we imagine that it is composed of a large number of very small parts all connected to each other. Each part will have some small mass, δm, and will be some distance, d, from the axis about which the body is spinning. (Here "δ" means "small".) The moment of inertia of each little mass about this axis equals the *mass times the square of its distance from the axis* (Figure 3.43).

$$\text{moment of inertia of } \delta m = \delta m \, d^2$$

The total moment of inertia of the spinning body is found by adding up the moments of inertia of each small part.

$$\text{moment of inertia of object} = \text{sum of moments of inertia of each part}$$

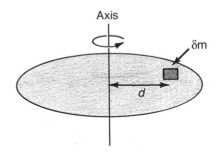

FIGURE 3.43
The moment of inertia of each part of the spinning body equals the mass of that small part times the square of the distance from the axis of rotation.

FIGURE 3.44
The moment of inertia is smaller when the mass is distributed closer to the axis. Jeff finds that it is easier to twist the barbell back and forth about a vertical axis (his arm) when the weights are closer to the axis. (Right) The barbell's mass is the same, but its moment of inertia is larger.

Moment of inertia is the rotational analog of mass. Objects with larger moments of inertia are more difficult to start spinning or to stop once they are spinning, just as bodies with greater mass are more difficult to speed up or slow down. However, two bodies with the same mass will not, in general, have the same moment of inertia. When much of a body's mass is far from the axis of rotation, its moment of inertia will be large. There are many parts that have large values of d, so $\delta m\, d^2$ is a big number. If, on the other hand, most of the mass is close to the axis, the momentum of inertia will be small because the values of d are small. A barbell with the weights positioned near the ends has a much larger moment of inertia than a barbell with the weights near the middle (Figure 3.44). With this latter configuration it is much easier to twist the barbell back and forth.

Even the same object will have different moments of inertia when it rotates about different axes. If you twirl a baton in the usual fashion (about an axis through its middle), its moment of inertia is smaller than if you twirl it about an axis through one end (Figure 3.45). In the latter case parts of the baton are much farther from the axis of rotation and consequently their contribution to the moment of inertia is quite large (remember that the distance is squared).

The concepts of angular speed and moment of inertia combine to give the angular momentum of a spinning body:

spin angular momentum = moment of inertia × angular speed

This is the rotational analog of linear momentum, which equals mass times velocity. The law of conservation of angular momentum in the case of spin angular momentum requires that there be no outside forces acting to increase or to decrease the spinning object's angular speed.

The law of conservation of angular momentum tells us that a spinning body will *spin faster if its moment of inertia is reduced.* An ice skater may start spinning with arms extended. Pulling the arms in will make the skater spin faster. With the arms closer to the axis or rotation, the moment of inertia of the skater is about one third as large as with the arms extended, so the angular speed is tripled. A diver doing somersaults will pull arms and legs in close to the body to reduce the moment of inertia, thereby increasing the spin rate. Figure 3.46 shows

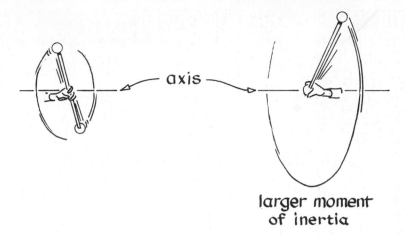

FIGURE 3.45
The moment of inertia of the baton is larger when it spins about an axis through one end than when the axis is through the center.

someone illustrating this effect by holding a weight in each hand at arm's length and spinning around on a stool. When the weights are pulled in close to the body, the combined moment of inertia of the person and the weights is reduced. Since the angular momentum stays constant, the person spins faster. If the weights are extended again, the angular speed will decrease.

An object can have both spin angular momentum and orbital angular momentum. The earth spins about its axis as its moves in its orbit about the sun. The total angular momentum of such a system is the sum of the spin and the orbital angular momenta. The earth-moon system presents an interesting situation. The moon has orbital angular momentum as it moves in its orbit, and the earth has spin angular momentum. As discussed in Section 2.10, the moon's gravitational pull on the earth's oceans causes the tides. The daily movement of the oceans, though it is very small, is subject to friction. This causes a very slow but *steady decrease in the earth's angular speed.* As the earth's spin angular momentum decreases, the moon's orbital angular momentum increases because the total angular momentum is conserved. The moon's orbit is slowly getting larger—it is moving away from the earth.

FIGURE 3.46
The person spins faster when the weights are pulled in closer to the body because the moment of inertia is smaller.

Pulsars are astronomical objects that emit a regular sequence of pulses of radio energy. The pulse period (i.e., the interval of time between one pulse and the next) is usually quite short (less than 1 second for most pulsars), and it is generally accepted that pulsars are rapidly rotating *neutron stars*.

A neutron star is a remnant of a once-normal star that may have ended its life in a brilliant flash called a "supernova explosion." Though small (about 20 kilometers across), neutron stars are very dense, with 1 teaspoon of neutron star stuff easily weighing a billion tons. Although the mechanism by which pulsars produce the radio energy we receive is not well understood, it is widely believed that what we are observing is akin to a "lighthouse effect." Radio waves are being beamed out into space from what might be termed "radio hot spots" on or near the surface of the neutron star, and as these hot spots are rotated across our line of sight, we receive regularly spaced, short bursts of radio radiation (Figure 3.47).

When pulsars were first discovered in 1967, their most prominent feature was the precision with which the pulses were spaced. Indeed, it was originally hoped that these objects could be used as extremely accurate timekeepers, surpassing even atomic clocks in their exactness. However, after continued observation, scientists found that these "astronomical clocks" were slowing down; their periods (the intervals between "ticks") were gradually increasing. We now think that what is happening is that the neutron star associated with a pulsar is losing energy through interactions with its environment and rotating more and more slowly with time. The analogy with a toy top slowly spinning down because of frictional losses to the table on which it rotates is perhaps not a bad one to have in mind here. The point is, astronomers expected a continual decay in the rotation rates of these compact stars and a corresponding lengthening of the periods of pulsars.

But nature is full of surprises. In 1969 a pulsar in the constellation of Vela suddenly sped up, that is, its period decreased! Afterwards it once again resumed its steady decline in spin at the same rate as

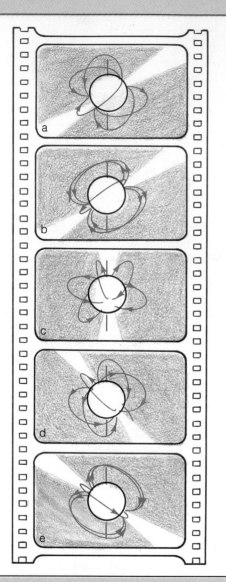

FIGURE 3.47
The "lighthouse" model of a pulsar. Radio "hot spots" associated with the strong magnetic field of the neutron star emit copious quantities of radio radiation. As these hot spots are carried across our line of sight by the spin of the neutron star, we receive sharp pulses of radio energy.

before. This was not an isolated event. In 1971 and again in 1975, similar speed-ups were observed (Figure 3.48). These brief events in which a pulsar increases its spin are called "glitches," and their explanation may well have its root in the law of conservation of angular momentum.

Current models for the structure of neutron stars suggest that they may have solid, crystalline crusts a few hundred meters thick on which small irregular features a few centimeters high exist. Imagine now what would happen if a "starquake" occurred and this crust fractured and settled by a small amount. (Calculations show that crustal collapse of only about 1 millimeter would be needed to account for the "glitches" if this model is correct.) The neutron star matter would then be slightly more concentrated toward the axis about which it is rotating, and consequently its moment of inertia would be smaller. Insofar as the neutron star can be considered an isolated system, the law of conservation of angular momentum applies to it, and it can be analyzed in much the same way as is an ice skater pulling in her arms while performing a pirouette. In particular, in order for its spin angular momentum to remain constant as the neutron star material collapses inward, its angular speed must increase. Naturally, increasing the rate of spin produces a shorter interval of time between successively observed radio beams and a decrease in the pulsar period. Since the redistribution of neutron star crustal material happens quickly and at irregular intervals, the nature of the glitches is neatly explained. They are expected to be observed for some time in the youngest (most rapidly rotating) pulsars like the one in Vela.

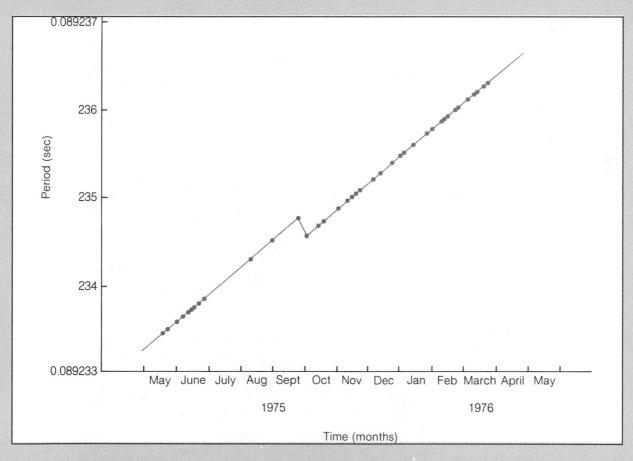

HISTORICAL NOTES

Most of the principles discussed in this chapter were developed piecemeal by a host of physicists and mathematicians in the seventeenth, eighteenth, and nineteenth centuries. They are extensions of Newton's mechanics, not fundamentally new formulations. As mentioned near the end of Section 3.2, the law of conservation of linear momentum has its roots in Newton's laws of motion. The ideas of work and potential energy arise naturally from Newton's second law of motion when the force on an object depends only on its position.

Before Newton's *Principia,* the importance of momentum and kinetic energy had been discovered experimentally. Galileo used the product of weight and velocity as momentum, a quantity that is nearly the same as our linear momentum, since weight and mass are proportional. In 1665 the Dutch physicist Christian Huygens reported that the quantity mv^2 was conserved during the collision of two hard balls. (Huygens later acquired fame for his work in optics. His ex-

FIGURE 3.49
Archimedes.

planation of the fundamental nature of light was contrary to Newton's but eventually was proven to be correct.) The actual use of the word "energy," the identification of kinetic energy as being equal to one half the mass times the speed squared, and the statement of the law of conservation of energy came about around the middle of the nineteenth century.

The simple machines discussed in Section 3.3 were first analyzed by the famous Greek mathematician and scientist Archimedes (287 BC?–212 BC). A citizen of Syracuse, Sicily, Archimedes (Figure 3.49) explained how a force could be amplified using a lever, a combination of pulleys, or other simple machines. The ancient historian Plutarch described how he demonstrated the usefulness of pulleys by moving a large, dry-docked ship by himself. This so impressed King Hiero of Syracuse that he persuaded Archimedes to construct machines to defend the city. These proved to be quite useful when the Romans attacked Syracuse by land and sea. Plutarch includes this account in *Life of Marcellus:*

> Archimedes began to ply his engines, and shot against the land forces of the assailants all sorts of missiles and immense masses of stones, which came down with incredible din and speed: nothing whatever could ward off their weight, but they knocked down in heaps those who stood in their way, and threw their ranks into confusion. At the same time huge beams were suddenly projected over the ships from the walls, which sank some of them with great weights plunging down from on high: others were seized at the prow by iron claws, or beaks like the beaks of cranes, drawn straight up into the air, and then plunged stern foremost into the depths, or were turned round and round by means of enginery within the city, and dashed upon the steep cliffs that jutted out beneath the wall of the city, with great destruction of the fighting men on board, who perished in the wrecks.

Archimedes' machines forced Marcellus, the Roman general, to abandon the direct assault in favor of a long siege. Two years later Syracuse fell, and Archimedes was killed by the conquerors.

Archimedes' most famous discovery was the principle of buoyancy that is named after him (Section 4.5). We should also note that he was one of the first to engage in two pursuits that have been central to physics since that time. First, as an engineer he used his discoveries to construct useful devices. Second, he was a scientific consultant for the military. To this day the military is a primary source of funding for research in physics and other sciences. Many great discoveries in physics and engineering, some quite terrifying, have come from efforts to devise offensive and defensive weapons of war.

From the two forms of mechanical energy introduced in this chapter, the concept of energy has been expanded to encompass many other diverse forms. A major part of the growth of science and technology in the last two centuries has been the discovery and application of new forms of energy. At the beginning of this century, Albert Einstein showed that the concept of energy even includes matter: matter can be converted into energy and vice versa (refer to Chapter 11). The energy shortage experienced by the United States in the 1970s and the resulting increase in energy costs quickly illustrated how dependent our society is on devices that require energy sources. Energy is now a necessity of contemporary society, almost equal in importance to food, labor, and raw materials. Energy has come a long way from its simple beginnings.

SUMMARY

Conservation laws are powerful tools for analyzing physical systems, particularly those in mechanics. Their main advantage is that it is not necessary to know the details of what is going on in the system at each instant. The use of conservation laws is based on a "before-and-after" approach: the total amount of the conserved physical quantity *before* an interaction is equal to the total amount *after* the interaction. Linear momentum, energy, and angular momentum are physical quantities that are defined and used mainly because they are conserved in isolated systems.

The main application of the law of conservation of linear momentum is to collisions. The total linear momentum before a collision equals the total linear momentum after the collision if the system is isolated. This applies to all collisions, both elastic and inelastic.

Work is done whenever a force acts through a distance in the same direction as the force. To be able to do work, a device or a person must have energy. The act of work involves the transfer of energy from one thing to another, the transformation of energy from one form to another, or both. In mechanics the main forms of energy are kinetic energy, potential energy, and heat energy. There are many other forms of energy corresponding to different sources of work. Any form of energy can be used to do work if a suitable conversion device is available. The law of conservation of energy tells us that in all such conversions, the total amount of energy remains constant.

Power is the rate of doing work or using energy. It is the measure of how fast energy is transferred or transformed.

Angular momentum is a conserved quantity in rotational motion.

SUMMARY OF IMPORTANT EQUATIONS

EQUATION	COMMENTS
FUNDAMENTAL EQUATIONS	
linear momentum $= mv$	Definition of linear momentum
$F = \dfrac{\Delta(mv)}{\Delta t}$	Alternate form of Newton's second law
Work $= Fd$	Definition of work
$KE = \frac{1}{2}mv^2$	Kinetic energy
$PE = mgd$	Gravitational potential energy
$P = \dfrac{\text{Work}}{t} \quad P = \dfrac{E}{t}$	Definition of power
ang. mo $= mvr$	Orbital angular momentum
SPECIAL CASE EQUATIONS	
$v = \sqrt{2gd}$	Speed after falling a distance d
$d = \dfrac{v^2}{2g}$	Vertical height reached given initial speed v

QUESTIONS

1. What is a conservation law? What is the basic approach taken when using a conservation law?

2. Why is the alternate form of Newton's second law of motion given in this chapter the more general form?

3. Could the linear momentum of a turtle be greater than the linear momentum of a horse? Explain why or why not.

4. An astronaut working with many tools some distance away from a spacecraft is stranded when the "maneuvering unit" malfunctions. How can the astronaut return to the spacecraft by sacrificing some of the tools?

5. For what type of interaction between bodies is the law of conservation of linear momentum most useful?

6. During a head-on collision of two automobiles, the occupants are decelerated rapidly. Use the idea of work to explain why an air bag that quickly inflates in front of an occupant reduces the likelihood of injury.

7. When climbing a flight of stairs, do you do work on the stairs? Do the stairs do work on you? What happens?

8. Identify as many different forms of energy as you can that are around you at this moment.

9. When you throw a ball, the work you do equals the kinetic energy the ball gains. If you do *twice* as much work when throwing the ball, does it go *twice as fast?* Explain.

10. How can the potential energy of something be negative?

11. Identify the energy conversions taking place in each of the following situations. Name all of the relevant forms of energy that are involved.
    a) A camper rubbing two sticks together to start a fire.
    b) An arrow shot straight upward, from the moment the bowstring is released by the archer to the moment when the arrow reaches its highest point.
    c) A nail being pounded into a board, from the moment a carpenter starts to swing a hammer to the moment when the nail has been driven some distance into the wood by the blow.
    d) A meteoroid entering the earth's atmosphere.

12. Solar-powered spotlights have batteries that are charged by solar cells during the day and then operate lights at night. Describe the energy conversions in this entire process, starting with the sun's nuclear energy and ending with the light from the spotlight being absorbed by the surroundings. Name all of the forms of energy that are involved.

13. Truck drivers approaching a steep hill that they must climb often increase their speed. What good does this do, if any?

14. If you hold a ball at eye level and drop it, it will bounce back, but not to its original height. Identify the energy conversions that take place during the process, and explain why the ball does not reach its original level.

15. A ball is thrown straight upward on the moon. Is the maximum height it reaches less than, equal to, or greater than the maximum height reached by a ball thrown upward on the earth with the same initial speed (no air resistance in both cases)? Explain.

16. Explain the distinction between elastic and inelastic collisions. Give an example of each.

17. Cart A and B stick together whenever they collide. The mass of A is twice the mass of B. How could you roll the carts towards each other in such a way that they would be stopped after the collision? (Assume there is no friction.)

18. A person runs up several flights of stairs and is exhausted at the top. Later the same person walks up the same stairs and does not get exhausted. Why is this? Ignoring air resistance, does it take more work or energy to run up the stairs than to walk up?

19. Explain the difference between *orbital* angular momentum and *spin* angular momentum.

20. Why do divers executing midair somersaults pull their legs in against their bodies?

21. A solid iron ball and a larger hollow iron ball have the same mass. Which has the larger moment of inertia about an axis through the center. Why?

PROBLEMS

1. In Section 2.4 we computed the force needed to accelerate a 1,000-kg car from 0 to 27 m/s in 10 s. Compute the force using the alternate form of Newton's second law. The change in momentum is just the car's momentum when going 27 m/s minus its momentum when going 0 m/s.

2. A runner with a mass of 80 kg accelerates from 0 to 9 m/s in 3 s. Find the net force on the runner using the alternate form of Newton's second law.

3. A pitcher throws a 0.5-kg ball of clay at a 6-kg block of wood. The clay sticks to the wood on impact, and their velocity afterwards is 3 m/s. What was the original speed of the clay?

4. A 3,000-kg truck runs into the rear of a 1,000-kg car that was stationary. The truck and car are locked together after the collision and move with speed 9 m/s. What was the speed of the truck before the collision?

5. A 50-kg boy on roller skates moves with a speed of 5 m/s. He runs into a 40-kg girl on skates. Assuming they cling together after the collision, what is their speed?

6. Two persons on ice skates stand face to face and then push each other away. Their masses are 60 kg and 90 kg. Find the ratio of their speeds immediately afterwards. Which person has the higher speed?

7. A loaded gun is dropped on a frozen lake. The gun fires, with the bullet going horizontally in one direction and the gun sliding on the ice in the other direction. The bullet's mass is 0.02 kg and its speed is 300 m/s. If the gun's mass is 1.2 kg, what is its speed?

8. A running back with a mass of 80 kg and a speed of 8 m/s collides with, and is held by, a 120-kg defensive tackle going in the opposite direction. How fast must the tackle be going before the collision for their speed afterwards to be zero?

9. A motorist runs out of gas on a level road 200 meters from a gas station. The driver pushes the 1,200-kg car to the gas station. If a 150-N force is required to keep the car moving, how much work does the driver do?

10. In Figure 3.10, the rock weighs 100 lb and is lifted 1 ft by the lever.
  a) How much work is done?
  b) The other end of the lever is pushed down 3 ft while lifting the rock. What force had to act on that end?

11. A weight lifter raises a 100-kg barbell to a height of 2.2 meters. What is the barbell's potential energy?

12. A microwave antenna with a mass of 80 kg sits atop a tower that is 50 m tall. What is the antenna's potential energy?

13. A windsurfer and sailboard have a combined mass of 90 kg. What is their kinetic energy when they are going 10 m/s?

14. While in orbit about the earth, a space shuttle's mass is 80,000 kg and its speed is 7,900 m/s. What is its kinetic energy?

15. The kinetic energy of a motorcycle and rider is 60,000 J. If their total mass is 300 kg, what is their speed?

16. In compressing the spring in a toy dart gun, 0.5 J of work is done. When the gun is fired, the spring gives its potential energy to a dart with a mass of 0.02 kg.

a) What is the dart's kinetic energy as it leaves the gun?
b) What is the dart's speed?

17. The tops of the World Trade Center towers in New York City are 412 m above the street. If a rock could fall from that height without air resistance, what would its speed be just before it hit the ground?

18. A student drops a water balloon out of a dorm window 12 m above the ground. What is its speed when it hits the ground?

19. A child on a swing has a speed of 7.7 m/s at the low point of the arc. How high will the swing be at the high point?

20. The cliff divers at Acapulco, Mexico jump off a cliff 26.7 m above the ocean. Ignoring air resistance, how fast are the divers going when they hit the water?

21. In the *Fallturm* at Bremen, Germany (see page 2), the 300-kg test chamber falls freely from a height of 110 m.
  a) What is the chamber's potential energy at the top of the tower?
  b) How fast is it going when it reaches the bottom of the tower? (You may want to convert it to mph for comparison to highway speeds.)

22. The fastest that a human has run is about 12 m/s.
  a) If a pole vaulter could run this fast and convert all of his or her kinetic energy into gravitational potential energy, how high would he or she go?
  b) Compare this height with the world record in the pole vault.

23. A bicycle and rider approach a hill going 10 m/s. Their total mass is 80 kg.
  a) What is the kinetic energy?
  b) If the rider coasts up the hill without pedaling, at what point will the bicycle come to a stop?

24. Compute how much kinetic energy was "lost" in the collision in Problem 4.

25. Compute how much kinetic energy was "lost" in the collision in Problem 5.

26. A 1,000-watt motor powers a hoist used to lift cars at a service station.
  a) How much time would it take to raise a 1,500-kg car 2 meters?

b) If it is replaced with a 2,000-W motor, how long would it take?

27. How long does it take a worker producing 200 W of power to do 10,000 J of work?

28. An elevator is able to raise 1,000 kg to a height of 40 m in 15 s.
   a) How much work does the elevator do?
   b) What is the elevator's power output?

29. A professor's little car can climb a hill in 10 s. The top of the hill is 30 m higher than the bottom, and the car's mass is 1,000 kg. What is the power output of the car?

30. In the annual Empire State Building race, contestants run up 1,575 steps to a height of 1,050 ft. In 1983, the winner of this race was Al Waquie, with a time of 11 min and 36 s. Mr. Waquie weighed 108 lb.
   a) How much work was done?
   b) What was the average power output (in ft-lb/s and in hp)?

31. A bicyclist rides to the top of Mt. Nebo, Arkansas in 23 minutes. The vertical height that the bicyclist climbs is 360 m and the total mass of bicycle and rider is 85 kg. What was the bicyclist's power output?

32. The moment of inertia of a diver in the "layout" position (body straight, arms extended over head) is about five times larger than when the diver is in the "tuck" position (arms wrapped around knees at the chest). At the beginning of a high dive the angular speed of a diver in the layout position is 0.5 revolutions per second. When the diver goes into the tuck position, what is the angular speed?

CHALLENGES

1. Rank the following three collisions in terms of the extent of damage that the car would experience. Explain your reasons for ranking the collisions as you did.
   a) A car going 10 m/s striking an identical car that was stationary.
   b) A car going 10 m/s running into an immovable concrete wall.
   c) A head-on collision between identical cars, both going 10 m/s.

2. The system depicted in Figure 3.7 can be used to measure the mass of a body. Specify what quantities would have to be known and what would have to be measured to do this. Suggest how these measurements might be performed.

3. A bullet with a mass of 0.01 kg is fired horizontally into a block of wood hanging on a string. The bullet sticks in the wood and causes it to swing upward to a height of 0.1 m. If the mass of the wood block is 2 kg, what was the initial speed of the bullet?

4. In a head-on, inelastic collision, a 4,000-kg truck going 10 m/s east strikes a 1,000 kg car going 20 m/s west.
   a) What is the speed and direction of the wreckage?
   b) How much kinetic energy was lost in the collision?

5. A person on a swing moves so that the chain is horizontal at the turning points. Show that the centripetal acceleration of the person at the low point of the arc is exactly 2 g, regardless of the length of the chain. This means that the force on the chains at the low point is equal to *three times* the person's weight. (Hint: the vertical distance between the turning point and the low point equals the length of the chain.)

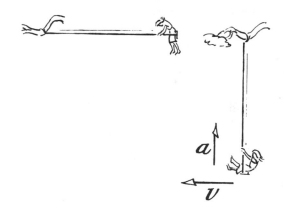

6. Assume that as a car brakes to a stop it undergoes a constant acceleration (deceleration). Explain why the stopping distance becomes *four times* as large if the initial speed is *doubled*.

7. The "shot" used in the shot-put event is a metal ball with a mass of 7.3 kg. When thrown in Olympic competition, it is accelerated to a speed of about 14 m/s. As an approximation, let's say that the athlete exerts a constant force on the shot while throwing it and that it moves a distance of 3 m while accelerating.
   a) What is the shot's kinetic energy?
   b) Compute the force that acts on the shot.
   c) It takes about 0.5 s to accelerate the shot. Compute the power required. Convert your answer to horsepower.

8. At the point in its orbit when it is closest to the sun, Halley's Comet travels 54,500 m/s. When it is at its most distant point,

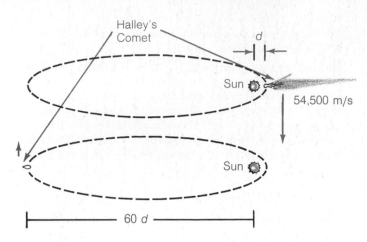

the distance between it and the sun is about 60 times the distance when it is at its closest point. What is the speed of the comet at the distant point?

9. A diver goes off a 10-meter platform and tries to execute several somersaults. Using the information in Problem 32 (above), compute how many somersaults (complete revolutions) the diver will make before hitting the water.

SUGGESTED READINGS

Bartlett, Albert A., and Hord, Charles W. "The Slingshot Effect: Explanations and Analogies." *The Physics Teacher* 23, no. 8 (November 1985):466–473.

Brancazio, Peter J. *Sport Science.* New York: Simon and Schuster, 1984. Chapter 4 "Rotating Bodies," Chapter 5 "Energy," and Chapter 6 "Collisions." Includes an analysis of Bob Beamon's record long jump in Mexico City.

Cohen, Morris R., and Drabkin, I. E. A Source Book in Greek Science. New York: McGraw-Hill, 1948. Good source on Archimedes and his work.

Drela, Mark, and Langford, John S. "Human-Powered Flight." *Scientific American* 253, no. 5 (November 1985):144–151. Along with drawings and specifications of different aircraft it includes a discussion of human power output.

Langford, John. "Triumph of Daedulus." *National Geographic* 174, no. 2 (August 1988):191–199. Describes the 72-mile flight of pedal-powered *Daedulus* from Crete to Santorin on 23 April 1988.

National Aeronautics and Space Administration. *The Voyager Flights to Jupiter and Saturn,* publication no. 585–182. Washington, D.C., GPO, 1982. Includes many beautiful color photos of the planets and their satellites.

Schefter, Jim. "Pedal Powered Choppers." *Popular Science* 232(5):80–118. Describes race to build the first successful human-powered helicopter.

OUTLINE

The Atlantis II sightseeing submarine, Bridgetown, Barbados.

PHYSICS OF MATTER

PROLOGUE: SUBS

In the late 1980s a new type of tourist excursion boat became quite popular—the submarine. Dozens of passenger subs are operating at resort islands all over the world. Forty people or more board specially built submarines with plenty of viewing ports. They submerge for close-up views of the fish and other natural life in the ocean.

The submarine has a surprisingly long history. Legend has it that Alexander the Great, in the fourth century BC, was lowered into the sea in a glass barrel to observe sea life. Many other familiar names in history, including Aristotle, Roger Bacon, Leonardo da Vinci, and Edmund Halley, wrote about or designed submersible devices. The first reports of a true submarine date to the 1620s and describe an oar-powered submarine operated in the Thames River. (King James I reportedly took a ride in it.) In the following centuries other human-powered submarines used propellers cranked by hand, pedaled with feet, or turned by the crew on a treadmill.

Throughout history submarines have been designed primarily for military use. The first documented attack (unsuccessful) by a submarine occurred in 1776. An American submarine named the *"Turtle"* (shaped like a coconut and crewed by one man) attempted to sink the British frigate *"Eagle"* by screwing a time bomb into its hull. The *"Eagle's"* copper sheathing proved impenetrable, and the bomb was allowed to sink. Submarine technology developed rapidly through the 19th and 20th centuries to the point that submarines played major roles in both world wars. Diesel engines powered the submarines when surfaced, and charged huge banks of batteries which propelled them when submerged. The advent of nuclear power plants in the 1950s provided the submarine with a nearly perfect propulsion system, since they do not require a steady supply of air, such as the diesels, and they can go for many *years* without refueling. Nuclear submarines have sailed under the polar ice at the North Pole and have circumnavigated the globe without surfacing.

Deep-diving submarines have been built for use in scientific research in oceanography and related fields as well as for salvaging operations on the ocean floor. A famous example of such underwater work occurred in 1986 when the "Alvin" explored and photographed the sunken *"Titanic"* at a depth of over 3 kilometers (about 2 miles).

Several of the basic concepts presented in this chapter are critical to the operation of submarines. Ballast tanks inside a submarine are used to change its buoyancy so it can sink or rise. When air in a ballast tank is allowed to escape, seawater rushes in to replace it, and the submarine gains enough weight to sink below the surface. Releasing compressed air from high-pressure air tanks into the ballast tanks expels the water, reducing the submarine's weight so it can rise upward. Oxygen for the crew to breathe is produced by using electricity to decompose water into oxygen and hydrogen.

The most serious threat to a submarine is the enormous forces that the water pressure causes on the submarine's outer surface. At a modest depth of 30 meters (about 100 feet) in seawater, where excursion submarines might operate, the force on a part of the submarine's hull the size of this page is about 16,000 newtons (3,600 pounds). At 300 meters, a typical depth for modern military submarines, the force on the same area is 10 times larger, approximately 18 tons. If the hull is not strong enough the submarine will be crushed by the sea, much like an aluminum can is crushed by the atmosphere when the air inside it is removed (Figure 4.18). The deepest submarine dive ever was to a depth of 10,900 meters (6.78 miles), where the force on each page-sized section of the hull was over 6 million newtons (680 tons). Such deep-diving submarines have crew compartments shaped like a sphere and made of super-strong materials such as titanium.

Pressure, the law of fluid pressures, and Archimedes' principle are some of the main topics of this chapter. They all apply directly to the operation of a submarine.

Up to this point we have been studying the physics of "objects." We have not been interested in the structure or composition of things, only in how they move in different situations. This chapter consists of two main parts. In the first part we consider the structure and properties of matter in general. In the rest of the chapter we discuss the physics of fluids, liquids and gases. The important concepts of pressure and density are introduced to better handle fluids. The variation of pressure with depth in a fluid and buoyancy are considered in detail. Sections 4.6 and 4.7 deal with useful principles concerning pressure changes in fluids at rest and in fluids in motion, respectively.

4.1   MATTER: PHASES, FORMS, AND FORCES

The subject of this section is matter—anything that has mass and occupies space. The earth, the water we drink, the air we breathe, our bodies, everything we touch is composed of matter. Obviously matter exists in different forms that have different physical properties. We can classify matter into four categories: *solid, liquid, gas,* and *plasma.* These are called the four *phases,* or *states,* of matter. (In physics, "gas" does *not* refer to gasoline, and "plasma" does *not* refer to the liquid part of blood.) Briefly we can distinguish the four phases as follows:

SOLIDS   Rigid; retain their shape unless distorted by a force. Examples: rock, wood, plastic, iron

LIQUIDS   Flow readily; conform to the shape of a container; have a well-defined boundary (surface); have higher densities than gases. Examples: water, beverages, gasoline, blood

GASES   Flow readily; conform to the shape of a container; do not have a well-defined surface; can be compressed (squeezed into a smaller volume) readily. Examples: air, carbon dioxide, helium

PLASMAS   Have the properties of gases but also conduct electricity; commonly exist at higher temperatures. Examples: gases in operating fluorescent, neon, and vapor lights; (see Figure 4.1) matter in the sun and stars.

Nearly all of the matter in our everyday experience appears as a solid, liquid, or gas. Traditionally these have been referred to as the three states of matter, with plasmas being a special "fourth" state of matter. Although plasmas are rare on earth, most of the matter in the universe is in the form of plasmas in stars. In the last 40 years the study of plasmas has grown to be one of the major subfields of physics because of the interest in nuclear fusion (the topic of Section 11.7). Nuclear fusion is the source of energy for stars, the sun included. One of the main goals in plasma physics is to artificially produce starlike plasmas in which fusion can occur.

There are many substances that do not fit easily into one of these slots. Granulated sugar and salt flow readily and take the shape of a container. But they are considered to be solids because a single granule of each does fit the description of a solid and because both can be crystallized into larger, solid chunks. Tar and molasses do not flow readily, particularly when they are cold, and so they act somewhat like solids. But given time they do flow and take the shape of their container; they are considered liquids.

Many substances are composites of matter in two different phases. Styrofoam behaves like a solid but is composed mostly of a gas trapped in millions of tiny, rigidly connected bubbles. The water in many rivers carries along tiny, solid particles that will settle if the water is allowed to stand. The mists that fill a shower stall and comprise fog consist of millions of small, round droplets of water

FIGURE 4.1
Fluorescent lights use glowing plasmas.

FIGURE 4.2
The phase of any substance depends on its temperature.

mixed in with the air. Apples and potatoes are solid but contain a great deal of liquid.

Another factor that complicates our neat classification of matter is the fact that the phase of a given substance can change with temperature and pressure. Water is a good example. Normally a liquid, water becomes a solid (ice) when cooled below 0°C (32°F; Figure 4.2). Under normal pressure water becomes a gas when its temperature is raised above 100°C (212°F). Even at room temperature water can be made to boil if the air pressure is reduced. Propane, carbon dioxide, and many other gases can be forced into the liquid phase at room temperature by increasing the pressure. Refrigerators and air conditioners depend on the pressure-induced liquification of gaseous Freon. (More on this in Section 5.7.)

To simplify our discussion in this chapter, we will consider only matter in one "pure" phase: solid like a rock, liquid like pure water, and gaseous like air. When we say that a substance exists in a particular phase, we mean its phase at normal room temperature and pressure (unless stated otherwise).

The phases of matter refer to the macroscopic (external) form and properties of matter. These in turn are determined by the microscopic (internal) composition of matter. Around 2,500 years ago some Greek philosophers theorized that all matter as we see and experience it is composed of tiny, indivisible pieces. It turns out that this is true up to a point: diamond, water, and air are all composed of extremely tiny "building blocks" or units that are the smallest entities that retain the identity of the substance. (But, as we shall see, the building blocks themselves are composed of still smaller particles.) Water is a liquid, and diamond is a solid because of the properties of the small units that comprise each. We can reclassify all substances by the nature of their intrinsic composition. We will proceed in order of increasing complexity.

The chemical *elements* represent the simplest and purest forms of matter. At this time, scientists have identified 109 different elements and have given names

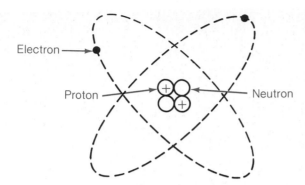

FIGURE 4.3
Simplified diagram of an atom. An atom consists of electrons in orbit about a compact nucleus, which in turn contains protons and neutrons. All atoms of a particular element have the same number of protons. This number is the element's atomic number. (Not drawn to scale.)

to 103 of them. Some of the common substances used in our society are elements, including hydrogen, helium, carbon, nitrogen, oxygen, neon, gold, iron, mercury, and aluminum. Each element is composed of a large number of incredibly small particles called *atoms.* There are 109 different atoms, one for each of the different known elements. Only about 90 of the elements exist naturally in the matter around us. The others are artificially produced in laboratories. The majority of the elements are quite rare and have names familiar only to chemists and other such scientists.

The atom is not an indivisible particle: it has its own internal structure. Every atom has a very dense, compact core called the *nucleus,* which is surrounded by one or more particles called *electrons.* The nucleus itself is composed of two kinds of particles, *protons* and *neutrons*[1] (Figure 4.3). The protons and electrons have equal but opposite electric charges and attract each other. The electrons, much lighter than protons and neutrons, orbit about the nucleus under the centripetal force exerted by the protons.

The number of protons in an atom determines its identity (which element it is). For example, atoms that have 2 protons are atoms of helium, those with 8 protons are atoms of oxygen, those with 79 protons are atoms of gold, and so on. This is called the element's *atomic number.* Each element is also given an abbreviation called its *chemical symbol.* Table 4.1 contains the chemical symbols and atomic numbers of some familiar elements. It also includes the phase of each element at room temperature and pressure. The Periodic Table of the Elements in Appendix B shows some of the properties of each of the elements. (We will take a closer look at the structure of atoms in Chapters 7 and 10.)

The next important class of matter consists of chemical *compounds.* Compounds are similar to elements in that they also have basic, identical building blocks. These are actually combinations of two or more atoms that are called *molecules.* There are millions of different compounds, including many common substances: water, salt, sugar, alcohol. Each molecule of a compound consists of two or more atoms that are held together by electrical forces. For example, each molecule of water consists of two atoms of hydrogen attached to one atom of oxygen (Figure 4.4). Each compound can be represented by a "formula"—a shorthand notation showing both the kinds and numbers of atoms in each molecule. You are probably familiar with the formula for water, H_2O. Some others are $NaCl$ (salt), CO_2 (carbon dioxide), CO (carbon monoxide), $C_6H_{12}O_6$ (sugar),

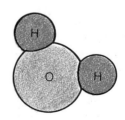

FIGURE 4.4
Water molecules consist of two hydrogen atoms attached to an oxygen atom. This structure is common to all molecules of water, ice, and steam.

---

[1]Protons and neutrons are themselves composed of smaller particles called quarks. More on this in Chapter 12.

TABLE 4.1   SOME COMMON CHEMICAL ELEMENTS

Element	Symbol	Atomic Number	Phase
Hydrogen	H	1	Gas
Helium	He	2	Gas
Carbon	C	6	Solid
Nitrogen	N	7	Gas
Oxygen	O	8	Gas
Neon	Ne	10	Gas
Sodium	Na	11	Solid
Aluminum	Al	13	Solid
Silicon	Si	14	Solid
Chlorine	Cl	17	Gas
Calcium	Ca	20	Solid
Iron	Fe	26	Solid
Cobalt	Co	27	Solid
Nickel	Ni	28	Solid
Copper	Cu	29	Solid
Zinc	Zn	30	Solid
Silver	Ag	47	Solid
Barium	Ba	56	Solid
Gold	Au	79	Solid
Mercury	Hg	80	Liquid
Lead	Pb	82	Solid
Uranium	U	92	Solid

and C_2H_5OH (ethyl alcohol). The study of how the atoms combine to form molecules is a major part of chemistry.

Some elements in the gas phase are also composed of molecules. The oxygen in the air we breathe is an element, but most of the oxygen atoms are paired up to form O_2 molecules. Ozone, O_3, is a rarer form of oxygen that is present in air pollution and also in the ozone layer about 25 miles above the earth's surface. Each ozone molecule is composed of three oxygen atoms. Nitrogen gas also exists in the form of molecules, N_2. Helium and neon are both gaseous elements whose atoms do not form molecules; they remain separate.

Many substances, such as air, stone, and seawater, are composed of two or more different compounds or elements that are physically mixed together. These are classified as *mixtures* and *solutions*. Air consists of dozens of different gases that are mixed together. Table 4.2 shows the composition of clean, dry air. The air we breathe also contains water vapor and pollutants (like carbon monoxide). The amount of such gases present in the air varies considerably from place to

TABLE 4.2 COMPOSITION OF CLEAN, DRY AIR	
Gas	Percent Composition (By volume)
Nitrogen (N_2)	78.1
Oxygen (O_2)	20.9
Argon (Ar)	0.93
Carbon dioxide (CO_2)	0.03
Other gases (Ne, He, and so on)	0.04

place and from day to day. The air in the Sahara Desert does not have quite the same composition as the air in Los Angeles.

A mixture of two elements is not the same as a compound. If you mix hydrogen and oxygen, you simply have a gaseous mixture. The individual hydrogen atoms and oxygen molecules remain separate. If you ignite the mixture, the hydrogen and oxygen atoms will combine to form water molecules. Energy is released explosively in the form of heat (fire) because the atoms have less total energy when they are bound to each other.

The example of hydrogen, oxygen, and water also illustrates that the properties of a compound (water) are usually quite different from the properties of the constituent elements (hydrogen and oxygen). Sodium (Na) is a solid that reacts violently with water. Chlorine (Cl) is a gas that is used to kill bacteria in drinking water. If either element were ingested alone, it could be fatal. But when sodium and chlorine are combined chemically, the result is salt (NaCl), a necessary dietary substance (Figure 4.5).

The basic unit of life is the cell, and each cell is composed of many different compounds. Many of these contain billions of atoms in each molecule, the DNA molecule being an important example. The main constituents of such "organic" molecules are hydrogen, carbon, oxygen, and nitrogen. Many other elements are present in smaller amounts. For example, your bones and teeth contain calcium.

FIGURE 4.5
Separately, chlorine and sodium are poisonous. When chemically combined, they form salt.

FIGURE 4.6
An atom is smaller than a ball in about the same proportion that a ball is smaller than the entire earth.

So atoms are basic to compounds as well as to elements. It is difficult to imagine how small atoms are and just how many there are in common objects. Atoms are about one 10 millionth of a millimeter in diameter. A ball 1 inch in diameter is about midway between the size of an atom and the size of the earth (Figure 4.6). In other words, if a 1-inch ball were expanded to the size of the earth, each atom in the ball would expand to about 1 inch in diameter.

Since atoms are so small, huge numbers of them are present in anything large enough to be seen. There are about 100 thousand billion billion (one followed by 23 zeros) atoms in each of your fingernails. Even the smallest particles that can be seen with the naked eye contain far more atoms than there are people on earth.

Atoms are nearly indestructible—only nuclear reactions affect them—and they are continually recycled. The earth and everything on it are believed to be composed of the debris of stars that exploded billions of years ago. Chemical processes such as fire, decay, and growth result in atoms being combined with and dissociated from other atoms. A single carbon atom in your earlobe may once have been part of a dinosaur, a redwood tree, a rose, or Leonardo da Vinci.

The constituent particles of matter, atoms and molecules, exert electrical forces on each other. ("Static cling" is an example of an electrical force.) The nature of these forces determines the properties of the substance. The forces between atoms in an element depend on the configuration of the electrons in each atom. In a compound, the size and shape of the molecules, as well as the forces between the molecules, affect its observed form and properties. We can relate the three common phases of matter to the interparticle forces as follows:

> **SOLIDS** Attractive forces between particles are very strong; the atoms or molecules are rigidly bound to their neighbors and can only vibrate.
>
> **LIQUIDS** The particles are bound together, though not rigidly; each atom or molecule can move about relative to the others but is always in contact with other atoms or molecules.
>
> **GASES** Attractive forces between particles are too weak to bind them together; atoms or molecules move about freely with high speed and are widely separated; particles are in contact only when they collide.

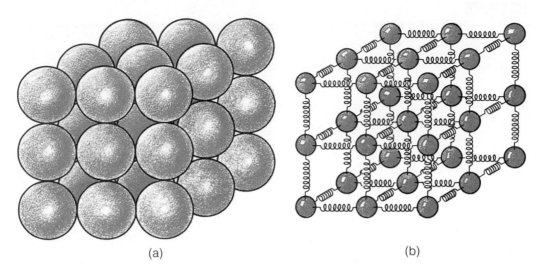

(a)   (b)

FIGURE 4.7
(a) The atoms or molecules in a crystalline solid are arranged in a regular three-dimensional pattern, similar to the way rooms are arranged in a large building. The atoms or molecules exert forces on each other. (b) A crystal behaves much like an array of particles that are connected to each other by springs.

A standard model for representing the interparticle forces in a solid consists of each atom connected to each neighbor by a spring (Figure 4.7). The atoms are free to oscillate like a mass on a spring. The vibration of the atoms or molecules is related to the temperature, as we will see in Chapter 5.

Often the atoms or molecules in a solid form a regular geometric pattern called a *crystal* (Figure 4.8). Salt is a crystalline solid in which the sodium and chlorine atoms alternate with each other. Carbon actually has two different crystalline forms. Graphite (soft and black, used in pencil lead) and diamond (hard and clear) are both pure carbon, but the geometric arrangement of their carbon atoms is different. In solids that do not have a crystal structure, called *amorphous*

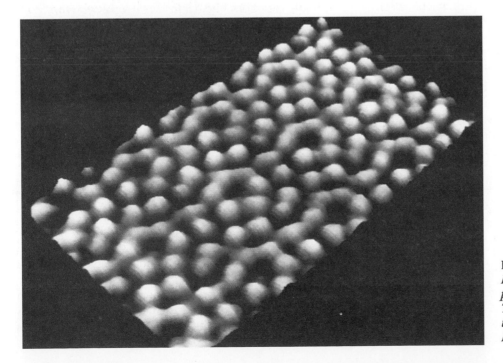

FIGURE 4.8
Image of the surface of silicon, produced with a Scanning, Tunneling Microscope (STM) at the IBM Thomas J. Watson Research Center.

What's in a name? Well, for the elements, names contain clues to their chemical and physical properties, to their places and means of discovery, to the origins of their chemical symbols, and to the scientists who figured prominently in their isolation. In short, a great deal can be learned about the chemical elements from a study of their names.

Consider the element bromine (Br). The word "bromine" derives from the Greek word *bromos*, meaning "stench"—an apt description for this reddish brown liquid that vaporizes easily at room temperature to produce a red gas with a strongly disagreeable odor. The element dysprosium (Dy) also takes its name from a Greek word, *dysprositos*, meaning "hard to get at." This, too, seems an appropriate term to apply to an element first discovered in 1886 but not isolated in a relatively pure form until about 1950. Or take the element iridium (Ir). Its name derives from the Latin word for rainbow, *iris*. This is also a rather nice description of an element whose crystalline salts are brilliantly colored: hydrated iridium tribromide is olive green, iridium tetrachloride is a rich brown, and iridium carbonyl is bright yellow, for example.

A number of elements obtain their names from their places of discovery. Probably the best examples are the elements erbium (Er), terbium (Tb), ytterbium (Yb), and yttrium (Y), all of which take their names from the village of Ytterby in Sweden near Stockholm (Figure 4.9). Ytterby is the site of a quarry that yielded many unusual minerals containing these and other uncommon elements. Further examples of elements named after places first associated with their discovery are californium (Cf), named for the state and university in which it was first produced by a team of Berkeley scientists

in 1950, and ruthenium (Ru), whose name comes from the Latin word for Russia *(Ruthenia)* and commemorates its discovery in platinum ores taken from the Ural Mountains. A final example of this type is helium (He), meaning "the sun"; it serves to remind us that the element was first detected in 1868 in the outer atmosphere of the sun during a solar eclipse!

The symbols of most of the chemical elements bear a direct correspondence to their names, usually consisting of the first letter and, at most, one additional letter from the name. For many of the common elements that have been known for centuries, this is not the case. Classic examples include iron (Fe), gold (Au), mercury (Hg), and copper (Cu). For such elements the symbols often originate from their Latin equivalents and date from the time hundreds of years ago when Latin was the language of scholarship the world over. *Iron* is an Anglo-Saxon word whose Latin equivalent is *ferrum*, hence the symbol Fe. Similarly, gold derives from an Anglo-Saxon word of the same spelling, but its symbol, Au, is taken from the Latin word *aurum* meaning "shining dawn." (Who can forget James Bond's archenemy Auric Goldfinger?) Mercury, named after the fleet-footed Roman god, takes its symbol, Hg, from the Latin word *hydrargyrum*, which translates to "liquid silver"—a perfect description of this shiny, liquid metal. Lastly, the symbol for copper, Cu, comes from the Latin *cuprum;* this word is a shortened corruption of the phrase *aes Cyprium*, meaning "metal from Cyprus." The Romans obtained virtually all of their supply of copper from this Mediterranean island. Other instances of this type of symbol derivation include silver (Ag from the Latin *argentum*), lead (Pb from the Latin

solids, the atoms or molecules are "piled together" in random fashion. Glass is a good example of this.

In a liquid the forces between the particles are not strong enough to bind them together rigidly. The atoms or molecules are free to move around as well as vibrate (Figure 4.10a). This is similar to a collection of ball-shaped magnets *plumbum*—cf. our word "plumber"), and sodium (Na from the Latin *natrium*).

A few of the element names derive from the names of scientists who have contributed to the discovery or the isolation of the elements. Specific cases include mendelevium (Md), named after Dmitri Mendeleev, a Russian chemist who developed the periodic table of the elements and then predicted the existence of the elements gallium (Ga), germanium (Ge), and scandium (Sc) (as they are now called); curium (Cm), named for Pierre and Marie Curie, who discovered the elements polonium (Po), named for Mme Curie's native Poland, and radium (Ra); and gadolinium (Gd), after Finnish chemist Johan Gadolin, the discoverer of yttrium.

As mentioned in the text, 109 elements have been identified. At the time of this writing, names have been established for the first 103 elements and proposed for elements 104 and 105. The last two again commemorate the effect of physicists who have been involved in the production of the new elements. In particular, it has been suggested that element 104 be named either kurchatovium (Ku) in honor of Igor Kurchatov, late head of Soviet Nuclear Research, or rutherfordium (Rf), after Ernest Rutherford, a New Zealand physicist. Element 105 has been tentatively named hahnium (Ha) after German scientist Otto Hahn, who participated in many of the first successful experiments involving nuclear fission in the late 1930s. (More on this in Chapter 11.) Elements 106, 107, 108, and 109 have yet to be named.

In summary, there can be a lot in a name, especially if it is the name of a chemical element. Perhaps this short excursion into the lexicology of the

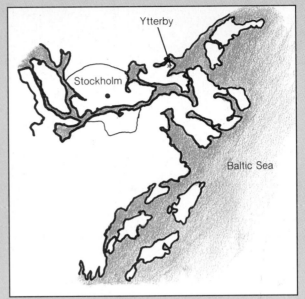

FIGURE 4.9
The elements erbium, terbium, ytterbium, and yttrium take their names from the village of Ytterby in Sweden.

elements will make you want to know more about this subject or will at least make it a little easier for you to remember the symbols of the more common elements that make up our environment. For those interested in pursuing this topic, a good place to start is the most recent edition of *The Handbook of Chemistry and Physics,* Section B, "The Elements and Inorganic Compounds."

clinging to each other. The forces between the particles are responsible for surface tension and the spherical shape of small drops.

Many compounds have an interesting intermediate phase between solid and liquid called the *liquid crystal* phase. The molecules have some mobility, as in a liquid, but they are arranged in a regular fashion, as in a solid. Liquid crystal

FIGURE 4.10
(a) The atoms or molecules in a liquid remain in contact with each other, but they are free to move about. (b) In a gas the atoms or molecules are not bound to each other. They move with high speed and interact only when they collide.

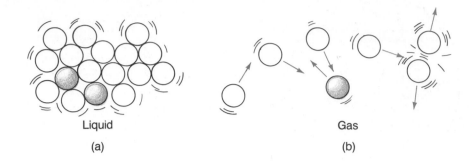

Liquid

(a)

Gas

(b)

displays (LCDs) in calculators and watches use electrically induced movement of the molecules to change the optical properties of the liquid crystal. (More on this in Section 9.1.)

In a gas, the atoms and molecules are widely separated and move around independently, except when they collide (see Figure 4.10b). Their speeds are surprisingly high: the oxygen molecules you are now breathing have an average speed of over 1,000 mph. As each molecule collides randomly with other molecules, its speed is sometimes increased and other times decreased.

Often the surface of a solid or a liquid forms a boundary for a gas. The surface of water in a glass, a lake, or an ocean forms a lower boundary for the air above. The inside of the walls of a tire are a boundary for the air inside (Figure 4.11). The high-speed atoms or molecules in the gas exert a force on the surface as their random motions cause them to collide with it. This is the basis for gas pressure, as mentioned in Section 3.6. In other words, the weight of a car is supported by the collisions of air molecules with the inner walls of the tires.

At normal temperatures and pressures, the average distance between the centers of the atoms or molecules in a gas is about 10 times that in a solid or liquid (Figure 4.12). There is a great deal of empty space in a gas, and that is why gases can be compressed easily.

Let us summarize the information in this section. The 109 different types of atoms (the elements) form the basis of matter as we experience it. Molecules are composed of two or more atoms "stuck" together to form the basic unit of compounds. The nature of the forces between the building blocks (atoms or molecules) determines whether the substance is a solid, a liquid, or a gas. Mixtures and solutions consist of different elements or different compounds mixed together.

FIGURE 4.11
The air molecules inside a tire collide with the walls and exert forces on it. If the molecules weren't moving, the tire would be flat.

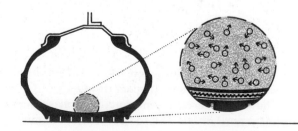

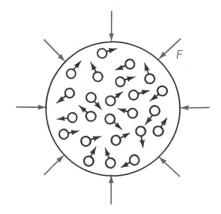

FIGURE 4.12
Gases can be compressed because the atoms and/or molecules are widely separated. Increasing the force (pressure) on the boundaries of a gas squeezes the particles closer together.

4.2   PRESSURE

Force and mass are two of the most important physical quantities in mechanics. When dealing with the motion of a single particle, all one needs to know is the particle's mass and the force acting on it. The effect of the force is determined by Newton's second law. In this section and the next, we introduce physical quantities that are "extensions" of force and mass. These quantities, *pressure* and *density,* are more useful for dealing with continuous, extended matter where surface area and volume are important.

Forces that are exerted by gases, liquids, and solids normally are spread over a surface. The force that the floor exerts on you when you are standing is distributed over the bottoms of your feet. A boat floating on water has an upward force spread over its lower surface (Figure 4.13). The air around a floating balloon

FIGURE 4.13
The water exerts an upward force that is spread over the bottom surface of the boat.

exerts forces on the balloon's surface. in situations like this, the physical quantity pressure is quite useful.

Note: p is the abbreviation for pressure, and P is the abbreviation for power.

> **PRESSURE**   The force that acts (perpendicularly) on a unit area of a surface. The force acting on a surface divided by its area.
>
> $$p = \frac{F}{A}$$

	Metric	English
Pressure *(p)*	pascal (Pa) (1 Pa = 1 N/m²)	pound per sq foot (lb/ft²)
		pound per sq inch (psi)
	millimeters of mercury (mm Hg)	inches of mercury (in Hg)

Pressure is a scalar: there is no direction associated with it.

The standard pressure units consist of a force unit divided by an area unit. The SI unit of pressure is the pascal, Pa, which equals 1 newton per square meter. The English unit, psi, is probably the most familiar to you. Air pressure in tires is measured in psi. For comparison:

$$1 \, psi = 6{,}890 \, Pa$$

This number is so large because a force of 1 pound on each square inch would produce a huge force on 1 square meter—a much larger area. The two pressure units involving mercury will be explained in Section 4.4. Another common unit of pressure is the *atmosphere (atm)*. It equals the average air pressure at sea level. (Later we will see that the air pressure is less at higher altitudes.) It is related to the other units as follows:

$$1 \, atm = 1.01 \times 10^5 \, Pa = 14.7 \, psi$$

This unit is like the g used as a unit of acceleration. We are all subject to the pressure of the air around us and to the acceleration of gravity, so it is natural to compare other pressures and accelerations to them. The old practice of using the length of a person's foot as a unit of distance is another example of taking a unit of measure provided by nature.

EXAMPLE 4.1   A 160-pound person stands on the floor. The area of each shoe that is in contact with the floor is 20 in². What is the pressure on the floor? Assuming the person's weight is shared equally between the two shoes, the force of one shoe is 80 pounds. So:

$$p = \frac{F}{A} = \frac{80 \, lb}{20 \, in^2}$$

$$p = 4 \, psi \quad \text{(standing on both feet)}$$

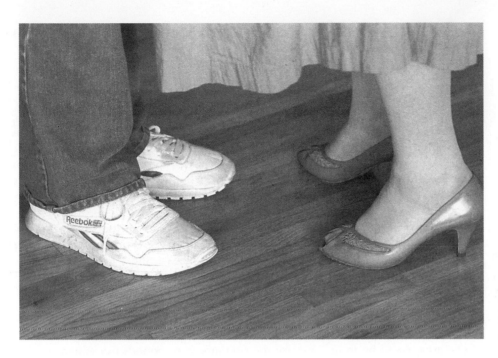

FIGURE 4.14
The pressure on the floor is much higher with narrow heels.

By Newton's third law of motion, the floor exerts an equal and opposite force on the shoes. So the pressure of the floor on the shoes is also 4 psi.

If the person stands on one foot instead, that shoe has all of the weight and the pressure is

$$p = \frac{160 \text{ lb}}{20 \text{ in}^2}$$

$$p = 8 \text{ psi} \quad \text{(standing on one foot)}$$

What if the person put on high-heeled shoes and balanced on the heel of one shoe (Figure 4.14). The bottom of the heel might measure 0.5 inches by 0.5 inches. The area is then 0.25 in^2. So the pressure in this case would be

$$p = \frac{160 \text{ lb}}{0.25 \text{ in}^2}$$

$$p = 640 \text{ psi} \quad \text{(balanced on a narrow heel)}$$

The same force causes much higher pressure when it acts over a smaller area. You can illustrate this nicely by squeezing a thumbtack between two fingers (Figure 4.15). The force that the point exerts on one finger is the same as the force that the head exerts on the other finger. The area of the point is much smaller than that of the head, so the pressure is much larger. These two examples show that pressure is a measure of how "concentrated" a force is.

There are many situations in which a liquid or a gas is under pressure and exerts a force on the walls of its container. The relationship between force and pressure can be used to determine the force on a particular area of the walls. Since pressure equals force divided by area, the pressure times the area equals the

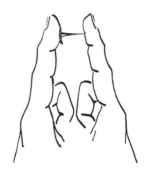

FIGURE 4.15
When a tack is squeezed between two fingers, the force on each finger is the same. However, the pressure that the point causes is much larger because its area is much smaller. This higher pressure causes pain in that finger.

force. In other words, the force on a surface equals the force on each square inch (the pressure) times the total number of square inches (the area).

$$F = pA$$

EXAMPLE 4.2

In the late 1980s there were several spectacular (and tragic) aircraft mishaps involving rapid loss of air pressure in the passenger cabins (Figure 4.16). (The causes included failure of a cargo door, outer skin rupture due to corrosion or cracks or both, and small bombs). Because the cabins are pressurized, there are large outward forces acting on windows, doors, and the aircraft skin.[2] Let's estimate the size of these forces.

The pressure inside a passenger jet cruising at high altitude (about 7,500 meters or 25,000 feet) is about 6 psi (0.41 atmosphere) greater than the pressure outside. What is the outward force on a window measuring 1 foot by 1 foot and on a door measuring 1 meter by 2 meters? The area of the window is

$$A = 1\,\text{ft} \times 1\,\text{ft} = 12\,\text{in} \times 12\,\text{in}$$

$$A = 144\,\text{in}^2$$

(The area has to be in in^2 because the pressure is in pounds per square *inch*.)

The force on the window is:

$$F = pA = 6\,\text{psi} \times 144\,\text{in}^2$$

$$F = 864\,\text{lb}$$

A rather small pressure causes a large force on the window. For the door we use SI units:

---

[2]One may ask why aircraft do not fly at low altitudes so that cabins would not need to be pressurized. There are several reasons why jets fly at 10,000 meters or higher. (1) If something goes wrong and an aircraft starts losing altitude, the pilot has more time to recover if it is very high to begin with. More than once an aircraft was saved after it had dived over 7,000 meters. (2) Because the air is thinner at high altitudes, the force of air resistance on an aircraft is smaller. Consequently it can travel faster and use less fuel than when flying at low altitudes. (3) The air is usually much "smoother" at high altitudes. There is much less turbulence, so passengers are more comfortable.

FIGURE 4.16
A large section of the aluminum skin on this older, heavily-used 737 jet ripped away while the aircraft was flying at 7,300 meters (24,000 feet) above sea level. One person was killed and dozens were injured but the aircraft landed safely. (April 1988)

$$p = 6\,\text{psi} = 6 \times 1\,\text{psi} = 6 \times 6{,}890\,\text{Pa}$$

$$p = 41{,}340\,\text{Pa}$$

$$A = 1\,\text{m} \times 2\,\text{m} = 2\,\text{m}^2$$

Consequently the force on the door is:

$$F = pA = 41{,}340\,\text{Pa} \times 2\,\text{m}^2$$

$$F = 82{,}680\,\text{N}$$

The force on the door is greater than the *weight of 4 small automobiles, or 100 people.*

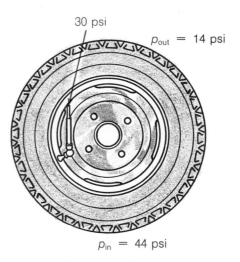

FIGURE 4.17
Absolute pressure inside of 44 psi produces a gauge pressure of 30 psi.

Pressure is a relative quantity. When you test the air pressure in a tire, you are comparing the pressure of the air inside the tire with the pressure outside the tire. When a tire is flat, there is still air inside it, but the pressure inside is the same as the atmospheric pressure outside.

For example, let us say that the atmospheric pressure is 14 psi. A tire is tested, and the air pressure gauge shows 30 psi. This means that the air pressure inside the tire is 30 psi higher than the air pressure outside the tire. So the actual pressure on the inner walls of the tire is $30 + 14 = 44$ psi. This is sometimes referred to as the *absolute pressure.* The pressure relative to the outside air (30 psi in this case) is then called the *gauge pressure* (Figure 4.17).

We can look at this another way. The pressure inside the tire, 44 psi, causes an outward force of 44 pounds on each square inch of the tire wall. The air pressure outside the tire causes an inward force of 14 pounds on each square inch. The net force due to air pressure on each square inch of the tire is then 30 pounds.

The gauge pressure of the air in a tire changes if the outside air pressure changes. What happens if the car is driven into a chamber, and the air pressure in the chamber is increased from 14 psi to 44 psi? The gauge pressure in the tires will then be zero, and the tires will go flat even though no air has been removed from them. When the air pressure is reduced to 14 psi, the tires will expand again to their normal shape.

In the previous example of the pressurized aircraft, the 6 psi is the gauge pressure. The absolute pressure inside the cabin might be 11 psi, and the air pressure outside might be 5 psi. The atmospheric pressure at 26,000 feet is about 5 psi.

DO-IT-YOURSELF PHYSICS

Put a small amount of water (a tablespoon is sufficient) into an aluminum can, and heat it to boiling. Put a glove on, and quickly plunge the can upside down into a sink with cold water in it. The steam in the can quickly condenses, causing the pressure to drop. The air pressure on the outside of the can is no longer balanced by the pressure inside—the can is crushed by the air (see Figure 4.18).

FIGURE 4.18
A can is crushed by the atmosphere if the pressure inside is reduced.

The standard pen-shaped tire-pressure tester illustrates nicely some of the physics that we have considered so far. It consists of a hollow tube (cylinder) fitted with a piston that can slide back and forth in the cylinder (Figure 4.19). Air from the tire enters the cylinder at the left end and pushes on the piston. The right end allows air from the outside to push on the other side of the piston. If the pressure inside the tire is greater than the outside air pressure, there is a net force to the right on the piston. A spring placed behind the piston is compressed by this net force. The greater the net force on the piston, the greater the compression of the spring. A calibrated shaft extends from the right side of the piston and out of the right end. When the piston is pushed to the right, the shaft protrudes from the right end the same distance that the spring is compressed. The length of shaft showing indicates the gauge pressure in the tire, since that is what causes the force on the piston.

We conclude this section with one last important note about pressure. Since gases are compressible, the volume of a gas can be changed (Figure 4.12). Whenever the volume of a fixed amount of gas is changed, the pressure in the gas changes also. Increasing the volume of a gas reduces the pressure. Decreasing the volume increases the pressure. To understand why this is the case, recall from the previous section that it is the collisions of the atoms and molecules in a gas with a surface that cause gas pressure. If the volume is decreased, the particles are squeezed together so there are more of them in the vicinity of the boundaries. More collisions occur each second, which means more force on the boundaries and higher pressure. The opposite happens when the volume is increased.

The temperature of a gas influences the speed of the atoms or molecules. Consequently, the pressure is also affected by the gas temperature. When the temperature of a given quantity of gas is kept constant, the pressure, p, is related to the volume, V, as follows:

$$pV = a \text{ constant} \qquad \text{(gas at fixed temperature)}$$

This means that the *volume of a gas is inversely proportional to the pressure.* If the pressure is doubled, the volume is halved. In Chapter 5, we will see what effect temperature has on pressure and volume.

4.3   DENSITY

Pressure is an extension of the idea of force. Similarly, *mass density* is an extension of the idea of mass. Just as pressure is a measure of the concentration of force, density is a measure of the concentration of mass.

D is the abbreviation for mass density, and d is the abbreviation for distance.

> **MASS DENSITY**   The mass per unit volume of a substance. The mass of a quantity of a substance divided by the volume it occupies.
>
> $$D = \frac{m}{V}$$

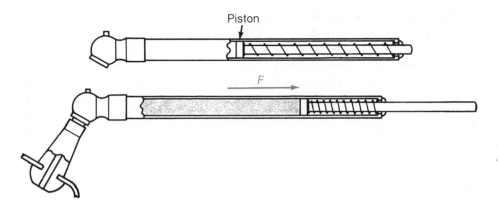

FIGURE 4.19
In a common tire-pressure tester, the pressure of the air in the tire pushes the piston to the right and compresses the spring. The higher the pressure, the greater the force and the greater the compression of the spring.

	Metric	English
Mass density D	kg per cubic meter (kg/m³)	slug per cubic foot
	gram per cubic centimeter (g/cm³)	

To find the mass density of a substance, one measures the mass of a part or "sample" of it and divides by the volume of that part or sample. It doesn't matter how much is used: the greater the volume, the greater the mass.

The dimensions of a rectangular aquarium are 0.5 meter by 1 meter by 0.5 meter. The mass of the aquarium is 250 kilograms larger when it is full of water than when it is empty (Figure 4.20). What is the density of the water? First, the volume of the water is:

$$V = l \times w \times h = 1\,\text{m} \times 0.5\,\text{m} \times 0.5\,\text{m} = 0.25\,\text{m}^3$$

EXAMPLE 4.3

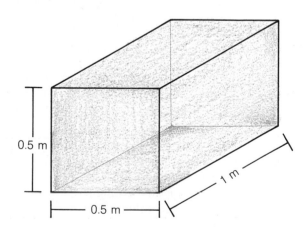

FIGURE 4.20
The mass of the water in this aquarium is 250 kilograms. Using this and the volume of the aquarium, we can compute the mass density of water.

So:

$$D = \frac{m}{V} = \frac{250 \text{ kg}}{0.25 \text{ m}^3}$$

$$D = 1{,}000 \text{ kg/m}^3 \qquad (\text{mass density of water})$$

If a tank with twice the volume were used, the mass of the water in it would be twice as much, and the density would be the same. The mass density of any amount of pure water is 1,000 kilograms/cubic meter.

When the same tank is filled with gasoline, the mass of the gasoline is found to be 170 kilograms. Therefore the density of gasoline is:

$$D = \frac{170 \text{ kg}}{0.25 \text{ m}^3} = 680 \text{ kg/m}^3$$

Except for small changes caused by changes in temperature or pressure, the mass density of any pure solid or liquid is constant. It is an identifying trait of that substance. The mass densities of pure gases, on the other hand, vary greatly with changes in temperature or pressure. We have seen that doubling the pressure on a gas halves the volume. That would double the mass density as well. In Chapter 5 we will describe how changing the temperature of a gas also alters its volume and therefore its density. By convention, the tabulated densities of gases are given at the *standard temperature and pressure (STP): 0°C temperature and 1 atmosphere pressure.*

The mass density of each element or compound is fixed. Water, lead, mercury, salt, oxygen, gold, and so on all have unique mass densities that have been measured and catalogued. The density of a mixture containing two or more substances depends on the density and percent content of each component. Most metals in common use are alloys, consisting of two or more metallic elements and other elements like carbon. For example, 14-carat gold is only about 58% gold. However, the mass density of a mixture with a particular composition is constant.

Table 4.3 lists the mass densities (column 2) of several common substances. The values for mixtures can vary and so are merely representative. The mass densities of the gases are for standard conditions.

Having a list of the densities of common substances is quite useful for three reasons. First, one can use mass density to help *identify* a substance. For example, one can determine whether or not a gold ring is solid gold by measuring its density and comparing it to the known value. Second, density measurement is used routinely to determine how much of a particular substance is present in a mixture. The coolant in an automobile radiator is usually a mixture of water and antifreeze. These two liquids have different densities (Table 4.3), and so the density of a mixture depends on the ratio of the amount of water to the amount of antifreeze. The higher the density, the greater the antifreeze content and the lower the freezing temperature of the coolant (Figure 4.21). By simply measuring the coolant density, one can determine the coolant's freezing temperature. If you have donated blood to a blood bank, part of the screening included checking to see if the hemoglobin content of your blood was high enough. This is done by determining if the blood's density is greater than an accepted minimum value.

Third, one can *calculate the mass* of something if one knows what its volume is. The mass of a substance equals the volume that it occupies times its mass density.

FIGURE 4.21
Ruth determines the freezing point of the car's coolant by measuring the coolant's density. This indicates the relative amount of antifreeze in the mixture.

TABLE 4.3 DENSITIES OF SOME COMMON SUBSTANCES

Substance	Mass density (D)	Weight density (D_w)	Specific Gravity
Solids			
Juniper wood	560 kg/m³	35 lb/ft³	0.56
Ice	917	57.2	0.917
Ebony wood	1,200	75	1.2
Concrete	2,500	156	2.5
Aluminum	2,700	168	2.7
Diamond	3,400	210	3.4
Iron	7,860	490	7.86
Brass	8,500	530	8.5
Nickel	8,900	555	8.9
Copper	8,930	557	8.93
Lead	11,340	708	11.34
Uranium	19,000	1,190	19
Gold	19,300	1,200	19.3
Liquids			
Gasoline	680	42	0.68
Ethyl alcohol	791	49	0.791
Water (pure)	1,000	62.4	1.00
Seawater	1,030	64.3	1.03
Antifreeze	1,100	67	1.1
Sulfuric acid	1,830	114	1.83
Mercury	13,600	849	13.6
Gases (at 0°C and 1 atm)			
Hydrogen	0.09	0.0056	0.00009
Helium	0.18	0.011	0.00018
Air	1.29	0.08	0.00129
Carbon dioxide	1.98	0.12	0.00198

$$m = V \times D$$

The mass of water needed to fill a swimming pool can be computed by measuring the volume of the pool. Let us say a pool is going to be built that will be 10 meters wide, 20 meters long, and 3 meters deep. How much water will it hold? The volume of the pool will be

EXAMPLE 4.4

$$V = l \times w \times h = 20\,m \times 10\,m \times 3\,m = 600\,m^3$$

$$m = V \times D = 600 \text{ m}^3 \times 1{,}000 \text{ kg/m}^3$$

$$m = 600{,}000 \text{ kg}$$

That is a lot of water.

In some cases it is practical to use another type of density called weight density. It is used commonly in the English system of units because weight is in more common use than mass.

> WEIGHT DENSITY   The weight per unit volume of a substance. The weight of a quantity of a substance divided by the volume it occupies:
>
> $$D_w = \frac{W}{V}$$

	Metric	English
Weight density D_w	newton per cubic meter (N/m^3)	pound per cubic foot (lb/ft^3)
		pound per cubic inch (lb/in^3)

The weight density of a substance is equal to the mass density times the acceleration of gravity. This is because the weight of a substance is just g times its mass.

$$W = m \times g \quad \rightarrow \quad D_w = D \times g$$

Column 3 in Table 4.3 shows representative weight densities in English units. These can be used in the same ways that mass densities are used. One interesting exercise is to compute the weight of the air in a room. (Often people do not realize that air and other gases do have weight.)

EXAMPLE 4.5   A college dormitory room measures 12 feet wide by 16 feet long by 8 feet high. What is the weight of air in it under normal conditions? The weight is related to the volume by the equation:

$$W = D_w \times V$$

But

$$V = l \times w \times h = 16 \text{ ft} \times 12 \text{ ft} \times 8 \text{ ft}$$

$$V = 1{,}536 \text{ ft}^3$$

So the weight is

$$W = D_w \times V = 0.08 \text{ lb/ft}^3 \times 1{,}536 \text{ ft}^3$$

$$W = 123 \text{ lb}$$

(This is for sea level pressure and 0°C temperature. At normal temperature, 20°C, the weight would be 115 pounds.)

Traditionally the term "density" has referred to mass density for those using the metric system and to weight density for those using the English system. When simply comparing the densities of different substances, yet another quantity is often used, the *specific gravity*. The specific gravity of a substance is simply the ratio of its density to the density of water. Column 4 of Table 4.3 shows the specific gravities of the substances. Diamond is 3.4 times as dense as water, so its specific gravity is 3.4 (Figure 4.22). This means that a certain volume of diamond will have 3.4 times the mass and 3.4 times the weight of an equal volume of water.

Water is used as the reference for specific gravities simply because it is such an important and yet common substance in our world. The density of water is another example of a "natural" unit of measure like the atmosphere (atm) and the acceleration of gravity (*g*). As a matter of fact, the original definition of the unit of mass in the metric system, the gram, was based on the density of water (see the *Physics Potpourri,* "The Metric System," in Section 1.2). The gram was defined to be the mass of 1 cubic centimeter of water at 0°C. This is why the density of water is exactly 1,000 kilograms/cubic meter.

One final note on the concept of density in general: we have only considered *volume* density—the mass or weight per unit volume. For matter that is primarily two dimensional, such as carpeting or a drumhead, it is often convenient to use *surface* density—the mass or weight per unit area. Similarly, for string, ropes, and cables, one can use a *linear* density—the mass or weight per unit length. The basic concept of density is that it is a measure of the concentration of matter. It relates mass or weight to physical size.

FIGURE 4.22
The specific gravity of diamond is 3.4 because it is 3.4 times as dense as water. The Hope diamond is shown here.

LEARNING CHECK

1. The three common phases of matter on earth are _____ , _____ , and _____ .

2. The molecule is the basic unit of all
   a) elements.
   b) compounds.
   c) solutions.
   d) mixtures.

3. If the same sized force is made to act over a smaller area
   a) the pressure is decreased.
   b) the pressure is not changed.
   c) the pressure is increased.
   d) it depends on the shape of the area.

4. The pressure difference between the inside and the outside of a tire is the _____ pressure.

5. The mass density of an object equals its _____ divided by its _____ .

4.4   FLUID PRESSURE AND GRAVITY

A fluid is any substance that flows readily. All gases and liquids are fluids, as are plasmas. Granulated solids such as salt or grain can be considered fluids in some situations because they can be poured and made to flow.

Fluids are very important to us: without air and water, life as we know it would be impossible. In the remainder of the chapter, we will discuss some of the properties of fluids in general. Usually the statements will apply to both liquids and gases, but the fact that gases are compressible and liquids are not is sometimes important.

We live in a sea of air—the atmosphere. Though we are usually not aware of it, the air exerts pressure on everything in it. This pressure varies with altitude and can cause one's ears to "pop" when riding in a fast elevator. When swimming under water, the same phenomenon can occur if you go deeper under the surface. In both the atmosphere and under water, the pressure is caused by the force of gravity. Without gravity, air and water would not be pulled to the earth. They would simply float off into space, just as we would. The pressure in any fluid increases as you go deeper in it and decreases when you rise through it.

Before considering the exact relationship between depth and pressure, we state two general properties of pressures in fluids. First, *fluid pressures act in all directions.* When you put a hand under water, the pressure acts not only on the top of your hand but on the sides and bottom too. Second, *the force of gravity causes the pressure in a fluid to vary with depth only,* not with horizontal position. In liquids, the pressure depends on the vertical distance from the surface and is independent of the shape of the container.

One can illustrate both of these principles by filling a rubber boot with water. When holes are punched in the boot, the pressure causes water to run out (Figure 4.23). Water will run out of holes in the top, sides, and bottom of the toe because the pressure acts in all directions on the inner surface. Water comes out faster from holes that are farther below the surface because the pressure is greater. The speed is the same for all holes that are at the same level, regardless of their orientation (up, down, or sideways) and their location (toe, heel, etc.). This is because the pressure in a particular fluid depends only on the depth, not on the lateral position.

The following law explains how the pressure in a fluid is related to gravity.

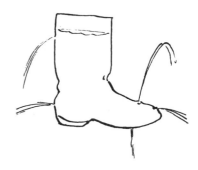

FIGURE 4.23
A rubber boot is filled with water. The pressure of the water acts on all parts of the inner surface of the boot and forces water out of any hole punched in the boot. The speed of the water coming out of a hole depends on how far the hole is below the surface of the water.

> LAW OF FLUID PRESSURE   The (gauge) pressure at any depth in a fluid at rest equals the weight of the fluid in a column extending from that depth to the "top" of the fluid divided by the cross-sectional area of the column.

This law is not so much a prescription for determining the pressure in a fluid as it is a description of what causes pressure in a fluid. For liquids, we can use it to derive the simple relationship between pressure and depth. Let us say a tank is filled with a liquid so that the bottom is some distance, h, below the liquid's surface. On the bottom, we look at a rectangular area that has length, l, and width, w (Figure 4.24). All of the liquid directly above the rectangle is in a column with dimensions l by w by h. The weight of this liquid pushes down on the rectangle. This causes a pressure on the bottom that is equal to the weight of the liquid in the column divided by the area of the rectangle.

$$p = \frac{F}{A} = \frac{\text{weight of liquid}}{\text{area of rectangle}}$$

It does not matter what the actual area is: a larger rectangle will have a proportionally larger amount of liquid in the column above. The actual height of the column of liquid is what determines the pressure. We compute the pressure using the fact that the weight of the liquid equals the weight density, D_w, of the liquid times the volume, V, of the column.

$$F = \text{weight of liquid} = D_w \times V = D_w \times l \times w \times h$$

$$A = l \times w$$

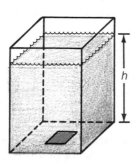
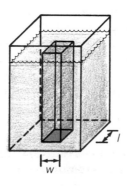

FIGURE 4.24
A tank is filled with a liquid to a depth h. Any rectangular area on the bottom supports the weight of all of the fluid directly above it. So the pressure on the bottom equals the weight of the liquid in the column divided by the area of that rectangle.

So:

$$p = \frac{F}{A} = \frac{D_w \times l \times w \times h}{l \times w}$$

$$p = D_w h \qquad \text{(gauge pressure in a liquid)}$$

Or, since $D_w = Dg$:

$$p = Dgh$$

This gives the pressure due to the liquid above. If there is also pressure on the liquid's surface (like atmospheric pressure), this will be passed on to the bottom as well. Also there is nothing special about the bottom. At some level above the bottom, the liquid above exerts a force, and therefore, pressure, on the liquid below. We can summarize our result as follows:

> In liquids, the absolute pressure at a depth h is greater than the pressure at the surface by an amount equal to *the weight density of the liquid times the depth*.

EXAMPLE 4.6   Let us calculate the gauge pressure at the bottom of a typical swimming pool— one that is 10 feet (3.05 meters) deep. The gauge pressure at that depth, using Table 4.3, is

$$p = D_w h = 62.4 \text{ lb/ft}^3 \times 10 \text{ ft}$$

$$p = 624 \text{ lb/ft}^2$$

To convert this to psi, we use the fact that 1 square foot equals 144 square inches.

$$p = 624 \frac{\text{lb}}{\text{ft}^2} = 624 \frac{\text{lb}}{144 \text{ in}^2}$$

$$p = 4.33 \text{ lb/in}^2 = 4.33 \text{ psi} \qquad \text{(gauge pressure)}$$

(The absolute pressure is this pressure plus the atmospheric pressure, 14.7 psi at sea level. Absolute pressure = 4.33 psi + 14.7 psi = 19.03 psi.)

At a depth of 20 feet, the gauge pressure would be twice as large, 8.66 psi. If we do this calculation in SI units:

$$p = Dgh = 1,000 \text{ kg/m}^3 \times 9.8 \text{ m/s}^2 \times 3.05 \text{ m}$$

$$p = 29,900 \text{ Pa}$$

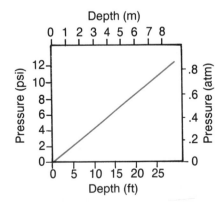

FIGURE 4.25
This is a graph of the (gauge) pressure under water versus the depth. The pressure increases 0.433 psi with every foot of depth. All liquids have the same shape graph, but the slope is greater if the density is higher.

The general result for the increase in pressure with depth in water is

$$p = 0.433 \text{ psi/ft} \times h \qquad \text{(for water, } h \text{ in ft, } p \text{ in psi)}$$

For every 10 feet of depth in water, the pressure increases 4.33 psi. Figure 4.25 shows a graph of the pressure under water versus the depth. It is a straight line, since the pressure is proportional to the depth. In seawater the density is slightly higher, and the pressure increases 4.47 psi for every 10 feet. Submarines and other devices that operate under water must be designed to withstand high pressures. For example, at a depth of 300 feet in seawater the pressure is 134 psi ($p = 0.447 \times 300$). The force on each square foot of a surface is over 9 tons.

At what depth in pure water is the gauge pressure 1 atm?

$$p = 0.433 \, \text{psi/ft} \times h$$

$$14.7 \, \text{psi} = 0.433 \, \text{psi/ft} \times h$$

$$\frac{14.7 \, \text{psi}}{0.433 \, \text{psi/ft}} = h$$

$$h = 33.9 \, \text{ft} = 10.3 \, \text{m}$$

EXAMPLE 4.7

Since mercury is 13.6 times as dense as water, the gauge pressure is 1 atmosphere at a depth of:

$$h = \frac{33.9 \, \text{ft}}{13.6} = 2.49 \, \text{ft} = 29.9 \, \text{in} = 0.76 \, \text{m} \, (\text{Figure 4.26.})$$

Fluid Pressure in the Atmosphere

So the law of fluid pressures has a simple form in liquids. For gases, things are a bit more complicated. We know that the density of a gas depends on the pressure. At greater depths in a gas, the increased pressure causes increased density. The total weight of a vertical column of gas can be computed only with the aid of calculus.

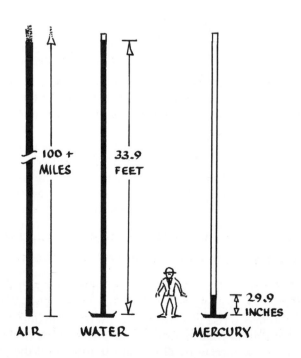

100 +
MILES

33.9
FEET

29.9
INCHES

AIR   WATER   MERCURY

FIGURE 4.26
The pressure at the bottom of each column is 1 atmosphere.

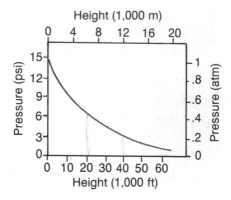

FIGURE 4.27
This is a graph of the absolute air pressure versus height above sea level. The graph is not a straight line because the density decreases with altitude. The pressure never quite goes to exactly zero.

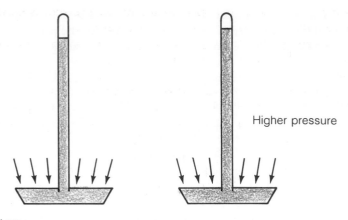

Higher pressure

FIGURE 4.28
A mercury barometer. If there is no air in the tube, the air pressure on the mercury in the bowl forces the mercury to rise up the tube. The pressure at the bottom of the tube equals the air pressure. The higher the air pressure, the higher the mercury.

The earth's atmosphere is a relatively thin layer of a mixture of gases. The decrease in air pressure with altitude is further complicated by variations in the temperature and composition of the gas. The heating of the air by the sun, the rotation of the earth, and a number of other factors cause the air pressure at a given place to vary slightly from day to day or even from hour to hour. Also the pressure is not always the same at points with the same altitude.

In spite of the complexity of the situation, we can make some statements about the general variation of air pressure with altitude. Since the atmosphere is a gas, there is no upper surface or boundary. The air simply gets thinner and thinner. At 9,000 meters (30,000 feet) above sea level, the density of the air is only about 35% of that at sea level. At this altitude, the average person cannot remain conscious because there is not enough oxygen in each breath. At about 160 kilometers (100 miles) above sea level, the density is down to 1 billionth of the sea-level density. This is often regarded as the effective upper limit of the atmosphere. Spacecraft can remain in orbit at this altitude because the thin air causes very little air resistance.

Figure 4.27 is a graph of the air pressure versus altitude for the lower atmosphere. (When comparing this graph with Figure 4.25, remember that the depth in a liquid is measured downward from the surface, while the height in the atmosphere is measured upward from sea level.) The pressure rapidly decreases with increasing height at low altitudes where the density is still fairly high. Remember that the pressure at any elevation depends on the weight of all of the air above. At 9,000 meters, the pressure is about 0.3 atmosphere. This means that only 30% of the air is above 9,000 meters. Even though the atmosphere extends to around 100 miles up, most of the air (70%) is less than 6 miles up.

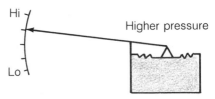

Higher pressure

FIGURE 4.29
Schematic of an aneroid barometer. With higher atmospheric pressure, the top of the can is pushed in more, causing the pointer to indicate higher pressure.

Air pressure is measured with a *barometer.* The simplest type is the mercury barometer. It consists of a vertical glass tube with its lower end immersed in a bowl of mercury. All of the air is removed from the tube, so there is no air pressure acting on the surface of the mercury in the tube (see Figure 4.28). The air pressure on the mercury in the bowl is transmitted to the mercury in the tube. This forces the mercury to rise up in the tube. (When drinking through a straw, you reduce the pressure in your mouth and the atmospheric pressure forces the drink up the straw.) The mercury will rise in the tube until the pressure at the base of the tube equals the air pressure. Hence the air pressure is measured by simply measuring the height of the column of mercury.

When the air pressure is 1 atmosphere (14.7 psi), the column of mercury is 760 millimeters (29.9 in) long. At lower pressures the column is shorter, so we can use the length of the column of mercury as a measure of the pressure (760 millimeters of mercury equals 1 atmosphere). Other liquids will work, but the tube has to be much longer if the liquid's density is small. For example, it would take a column of water 10.3 meters (33.9 feet) long to produce, and therefore measure, a pressure of 1 atmosphere (Figure 4.26).

This unit of measure is not limited to measuring air pressure. Blood pressure is also given in millimeters of mercury. If your blood pressure is 100 over 60, it means that the pressure drops from 100 millimeters of mercury during each heartbeat to 60 millimeters of mercury between beats.

A more portable type of barometer consists, more or less, of a very short metal can with the air removed from the inside (see Figure 4.29). The air pressure causes the ends to be squeezed inward. The higher the pressure, the greater the distortion of the ends. A pointer is attached to one end and moves along a scale when the distortion of the end changes. This is called an *aneroid barometer.*

The variation of air pressure with height is used to measure the altitude of aircraft. An *altimeter* is simply an aneroid barometer with a scale that registers altitude instead of pressure (Figure 4.30). For example, if a regular barometer in an airplane indicated the pressure was 0.67 atmosphere, one could infer that the altitude was 10,000 feet. Each pressure reading on the barometer would correspond to a different altitude.

A more sophisticated instrument is used to measure the vertical speed of an aircraft—how fast it is going up or down. This device, called a *vertical airspeed indicator* in airplanes (refer to Figure 4.30) and a *variometer* in gliders, senses changes in the air pressure. When the aircraft is going up, the measured air pressure decreases. The instrument converts the rate of change of the air pressure into a vertical speed.

The law of fluid pressures applies to granulated solids in much the same way that it does to liquids. The walls of storage bins and grain silos are reinforced near the bottom, because the pressure is higher there. Again, it is the force of gravity pulling on the material above that causes the pressure.

FIGURE 4.30
An altimeter (upper) measures altitude by measuring the air pressure. A vertical airspeed indicator (lower) measures vertical speed by measuring how rapidly the air pressure is increasing or decreasing.

DO-IT-YOURSELF PHYSICS

Fill a glass with water, hold a piece of paper on the top, and invert it. (You might want to do this over a sink.) When you remove your hand from the piece of paper, why doesn't the water fall out?

4.5 ARCHIMEDES' PRINCIPLE

The force of gravity causes the pressure in a fluid to increase with depth. This in turn causes an interesting effect on substances partly or totally immersed in a fluid. In some cases the substance floats—wood or oil on water, blimps, or hot-air balloons in air. In other cases the substance seems lighter—a 100-pound rock can be lifted with a force of about 60 pounds when it is under water. Obviously some force acts on the foreign substance to oppose the downward force of gravity (weight). This force is called the *buoyant force*.

> BUOYANT FORCE   The upward force exerted by a fluid on a substance partly or completely immersed in it.

This force acts on anything immersed in any gas or liquid. As long as there are no other forces acting on the substance, there are three possibilities (Figure 4.31). If the buoyant force is *less than* the weight of the substance, it will *sink*. If the buoyant force is *equal* to the weight, the substance will *float*. If the buoyant force is *greater than* the weight of the substance, it will *rise* upward. Examples of each case (in the same order) are a rock in water, a piece of wood floating on water, and a helium-filled balloon rising in air. Both the rock and the balloon experience a net force because the weight and the buoyant force do not cancel each other. This net force causes each to accelerate momentarily until reaching a terminal velocity.

At this point we might pose two questions. First, what causes the buoyant force? Second, what determines the size of the buoyant force? The answer to the first question can be arrived at rather simply in light of the previous section. Consider an object that is completely immersed (Figure 4.32). Since its bottom surface is deeper in the fluid than its top surface, the pressure on the bottom surface will be greater. Hence the *upward force* on its bottom surface caused by this pressure is *greater* than the *downward force* on its top surface. In short the difference in fluid pressure acting on the surfaces of the object causes a net upward force. (The forces on the sides of the object are equal and opposite, so they cancel out.)

When something floats on the surface of a liquid, only its lower surface experiences the fluid pressure of the liquid. This causes an upward force as before.

As for the second question, the size of the buoyant force is determined by a law formulated in the third century BC.

F_b = Buoyant force

FIGURE 4.31
The rock sinks because the buoyant force on it is less than its weight. The wood floats because the buoyant force on it is equal to its weight. The helium balloon rises upward because the buoyant force on it is greater than its weight.

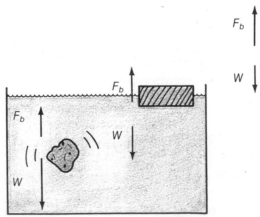
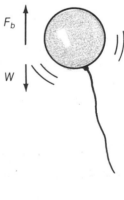

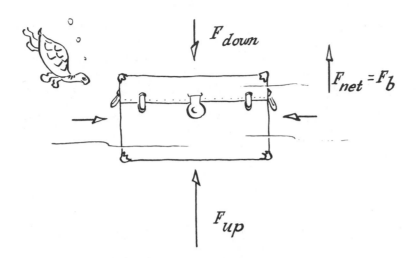

FIGURE 4.32
*The fluid pressure causes a force
on each surface of an immersed
object. The pressure on the lower
surface is greater than the
pressure on the upper surface.
Consequently the upward force
on the bottom is greater than the
downward force on the top. The
net upward force is the buoyant
force.*

> ARCHIMEDES' PRINCIPLE   The buoyant force acting on a substance in a fluid at rest is equal to the weight of the fluid displaced by the substance.
>
> $$F_b = \text{Weight of displaced fluid}^3$$

When a piece of wood is placed in water, it displaces some of the water: part of the wood occupies the same space or volume that was formerly occupied by water. The weight of this displaced water equals the buoyant force acting on the wood. Obviously any object that is completely submerged in a fluid will displace a volume of fluid equal to its own volume.

If the buoyant force on an object is less than its weight, it sinks, but the *net* downward force is reduced. Figure 4.33 shows an object hanging from a scale. Its weight is 10 newtons. As the object is lowered into a beaker of water, it displaces some of the water. The scale reading is reduced by an amount equal to the buoyant force. When the scale shows 6 newtons, the buoyant force is 4 newtons.

$$\text{Scale reading} = \text{weight} - \text{buoyant force}$$

$$6\,\text{N} = 10\,\text{N} - 4\,\text{N}$$

---

[3](For the mathematically inclined) In the simple case of a box-shaped object immersed (but level) in a liquid, we can prove Archimedes' principle. Assume the box's dimensions are l, w, and h. The downward force on the top of the box is:

$$F_{\text{top}} = p_{\text{top}} \times A_{\text{top}} = p_{\text{top}} \times l \times w$$

Here p_{top} is the pressure at the top of the box. At the bottom of the box the pressure is higher because it is deeper in the fluid. Specifically the pressure there, p_{bottom}, equals:

$$p_{bottom} = p_{\text{top}} + (D_w \times h)$$

where D_w is the weight density of the fluid. Therefore the upward force on the bottom is:

$$F_{bottom} = p_{\text{bottom}} \times l \times w = (p_{\text{top}} + (D_w \times h)) \times l \times w$$

$$F_{bottom} = (p_{\text{top}} \times l \times w) + (D_w \times h \times l \times w)$$

The buoyant force F_b is just the upward force minus the downward force:

$$F_b = F_{bottom} - F_{\text{top}} = (p_{\text{top}} \times l \times w) + (D_w \times h \times l \times w) - (p_{\text{top}} \times l \times w)$$

$$F_b = D_w \times h \times l \times w = D_w \times V = W$$

$$F_b = \text{Weight of fluid displaced}$$

FIGURE 4.33
The weight of an object is 10 newtons. When the object is immersed in water, the scale reads a smaller force because of the upward buoyant force. The buoyant force is equal to the weight of the water that the object displaces.

The weight of the water that spills over the side of the beaker is also 4 newtons. If the object is lowered farther into the water, the buoyant force will increase, and the scale reading will decrease.

One important thing to notice here is that the buoyant force acting on a substance doesn't depend on what the substance is, only on how much fluid it displaces. *Identical balloons* filled to the same size with helium, air, and water all have the *same buoyant force* acting on them. However, only the helium-filled balloon floats in air because its weight is less than the buoyant force. Also, the weight of a substance alone does not determine whether or not it will float. A tiny pebble will sink in water, but a 2-ton log will float.

The key here is density. Let's take the simplest case of a solid object submerged in a liquid. The volume of the liquid that is displaced by the object is, of course, equal to the object's volume. So:

$$\text{Weight of object} = \text{weight density (of object)} \times \text{volume}$$

$$W = D_w(\text{object}) \times V$$

The buoyant force is

$$\text{Buoyant force} = \text{weight of displaced fluid} \qquad (\text{Archimedes' principle})$$

$$F_b = \text{weight density (of fluid)} \times \text{volume}$$

$$F_b = D_w(\text{fluid}) \times V$$

From this we conclude that if the weight density of the object is greater than the weight density of the fluid, the object's weight is greater than the buoyant force, and the object sinks. If its density is less than the fluid's density, it floats. By simply comparing the densities (or specific gravities) of the substance and the fluid, one can determine whether or not the substance will float. *Any substance with a smaller density than that of a given fluid will float in (or on) that fluid.*

In Table 4.3 we see that juniper, ice, gasoline, ethyl alcohol, and all of the gases will float on water because their densities are less than that of water. (Actually the alcohol would simply mix in with the water. In the case of a fluid being immersed in another fluid, it is best to imagine the first being enclosed in a balloon or some other container that has negligible weight but keeps the fluids from mixing.) The density of mercury is so high that everything in the table except gold and uranium will float on it (Figure 4.34). Hydrogen and helium both float in air, while carbon dioxide sinks.[4]

DO-IT-YOURSELF PHYSICS

(You have to do this one outside—and no smoking!) Fill a glass about two thirds full with equal amounts of water and gasoline. These two liquids do not mix, and you can clearly see that the gasoline does float on the water. Now put an ice cube into the glass. The ice sinks in the gasoline but floats on the water.

[4]A deadly example of the heavier-than-air nature of carbon dioxide occurred near Lake Nios in Cameroon, West Africa in August 1986. As a result of a small volcanic eruption or earthquake beneath the lake, a large volume of trapped CO_2 (in addition to carbon monoxide and hydrogen sulfide) was released; the dense CO_2 displaced the air along the lakeshore and caused the tragic deaths of over 1,700 persons by suffocation.

FIGURE 4.34
An iron ball floats in mercury because the density of mercury is greater than the density of iron (almost double).

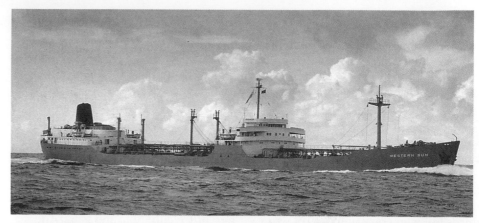

FIGURE 4.35
A loaded ship rides lower in the water because it must displace more water.

Ships and blimps float, even though they are constructed out of materials with higher densities than the fluids in which they float. They are composites of substances: most of the volume of a ship is occupied by air, and most of the volume of a blimp is occupied by helium. In both cases the *average density* of the object is less than the density of the fluid.

As a ship is loaded with cargo, it sinks lower into the water. This makes it displace more water, thereby increasing the buoyant force and countering the added weight (see Figure 4.35).

Archimedes' principle is routinely used to measure the densities or specific gravities of solids that sink. The object is hung from a scale, and the reading is recorded. (It does not matter whether the scale gives mass or weight. Here we assume it reads weight.) Then the object is completely immersed in water, and the new scale reading is recorded (see Figure 4.36). The *difference* in these two readings equals the weight of the displaced water, by Archimedes' principle.

W of displaced water = scale reading out of water − scale reading in water

W of displaced water = scale (out) − scale (in)

So the weight of the object is known, and the weight of an *equal volume of water* is also known. Density is weight divided by the volume. Since the volumes are the same, the density of the object divided by the density of the water *equals* the weight of the object divided by the weight of the water.

FIGURE 4.36
The specific gravity of a solid is measured by comparing the scale reading when it is out of water to the scale reading when it is completely submerged. The difference between the two readings is equal to the weight of the displaced water.

$$\frac{\text{Density of object}}{\text{Density of water}} = \frac{\text{scale reading out of water}}{\text{scale (out)} - \text{scale (in)}}$$

This ratio is just the specific gravity of the object.

$$\text{Specific gravity} = \frac{\text{scale (out)}}{\text{scale (out)} - \text{scale (in)}}$$

In Section 4.3, we describe how the density of the coolant in an automobile engine indicates the antifreeze content. The density is measured with a simple device that employs Archimedes' principle. Figure 4.37 shows an antifreeze tester that consists of a narrow glass tube containing five balls. The balls are specially made so that each one has a slightly higher density than the one above it. The coolant is drawn up into the glass tube with a suction bulb. Each ball will float if its density is less than the density of the coolant. The number of balls that float depends on the density of the coolant. If the antifreeze content is high, the coolant's density is high, and all of the balls float.

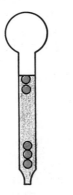

Higher-density coolant

FIGURE 4.37
When radiator coolant is drawn up into the antifreeze tester, the number of balls that float depends on the coolant's density. The density indicates the antifreeze content and therefore the freezing temperature of the coolant.

To check the hemoglobin content of blood, a drop of it is placed in a liquid that has the correct minimum density. If the blood sinks, its density is greater than that of the liquid, and the hemoglobin content is high enough for the blood to be accepted.

Now we will give a couple of examples using Archimedes' principle.

EXAMPLE 4.8   A contemporary Huckleberry Finn wants to construct a raft by attaching empty, plastic, 1-gallon milk jugs to the bottom of a sheet of plywood. The raft and passengers will have a total weight of 300 pounds. How many jugs are required to keep the raft afloat on water?

The buoyant force on the raft must be at least 300 pounds. Consequently the raft must displace a volume of water that weighs 300 pounds.

$$F_b = \text{weight of water displaced}$$

$$= D_w(\text{water}) \times \text{volume of water displaced}$$

$$300 \text{ lb} = 62.4 \text{ lb/ft}^3 \times V$$

So:

$$V = \frac{300 \text{ lb}}{62.4 \text{ lb/ft}^3}$$

$$V = 4.8 \text{ ft}^3$$

To keep 300 pounds afloat, 4.8 cubic feet of water must be displaced. One cubic foot equals 7.48 gallons. So $4.8 \times 7.48 = 36$ one-gallon jugs would be enough.

EXAMPLE 4.9   Before the Hindenberg disaster, blimps, zeppelins, and balloons were filled with hydrogen. Now helium is used. Let us compare the two gases in terms of their buoyancy in air. Each cubic foot of hydrogen gas weighs 0.0056 pound (Table 4.3). When in an airship, this hydrogen will displace a cubic foot of air. So each cubic foot of hydrogen has a buoyant force of 0.08 pound, the weight of one cubic foot of air. Therefore the net force on each cubic foot of hydrogen gas is

$$\text{Net force} = 0.08 \text{ lb} - 0.0056 \text{ lb}$$

$$F = 0.0744 \text{ lb} \quad (\text{hydrogen})$$

Each cubic foot of hydrogen can lift 0.0744 pound (see Figure 4.38). If helium is used instead, the buoyant force is still the same, but the weight of each cubic foot of helium is 0.011 pound. So:

$$\text{Net force} = 0.08 \text{ lb} - 0.011 \text{ lb}$$

$$F = 0.069 \text{ lb} \quad (\text{helium})$$

Each cubic foot of hydrogen can lift 8% more than a cubic foot of helium. A balloon filled with hydrogen can lift 8% more than an identical balloon filled with helium. However, the big factor that tilts the scale in favor of helium is that it does not burn, as hydrogen does.

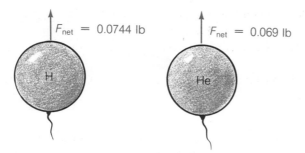

$F_{net} = 0.0744$ lb $F_{net} = 0.069$ lb

FIGURE 4.38
One cubic foot of hydrogen can lift 0.0744 pounds. One cubic foot of helium can lift 0.069 pounds.

The air exerts a buoyant force on everything in it, not just balloons. This force, 0.08 pound for each cubic foot, is so small that it can be ignored except when dealing with other gases. For example, the volume of a person's body is typically around 2 or 3 cubic feet. (The density of the human body is about the same as that of water. That is why we just barely float in water. So the approximate volume of your body is just your weight divided by the weight density of water.) This means that the buoyant force on such a person due to the air is between 0.16 and 0.24 pound.

4.6 PASCAL'S PRINCIPLE

When a force acts on a surface of a solid, the resulting pressure is "transmitted" through the solid only in the original direction of the force. When you sit on a stool, your weight causes pressure in the legs that is transmitted to the floor. This pressure doesn't act to the side of the legs, only downward.

In a fluid, any pressure caused by a force is transmitted everywhere throughout the fluid and acts in all directions. When you squeeze a tube of toothpaste, the pressure is passed on to all points in the toothpaste. An inward force on the sides of the tube can cause the toothpaste to come out of the end. This property of fluids should be familiar to you and might even be used to distinguish fluids from solids. Pascal's principle is a formal statement of this phenomenon.

> **PASCAL'S PRINCIPLE**   Pressure applied to an enclosed fluid is transmitted undiminished to all parts of the fluid and to the walls of the container.

This property of fluids is exploited in widely used hydraulic systems. Hydraulic jacks and the brakes systems in automobiles are common examples. The basic components of such systems are piston and cylinder combinations. Figure 4.39 shows a small piston and cylinder on the left connected via a tube to a larger piston and cylinder on the right. (The effect of gravity can be ignored.) A liquid is in both cylinders and the connecting tube. When a force acts on the left piston, the resulting pressure is passed on throughout the liquid. This pressure causes a force on the right piston. Since this piston is larger than the one on the left, the force $(F = pA)$ will also be larger. For example, if the area of the right piston is five times that of the left piston, the force will be five times as large. This system behaves like a lever (Figure 3.10). A small force acting at one place causes a larger

FIGURE 4.39
The pressure caused by the force on the small piston is transmitted throughout the fluid and acts on the larger piston. The resulting force on this piston is larger than the force on the smaller piston.

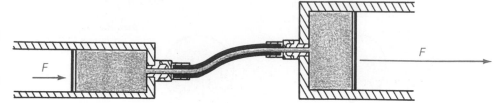

force at another place. As with the lever, the smaller piston will move a correspondingly greater distance than the larger piston. The work done would be the same for both pistons.

In automotive brake systems, the brake pedal is connected to a piston that slides in the "master" cylinder. This cylinder is connected via tubing to "wheel" cylinders on the brakes on the wheels (Figure 4.40). (Actually there are two cylinders in tandem in the master cylinder. Each is connected to two wheels so that if one subsystem fails, the other two wheels will still have braking power.) Brake fluid fills the cylinders and tubing. When the brake pedal is pushed, the piston produces pressure in the master cylinder that is transmitted to the wheel cylinders. The piston in each wheel cylinder is attached to a mechanism that applies the brakes. In disc brakes, used on the front wheels of most cars, the piston squeezes disc pads against the sides of a rotating disc attached to the wheel. This action is very similar to that of rim brakes on bicycles. The mechanism in drum brakes, used on the rear wheels of most cars, is a bit more complicated.

Using a hydraulic brake system serves two purposes. By using wheel cylinders with larger diameters than those of the master cylinder, there is a mechanical advantage: the forces on the wheel cylinder pistons are larger than the force applied to the master cylinder piston. Also this is an efficient way to transmit a force from one location in the car to four other locations.

4.7   BERNOULLI'S PRINCIPLE

In this section we present a simple principle that applies to *moving fluids.* Water flowing in a stream or through pipes to a water faucet, air moving through heating ducts or as a cool breeze in the summer, blood flowing through your arteries and veins—these are all examples of fluids flowing. It is common for each of these fluids to speed up and slow down as it flows. Accompanying any change in the speed of the fluid is a change in pressure of the fluid. This is stated in the following principle, named after the Swiss physicist and mathematician Daniel Bernoulli.

> BERNOULLI'S PRINCIPLE   For a fluid undergoing steady flow,[5] the *pressure is lower* where the fluid is *flowing faster.*

This principle is based on the conservation of energy. A fluid under pressure has what can be called *pressure potential energy.* The higher the pressure, the

---

[5]Steady flow means that there is no random swirling of the fluid and that there are no outside forces increasing or decreasing the rate of flow.

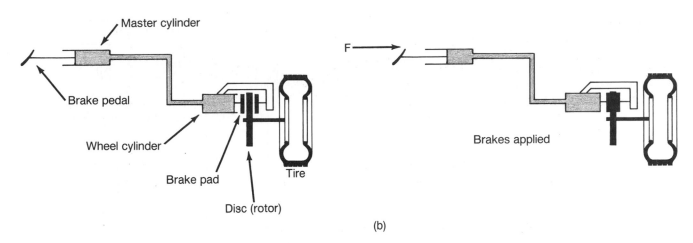

Master cylinder

Brake pedal

Wheel cylinder

Brake pad

Disc (rotor)

Tire

F

Brakes applied

(b)

FIGURE 4.40
(a) Simplified diagram of an automotive hydraulic brake system. Only one wheel, equipped with disc brakes, is shown. (b) When the brake pedal is pushed, the pressure increase in the master cylinder is passed on through the fluid to the wheel cylinder. The piston is pushed outward, and the brake pads squeeze the disk.

greater the potential energy of any given volume of fluid. (When you open a faucet, water rushes out as the potential energy due to pressure is converted into kinetic energy of running water. Low water pressure makes the water come out slowly, with low kinetic energy, because it starts with low potential energy.) Moving fluids have both kinetic and potential energy. When a fluid speeds up, its kinetic energy increases. Its total energy remains constant so its potential energy and therefore its pressure decrease.

One of the best examples of Bernoulli's principle is that depicted in Figure 4.41. Water flowing through a pipe passes through a smaller section spliced into the pipe. The water speeds up when it enters the narrow region, then slows down when it reenters the wide region. This occurs because the volume of fluid passing through each part of the pipe each second is the same. Where the cross-sectional area is smaller, the fluid must flow faster if the same number of cubic inches of fluid is to get through each second. (You've seen this already if you've ever used your thumb to partially block water coming out of the end of a garden hose.) Bernoulli's principle tells us that the pressure in the moving water is *smaller in the narrow section* than in the wide sections upstream and downstream. Pressure gauges placed in the pipe show this to be so. This is one of those rare situations in which physical fact runs counter to one's intuition. On first thought most people would predict that the pressure should be higher in the narrow part of the pipe because the fluid is "squeezed into a smaller stream." But that is not the case. The pressure is actually lower in the narrow part.

Carburetors on gasoline engines used in lawn mowers, some cars,[6] and dozens of other powered devices exploit Bernoulli's principle. The carburetor's function is to mix gasoline droplets into the air that enters the engine. The passageway of the air is narrowed so that the air speeds up and the pressure decreases (see Figure 4.42). A small tube is placed with one end in the narrow part of the airway and the other in a bowl filled with gasoline. When air flows through the carburetor, the reduced pressure causes gasoline to be forced upwards in the tube, just like cola in a soda straw. This arrangement is convenient because the flow of gasoline automatically stops when the engine is turned off and the air stops flowing.

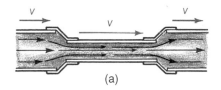

(a)

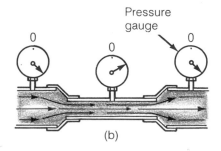

(b)

FIGURE 4.41
(a) The water travels faster in the narrow section of the pipe. (b) The pressure is lower *where the water is* moving *faster.*

---

[6]Increasingly carburetors on cars are being replaced by *fuel injection systems.* These inject the gasoline directly into the air entering the engine much like a toy water pistol squirts water into the air.

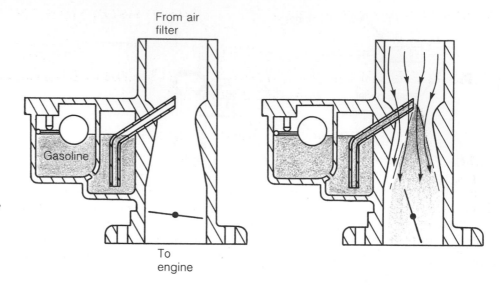

From air
filter

Gasoline

To
engine

FIGURE 4.42
Simplified diagram of a carburetor. Lower pressure in the moving air allows atmospheric pressure to force gasoline up the small tube and into the airstream.

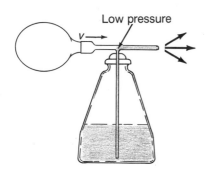

Low pressure

v

FIGURE 4.43
Perfume atomizers use Bernoulli's principle. Air is forced to move through the horizontal tube when the bulb is squeezed. In the narrow section of this tube the air has to speed up so the pressure is lower. Normal air pressure on the perfume in the bottle forces it to rise upward into the stream.

A very similar system is used in atomizers on perfume bottles. When a bulb is squeezed, air moves through a horizontal tube that has a narrow segment (Figure 4.43). The air speeds up in this section, and so the air pressure is lower. A small tube runs from this section down into the perfume. Since the air pressure is reduced, the normal air pressure acting on the surface of the perfume forces the liquid to rise upward in the tube and to enter the moving air.

DO-IT-YOURSELF PHYSICS

Hold the top of a piece of paper horizontally just below your lips, so that the paper hangs limp (see Figure 4.44). Blow hard over the top of the paper. The moving air above the paper is at a lower pressure, so the paper is lifted up by the higher pressure below it.

The next time you are drinking with a straw, blow hard *across* the top of the straw, as when you blow across a bottle to produce sound. The drink will rise up in the straw because of the lower pressure in the moving air.

FIGURE 4.44

HISTORICAL   NOTES

The Atomic Theory of Matter

The nature of matter appears to have been one of the earliest subjects of human speculation. The Greek philosopher Democritus, who lived about a century before Aristotle, developed the theory that all matter is composed of particles too small to be seen. He called these particles "atoms," from the Greek word for indivisible, since he thought them to be the final result of any repeated subdivision of matter. The atomic theory as expounded by Democritus and other Greek philosophers was not accepted by Aristotle. For this and other reasons, the atomic theory was abandoned for more than 2,000 years.

The successes of Galileo and Newton opened a new era of experimentation and advancement in all of the sciences. The development of chemistry in the 17th and 18th centuries revived the atomic theory because it could explain many of the new discoveries. Robert Boyle (1626–1691), an Irish-born physical scientist, identified "elements" as those substances that could not be further decomposed by chemical analysis. A century later the French chemist Antoine Lavoisier (1743–1794) refined the concept of the element and showed that mass is conserved during chemical reactions (see Figure 4.45). When paper burns, the total

FIGURE 4.45
Lavoisier with his wife Marie, who was also his assistant. (Painting by Jacques Louis David.)

FIGURE 4.46
English scientist John Dalton developed the concepts of atoms and molecules, and their relationship to elements and compounds.

mass of the combustion products, ashes and smoke is the same as the total mass of the paper and oxygen that were combined during the burning. (Lavoisier was guillotined during the French Revolution because he worked for King Louis XVI.) During the 18th century, several of the elements were correctly classified: iron, gold, mercury, carbon, and oxygen, to name a few. Salt and several other compounds were mistakenly identified as elements because they could not be decomposed into their constituent elements with the techniques then available.

In 1808, John Dalton (1766–1844), an English schoolteacher, presented what was essentially the modern atomic theory (see Figure 4.46). He correctly stated that the properties of each element are determined by the properties of its atoms, the atoms of different elements have different masses, and two or more different atoms in combination form the basic units of chemical compounds. The actual structure of atoms and the fact that they aren't indestructible were only determined during the last 100 years.

Fluids

The scientific study of fluids probably began with Archimedes' discoveries about buoyancy. In his book *On Floating Bodies,* he proves the famous principle that immortalized his name. Apparently Archimedes' interest in the topic began when King Hiero of Syracuse presented him with a problem. The king had given a certain weight of gold to a craftsman to be made into a crown. The finished product was presented to the king, but he suspected that some of the gold had been replaced by an equal weight of silver. King Hiero asked Archimedes to find a way to determine the composition of the crown without damaging it. The story goes that Archimedes thought of the solution when he noticed how water was displaced as he sank into a bathtub. He supposedly rushed home naked, shouting, "Eureka!" ("I have found it!").

Historians do not agree on the exact method used by Archimedes to prove that there was indeed some silver in the crown. He could have compared the volumes of the crown, an equal weight of gold, and an equal weight of silver by measuring the water each displaced when submerged. Or he could have compared the weights of the three objects when immersed in water (as in Figure 4.36). Either method could have been used to determine the amount of silver in the crown.

The concept of specific gravity was developed in Arabia approximately 1,000 years ago. Arabian scientists also described how to use Archimedes' principle to measure the specific gravities of solids.

The French mathematician Blaise Pascal (1623–1662) made several discoveries about the nature of pressure in liquids, the most famous being the principle named after him (Section 4.6). He described how this principle implies that a force can be "amplified" using combinations of different-sized pistons and cylinders (Figure 4.39). Pascal also showed experimentally that the pressure in a liquid depends only on the depth and not on the shape of the container. The pressure at the bottom of an inverted cone filled with water is the same as the pressure at the bottom of a narrow tube filled with water to the same level.

For 2,000 years after Aristotle's time, the philosophical principle *horror vacui*—that "nature abhors a vacuum"—obscured the understanding of air pressure. It was used to explain why a liquid could be drawn up into a tube by removing some of the air. When a soda straw is being used, the drink is supposedly pulled into the straw because nature does not allow the creation of a vacuum. Galileo was puzzled by reports that a suction pump could not raise water in a pipe higher than about 33 feet. This indicated to him that there seemed to be a limit to *horror vacui*.

The invention of the mercury barometer by the Italian mathematician Evangelista Torricelli (1608–1647), a student of Galileo, proved to be the key to discovering the nature of atmospheric pressure. Torricelli correctly suggested that it is the pressure of the "sea of air" (the atmosphere) that forces liquids to rise in a tube when air is removed from it. Pascal learned of Torricelli's work and performed many experiments with his own barometers (see Figure 4.47). (In one of them he used red wine in a glass tube 46 feet long.) Pascal reasoned that the column of mercury in a barometer should be shorter at higher elevations. Yet he could not detect a difference in the length when he moved his barometer to the top of a church tower. Pascal's brother-in-law took a barometer to the Puy de Dôme, a mountain in south central France, and did observe the effect.

The German physicist Otto von Guericke (1602–1686) performed several experiments that dramatically illustrated that atmospheric pressure can cause large forces. In one such experiment he filled a copper sphere with water, sealed it, and then pumped out the water. The unbalanced air pressure on the outside caused the sphere to implode. Von Guericke invented air pumps and used them to produce near vacuums. In 1654 he demonstrated his *Magdeburg hemispheres* to Emperor Ferdinand III. He fitted two hemispheres together and pumped the air out of the spherical cavity thus formed (see Figure 4.48). The force of the air pressure acting on the two hemispheres was so great that eight horses pulling on each one were unable to separate them.

Robert Boyle used a modification of the barometer to discover the relationship between pressure and volume in a gas (see the end of Section 4.2). Boyle's apparatus was a U-shaped glass tube partially filled with mercury. One side was sealed off so that some air was trapped above the mercury. By pouring more

FIGURE 4.47
A large barometer built by Pascal in Rouen, France to measure air pressure.

FIGURE 4.48
Otto von Guericke demonstrating that the air is stronger than horses.

mercury into the other side or by removing some, Boyle was able to vary the pressure on the trapped air. The difference in height between the mercury in the two sides indicated the pressure on the air. Boyle showed that the volume occupied by the trapped air was inversely proportional to the pressure on it. This law, $pV = $ a constant, is known as *Boyle's law.*

SUMMARY

The matter that makes up the material universe can be classified in different ways. In terms of the external, observable properties of matter, we can identify four phases: solid, liquid, gas, and plasma. In terms of the submicroscopic composition, there are elements, compounds, and combinations of these. The nature of the constituent particles—atoms and molecules—and the forces that act between them determine the physical properties of a given element or compound.

Pressure, mass density, and weight density are extensions of the concepts of force, mass, and weight. Pressure is a measure of the "concentration" of any force that is spread over an area. Mass density is the mass of a substance per unit volume.

Liquids and gases are fluids: they flow readily and take the shapes of their containers. The force of gravity causes the pressure in a fluid to increase with depth. The law of fluid pressures states how the pressure at any point in a fluid is determined by the weight of the fluid above. In liquids the pressure is proportional to the distance below the liquid's surface. Gases are compressible, and so the pressure increases with depth in a different way, since the density is not constant.

Anything partly or completely immersed in a fluid experiences an upward buoyant force. By Archimedes' principle, this force is equal to the weight of the fluid that is displaced. Ships, boats, submarines, blimps, and hot-air balloons all rely on this principle. Archimedes' principle is also routinely used to measure the densities of solids and liquids.

Pascal's principle is a statement of the rather obvious fact that any additional pressure in a fluid is passed on uniformly throughout the fluid.

Bernoulli's principle is exploited by carburetors, atomizers, and other devices. It states that the pressure in a moving fluid decreases when the speed increases and vice versa.

SUMMARY OF IMPORTANT EQUATIONS

EQUATION	COMMENTS
FUNDAMENTAL EQUATIONS	
$p = \dfrac{F}{A}$	Definition of pressure
$D = \dfrac{m}{V}$	Definition of mass density
$D_w = \dfrac{W}{V}$	Definition of weight density
$F_b = W_{\text{Fluid displaced}}$	(Archimedes' principle)
SPECIAL CASE EQUATIONS	
$p = D_w h = Dgh$	Pressure (gauge) at a depth, h, in a liquid
$p = 0.433 \text{ psi/ft} \times h$	Pressure (gauge) in psi under water at a depth, h, in feet

QUESTIONS

1. Describe the four states of matter. Compare their external, observable properties. Compare the nature of the forces between atoms or molecules (or both) in the solid, liquid, and gas phases.

2. Identify some of the elements that exist in pure form around you.

3. What is the difference between a mixture of two elements and a compound formed from the two elements?

4. If you classify everything around you as an element, a compound, or a mixture, which category would have the largest number of entries?

5. Why can gases be compressed much more readily than solids or liquids?

6. Use the concept of pressure to explain why snowshoes are better than regular shoes for walking in deep snow.

7. Explain the difference between gauge pressure and absolute pressure.

8. How can you use the volume of some pure substance to calculate its mass?

9. The mass density of a mixture of ethyl alcohol and water is 950 kg/m^3. Is the mixture mostly water, mostly alcohol, or about half and half? What is your reasoning?

10. Would the weight density of water be different on the moon than it is on earth? What about the mass density?

11. The way pressure increases with depth in a gas is different from the way it does in a liquid. Why?

12. If the acceleration of gravity on the earth suddenly increased, would this affect the atmospheric pressure? Would it affect the pressure at the bottom of a swimming pool? Explain.

13. If the earth's atmosphere warmed up and expanded to a larger total volume but its total mass did not change, would this affect the atmospheric pressure at sea level? Would this affect the pressure at the top of Mt. Everest? Explain.

14. Is there a pressure variation (increase with depth) in a fuel tank on a spacecraft in orbit? Why or why not?

15. Explain how a barometer can be used to measure altitude.

16. Why does the buoyant force always act upward?

17. What substances would sink in gasoline but float in water?

18. It is easier for a person to float in the ocean than in an ordinary swimming pool. Why?

19. In "The Unparalleled Adventure of One Hans Pfaall" by Edgar Allen Poe, the hero discovers a gas whose density is "37.4 times" less than that of hydrogen. How much better at lifting would a balloon filled with the new gas be compared to one filled with hydrogen?

20. A brick is tied to a balloon filled with air and is then tossed into the ocean. As the balloon sinks, the buoyant force on it decreases. Why?

21. What is Pascal's principle?

22. How does a car's brake system make use of Pascal's principle?

23. What useful thing happens when the speed of a moving fluid increases?

24. The pressure in the air along the upper surface of an aircraft's wing (in flight) is lower than the pressure along the lower surface. Compare the speed of the air flowing over the wing to that of the air flowing under the wing.

25. How does a perfume atomizer make use of Bernoulli's principle?

PROBLEMS

1. A grain silo is filled with 2 million pounds of wheat. The area of the silo's floor is 400 sq ft. Find the pressure on the floor in pounds per square foot and in psi.

2. A bicycle tire pump has a piston with area 0.44 in.^{2}. If a person exerts a force of 30 lb on the piston while inflating a tire, what pressure does this produce on the air in the pump?

3. A large truck tire is inflated to a gauge pressure of 80 psi. The total area of one sidewall of the tire is 1,200 in.^{2}. What is the outward force on the sidewall due to the air pressure?

4. The water in the plumbing in a house is at a gauge pressure of 300,000 Pa. What force does this cause on the top of the tank inside a water heater if the area of the top is 0.2 square meters?

5. A metal can has dimensions 8 in. by 4 in. by 10 in. high. All of the air inside the can is removed with a vacuum pump. Assuming normal atmospheric pressure outside the can, find the total force on one of the 8-by-10-in. sides.

6. A viewing window on the side of a large tank at a public aquarium measures 50 in. by 60 in. The average gauge pressure due to the water is 8 psi. What is the total outward force on the window?

7. A large chunk of metal has a mass of 393 kg, and its volume is measured to be 0.05 m³.
a) Find the metal's mass density and weight density in SI units.
b) What kind of metal is it?

8. A small statue is recovered in an archaeological dig. Its weight is measured to be 96 lb and its volume 0.08 ft³.
a) What is the statue's weight density?
b) What substance is it?

9. A large tanker truck can carry 20 tons (40,000 lb) of liquid.
a) What volume of water can it carry?
b) What volume of gasoline can it carry?

10. A manufacturer orders 10,000 kg of copper ingots. What volume does the copper occupy?

11. A large balloon used to sample the upper atmosphere is filled with 900 m³ of helium. What is the mass of the helium?

12. A certain part of an aircraft engine has a volume of 1.3 ft³.
a) Find the weight of the piece when it is made of iron.
b) If the same piece is made of aluminum, find the weight and determine how much weight is saved by using aluminum instead of iron.

13. Find the gauge pressure at the bottom of a swimming pool that is 12 ft deep.

14. The depth of the Pacific Ocean in the Mariana Trench is 36,198 ft. What is the gauge pressure at this depth?

15. Calculate the gauge pressure at a depth of 300 m in seawater.

16. A storage tank 30 m high is filled with gasoline.
a) Find the gauge pressure at the bottom of the tank.
b) Calculate the force that acts on a square access hatch at the bottom of the tank that measures 0.5 meter by 0.5 meter.

17. The highest point in North America is the top of Mt. McKinley in Alaska, 20,320 ft above sea level. Using the graph in Section 4.4, find the approximate air pressure.

18. The Concord, a supersonic passenger jet, flies at an altitude of 15,000 m. What is the approximate air pressure at that height?

19. An ebony log with volume 12 ft³ is submerged in water. What is the buoyant force on it?

20. An empty storage tank has a volume of 1,500 ft³. What is the buoyant force exerted on it by the air?

21. A piece of concrete measures 3 ft by 2 ft by 0.5 ft.
a) What is its weight?
b) Find the buoyant force that acts on it when it is submerged in water.
c) What is the net force on the concrete piece when it is under water?

22. A juniper wood plank measuring 0.25 ft by 1 ft by 16 ft is totally submerged.
a) What is its weight?
b) What is the buoyant force acting on it?
c) What is the size and the direction of the net force on it?

23. The volume of an iceberg is 100,000 ft³.
a) What is its weight, assuming it is pure ice?
b) What is the volume of seawater it displaces when floating? (Hint: You know what the weight of the seawater is.)
c) What is the volume of the part of the iceberg out of the water?

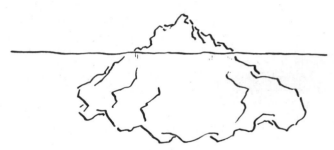

24. A boat (with a flat bottom) and its cargo weigh 5,000 N. The area of the boat's bottom is 4m². How far below the surface of the water is the boat's bottom when it is floating in water?

25. A scales reads 100 N when a piece of aluminum is hanging from it. What does it read when it is lowered so the aluminum is submerged in water?

26. A rectangular block of ice with dimensions 2 m by 3 m by 0.2 m floats on water. A person weighing 600 N wants to stand on the ice. Would the ice sink below the surface of the water?

CHALLENGES

1. When exactly 1 cup of sugar is dissolved in exactly one cup of water, *less than* 2 cups of solution result. Why?

2. The cabin pressure inside a space shuttle is maintained at 1 atm. An air lock—a tube 63 in. in diameter with hatches at both ends—allows astronauts in space suits to leave the shuttle for space walks. When the inner hatch is open, what is the total force on the other end of the air lock?

3. What would be the mass density, weight density, and specific gravity of aluminum on the moon? The acceleration of gravity there is 1.6 m/s^2.

4. The mass of the earth is 6×10^{24} kg, and its radius is 6.4×10^6 m. What is the average mass density of the earth?

5. Two swimming pools are 8 ft deep, but one measures 20 ft by 30 ft, and the other measures 40 ft by 60 ft. Identical drain valves at the bottom of each pool are 10 in.^{2} in area. Compare the force on each valve.

6. Scuba divers take their own supply of air with them when they go under water. Why couldn't they just take a long hose with them from the surface and breathe through it?

7. A motorist driving through Colorado checks the tire pressure in Denver (elevation 5,000 ft) then again at the Eisen-

hower Tunnel (elevation 11,000 ft). Would the pressures be the same? What two main factors affect the tire pressure as the car climbs?

8. A glass contains pure water with ice floating in it. After the ice melts, will the water level be higher, lower, or the same? (Ignore evaporation.)

9. At one point in the novel *Slapstick,* by Kurt Vonnegut (Delacorte Press, 1976), the force of gravity on earth suddenly "increased tremendously." The result:

   . . . elevator cables were snapping, airplanes were crashing, ships were sinking, motor vehicles were breaking their axles, bridges were collapsing, and on and on.

   Would a ship really sink if the force of gravity were increased?

10. A brick rests on a large piece of wood floating in a bucket of water. The brick slides off and sinks. Does the water level in the bucket go up, go down, or stay the same?

11. As a helium balloon rises up in the air, work is done on it against the force of gravity. What is doing the work? What energy transfer or transformation is taking place?

SUGGESTED READINGS

Brancazio, Peter J. *Sport Science.* New York: Simon and Schuster, 1984. Chapter 10, particularly the last two sections, takes a close look at the role of Bernoulli's principle in sports.

Cajori, Florian. *A History of Physics.* New York: Dover, 1962 (originally published by Macmillan in 1929). The chapter on 17th century physics describes the research of Pascal, Torricelli, and others on the properties of gases.

Fisher, Arthur. "Seeing Atoms." *Popular Science* 234, no. 4 (April 1989):102–107. Describes the scanning tunneling electron microscope, with several pictures.

Gamov, George. *Biography of Physics.* New York: Harper and Row, 1961. Chapter 1 has a lot about Archimedes, including an account of the king's crown.

Scott, Arthur F. "The Invention of the Balloon and the Birth of Modern Chemistry." *Scientific American* 250, no. 1 (January 1984):126–137. An account of the beginnings of ballooning and its relationship to the study of the nature of matter.

Walker, Jearl. *The Flying Circus of Physics with Answers.* New York: Wiley, 1977. "A collection of problems and questions about physics in the real, everyday world." Chapter 4 addresses dozens of questions about fluids.

Weast, Robert C., ed. *Handbook of Chemistry and Physics,* 66th ed. Boca Raton, Florida: CRC Press. Gives a short history of each element.

CHAPTER

5

OUTLINE

196

TEMPERATURE AND HEAT

PROLOGUE: SOLAR ENERGY

Cooking, drying hair, staying warm in cold weather, melting raw materials in factories—these and many other processes that we take for granted require heat energy. Where does this energy come from? By tracing the origins of the energy that we use, we find that most of it comes from the sun. We use some *solar energy* (the energy in sunlight) directly. For example, sunshine entering windows on the sunny sides of buildings help heat them in the winter. But mostly we use solar energy that has been first converted into other forms of energy.

Trees and other plants convert solar energy into chemical energy, a form of potential energy. When you build a campfire to cook food, the potential energy stored in the wood is released as heat energy. Oil, coal, natural gas, and other fossil fuels—our main source of energy at this time—come from the remains of plants that stored solar energy hundreds of millions of years ago. Cooking with natural gas and using a car's heater to stay warm involve conversion of this chemical energy into heat energy.

What about an electric toaster or hair dryer? Most of the electrical energy produced in the United States comes from power plants that burn coal or other fossil fuels, use the heat to boil water into steam, and then use the steam to turn an electrical generator. In some parts of the world a great deal of electrical energy comes from hydroelectric power stations (described in Section 3.5). They convert gravitational potential energy of water behind a dam into electrical energy. But it is the sun that gives the water its potential energy. Sunlight warms the oceans and lakes, causing water to evaporate into the air. Moisture-laden air, also warmed by the sun, rises upwards and forms clouds. Whenever the density of water vapor in the clouds becomes too great, drops form and fall as rain. Streams and rivers carry the runoff to the reservoirs behind the dams.

Fossil fuels and hydroelectric power are both limited. Sooner or later the supply of coal and crude oil will be used up, and there are only so many rivers that can be dammed. A hundred years from now most of the energy used by our great, great, grandchildren will likely come from other sources. When seeking alternatives to our current energy supplies, an obvious place to look is up—toward the sun. For the contiguous 48 states, an average of 200 joules of solar energy is delivered to every square meter each second (this average includes reductions due to nighttime, cloudiness, and winter). An enormous amount of energy is available for the taking.

Solar cells, which covert solar energy directly into electrical energy, have been in use for decades, first in the space program to supply electricity to satellites in orbit, then in devices such as calculators, radios, outside lighting, and so on. Recent breakthroughs in cell efficiency and production costs suggest that solar cells may be competitive with conventional power plants in the near future.

Solar One, near Barstow, California, is a solar-steam-electric power plant that generated electricity from 1982 to 1988. Sunlight reflected off 1,818 computer-controlled mirrors (heliostats) that moved with the sun and directed its energy on a boiler at the top of a tower. The boiler, glowing white hot, produced steam that ran a conventional turbine and generator. Some of the steam was used to heat a huge tank containing rocks, sand, and a special type of oil. This stored heat energy was used to produce steam when clouds blocked the sun and for a time in the evening after the sun had set.

Wind energy, used for centuries to mill grain and to pump water, is also used to generate electricity. Large "wind farms" located in regions with strong, steady winds have hundreds of individual generators turned by propellers. They convert the kinetic energy of the moving air into electrical energy. And how does the air get its kinetic energy? It comes from the sun. Uneven heating of the earth's surface causes air to rise at some places, thereby lowering the atmospheric pressure at those locations. Wind is simply moving air pushed by the surrounding higher-pressure air toward the regions of lowered pressure.

There are other potential ways to tap solar energy, such as using the temperature difference between different depths in the oceans. Much more emphasis will be placed on solar energy and other renewable sources of energy in the coming decades.

Many of the key concepts presented in this chapter, including radiation, convection, phase transitions, and heat engines, find direct applications in the different methods of using solar energy.

In the last two chapters we saw that temperature and heat energy are very important in physics. The phase of any material depends on its temperature. The density, pressure, and volume of a gas can vary with the temperature of the gas. Heat energy is an important consideration in energy conversions, as we discussed in Chapter 3.

In this chapter we take a formal look at temperature and heat. First we consider temperature—what it is, how it is measured, and how matter is affected when it changes. Then we elaborate on the idea of heat energy by introducing the concepts of heat and internal energy and showing how they relate to temperature and changes of phase. The chapter concludes with a discussion of heat engines and heat movers.

5.1   TEMPERATURE

Temperature is perhaps the most commonly used physical quantity in our daily lives, next to time. We might loosely define temperature as a measure of hotness or coldness. But the concepts of hot and cold are themselves rather vague, subjective, and relative. In the summer an air temperature of 70°F feels cool, while the same temperature in the winter feels warm.

The idea of temperature is quite distinct from that of heat. For example, when one drop is removed from a cup of hot water, both the drop and the cup have the *same* temperature, but the amount of heat energy associated with each is very different (Figure 5.1). Placing the drop in the palm of your hand would not have nearly the same effect (pain) as pouring the cup of water into it.

All thermometers depend on some physical property that changes with temperature. The most common type exploits the fact that a liquid, usually mercury or red-colored alcohol, will expand or contract when heated or cooled, rising or falling in a glass tube as the temperature varies. Some thermometers are based on other temperature-dependent physical properties, including the volume of solids, the pressure or volume of a gas, the electrical properties of metals, the amount and frequency of radiated energy, and the speed of sound in a gas.

There are three different temperature scales in common use: Fahrenheit, Celsius, and Kelvin. The normal freezing and boiling temperatures of water—called the *phase transition* temperatures—are used to compare the three scales. In the Fahrenheit scale, the boiling point of water under a pressure of 1 atmosphere is 212°—designated 212° F. The freezing point of water under this pressure is 32° F. So there are 180 units, called *degrees,* that separate the two temperatures. The Celsius scale, formerly called the centigrade scale, is metric based; it uses 100 degrees between the freezing and boiling points of water. Zero degrees Celsius—designated 0° C—is the freezing temperature, and 100° C is the boiling temperature.

Most of us spend our lives subjected to temperatures within the range of −60 to 120° F (−51 to 49° C). Much higher temperatures exist in common places: the interiors of stoves and automobile engines, the filaments of light bulbs, the flame of a candle, and so on. The sun's surface temperature is about 10,000° F (5,700° C), and its interior is at about 27,000,000° F (15,000,000° C). Temperatures this high have been produced on the earth in experiments with plasmas and in nuclear explosions. There is no upper limit on temperature.

At the other extreme, there *is* a limit on cold temperatures. The coldest possible temperature, called *absolute zero,* is −459.67° F (−273.15° C). Since it is impossible to go below this temperature, the Kelvin scale is a convenient one because it uses absolute zero as its starting point (zero). (This scale is also referred to as the "absolute temperature scale.") The size of the unit in the Kelvin scale is the same as that in the Celsius scale, except it is called a Kelvin instead

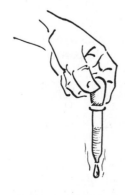 

FIGURE 5.1
The cup of water and the drop have the same temperature, but the cup of water can transfer much more heat energy to its surroundings.

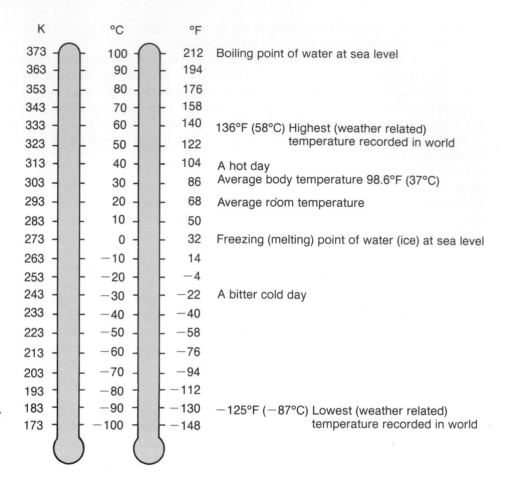

K	°C	°F	
373	100	212	Boiling point of water at sea level
363	90	194	
353	80	176	
343	70	158	
333	60	140	136°F (58°C) Highest (weather related)
323	50	122	temperature recorded in world
313	40	104	A hot day
303	30	86	Average body temperature 98.6°F (37°C)
293	20	68	Average room temperature
283	10	50	
273	0	32	Freezing (melting) point of water (ice) at sea level
263	−10	14	
253	−20	−4	
243	−30	−22	A bitter cold day
233	−40	−40	
223	−50	−58	
213	−60	−76	
203	−70	−94	
193	−80	−112	
183	−90	−130	−125°F (−87°C) Lowest (weather related)
173	−100	−148	temperature recorded in world

FIGURE 5.2
Comparison of the three temperature scales. The Celsius and Kelvin scales have the same sized unit or degree. The Fahrenheit degree is five-ninths the Celsius degree.

of a degree. Any temperature in the Kelvin scale equals the corresponding Celsius value plus 273.15. The normal boiling and freezing temperatures of water are 273.15 K and 373.15 K, respectively.

Figure 5.2 shows a comparison of the three temperature scales. Note that the Fahrenheit and Celsius scales agree at −40°. Table 5.1 lists some representative temperatures. As a physical quantity, temperature is represented by T.

Unfortunately, T is used to represent both temperature and period (Section 1.2).

What determines the temperature of matter? In other words, what is the difference between a cup of coffee when it is hot (200° F) and the same cup of coffee when it is cold (70° F)? The atoms and molecules that compose matter have kinetic energy. In gases, they move about randomly with high speed. In liquids and solids, they vibrate much like a mass on a spring or an object oscillating in a hole (Section 3.5). At *higher temperatures* the atoms and molecules in matter *move faster* and have *higher kinetic energy*. In particular,

> The Kelvin temperature of matter is proportional to the average kinetic energy of the constituent particles.
>
> Kelvin scale temperature ∝ average *KE* of atoms and molecules

Because of collisions between the particles, during which energy is exchanged, the particles do not all have exactly the same kinetic energy at each instant, nor does the energy of a given particle stay exactly the same from one moment to the

TABLE 5.1 REPRESENTATIVE TEMPERATURES IN THE THREE TEMPERATURE SCALES			
Description	°F	°C	K
Absolute zero	−459.67	−273.15	0
Helium boiling point	−452	−268.9	4.25
Nitrogen boiling point	−320.4	−195.8	77.35
Oxygen boiling point	−297.35	−182.97	90.18
Alcohol freezing point	−175	−115	158
Mercury freezing point	−37.1	−38.4	234.75
Water freezing point	32	0	273.15
Normal body temperature	98.6	37	310.15
Water boiling point	212	100	373.15
"Red hot" (approx.)	800	430	700
Aluminum melting point	1220	660	933
Iron melting point	2797	1536	1809
Sun's surface (approx.)	10,000	5700	6000
Sun's interior (approx.)	27×10^6	15×10^6	15×10^6
Highest laboratory temperature	410×10^6	230×10^6	230×10^6

next. But the *average* kinetic energy of all of the particles is constant as long as the temperature stays constant.

So when a cup of coffee is hot, the molecules in it have higher average kinetic energy than when it is cold. If you put your finger into hot coffee, the atoms and molecules in the coffee pass on their higher kinetic energy to the atoms and molecules in your finger by way of collisions: your finger is warmed.

This fact—that temperature depends on the average kinetic energy of atoms and molecules—is an important one and should help you understand many of the phenomena we will discuss in this chapter. In gases, higher kinetic energy means that the atoms and molecules move about with higher speeds. For liquids and solids, the molecules vibrate through a greater distance like a pendulum swinging through a larger arc. You may recall that when particles oscillate like this, they also have potential energy. This is the case here, but the potential energy of "bound" atoms and molecules in liquids and solids is not directly related to the temperature. This potential energy is important when a substance undergoes a change of phase—like freezing or boiling. More on this later.

This principle also accounts for the existence of an absolute zero. At colder temperatures the average kinetic energy of the particles is smaller. If they stopped moving altogether, the average kinetic energy would be zero. This would be the lowest possible temperature—absolute zero (see Figure 5.3). The word "would" is used because, as it turns out, absolute zero can never be reached. The atoms and molecules in a substance cannot be completely stopped. Researchers do get very close to absolute zero—within a millionth of a degree—but they cannot reach it exactly.

At very low temperatures many substances acquire quite unusual properties. Plastic and rubber become as brittle as glass. Below about 2 K, helium is a

PHYSICS POTPOURRI   To Breathe or Not to Breathe, That Is the Question.

As mentioned in Chapter 4, we live in a "sea" of air that provides us with one of the basic ingredients of sustained life: oxygen. Indeed, had the earth been unable to retain an atmosphere—for more than several billion years—the evolution of life as we know it could not have taken place at all. But what physical conditions or quantities determine whether or not a planet is capable of holding onto an appreciable atmosphere? More specifically, why does the earth possess an atmosphere, while Mercury and the moon do not (see Figure 5.4)? Is the composition of such an atmosphere related to these physical conditions? That is, is it possible to understand, in terms of our physics, why the atmosphere of Jupiter is primarily hydrogen, while that of the earth is made up principally of heavier gases like nitrogen and oxygen (see Table 4.2)? Let us see.

Every planetary body in our solar system receives some radiation from the sun: those closer to the sun receive more radiation and hence have higher surface temperatures than those farther away. Given our definition of temperature, it is clear that atoms and molecules in the atmosphere of a planet near the sun will generally have higher average kinetic energies (and higher average speeds) than those in the atmosphere of a planet far from the sun. The ability of a planet to retain these atmospheric atoms and molecules depends primarily on the strength of its gravitational field. If the average speeds of the atmospheric particles are high and the planet's surface gravity is low, the atmosphere will gradually escape. Conversely, if the average speeds of the atmospheric particles are low and the surface gravity is high, the atmosphere will be retained. This is why Mercury and the moon possess no atmospheres. Mercury is a small planet very close to the sun: it is hot (having a surface temperature over 600 K) and has a low surface gravity (one g on Mercury equals 3.7 m/s^2 as opposed to 9.8 m/s^2 on the earth). Any atmosphere it may have had once escaped long ago because its particles were far too energetic to have been held in by the planet's weak gravity. Similarly, the moon's atmosphere must have dissipated billions of years ago despite its cooler temperature (about 300 K, like that of the earth) because its surface gravity is only about one sixth that of the earth.

For planets with moderate temperatures and surface gravities, some gases can be retained, while others cannot. It is possible to predict which gases will be retained in a given circumstance by comparing the average molecular speed of a gas with the escape velocity of the planet. The *escape velocity* (as discussed in Section 3.5) is simply the minimum speed needed to overcome the gravitational pull of the planet. The higher the surface gravity of

FIGURE 5.3
At lower temperatures, atoms and molecules have lower average kinetic energy (KE). *At absolute zero, their kinetic energies would be zero: they would be stationary.*

400 K   200 K   0 K

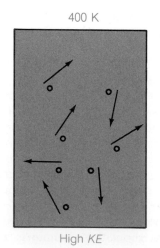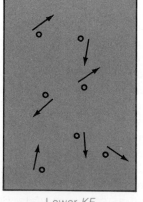

High *KE*   Lower *KE*   Zero *KE*

a planet, the greater its escape velocity. In general, a planet will retain a particular gas in its atmosphere only if its escape velocity is at least six times larger than the average molecular speed of that gas.

Only under these circumstances will the planet's surface gravity be high enough to prevent the gas from leaking off into space.

continued on next page

FIGURE 5.4
The planet Earth (a) has an atmosphere, but the planet Mercury (b) does not.

"superfluid" liquid; it flows without friction. Some materials become "superconductors"; they conduct electricity without resistance (more on this in Chapter 7).

The nature of matter is also different at very high temperatures. Above about 6,000 K, the kinetic energies of atoms are so high that they cannot bind together. Hence there can be no solids or liquids—or even molecules. Above about 20,000 K, the electrons break free from atoms, and only plasmas can exist.

5.2   THERMAL EXPANSION

Thermal expansion is an important phenomenon that is exploited by the common types of thermometers and by a variety of other useful devices. In almost all cases substances that are not constrained expand when their temperatures increase.

Figure 5.5 displays the rule graphically. Surface temperatures (in Kelvins) for the planets are plotted along the horizontal axis, and their escape velocities *(divided by 6)* are plotted on the vertical axis. Superimposed on this plot are lines for different gases representing the average molecular speeds of these gases at each temperature. Several things can be noted: (1) For a given gas, as the temperature increases (far left), the average molecular speed increases. This is to be expected, of course, on the basis of what is meant by the temperature of a gas. (2) At a given temperature the speeds of the lighter atoms and molecules, like hydrogen and helium, are greater than those of the heavier ones, like nitrogen and carbon dioxide. This, too, is easily understood in terms of what we know about the physics of gases. For a fixed temperature, the average kinetic energy of, say, a helium atom will be equal to that of a carbon dioxide molecule. But the kinetic energy of any particle is proportional to the product of its mass and its velocity squared. For a particular value of kinetic energy, higher masses correspond to smaller velocities. Hence at any temperature the more massive CO_2 molecule will be moving more slowly than the lighter He atom. (3) For a planet to retain a particular gas, its position on the graph must lie above the line for that gas. Notice that the points corresponding to the planets Jupiter and Saturn lie above all the lines. These planets are so massive and so cold (being far from the sun) that they can retain all the gases considered here in their atmospheres—even hydrogen, which has the lightest atoms. Earth, on the other hand, can hold onto the heavier gases like nitrogen and oxygen but not hydrogen. And, as discussed above, the moon and the planet Mercury are incapable of holding onto any of the common atmospheric constituents.

The analysis presented here has ignored many factors that complicate the question of planetary atmosphere retention (such as interactions between planetary atmospheres and the "solar wind," which is a continuous stream of electrons and other charged particles that leave the sun and propagate outwards into the interplanetary medium) but is correct in its broad outline. The important point to keep in mind is that, armed only with the physics we have developed so far, we are in a position to make some important predictions about the likely existence of atmospheres on other planetary bodies within the solar system. Such predictions figure prominently in attempts to identify locations beyond the earth where life may have evolved.

FIGURE 5.5
The dots indicate the surface temperatures and escape velocities (divided by six) for the moon and eight of the planets. The solid lines show the variation of molecular speed as a function of temperature for several common atmospheric gases.

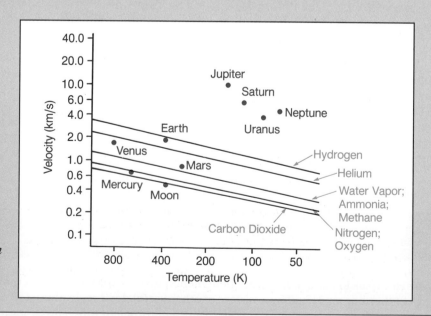

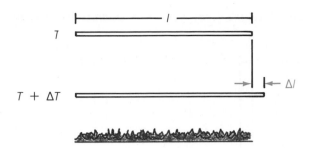

FIGURE 5.6
A metal rod has length l *when the temperature is* T. *When the temperature is increased by an amount* ΔT, *the length of the rod increases by a proportional amount* Δl.

The air in a balloon expands when heated, the mercury in a thermometer expands upward in the glass tube when placed in a hot liquid, and bridges become longer in the summer. If the substance is constrained sufficiently, it will not expand; but forces or pressures will be created in response to the constraint. For example, if an empty pressure cooker is sealed and then heated, the air inside is prevented from expanding. But the pressure inside increases and causes larger and larger outward forces on the inner surfaces of the pressure cooker.

We can see qualitatively why this expansion occurs. At higher temperatures the atoms and molecules in a solid or a liquid vibrate through a larger distance and so push each other apart slightly. In gases they move with higher and higher speed as the temperature rises. A balloon will expand when heated because the higher-speed air molecules will cause higher pressure and push the balloon's surface outward as they collide with it. In both, the expansion occurs in all three dimensions. For example, as a brick is heated, its length, width, and thickness all increase proportionally.

We can use basic mathematics to predict the amount of expansion that occurs. Let us first consider the simplest case—the thermal expansion of a long, thin solid such as a metal rod. The main expansion will be an increase in its length, *l* (see Figure 5.6). This increase, designated Δ*l*, depends on three factors:

1. The *original length, l.* The longer the rod is to begin with, the greater the change in length will be.

2. The *change in temperature,* designated Δ*T.* The larger the increase in temperature, the greater the increase in length.

3. The *substance.* For example, the increase in length of an aluminum rod will be more than twice that of an identical iron rod under the same conditions.

Point 3 can be tested through experimentation. The expansions of different solids are measured under similar conditions. The results are used to assign a *coefficient of linear expansion* to each material. This is a physical property of each substance, much like mass density or weight density. It is represented by the Greek letter alpha, α. Since aluminum expands more than iron under the same circumstances, the coefficient of linear expansion of aluminum is larger than that of iron (Figure 5.7).

The equation that gives the change in length in terms of the change in temperature and the coefficient of linear expansion is

$$\Delta l = \alpha \, l \, \Delta T$$

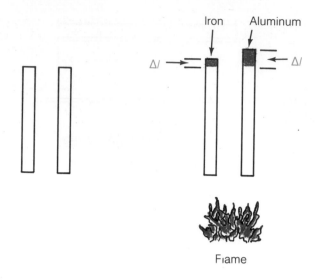

FIGURE 5.7
Aluminum expands more than iron when heated the same amount; its coefficient of linear expansion is larger.

The change in length is proportional to the change in temperature and to the original length. The coefficient of linear expansion, α, is the constant of proportionality. Table 5.2 lists α for several different solids. In the equation, the units of l and Δl must be the same. The units of temperature, usually °C, are canceled by the units of α, which are their reciprocal; that is to say, the units of alpha may be expressed as 1/°C.

EXAMPLE 5.1

The center span of a steel bridge is 1,200 meters long on a winter day when the temperature is $-5°$ C. How much longer is the span on a summer day when the temperature is $35°$ C?

First, the change in temperature is the final temperature minus the initial temperature:

$$\Delta T = 35 - (-5) = 35 + 5 = 40°\ C$$

From Table 5.2, the coefficient of linear expansion for steel is

$$\alpha = 12 \times 10^{-6}/°C$$

So:

$$\Delta l = \alpha l\ \Delta T$$
$$= (12 \times 10^{-6}/°C) \times 1{,}200\ m \times 40°\ C$$
$$= (12 \times 10^{-6}/°C) \times 48{,}000\ m\text{-}°C$$
$$= 576{,}000 \times 10^{-6}\ m$$
$$\Delta l = 0.576\ m$$

TABLE 5.2 SOME COEFFICIENTS OF LINEAR EXPANSION	
Solid	$\alpha\ (\times\ 10^{-6}/°C)$
Aluminum	25
Brass or bronze	19
Brick	9
Copper	17
Glass (plate)	9
Glass (Pyrex)	3
Ice	51
Iron or steel	12
Lead	29
Quartz (fused)	0.4
Silver	19

This is a considerable change in length and must be allowed for in the bridge design. Expansion joints, which act somewhat like loosely interlocked fingers, are placed in bridges, elevated roadways, and other such structures to allow thermal expansion to occur (Figure 5.8).

FIGURE 5.8
Expansion joints allow for the thermal expansion of bridges and other elevated roadways. Each end of a section of the roadway is connected to a metal "comb." The teeth of the two combs fit together and can move back and forth as the lengths of the sections change. The thermometer is 30 centimeters (1 foot) wide.

The equation also works when the temperature decreases. When this occurs, the change in temperature is negative, and so the change in length is also negative: the solid becomes shorter. Another way of saying this is that thermal expansion is a reversible process. If something becomes longer when heated, it will get shorter when cooled. Even though it is not mentioned each time, most of the phenomena discussed in this chapter are *reversible*. If something happens when the temperature increases, the reverse will happen when the temperature decreases.

The *bimetallic strip* is an ingenious and widely used application of thermal expansion. As the name implies, a bimetallic strip consists of two strips of different metals bonded to one another (Figure 5.9). The two metals have different coefficients of linear expansion, so they expand by different amounts when heated. The result is that the bimetallic strip bends—one way when heated and the other way when cooled. The greater the change in temperature, the greater the bending. For example, if brass and iron are used, the brass will expand and contract more than the iron will. The brass will be on the outside of the curve when the strip is hot and on the inside of the curve when it is cold.

Thermostats, thermometers, and choke-control mechanisms on some automobiles often contain a bimetallic strip that is curled into a spiral. The coil will either partly unwind or will wind up more tightly when the temperature changes. To make a thermometer, a pointer is attached to one end of the spiral, and the other end is fixed. As the temperature varies, the pointer moves over a scale that indicates the temperature (Figure 5.10). In thermostats, the movement of the end of the spiral is used to turn a switch on or off. The switch might turn on a heater, an air conditioner, a fire alarm, or the cooling unit in a refrigerator.

As mentioned, thermal expansion occurs in all three dimensions. The bridge becomes not only longer but also wider and thicker. The area of any surface increases, as does the volume of the solid. If a solid has a hole in it, thermal expansion will make it bigger, contrary to most people's intuition (Figure 5.11). This is because thermal expansion causes every point in a solid to move *away* from every other point. (It is similar to what happens when a photograph is enlarged.) A point on one side of a hole moves away from any point on the other side of the hole.

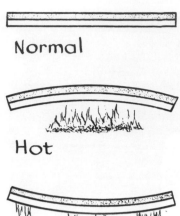

FIGURE 5.9
In this bimetallic strip, the metal composing the upper layer has a larger coefficient of thermal expansion than the metal composing the lower layer. When heated, the strip curves downward because the upper layer undergoes a greater change in length. When cooled, the strip curves upward.

FIGURE 5.10
This thermometer uses a bimetallic strip in the shape of a spiral. Even small temperature changes cause the pointer attached to the outer end of the spiral to rotate noticeably.

Liquids

The behavior of liquids is quite similar to that of solids. Since liquids do not hold a certain shape, it is only practical to consider the change in volume caused by thermal expansion. In general, liquids expand considerably more than solids. This means that when a container holding a liquid is heated, the level of the liquid will usually rise because the increase in volume of the liquid exceeds the increase in volume of the container. When a mercury thermometer is heated, the mercury in the tube and in the glass bulb at the bottom expands more than does the glass. If the glass expanded more than the mercury, the column would go down at higher temperatures instead of up.

At the beginning of this section we implied that there are exceptions to the general rule that matter expands when heated. The most important example of

FIGURE 5.11
The hole becomes larger when the object is heated.

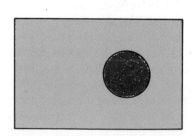

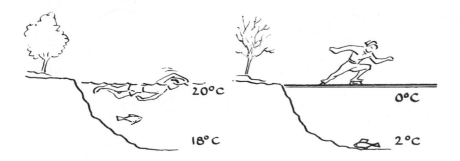

FIGURE 5.12
Above 4° C, the warmer water rises to the surface. Below 4° C, the cooler water rises to the surface.

this is water that is near its freezing temperature. Above 4°C (39°F), water expands when heated like ordinary liquids. But between 0°C and 4°C, water actually contracts when heated and expands when cooled. The volume of a given amount of water at 3°C is *less* than the volume of the same water at 1°C. This anomaly accounts for the fact that lakes, ponds, and other bodies of water freeze on top first. As long as the average water temperature is above 4°C, the warmer (less dense) water is buoyed to the surface, and the cooler water is at the bottom (see Figure 5.12). As the air gets cooler in autumn, the water at the surface is cooled. When the average temperature of the water is below 4°, the cooler water (closer to freezing) is now less dense and rises to the surface. Consequently the surface water freezes first because it is cooler than the water below, and it is in contact with the cold air.

Water is also unusual in that its density when in the solid phase (ice) is less than its density when in the liquid phase. Because of this, ice floats in water, whereas most solids sink in their own liquid.

Gases

The volume expansion of gases is larger than that of solids and liquids. Also, the amount of expansion does not vary with different gases (except at very low temperatures or very high pressures). Instead of relating the expansion to a change in temperature, it is simpler to state the relationship between the volume occupied by the gas and the temperature. In particular, as long as the pressure remains constant, *the volume occupied by a given amount of gas is proportional to the temperature (in Kelvins).*

$$V \alpha T \quad \text{(gas at fixed pressure)}$$

The temperature must be in Kelvins. When the temperature of a gas is doubled, the volume doubles. The volume of a balloon at 303 K (86°F) is about 10% larger than the volume of the same balloon at 273 K (32°F) (see Figure 5.13). The volume change of typical solids is less than 1% over the same temperature range. In liquids it is a maximum of 5%.

We can combine this with the pressure-volume relationship given at the end of Section 4.2. The result:

$$\frac{pV}{T} = \text{a constant}$$

The constant depends on the quantity of gas. A given amount of gas can have any combination of pressure, volume, and temperature, as long as the three values

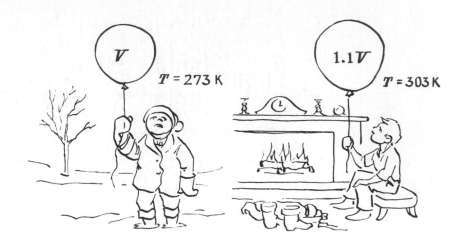

FIGURE 5.13
If the pressure in a gas is kept constant, the volume that the gas occupies is proportional to the Kelvin temperature. A 10% increase in temperature will cause the volume of a balloon to increase by 10%.

Refer back to Archimedes' Principle, Section 4.5.

satisfy the equation. This equation, known as the *ideal gas law,* expresses the interdependence of p, V, and T. For example, if the volume of the gas is fixed, the pressure will increase whenever the temperature increases. In other words, the pressure is proportional to the temperature (in Kelvins) as long as the volume stays constant. In the earlier example of heating air in a pressure cooker, the volume of the air inside would be nearly constant. (This is because the volume expansion of the metal is quite small compared to that of a gas.) So the pressure inside would increase proportionally with the temperature.

Regardless of which phase of matter is involved, changing the temperature of a substance will not change its mass or its weight. Since thermal expansion causes the volume to increase but the mass and weight stay the same, the *mass density and weight density decrease.* The mass density of a hot piece of iron is slightly less than the mass density of the same piece when it is cold. The mass density of the balloon referred to earlier is about 10% less when its temperature is 303 K than when its temperature is 273 K.

The reduction in density of a gas at constant pressure because of heating is used in hot-air balloons. The air in the balloon, which is basically a large bag with a small opening at the bottom, is heated with a burner (see Figure 5.14). The pressure inside remains equal to the atmospheric pressure because of the opening. Consequently the air expands, and its density decreases. The balloon can float in the air because it is filled with a gas (hot air) that has a smaller density than the surrounding fluid (cooler air). As the air inside cools, the balloon will sink toward the earth until the burner heats the air again.

5.3   THE FIRST LAW OF THERMODYNAMICS[1]

We have discussed what temperature is and how changes in temperature can affect the physical properties of matter—density in particular. The next item to

---

[1]The word "thermodynamics" derives from the Greek words *thermē*, meaning "heat," and *dynamikos,* meaning "powerful." *Thermodynamics* is that branch of physics which deals with the mechanical actions or relations of heat flow.

FIGURE 5.14
A hot-air balloon exploits thermal expansion.

consider is how the temperature of matter is changed. This will lead us back to the concept of energy.

There are two general ways to increase the temperature of a substance:

1. By exposing it to something that has a higher temperature

2. By doing work on it in certain ways

The first way is very familiar to you. When you heat something on a stove, warm your hands over a heater, or feel the sun warm your face, the temperature increase is caused by exposure to something that has a higher temperature: the stove, the heater, or the sun. As mentioned before, the atoms and molecules in the substance being warmed gain kinetic energy from those in the hotter substance. To raise the temperature of a substance, its atoms and molecules must gain kinetic energy.

Friction is the best example of raising the temperature of a substance by the second way (doing work on it). When something is heated by friction, work is done on it. This was discussed in more detail at the end of Section 3.4.

Another example of raising the temperature of a substance by doing work on it is the compression of a gas. When a gas is squeezed (quickly) into a smaller volume, its temperature increases (see Figure 5.15). In diesel engines air is compressed so much that its temperature is raised above the combustion temperature of diesel fuel. When the fuel is injected into the compressed hot air, it ignites.

The actual computation of the temperature rise of a gas when it is compressed a certain amount is beyond the scope of this text. But we can illustrate the type of heating that occurs by stating the results of one example: If air at $27°$ C is compressed to one twentieth of its initial volume in a diesel engine, its temperature will increase to over $700°$ C (Figure 5.16).

Both processes can be reversed to cause the temperature of a substance to decrease. The cooler air in a refrigerator lowers the temperature of a pitcher of tea. Air escaping from a tire is cooled as it expands.

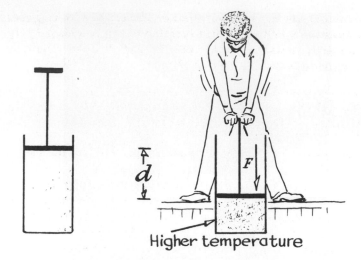

FIGURE 5.15
Work is done on a gas when it is compressed in a cylinder. This work causes the temperature of the gas to increase.

DO-IT-YOURSELF PHYSICS

Locate a bicycle tire pump and use it to inflate a flat tire. Then grab the lower end of the pump or the metal fittings on the hose; notice that they are warm. The work you do in compressing the air heats it.

Temperature depends on the average kinetic energy of atoms and molecules. So for matter to undergo a change in temperature, its atoms and molecules must gain energy (increase temperature) or lose energy (decrease temperature). In gases, the constituent particles have kinetic energy only. All of the energy given to the atoms and molecules acts to increase the temperature of the gas. Things are

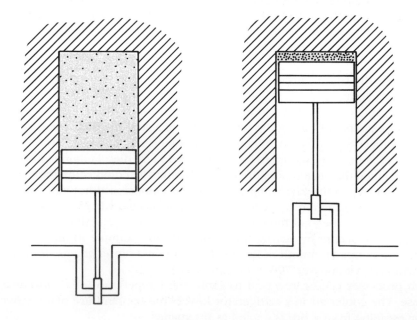

FIGURE 5.16
The air in a diesel engine has an initial temperature of 27° C. It is compressed by the piston until it occupies one twentieth of its original volume. This raises the temperature to 721° C.

different in solids and liquids. The atoms and molecules have kinetic energy *and* potential energy because they are bound to each other and oscillate. Energy given to the particles goes to increase both their *PE* and their *KE*. The concept of *internal energy* incorporates both forms of energy.

> **INTERNAL ENERGY** The sum of the kinetic energies and potential energies of all the atoms and molecules in a substance.[2]

Internal energy is represented by *U*. Its units of measure are the same as those of energy and work.

In gases, the internal energy is the total of the kinetic energies only: the atoms and molecules do not have potential energy (gravitational potential energy is not included in internal energy). In solids and liquids, both the kinetic energy and the potential energy of the particles contribute to the internal energy.

The concept of "heat energy" used in Section 3.4 and elsewhere is actually internal energy. The term heat energy was a useful compromise in nomenclature because it indicated that there is a form of energy related to our everyday concept of heat.

When something is heated, its internal energy increases. If this is accomplished by exposure to a hotter substance, we say that *heat* has flowed from the hotter substance into the cooler substance.

> **HEAT** The flow of internal energy. The form of energy that is transferred between two substances because they have different temperatures.

Heat is represented by *Q*. Its units are the same as those of work and energy. Traditionally the calorie, kilocalorie (also written Calorie), and British thermal unit (Btu) were used exclusively as units of heat. The joule, erg, and foot-pound were used for work and the other forms of energy. Now the joule is becoming the standard unit for heat as well.

The internal energy of something increases when heat flows into it and decreases when heat flows out of it (Figure 5.17). Heat represents a transfer of internal energy. In this respect it is much like work in mechanics. When work is

---

[2]Sometimes the concept of internal energy is expanded to include other forms of energy possessed by atoms and molecules. For example, batteries and gasoline have chemical energy that could be included in internal energy.

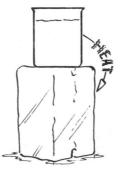

FIGURE 5.17
Heat flows into the water from the burner, increasing the water's internal energy. Heat flows out of the water into the ice, lowering the water's internal energy.

done, energy is transferred from one thing to another or is transformed from one form to another (Section 3.4). Both heat and work involve transfer: they don't exist in isolation. It is interesting that both can increase the internal energy of a substance.

The first law of thermodynamics is a formal summary of the preceding statements.

> **FIRST LAW OF THERMODYNAMICS**   The change in internal energy of a substance equals the work done on it plus the heat added to it.
>
> $$\Delta U = \text{Work} + Q$$

The work referred to in this law must be the type that transfers energy directly to atoms and molecules, such as compressing a gas. Work is done on an object when it is lifted, but this does not affect its internal energy—just its gravitational potential energy.

The work is positive if it is done *on* the substance and negative if the substance does work on something else. When gas in a cylinder is compressed by a piston, the work that is done is positive. If the piston is then released, the gas will expand and push the piston out. In this case the gas does work on the piston, and the work is negative (Figure 5.18). Similarly Q is positive when heat flows into the substance and negative when heat flows out of it. If you place a brick in a hot oven, heat flows into it, and Q is positive. Placing the brick in a freezer would result in a flow of heat out of it and a negative Q.

The first law of thermodynamics is nothing more than a restatement of the law of conservation of energy as it applies to thermodynamic systems. Work done or heat added to a substance is "stored" in it as internal energy. Aside from its theoretical significance, the first law of thermodynamics is an important principle used in the analysis of things like internal combustion engines and air conditioners.

This section began with a consideration of how the temperature of matter is changed. How does internal energy fit in with this? Temperature depends on the

FIGURE 5.18
When the piston compresses the gas, the work done on the gas is positive. When the gas pushes the piston back, it does work on the piston. In this case, the work done on the gas is negative.

Work done on the gas

Work done by the gas

kinetic energies of the atoms and molecules in a substance. But that is part of the internal energy of the substance. This means that at higher temperatures, a substance has higher internal energy.

The concept of internal energy is a bit abstract. Perhaps you are wondering why it is used at all, since temperature is determined only by the kinetic energies of particles. Internal energy is very useful when considering phase transitions. When water boils, for example, heat is added to it, but the temperature stays the same. This means that the average kinetic energy of the water molecules also stays the same. The heat added during a change in phase increases the internal energy by increasing only the *potential energies* of the molecules. The energy is used to break the bonds between the water molecules and free them. We will take a closer look at this in Section 5.6.

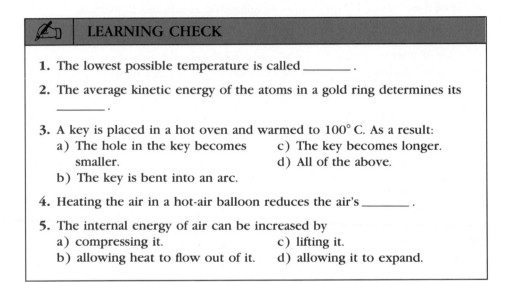

LEARNING CHECK

1. The lowest possible temperature is called _____ .

2. The average kinetic energy of the atoms in a gold ring determines its
   _____ .

3. A key is placed in a hot oven and warmed to 100° C. As a result:
   a) The hole in the key becomes smaller.
   b) The key is bent into an arc.
   c) The key becomes longer.
   d) All of the above.

4. Heating the air in a hot-air balloon reduces the air's _____ .

5. The internal energy of air can be increased by
   a) compressing it.
   b) allowing heat to flow out of it.
   c) lifting it.
   d) allowing it to expand.

5.4 HEAT TRANSFER

Transferring heat is the commoner of the two ways to change the temperature of something. Heat transfer occurs whenever there is a temperature difference between two substances or between parts of the same substance. In this section we discuss the three different mechanisms for heat transfer: *conduction, convection,* and *radiation.*

> CONDUCTION   The transfer of heat between atoms and molecules in direct contact.
>
> CONVECTION   The transfer of heat by buoyant mixing in a fluid.
>
> RADIATION   The transfer of heat by way of electromagnetic waves.

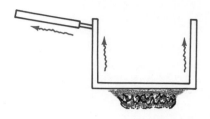

FIGURE 5.19
*Heat is conducted to the pan
from the flame in contact with it.
Conduction also takes place
within the pan: heat flows from
the hot bottom up the sides and
into the handle.*

Conduction

Conduction occurs when a pan is placed on a hot stove, when you put your hands into cold water, and when an ice cube comes into contact with warm air. The atoms and molecules in the warmer substance transfer some of their energy directly to the particles in the cooler substance. In these examples the conduction takes place across the boundary between the two substances, where the atoms and molecules collide with each other as they vibrate. Conduction also is responsible for the transfer of heat from one part of a solid to another part. (This also happens in fluids but is not as important as convection.) Even though only the bottom of a pan is in contact with a hot burner, the heat flows through the metal and soon raises the temperature of all parts of the pan (Figure 5.19). As the atoms and molecules on the bottom are heated, their constant jostling passes some of their increased speed onto neighboring atoms and molecules.

The ease with which heat flows within matter varies greatly. Materials through which heat moves slowly are called *thermal insulators.* Wool, Styrofoam, and bundles of fiberglass strands are all good insulators because they contain large amounts of trapped air or other gases. Conduction is poor within a gas because the atoms and particles are not in constant contact with each other. Iron, copper, and other metals are *thermal conductors:* heat flows readily through them. Concrete, stone, wood, and glass are between the two extremes. A vacuum completely prevents conduction because there are no atoms or molecules present.

Metals are good conductors of heat for the same reason that they are good conductors of electricity. Some of the electrons in the atoms in metals are free to move about from one atom to the next. It is the motion of these "conduction electrons" that constitutes an electric current (Chapter 7). These electrons can also carry internal energy from the warmer part of a metal to the cooler parts.

The conduction of heat within matter is similar to the flow of a fluid, except that nothing material moves from one place to another. The rate at which heat flows from a hot part of an object to a cold part depends on several things: the difference in temperature between the two places, the distance between them, the cross-sectional area through which the heat flows, and how good a thermal conductor the substance is.

Heat conduction is important in our daily lives. In cold weather we wear clothes that slow the conduction of heat from our warm bodies to the cold air. Handles on metal pans are often made of wood or plastic to reduce the conduction of heat from the burner to your hand. One reason carpets and rugs are used is that they feel warm when you step on them with bare feet. A rug is no warmer than the bare floor next to it, but it is a poorer conductor. When you step on the rug, very little heat is conducted away from your feet, so they stay warm. When you step on bare wood or tile, materials that are better thermal conductors, your feet are cooled more because heat is conducted from them more rapidly. (Figure 5.20).

Convection

Convection is the dominant mode of heat transfer within fluids. Whenever part of a fluid is heated, its density is decreased, and it rises. (Water below 4°C is an exception.) The result is a natural mixing of the fluid. Conduction can then occur between the warmer fluid and the cooler fluid around it.

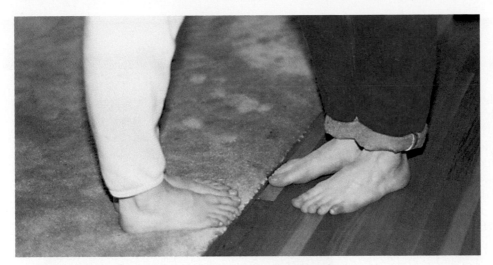

FIGURE 5.20
A rug feels warmer than bare floor because it cuts down on the conduction of heat from your feet.

A room with a wood stove or heater is warmed by convection (if no fans are used). The air that is heated rises to the ceiling and is replaced by air near the floor that moves to the heat source. The result is natural circulation of air along the ceiling, floor, and walls of the room (Figure 5.21). The warmed air near the ceiling cools when it contacts the walls and then sinks to the floor. The same type of circulation can occur in heated aquariums and swimming pools.

You may have heard the statement "heat rises." This is *not* physically correct. Heat is not something material that can rise or fall. "Heated air rises" or "heated fluids rise" are better statements for conveying the idea. Remember also that it is the denser surrounding fluid that pushes the warmer, less dense fluid upwards.

Mechanical mixing of a fluid is *not* convection. Stirring cool cream into hot coffee with a spoon is not convection because the mixing is not caused by

FIGURE 5.21
Convection causes a natural circulation of air in a room heated with a wood stove. Air heated by contact with the stove rises to the ceiling. Air cooled by contact with an outside wall sinks to the floor. This results in lateral movement of air along the ceiling and the floor.

Spacecraft photographs of the earth reveal a blue planet partially covered with a swirling pattern of brilliant white clouds (see Figure 5.4a). Similar photographs of the giant planet Jupiter show a colorful array of cloud patterns dominated by an alternating sequence of light-colored *zones* and darker, reddish-colored *belts* extending throughout the planet's midlatitudes and a large, reddish oval in the southern hemisphere called the *Great Red Spot* (Figure 5.22). At first glance it would seem that the forces controlling the weather patterns on these two planets must be different to produce such dissimilar cloud forms. And yet this may not be entirely so, as the Pioneer and Voyager missions to Jupiter have shown.

Winds on the earth blow from regions of high atmospheric pressure toward regions of low pressure. If the earth did not spin, air near the equator, heated by the sun's rays, would rise, creating a region of low pressure in its wake near the surface. Cooler air from the northern (or southern) midlatitudes would flow toward the equator to fill the void. The warmer, rising air would gradually cool and sink back near the polar ice caps, producing high-pressure areas and completing the driving mechanism of the convection cycle. In the absence of rotation, we would find that the prevailing winds on the earth would blow from the poles to the equator.

But the earth does rotate, and this rotation acts to break up this simple, two-component convection pattern into six smaller convection cells, three in the northern hemisphere and three in the southern. Figure 5.23 shows schematically the major convection zones in the earth's atmosphere and the wind patterns associated with the highs and lows

FIGURE 5.22
Voyager 1 photograph of cloud formations in the atmosphere of Jupiter. Comparison of this picture with Figure 5.4.a seems to suggest that weather patterns on earth and Jupiter are very different. But this may not be necessarily so.

produced by these cells. Notice that the winds blow in a counterclockwise direction around lows in the northern hemisphere and in a clockwise direction around the highs. The directions are reversed in the southern hemisphere.

The basic pattern of wind flow is the same on Jupiter. For example, the Great Red Spot is a high-pressure system, its reddish clouds being located

thermal buoyancy. Similarly, hot or cold air being blown into a room results in a heat transfer from the mixing of the fluid but is not convection.

Convection in the earth's atmosphere is a major cause of clouds, wind, thunderstorms, and other meteorological phenomena. White, puffy, cumulus clouds are formed when warm air rises into cooler air above.

Sea breezes—steady winds blowing into shore along coasts—are caused by convection. Sunshine warms the land more than the sea, so the air over the

FIGURE 5.23
The principal convection currents in the earth's atmosphere. As a result of the earth's rotation, six major convection cells encircle the earth. The surface winds associated with these cells are indicated by dashed arrows.

near the top of Jupiter's atmosphere. Careful examination of the cloud motions in and around the Great Red Spot shows that it rotates counterclockwise with a period of about 6 days. Thus the Great Red Spot is a large, long-lived (it has been observed since the mid-1600s!) cyclone-like disturbance in Jupiter's southern hemisphere. Other similar disturbances appear in Jupiter's southern hemisphere as *white ovals,* while in the northern hemisphere they can be found in the form of *brown ovals.* Whereas the white ovals are high-pressure regions,

the brown ovals are regions of lower-than-average atmospheric pressure. Such disturbances may be more common in the solar system than scientists once believed, particularly in the light of the Voyager 2 images of Neptune (Figure 5.24a).

Similarly, the Jovian zones are regions of generalized high pressure, while the belts are regions of lower atmospheric pressure. Because of Jupiter's rapid spin (a day on Jupiter is only 10 earth hours

continued on next page

ground is heated and rises upward.[3] This reduces the air pressure over the land, so the higher pressure over the sea forces air to move inland (Figure 5.25a). At night the land cools off more than the sea, so the process is reversed, and a *land*

---

[3]Only the surface of the ground is heated by the sun because the soil is a poor conductor. Heat given to the sea is quickly spread deeper below the surface by currents and convective mixing. We will also see in Section 5.5 that water requires an inordinate amount of heat to raise its temperature.

long), these highs and lows are stretched around the planet, producing its characteristic alternating bright and dark bands. Along the northern boundary of a zone in the northern hemisphere of Jupiter, the winds blow eastward, while along the southern boundary they blow westward, just as around highs on earth. Thus the direction of flow is the same around comparable regions in both atmospheres. Similar statements can be made about the directions of flow in regions bordering belts on Jupiter and lows on the earth. Again, spacecraft observations of the outer planets have revealed that flow patterns like these are recurrent throughout the solar system (see Figure 5.24b).

The structure and appearance of Jupiter's atmosphere above 45° (and below −45°) latitude become less ordered than those in the regions near the equator and may require some physical processes not commonly observed on earth. However, many of the weather patterns seen on Jupiter and the other large gaseous planets in the solar system now appear to be extensions of earthlike phenomena to conditions prevalent on these giant planets—an interesting and perhaps somewhat surprising result, by Jove!

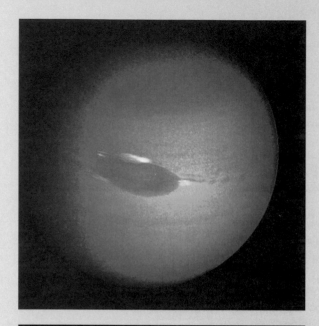

FIGURE 5.24
(a) Voyager 2 photograph of Neptune taken in August 1989 shows what has been labeled the Great Dark Spot, a vortex roughly the size of earth located 20 degrees south of Neptune's equator which slowly turns in a counterclockwise direction, completing one rotation in about 16 days.
(b) Voyager 2 photograph of Saturn showing alternating light and dark bands in its atmosphere. Although narrower and more subdued in color and detail, these structures are believed to be regions of alternating high and low pressure separated by boundaries of high winds.

breeze is produced. Air over the cooler ground sinks and forces air to move out to sea (Figure 5.25b).

A dramatic and useful form of convection often occurs as the sun warms the earth. Air in contact with hot ground is heated and expands upward. If conditions are favorable, the hot air will form an invisible "bubble" that breaks free from the

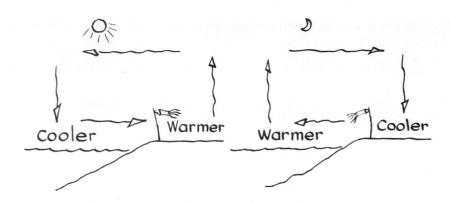

FIGURE 5.25
(a) Sea breeze. Warmed air rises from the heated land, causing cooler air to be drawn in from the sea. (b) The land cools at night, so the flow is reversed.

surface and rises upward into the air. (This is similar to the formation of steam bubbles on the bottom of a pan of boiling water.) These bubbles of rising heated air are called *thermals* (Figure 5.26). Hang glider and sailplane pilots and soaring birds such as eagles, hawks, and vultures seek out thermals and circle around in them. The upward speed of a typical thermal is around 5 meters per second (11 mph), so they provide a nice, free ride upward.

Radiation

Radiation is the transfer of heat via electromagnetic waves. We've all felt the warmth of the sun on our faces and the heat radiating from a hot fire. This is heat radiation—a type of wave related to radio waves and x-rays. (We will discuss electromagnetic waves in more detail in Chapters 8 and 9.) The sun's radiation passes through 150 million kilometers (93 million miles) of empty space and heats the earth and everything on it. This is the only one of the three types of heat transfer that can operate through a vacuum: without the sun's radiation the earth would be a cold, lifeless rock.

Infrared heat lamps warm things by emitting radiation. From a distance you can feel the heat from a camp fire or other heat source because the radiation from it warms your hands and face. Radiators used in steam-heating systems emit radiation that helps to heat rooms. (Actually when a radiator heats a room, all three types of heat transfer occur. Heat is conducted from the radiator to the air in

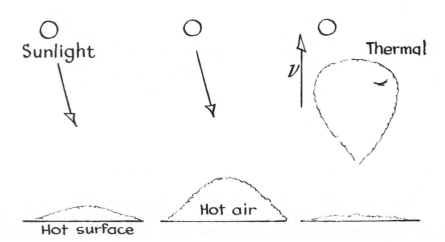

FIGURE 5.26
Thermals are rising "bubbles" of air formed on surfaces heated by the sun. They give a free upward ride to soaring birds and aircraft. Finding a thermal and then staying in it are a bit difficult because thermals are invisible.

As discussed above, energy can be moved or transported from one location to another by three mechanisms: conduction, convection, and radiation. Usually the bulk of the energy in any system is naturally transferred by the mechanism that is the most efficient. In a star like our own sun, energy is moved primarily by radiation throughout most of its interior but by convection during the last 20% or so of its journey from the sun's core. The specific mechanism at work depends upon the local physical conditions in the star.

The sun is an average star, one of over 10 billion contained in the Milky Way Galaxy. It is in what is called the *main sequence* phase of its lifetime. We might refer to this as the "adulthood" of the sun's life. In this period, energy is produced at the center, or core, of the star by nuclear fusion reactions involving protons (see Chapter 11 for more on fusion). This energy is transported from the core out to regions just beneath the solar surface—a distance of some 80% of the sun's radius—by radiation. Energy propagates from the core in the form of electromagnetic waves, working its way outward until it reaches what is known as the convection zone. This process is painfully slow, however. The electromagnetic waves travel only about 1 millimeter before they undergo collisions with particles in the solar interior that randomly alter their directions of motion; it requires about a million years for energy produced at the core to reach its outer parts.

When the energy arrives at the outer 20% of the sun's interior, radiation becomes less efficient as a transport mechanism than convection. The reason for this has to do with what is called the *opacity* of the gas. Essentially the opacity is a measure of how effectively the gas absorbs electromagnetic waves. Although the gas in the solar interior is fairly opaque, radiation is still more efficient than convection at moving the energy outward. However, in the very outermost parts of the sun, the gas temperatures are low enough that hydrogen atoms can exist. The formation of these atoms suddenly makes the gas extremely opaque—so much so that convection becomes the preferred mode of energy transport, and a convection zone develops. From here the energy finally bursts forth into space.

Photographs of the solar surface taken from high-altitude balloons, rockets, or satellites permit us to see the top of this convection zone (Figure 5.27). The rolling, bubbling gases can be clearly seen in the form of light and dark convection cells; such mottled photographs have been variously described as "granulation" or "salt-and-pepper" pat-

---

contact with it. This causes a convection current. The radiation emitted by the hot radiator warms the walls and other things in the room.)

FIGURE 5.28
A glowing light bulb transfers heat by both radiation and convection.

DO-IT-YOURSELF PHYSICS

Hold your hand a few inches from the side of a bare light bulb. The warmth you feel is from heat transferred by radiation. Place your hand the same distance directly above the bulb, and notice that it feels warmer (Figure 5.28). Now both radiation and convection of the heated air warm your hand.

You might think of the radiation as a "vehicle" or "carrier" of internal energy. Internal energy of atoms and molecules is converted into electromagnetic energy—radiation. The radiation then carries the energy through space until it is terns or as "oatmeallike" in texture. The brighter areas are warmer, updrafting material. The entire pattern continually changes as the cells rise and fall on time scales of about 10 minutes; each bubble has an irregular shape and is about 2,000 kilometers across—half the size of the United States! These are no small cauldrons of boiling brew!

Conduction does not play a significant role in the transport of energy in stars like the sun because the sun's material is gaseous and not extremely dense. Conduction does become an important mechanism in the very terminal stages of most star's lives, however. In this final phase of their evolution, all but the most massive stars end up as small (earth sized or less), dense (more than a billion times as dense as water) objects called *white dwarfs,* or *neutron stars* (see the *Physics Potpourri* "Starquakes," in Section 3.8). Because of the compactness of these endpoints of stellar evolution, their properties are in many ways similar to those of ordinary solid matter. It may not be too surprising then, given our discussion of heat flow in solids, to find that conduction is an important means by which these stars remove the last vestiges of energy from their cores to their surface.

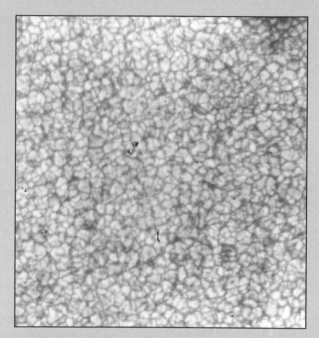

FIGURE 5.27
The surface of the sun.

absorbed by something. When absorbed, the energy in the radiation is converted into internal energy of the atoms and molecules of the absorbing substance.

Everything emits electromagnetic radiation: the sun, the earth, your body, this book. But the amount and type of radiation depend on the temperature of the emitter. The hotter something is, the more electromagnetic radiation it emits. Things below about 800° F (430° C) emit mostly infrared light, which we cannot see. Hotter substances emit *more* infrared light and also visible light. That is why we can see things in the dark that are "red hot" or "white hot." Things that are even hotter, like the sun at 10,000° F, emit more infrared and visible light but also emit ultraviolet light. We will take a closer look at heat radiation in Chapter 8.

So everything around you is emitting radiation and also absorbing the radiation emitted by other things. How does this lead to a net transfer of heat? Emission of radiation cools an object, while absorption of radiation warms it. If something absorbs radiation faster than it emits radiation, it is heated. Your face is warmed by the sun because it absorbs more radiation than it emits.

Rising energy costs in the 1970s and 1980s prompted many people to become more aware of heat transfer into and out of buildings. It costs money to add heat

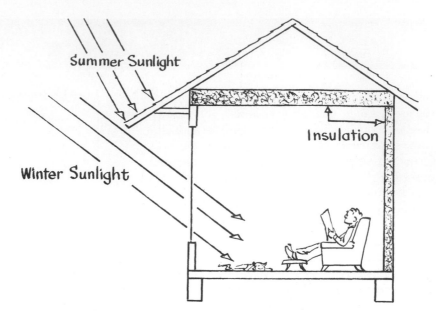

FIGURE 5.29
Heating and cooling costs for a home can be lowered by keeping in mind the mechanisms of heat transfer.

to buildings in the winter and to extract heat from them in the summer. These costs can be reduced by limiting the flow of heat between the inside and the outside of the building. Standard insulating procedures take into account all three mechanisms of heat transfer (Figure 5.29). Insulation in the walls and ceilings reduces conduction. More insulation is placed in the ceiling because of convection: the warmest air in the room is at the ceiling, so that is where heat would be transferred most rapidly out of the room during the winter. Window shades, blinds, and canopies can be used to keep direct sunlight out during the summer.

Thermos bottles are also designed to limit the flow of heat into or out of their contents. Instead of some kind of insulation, thermos bottles use a near vacuum between double-walled containers to almost eliminate heat flow due to conduction. The inner glass chamber is coated with silver or aluminum to reflect radiation.

5.5   SPECIFIC HEAT CAPACITY

Transferring heat to a substance or doing work on it increases its internal energy. In this section we describe how the temperature of the substance is changed as a result. To simplify matters, we assume that no phase transitions take place.

We might state the topic now under consideration in the form of a question: To increase the temperature of a substance by some amount ΔT, what quantity of heat, Q, must be added to it? (We could just as well ask how much work must be done on it.)

The amount of heat needed is proportional to the *temperature increase*. It takes twice as much heat to raise the temperature 20° C as it does to raise it 10° C. So:

$$Q \propto \Delta T$$

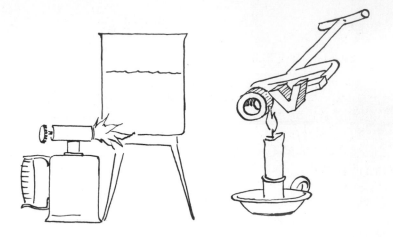

FIGURE 5.30
It takes about nine times as much heat to warm water 1° as it takes to warm the same mass of iron 1°.

Logically the amount of heat needed also depends on the *quantity* of the substance to which the heat is added. For a given increase in temperature, 2 kilograms of water will require two times as much heat as 1 kilogram. So:

$$Q \propto m$$

The quantity of heat required also depends on the *substance*. It takes more heat to raise the temperature of water 1° C than it does to raise the temperature of an equal mass of iron 1° C (Figure 5.30). As with thermal expansion (Section 5.2), a number can be assigned to each substance indicating the relative amount of heat needed to raise its temperature. This number, called the *specific heat capacity, C,* is determined experimentally for each substance. The larger the specific heat capacity of a substance, the greater the amount of heat needed to raise its temperature by a given amount.

$$Q \propto C$$

We can combine the three proportionalities into the following equation:

$$Q = Cm\,\Delta T$$

The amount of heat required equals the specific heat capacity of the substance times the mass of the substance, times the temperature increase. The SI unit of specific heat capacity is the joule per kilogram-degree Celsius. If the specific heat capacity of a substance is 1,000 J/kg-°C, it takes 1,000 joules of energy to raise the temperature of 1 kilogram of that substance 1° C. (The Kelvin is the same size unit as the °C and so can be used also.) Table 5.3 lists the specific heat capacities of some common substances.

TABLE 5.3 SOME SPECIFIC HEAT CAPACITIES	
Substance	C (J/kg-°C)
Solids	
Aluminum	890
Concrete	670
Copper	390
Ice	2000
Iron and steel	460
Lead	130
Silver	230
Liquids	
Gasoline	2100
Mercury	140
Seawater	3900
Water (pure)	4180

Let us compute how much energy it takes to make a cup of coffee or tea. Eight ounces of water has a mass of about 0.22 kilograms. How much heat must be added to the water to raise its temperature from 20° C to the boiling point, 100° C? The change in temperature is

EXAMPLE 5.2

$$\Delta T = 100 - 20 = 80° C$$

The specific heat capacity, *C,* of water is 4,180 (from Table 5.3). Therefore:

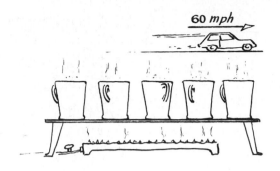

FIGURE 5.31
The energy needed to heat 5 cups of water to boiling is about the same as the energy needed to accelerate a small car to 60 mph.

$$Q = Cm\,\Delta T$$
$$= 4{,}180\ \text{J/kg-}°C \times 0.22\ \text{kg} \times 80°C$$
$$Q = 73{,}600\ \text{J}$$

It takes an enormous amount of energy to heat water. You may recall that in Example 3.6 we computed the kinetic energy of a small car traveling at highway speed. The answer was 364,500 joules. This much energy is only enough to bring about 5 cups of water to the boiling point from 20°C (Figure 5.31). Usually a relatively large amount of mechanical energy does not produce a large temperature change when it is converted into heat. Example 5.3 shows this another way.

EXAMPLE 5.3   A 5-kilogram concrete block falls to the ground from a height of 10 meters. If all of its original potential energy goes to heat the block when it hits the ground, what is its change in temperature?

There are two energy conversions. The block's gravitational potential energy, *PE = mgd*, is converted into kinetic energy as it falls. When it hits the ground, the kinetic energy is converted into heat in the inelastic collision (Figure 5.32). Actually this heat would be shared between the block and the ground, but we

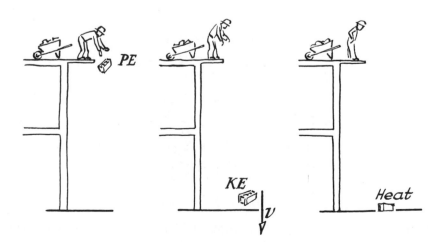

FIGURE 5.32
A concrete block falls 10 meters to the ground. If all of the block's energy is converted into heat that is absorbed by the block only, its temperature is raised 0.15° C.

assume that it all goes to the block. So the heat transferred to the block equals its original potential energy.

$$Q = PE = mgd = 5 \text{ kg} \times 9.8 \text{ m/s}^2 \times 10 \text{ m}$$

$$Q = 490 \text{ J}$$

The increase in temperature, ΔT, of the block is:

$$Q = Cm \, \Delta T$$

$$490 \text{ J} = 670 \text{ J/kg-}°\text{C} \times 5 \text{ kg} \times \Delta T$$

$$490 \text{ J} = 3,350 \text{ J/}°\text{C} \times \Delta T$$

$$\frac{490 \text{ J}}{3,350 \text{ J/}°\text{C}} = \Delta T$$

$$\Delta T = 0.15°\text{C}$$

At the end of Section 3.4 we discussed how mechanical energy is often converted into heat. At that time you may have wondered why you usually don't notice this heat when you slide something across a floor or drop a book on a table. It is simply because the quantities of energy or work that we typically deal with do not go far in changing the temperatures of the objects involved.

Let's consider one last example in which the amount of mechanical energy is so large that considerable heating does take place.

A satellite in low earth orbit experiences slight air resistance and eventually reenters the earth's atmosphere. As it moves downward through the increasingly dense air, the frictional force of air resistance converts the satellite's kinetic energy into heat. If the satellite is mostly aluminum and all of its kinetic energy is converted into heat, what would be its temperature increase?

EXAMPLE 5.4

Again, some of the heat goes to the air and some to the satellite. Just to get some idea of the potential heating, we assume that all of the heat from the friction flows into the satellite. We do not need to know the mass of the satellite, because m divides out, as we will see.

The heat transferred to the satellite equals its original kinetic energy. In Section 2.9 we computed the speed of a satellite in low earth orbit—7,900 meters/second. So:

$$Q = KE = \frac{1}{2} mv^2 = \frac{1}{2} \times m \times (7,900 \text{ m/s})^2$$

$$Q = (31,200,000 \text{ J/kg}) \times m$$

The increase in temperature caused by this much heat is:

$$Q = (31,200,000 \text{ J/kg}) \times m = Cm \, \Delta T$$

$$31,200,000 \text{ J/kg} = 890 \text{ J/kg-}°\text{C} \times \Delta T$$

$$\frac{31,200,000 \text{ J/kg}}{890 \text{ J/kg-}°\text{C}} = \Delta T$$

$$\Delta T = 35,000°\text{C}$$

Of course its temperature would not actually increase this much: the satellite would start to melt. But even if 90% of the heat went to the air, the remaining 10% would still be enough to melt the satellite. (This would make $\Delta T = 3{,}500°$ C.)

Refer back to Physics Potpourri, "Meteoritics," in Section 3.4.

The purpose of Example 5.4 was to show you why satellites and meteoroids usually disintegrate when they enter the earth's atmosphere. The friction from the air transforms their kinetic energy into enough heat to raise their surface temperatures by several thousand degrees. The objects simply start melting on the outside. Meteors can be seen at night because of their extreme temperatures: they leave behind a trail of hot, glowing air and meteoroid fragments.

Before the middle of the 19th century, the concepts of heat and mechanical energy were not connected to each other. It was thought that heat dealt with changes in temperature only and had nothing directly to do with mechanical energy. (The explanations of heating caused by friction were a bit strange as a result.) Water once again was used as a basis for defining a unit of measure: the calorie was defined to be the amount of heat needed to raise the temperature of 1 gram of water 1°C. Similarly, the British thermal unit, Btu, was defined to be the amount of heat needed to raise the temperature of 1 pound of water 1°F. This made the specific heat capacity of water equal to 1 cal/g-°C = 1 Btu/lb-°F.

In 1843, James Joule announced the results of his experiments in which a measurable amount of mechanical energy was used to raise the temperature of water by stirring it. The amount of heat given to the water equaled the mechanical energy expended. The result allowed Joule to calculate what was called the "mechanical equivalent of heat"—the relationship between the unit of energy and the unit of heat:

$$1 \text{ cal} = 4{,}186 \text{ J}$$

It was only later that the metric unit of energy was renamed in honor of Joule.

The "food calorie" that dieters count is actually the kilocalorie. So:

1 food calorie = 1,000 calories

It is used as a measure of the energy content of foods.

You may have noticed that the specific heat capacity of water is quite high, nearly twice as high as that of anything else in Table 5.3. This ability to absorb (or release) large amounts of heat is another property of water that adds to its uniqueness—and its usefulness. Water is used as a coolant in automobile engines, power plants, and countless industrial processes partly because it is plentiful and partly because its specific heat capacity is so high. Engine parts that are near where the fuel burns are exposed to very high temperatures. The metal would be damaged, or even melt, if there were no way of cooling the parts. Water is circulated to these areas of the engine, where it absorbs heat from the hotter metal, thereby cooling the parts. The heated water flows to the radiator, where it gives up the heat to the air.

5.6 PHASE TRANSITIONS

A phase transition or "change of state" occurs when a substance changes from one phase of matter to another. Table 5.4 lists the common phase transitions, the

TABLE 5.4 COMMON PHASE TRANSITIONS		
Name	Phases Involved	Effect on U
Boiling	Liquid to gas	Increases U
Melting	Solid to liquid	Increases U
Condensation	Gas to liquid	Decreases U
Freezing	Liquid to solid	Decreases U

FIGURE 5.33
The temperature of boiling water stays constant even though heat is flowing into it.

phases involved, and the effect of the transition on the internal energy of the substance.[4]

Let's say that a pan of water is placed on a stove and that heat is added to the water at some rate. This causes the temperature to rise steadily until the water starts to boil. Then the temperature stays the same ($100°$ C $= 212°$ F at sea level) even though heat is being added to the water. The heat is no longer increasing the kinetic energy of the water molecules: it is breaking the "bonds" that hold the molecules in the liquid state. The molecules are given enough energy to break free from the water's surface and become free molecules of steam (Figure 5.33).

Below the boiling temperature, the water molecules are bound to each other and so have negative potential energies. During boiling, each molecule in turn is given enough energy to break free of the bonds. The kinetic energies of the molecules stay the same during boiling and therefore the temperature of the water stays constant.

A similar process occurs when a solid melts. The atoms or molecules go from being rigidly bound to each other in the solid state to being rather loosely bound in the liquid state. As in boiling, the increase in internal energy that occurs during melting goes to increase *only* the potential energies of the atoms or molecules. The temperature of ice remains at $32°$ F ($0°$ C) while it is melting.

Condensation and freezing are simply the reverse processes of boiling and melting, respectively. Here the potential energies of the atoms and molecules decrease, while the kinetic energies and the temperature stay the same.

The temperature at which a particular phase transition occurs depends on the properties of the atoms or molecules in the substance—particularly the masses of the particles and the forces acting between them. When these forces are very strong, such as in salt, the melting and boiling temperatures are quite high. When the forces are very weak, such as in helium, the phase-transition temperatures are very low—near absolute zero.

The boiling temperature of each liquid varies with the pressure of the air (or other gas) that acts on its surface. When the pressure is 1 atmosphere, water boils at $100°$ C. At an elevation 3,000 meters (10,000 feet) above sea level, where the pressure is about 0.67 atmospheres, the boiling point of water is reduced to $90°$ C. That is why it takes longer to cook food by boiling at higher elevations. The temperature of the boiling water is lower, and so conduction of heat into the food is slower. If the pressure is *increased* to 2 atmospheres such as in a pressure cooker, the boiling point of water is $120°$ C (Figure 5.34). One type of nuclear

FIGURE 5.34
Higher pressure inside the pressure cooker makes the boiling point of water about $120°C$.

---

[4]Sublimation is a fairly rare phase transition in which a solid goes directly to a gas, bypassing the liquid phase. The atoms or molecules break free from the rigid forces, binding them in a crystal, and move off in the gas phase. Dry ice (solid carbon dioxide) undergoes sublimation at temperatures above $-78.5°$ C under 1 atmosphere of pressure. This means that carbon dioxide cannot exist in the liquid phase under normal pressure. Mothballs also undergo sublimation at room temperature.

power plant, called a *pressurized water reactor,* uses high pressure to keep water from boiling even at several hundred degrees Celsius.

This dependence of the boiling temperature on the pressure makes it possible to induce boiling or condensation simply by changing the pressure. For example, water exists as a liquid at 110° C when under 2 atmospheres of pressure. If the pressure is reduced to 1 atmosphere, the boiling temperature is lowered to 100° C, and the water begins to boil. In the same manner, steam at 110° C and 1 atmosphere pressure will start to condense if the pressure is increased to 2 atmospheres. As we will see in Section 5.7, such pressure-induced phase transitions are the basis for the operation of refrigerators and other "heat movers."

A specific amount of heat must be added or removed from a particular substance to complete a phase transition. For example, 334,000 joules of heat must be added to each kilogram of ice at 0° C to melt it. This quantity is called the *latent heat of fusion* of water. A much larger amount, 2,260,000 joules, must be added to each kilogram of water at 100° C to convert it completely into steam. This is the *latent heat of vaporization* of water. During the reverse processes, freezing and condensation, the same amounts of heat must be extracted from the water and steam, respectively.

EXAMPLE 5.5   Ice at 0° C is used to cool water from room temperature (20° C) to 0° C. How much water can be cooled by 1 kilogram of ice?

The ice absorbs heat as it melts, cooling the water in the process. The maximum amount of water it can cool to 0° C corresponds to all of the ice melting. So the question is, how much water will be cooled by 20° C when 334,000 joules of heat are extracted from it?

$$Q = Cm\,\Delta T$$

$$-334{,}000\,\text{J} = 4{,}180\,\text{J/kg·°C} \times m \times -20°\text{C}$$

$$-334{,}000\,\text{J} = -83{,}600\,\text{J/kg} \times m$$

$$\frac{-334{,}000\,\text{J}}{-83{,}600\,\text{J/kg}} = m$$

$$m = 4.0\,\text{kg}$$

Ice at 0° C can cool about four times its own mass of water from 20° C to 0° C—provided this takes place in a well-insulated container to minimize heat conduction into or out of the water.

These large latent heats of ice and water have common applications. The reason that ice is so good at keeping drinks cold is that it absorbs a large amount of heat while melting. As long as there is ice in the drink, the temperature remains near 0°C. One reason water is ideal for extinguishing fires is that it absorbs a huge amount of heat when it vaporizes. When poured on a hot, burning substance, the water absorbs heat as it boils, thereby cooling the substance (Figure 5.35).

Let's summarize the relationship between the temperature of H_2O, its phase, and its internal energy. A block of ice at a temperature of $-25°$ C is placed in a special chamber. The pressure in the chamber is kept at 1 atmosphere while heat is added to the ice at a fixed rate. Figure 5.36 shows a graph of the temperature of the H_2O versus the amount of heat added to it.

FIGURE 5.35
Exploiting water's high heat of vaporization.

Between points *a* and *b* on the graph, the heat added to the ice simply raises its temperature. At point *b,* the ice starts to melt, and the temperature stays fixed at 0°C until all of the ice is melted, point *c.* From *c* to *d* the heat that is added to the water goes to increase its temperature. Between *d* and *e* the water boils while the temperature remains at 100° C. As soon as all of the water is converted into steam, at *e,* the temperature of the steam starts to rise.

We could reverse the process by placing steam in the chamber and removing heat from it. The result would be like moving along the graph from right to left.

At temperatures below their boiling points, liquids can gradually go into the gas phase through a process known as *evaporation.* Water left standing will eventually "disappear" because of this. How can this phase transition occur at temperatures below the boiling point? Individual atoms or molecules in a liquid can go into the gas phase if they have enough energy. At temperatures below the boiling point, some of the atoms or molecules do have enough energy to do this. Even though the *average* energy of the particles is too low, some have more energy than the average, and some have less. Atoms or molecules with higher-than-average energy can break free from the liquid if they are near the surface.

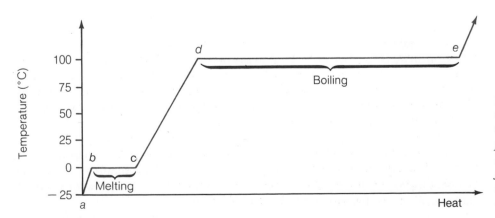

FIGURE 5.36
Graph of the temperature of H_2O versus the heat added to it. The phase transitions correspond to the two places where the graph is flat: the internal energy increases, but the temperature stays constant.

FIGURE 5.37
Fog in the cooler air in a valley. Water droplets form when the humidity exceeds the saturation density—when there is too much water vapor in the air.

Once in the air, the atoms or molecules can remain in the gas phase even though the temperature is below the boiling point.

Because of evaporation, water vapor is always present in the air. The amount varies with geographic location (proximity to large bodies of water), climate, and weather. *Humidity* is a measure of the amount of water vapor in the air.

> HUMIDITY   The mass of water vapor in the air per unit volume. The density of water vapor in the air.

The unit of humidity is the same as that of mass density. The humidity is generally in the range from about 0.001 kilograms per cubic meter (cold day in a dry climate) to about 0.05 kilograms per cubic meter (hot, humid day). Note that these densities are much less than the normal density of the air, 1.29 kilograms per cubic meter. Even in humid conditions, water vapor is only a small component of the air—less than 5%.

At any given temperature there is a maximum possible humidity, called the *saturation density.* This upper limit exists because the water molecules in the air "want" to be in the liquid phase. If there are too many molecules in the air, the probability is high that several will get close enough to each other for the attractive force to take over. They begin to form droplets and condense onto surfaces. This is what happens during the formation of fog, dew, or the mist in a shower (Figure 5.37). The saturation density is *higher* at higher temperatures because the water molecules are moving faster and are less likely to "stick" together when they collide. Hence, more molecules can be present in any volume of air without droplet formation. Table 5.5 lists the saturation density of water vapor in the air at several different temperatures.

TABLE 5.5   SATURATION DENSITIES OF WATER VAPOR IN THE AIR

Temperature		Saturation Density
(°C)	(°F)	(kg/m³)
−15	5	0.0016
−10	14	0.0022
−5	23	0.0034
0	32	0.0049
5	41	0.0068
10	50	0.0094
15	59	0.0128
20	68	0.0173
25	77	0.0228
30	86	0.0304
35	95	0.0396
40	104	0.0511

The reverse of evaporation also takes place: water molecules in the air near the surface of water can be deflected into the water and "captured." When the humidity is well below the saturation density at that temperature, evaporation occurs faster than reabsorption of water molecules. If the humidity increases to the saturation density, water molecules are reabsorbed into the water at the same rate that they evaporate. The water does not disappear. This is why damp towels do not dry well in humid environments.

The upshot of this is that it is not so much the humidity that is important as how close the humidity is to the saturation density. For this reason, *relative humidity* is a more useful quantity.

> RELATIVE HUMIDITY   The humidity expressed as a percentage of the saturation density.
>
> $$\text{Relative humidity} = \frac{\text{humidity}}{\text{saturation density}} \times 100\%$$

When the relative humidity is 40%, the air is "holding" 40% of the maximum amount of water vapor that it can hold.

What is the relative humidity when the humidity is 0.009 kilograms per cubic meter, and the temperature is 20°C?   **EXAMPLE 5.6**

From Table 5.5, the saturation density at 20°C is 0.0173 kilograms per cubic meter. Therefore:

$$\text{Relative humidity} = \frac{0.009 \text{ kg/m}^3}{0.0173 \text{ kg/m}^3} \times 100\%$$

$$\text{Relative humidity} = 52\%$$

The same humidity in air at 15°C would make the relative humidity 70%.

FIGURE 5.38
Water vapor in the cooled air near the glass condenses into drops on the surface.

When *air is cooled* and the water vapor content stays constant, the *relative humidity increases.* When *air is heated* and the humidity stays constant, the *relative humidity decreases.* This is why heated buildings often feel dry in the winter. Cold air from the outside enters the building and is heated. Unless water vapor is artificially added to this air with a humidifier, the relative humidity will be very low.

When the humidity is fairly high, water droplets form on the surfaces of cold objects such as a glass of ice water (see Figure 5.38). The air near the cold surface is cooled to a temperature at which the humidity exceeds the saturation density. The water vapor in the cooled layer of air near the surface condenses onto the surface. This process also occurs when car windows "fog over" in cold weather.

The process of evaporation cools a liquid. Atoms or molecules in the gas phase have more energy than when in the liquid phase. When water molecules evaporate, they take away some internal energy from the water, thereby cooling it. Here is another way of looking at it: since the molecules with the higher kinetic energies are the ones that evaporate, the average kinetic energy of the water molecules that remain is lowered.

Our bodies are cooled by the evaporation of perspiration. On hot days we feel hotter if the humidity is high because evaporation is inhibited. A breeze feels cool because it removes air near our skin that has a higher humidity due to perspiration. The drier air that replaces it allows for more rapid evaporation—and cooling.

5.7   HEAT ENGINES AND THE SECOND LAW OF THERMODYNAMICS

Along with the efforts made in the recent past to conserve energy used to heat and to cool buildings, much has been done to improve the efficiency of other ways that we use energy. In this section we will consider some of the basic theoretical principles that are involved in energy-conversion devices that use heat and mechanical energy.

Most of the energy used in our society comes from fossil fuels—coal, oil, and natural gas. Part of these fuels are burned directly for heating, such as in gas stoves and oil furnaces. But most of these fuels are used as the energy input for devices that are classified together as *heat engines.*

> **HEAT ENGINE**   A device that converts heat energy into mechanical energy or work. It absorbs heat from a hot source such as burning fuel, converts some of this energy into usable mechanical energy or work, and outputs the remaining energy as heat at some lower temperature.

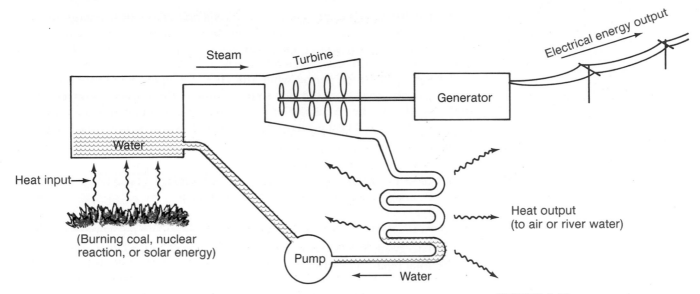

FIGURE 5.39
Simplified sketch of an electric power plant showing the energy (heat) inputs and ouputs.

Gasoline engines, diesel engines, jet engines, and steam-electric power plants are all heat engines. In gasoline and diesel engines, some of the heat from burning fuel is converted into mechanical energy. The remainder of the heat is ejected to the air from the exhaust pipe, the radiator, and the hot surfaces of the engine. Coal, nuclear, and the Solar One power plants produce electricity by using steam to turn a generator (see Figure 5.39). Heat from burning coal, fissioning nuclear fuel, or the sun is used to boil water. The steam is piped to a turbine (basically a propeller) that is given rotational energy by the steam. A generator connected to the turbine converts rotational energy into electrical energy. After the steam leaves the turbine, it is condensed back into the liquid phase by being cooled with river water or the air. Cooling towers, which function somewhat like automobile radiators, are used in the latter case. Most of the heat given to the water to boil it is ejected from the plant as waste heat when the steam is condensed.

Even though the actual inner workings of heat engines are quite complicated, from an energy point of view we can represent them with a simple schematic diagram (see Figure 5.40). Energy in the form of heat from some source is input

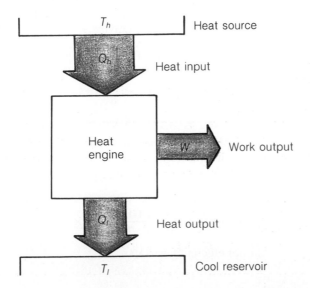

FIGURE 5.40
Diagram of a heat engine. The mechanism absorbs heat from the heat source, uses some of it to do work, and releases heat at some lower temperature.

into a mechanism. The mechanism converts some of the heat energy into mechanical energy and rejects the remainder. In this simplified representation of a heat engine, the heat source has some fixed (high) temperature, T_h, and the reservoir that absorbs the rejected heat has some fixed (lower) temperature, T_l.

During a given time period while the engine is operating, some quantity of heat, Q_h, is absorbed from the heat source. Some of this energy is converted into usable work, and the remainder of the original energy input is ejected as waste heat, Q_l. This wasted heat is unavoidable. The following law is a formal statement of this.

> **SECOND LAW OF THERMODYNAMICS**   No device can be built that will repeatedly extract heat from a source and deliver mechanical work or energy without ejecting some heat to a lower-temperature reservoir.

The energy efficiency of any device or process is just the usable output divided by the total input, times 100%.

$$\text{Efficiency} = \frac{\text{energy or work output}}{\text{energy or work input}} \times 100\%$$

If the efficiency of a device is 25%, then one fourth of the input energy is converted into usable form. The remaining three fourths is "lost"—released as waste heat.

For heat engines, the input energy is Q_h, and the output is the work. Therefore:

$$\text{Efficiency} = \frac{\text{Work}}{Q_h} \times 100\% \qquad \text{(heat engine)}$$

As we have seen, there are many different types of heat engines. Some of them use processes that are inherently more efficient than others. However, there is a theoretical upper limit on the efficiency of a heat engine. This maximum efficiency is called the *Carnot efficiency* after the French engineer Sadi Carnot. Carnot discovered that the efficiency of a perfect heat engine is limited by the temperatures of the heat source and of the low-temperature reservoir. In particular:

$$\text{Carnot efficiency} = \frac{T_h - T_l}{T_h} \times 100\%$$

(T_h and T_l must be in Kelvins.)

Real heat engines have friction, imperfect insulation, and other factors that reduce their efficiencies. But even if these were completely eliminated, a heat engine could not have an efficiency of 100%.

EXAMPLE 5.7   A typical coal-fired power plant uses steam at a temperature of 1,000° F (810 K). The steam leaves the turbine at a temperature of about 212° F (373 K). What is the theoretical maximum efficiency of the power plant? Here, T_h = 810 K and T_l = 373 K. So:

$$\text{Carnot efficiency} = \frac{810 - 373}{810} \times 100\%$$

$$\text{Carnot efficiency} = 54\%$$

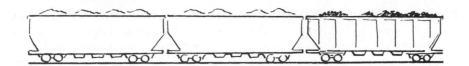

FIGURE 5.41
For every three railroad cars of coal burned at an electric power plant, the energy contained in about two of them goes to waste heat. This is a consequence of the second law of thermodynamics, as well as an imperfect heat engine.

This is the ideal efficiency. The actual efficiency of a coal power plant is usually around 35% to 40%.

Nearly two thirds of the energy input to a power planet is lost as waste heat (Figure 5.41). Not only is the energy not used, it causes thermal pollution by artificially heating the environment. The actual top efficiencies of the other heat engines in common use are even lower: diesel engine, 35%; jet engine, 23%; gasoline (piston) engine, 25%; Wankel (rotary) engine, 17%.

There are two general ways to try to improve the efficiency of a heat engine. One is to improve the process and reduce energy losses so that the efficiency gets closer to the Carnot efficiency. The other way is to increase the Carnot efficiency by raising T_h or lowering T_l. The efficiencies of steam-based heat engines have been improved a great deal through the use of higher-temperature steam.

Heat Movers

Refrigerators, air conditioners, and heat pumps are devices that act much like heat engines in reverse. They use an input of energy to cause heat to flow from a cooler substance to a warmer substance—opposite the natural flow from hot to cold. The purpose of refrigerators and air conditioners is to remove heat from an area and thereby cool it. Heat pumps are designed to heat an interior space in the winter as well as cool it in the summer. When heat energy is removed from one location by these devices, a larger amount of heat energy is ejected at a different location. A refrigerator removes heat from its interior and ejects heat into the room. We will call these devices *heat movers,* since the term describes what they do.

The mechanisms in the three devices are much the same. A gas, called the *refrigerant,* is forced to change phases in a cyclic process. The gas must have a fairly low boiling point and be condensed easily when the pressure on it is increased. Freon, CCl_2F_2 is the most commonly used refrigerant. The gas is compressed into the liquid phase by a pump and is forced to flow through a small opening called an *expansion valve* (Figure 5.42). The pressure on the other side of the valve is kept low so that the refrigerant instantly goes into the gas phase. This phase transition cools the refrigerant, and in the process it absorbs heat from the surroundings. The gas flows back to the pump, where it is recompressed into the liquid phase. During this phase transition it releases heat. Once the refrigerant is liquified, it flows through the expansion valve, and the cycle repeats. Energy must be supplied to the pump to make the process work.

Heat movers can be represented with a diagram similar to that of heat engines (Figure 5.43). Unlike heat engines, the mechanism in heat movers has an *input* of energy or work. During a given period of time, a quantity of heat, Q_l is absorbed from a cool reservoir with temperature T_l. An amount of work is done on the refrigerant by the pump, and an amount of heat Q_h is released to a warm reservoir

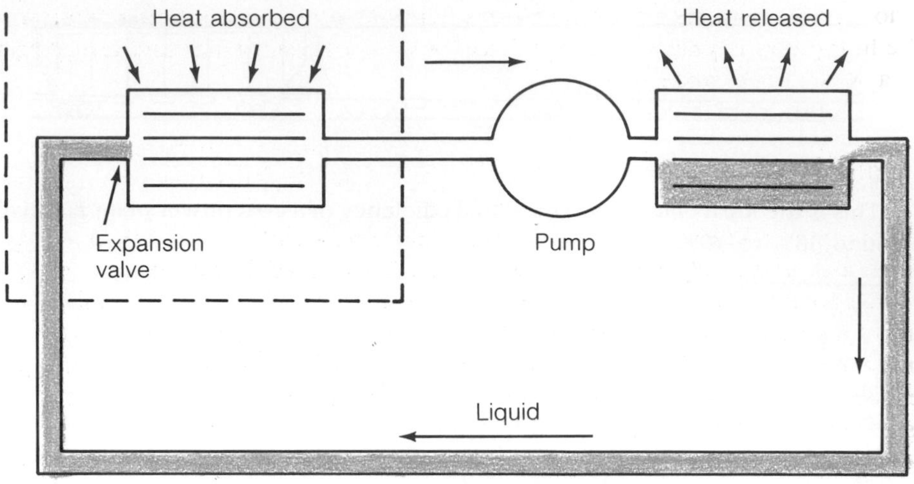

FIGURE 5.42
Refrigerators remove heat from their interiors by exploiting the phase transition of the refrigerant. As the refrigerant vaporizes, it absorbs heat from the refrigerator's interior. As it condenses, it releases heat to the air in the room.

with temperature T_h. In refrigerators the cool reservoir is the air in its interior. The warm reservoir is the air in the room. Air conditioners absorb heat from the air that they circulate inside a building, an automobile, and so on, and release heat to the warmer outside air. When in the "heating mode," heat pumps remove heat from the outside air, ground, or groundwater and release heat inside a building. In the "cooling mode" heat pumps reverse the heat flow: heat is removed from the inside of the building and is ejected to the outside.

The law of conservation of energy tells us that the amount of heat released, Q_h, is equal to the amount of heat absorbed, Q_l, plus the energy input to the mechanism. The relative values of Q_h, Q_l, and the energy input depend on the efficiency of the pumping mechanism, the difference between the two temperatures, T_h and T_l, and other factors. In the average refrigerator the amount of heat removed from the interior is about *three times* the amount of energy input. In the heating mode, the amount of heat delivered by a good heat pump is about *three times* the

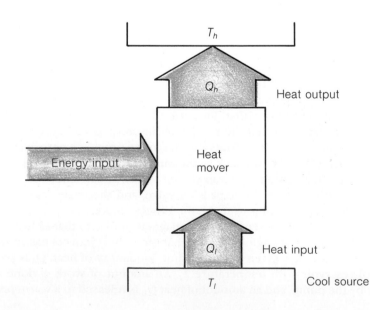

FIGURE 5.43
Diagram of a heat mover. A mechanism uses an energy input to extract heat from a cool source and to release heat at a higher temperature.

amount of energy consumed by the mechanism. Energy is not being "created": the heat pump is simply removing heat from a cooler substance and giving heat to a warmer substance.

Usable Energy

The output of a heat engine is low-temperature heat and useful work or energy. The latter eventually becomes heat as well because of friction and other processes. So the overall effect of a heat engine is to convert higher-temperature heat into lower-temperature heat. This heat, in turn, is no longer available to do as much useful work: Carnot tells us that lower temperatures would give us lower efficiencies. Even though energy is not "lost" or "destroyed" in a heat engine, it is *made unavailable* for similar use again.

This is a general result of energy transformations: the energy that remains is less usable or less organized than the original energy. Electricity is "high-quality" energy: it is very organized and can be used in many different ways. Light bulbs, heaters, and blow dryers convert electrical energy into lower-quality energy—heat (Figure 5.44). (Even the light from a light bulb eventually becomes heat.) There are many other examples of this. Even the high-quality energy in fine Swiss chocolate becomes low-quality heat after we eat it and exercise.

Observations like these lead us to the following conclusion: *natural processes tend to increase the disorder of systems.*

This tendency is not limited to energy. When you let the air out of a tire, you reduce the order of the system: low-pressure air is less useful and more disordered than high-pressure air. Refined and manufactured metal products, like cars, become less ordered as they rust and fall apart. A fallen tree in a forest decays into a more disordered state. This list goes on and on.

The physical quantity *entropy* is a measure of the *amount of disorder* in a system. Natural processes tend to *increase* the entropy of a system because they

FIGURE 5.44
Energy transformation that leaves energy in a less usable form.

increase the disorder. This is another way of stating the second law of thermodynamics.

The entropy of part of a system can be decreased temporarily at the expense of the rest of the system. As living things grow, they become more highly ordered, and their entropy decreases. But in so doing they increase the entropy of the food they consume. After they die, the entropy of the remains once again increases. As a car battery is charged, its entropy decreases, but the charging system is consuming energy and increasing entropy.

The term "energy shortage" is somewhat misleading. There is as much energy around now as ever before. But our society uses up high-quality, conveniently stored energy—fossil fuels, uranium, forests—and leaves low-quality heat. We are increasing the entropy of the world, leaving less *usable* energy available for future generations. By tapping renewable energy sources—solar, wind, geothermal, and tidal—we use continuous supplies of energy that would otherwise go to waste. Solar energy eventually becomes high-entropy heat, whether it is absorbed by the ground or by solar cells. If absorbed by solar cells, we get low-entropy electricity as an intermediate step.

> *The energies of our system will decay, the glory of the sun will be dimmed, and the earth, tideless and inert, will no longer tolerate the race which has for a moment disturbed its solitude.*
>
> Arthur James Balfour (1848–1930)

Temperature

The first primitive thermometer was invented by none other than Galileo in 1592. It used the expansion and contraction of air with temperature. A glass bulb equipped with a long, thin neck was partially filled with water and then inverted into a container of water (Figure 5.45). The water level would move up and down in the neck as changes in temperature caused the air in the bulb to expand and contract. The device was also affected by changes in atmospheric pressure, so it was a combination thermometer and barometer. Galileo did not equip his device with a temperature scale.

In 1657 a number of disciples of Galileo formed an "academy of experiment" in Florence and developed functional thermometers. These were based on the thermal expansion of alcohol in sealed tubes and were not affected by changes in atmospheric pressure. The academicians established temperature scales for their thermometers by choosing two fixed points and dividing the interval into a number of degrees.

The Florentine thermometers became quite popular in Europe and stimulated many scientists to attempt improvements. An astronomer in Paris named Ismaël Boulliau was the first to use mercury in a thermometer. Perhaps the most famous name in the history of the development of the modern thermometer is Gabriel

Daniel Fahrenheit (1686–1736), a manufacturer of meteorological instruments. Fahrenheit experimented extensively with both alcohol and mercury thermometers, as well as with fixed points and temperature scales. The scale that he eventually chose—our modern Fahrenheit scale—used as its fixed points the temperature of an ice-salt-water mixture and the temperature of the human body. Fahrenheit divided the interval into 96 steps and made the first fixed point 0 degrees. On this scale the temperature of melting ice was found to be 32° and that of boiling water (at normal atmospheric pressure) 212°. Fahrenheit made the important discovery that the boiling temperature of liquids varies with atmospheric pressure. This accounted for the confusion that other experiments encountered when trying to use the boiling point of a liquid as a fixed point on a temperature scale. Apparently there was some error in Fahrenheit's measurement of body temperature because the average is 98.6° F, not 96° F.

Several experimenters sought to establish a scale based on 100 degrees between fixed points—a "centigrade" scale. The Swedish astronomer Andreas Celsius (1701–1744) introduced a scale that used 100 degrees between the freezing and boiling points of water, but it read backwards: 100° was the freezing temperature and 0° was the boiling temperature. This scale was later inverted and is our modern Celsius scale. But no one temperature scale has ever quite gained universal acceptance. During the 18th century there were more than a dozen different scales in use.

The existence of an "absolute zero" was first indicated in experiments with air thermometers, like Galileo's. The volume of a fixed amount of a gas was known to decrease with temperature. One could simply graph the volume of the gas versus the temperature and extrapolate the straight line to the point where the volume would be zero (see Figure 5.46). The corresponding temperature would logically be the coldest possible temperature—absolute zero. It was known to exist even before it was possible to get within 200 degrees of it. This temperature is the zero point of the Kelvin scale, named after the British physicist Lord Kelvin, who first established an absolute temperature scale.

FIGURE 5.45
Galileo's "thermometer."

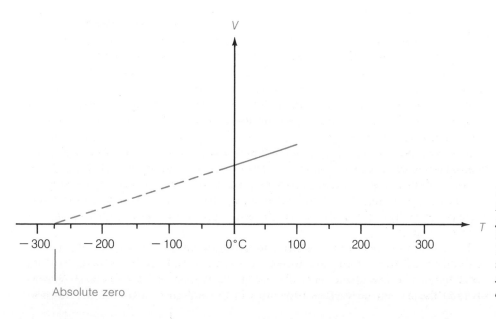

FIGURE 5.46
Lowering the temperature of a gas under constant pressure lowers the volume it occupies. The dotted line indicates that the volume would go to zero at some very low temperature—absolute zero.

FIGURE 5.47
*Benjamin Thompson, later
Count Rumford, at a cannon
factory.*

FIGURE 5.48
James Prescott Joule.

Heat

The development of the understanding of the basic nature of heat presents an interesting counterexample to the general progress of science. Originally Democritus and other Greeks thought that heat was some kind of material substance that flowed into objects when they were heated. This view was challenged by many leading scientists of the 17th century, including such familiar ones as Isaac Newton and Robert Boyle, who considered heat to involve some kind of internal motion. Although this was the correct approach, it was replaced in the 18th century by a reversion to the material model of heat. Joseph Black (1728–1799), a Scotch scientist, and Antoine Lavoisier were two of the strongest proponents of the *caloric theory* of heat. Heat supposedly was an invisible, massless fluid called "caloric." This model, though incorrect, was adequate for explaining many thermodynamic phenomena. Lavoisier measured the specific heat capacities of many substances, and Black successfully measured the quantity of heat needed to melt a given amount of ice. The science of thermodynamics progressed quite nicely during the next century, even though the basic concept of heat was wrong.

The Achilles' heel of the caloric theory was friction. How was heat, and therefore caloric, produced by friction? Among the first to dispute the caloric theory was Benjamin Thompson, later titled Count Rumford (Figure 5.47). Thompson (1753–1814), one of the first Americans to leave his mark in science, was born in Massachusetts and was employed there until he fled to Europe during the Revolutionary War. Thompson spent the rest of his life in England, Austria, Bavaria (where he was given his title), and finally France. He worked as a scientist, an industrialist, and in various capacities in the military. At one point Rumford supervised the construction of cannons at a factory in Munich. The cannon barrels were made by boring out the centers of solid iron cylinders. During this process friction between the drill bit and the cannon barrel produced heat. The caloric theory implied that the friction released caloric from the metal. Rumford performed many experiments and found that the amount of heat that could be produced when a dull bit was used was unlimited. How could the iron contain an infinite quantity of a material substance (caloric)? Rumford reasoned that the heat was produced by the action of the drill bit and that caloric did not exist.

Several experiments by the English brewer and amateur scientist James Prescott Joule (1818–1889) indicated that heat is a form of energy (Figure 5.48). His most famous experiment consisted of a paddle wheel that was made to churn water, thereby heating it, by the action of a falling weight. Joule measured the temperature change of the water and used its specific heat to relate the amount of work done by the falling weight to the heat gained by the water. The result is the famous "mechanical equivalent of heat" (Section 5.5). In other experiments Joule measured the heat produced by the compression of air and by electrical currents. This, along with the work of many other scientists, led to the first law of thermodynamics.

A great impetus to the science of thermodynamics was provided by attempts to improve the performance of steam engines. The SI unit of power is named after James Watt, a Scottish engineer who was responsible for several design changes. Nicolas Leonard Sadi Carnot (1796–1832) was a French engineer who introduced the theoretical analysis of heat engines (see Figure 5.49). Initially a follower of the caloric theory, Carnot was nonetheless able to determine what factors affect the efficiency of a heat engine. His analysis was later modified to include the fact that heat is associated with internal energy, not caloric.

The study of thermodynamics is at the forefront of physics even today. The 1982 Nobel Prize in physics was awarded to Kenneth G. Wilson for his theoretical work on phase transitions.

FIGURE 5.49
Sadi Carnot.

SUMMARY

Temperature is the basis of our sense of hot and cold. Physically the temperature of a substance is proportional to the average kinetic energy of its atoms and molecules. Three different temperature scales are in common use today: the Fahrenheit, Celsius, and Kelvin scales. The Kelvin temperature scale uses the lowest possible temperature, absolute zero, as its zero point.

In nearly all cases, matter expands when its temperature is raised. The amount of expansion depends on the substance and the temperature change. This property is exploited in mercury and alcohol thermometers and in bimetallic strips.

The internal energy of a substance is the total potential and kinetic energies of its atoms and molecules. It can be changed by doing work on the substance and by transferring heat to it. The first law of thermodynamics states that the change in internal energy of a substance equals the work done on it plus the heat added to it.

There are three ways that heat can be transferred from one place to another. Conduction is the transfer of heat via contact between atoms or molecules. Convection is the transfer of heat via the buoyant mixing in a fluid. Radiation is the transfer of heat via electromagnetic radiation. Often all three processes occur simultaneously.

Except during phase transitions, the temperature of matter increases whenever its internal energy increases. The specific heat capacity is a parameter of each substance that relates the mass and temperature increase to the heat added. During phase transitions the potential energies of the atoms and molecules change while the average kinetic energy remains constant. So the internal energy increases while the temperature stays constant. Evaporation occurs below the boiling temperature and is responsible for the water vapor present in the air.

Humidity and relative humidity are two measures of the water vapor content.

Heat engines are devices that use heat from a hot source to do work. They necessarily release heat at a cooler temperature in the process. The maximum efficiency of a theoretically "perfect" heat engine is determined by the two temperatures. Heat movers use an energy input to remove heat from a cool substance and to transfer it to a warmer substance.

Entropy is a measure of the disorder in a system. Heat engines do not "consume" energy, but they do increase entropy, thus reducing the energy's availability. Increasing the entropy (disorder) of the universe is a result of all natural processes.

SUMMARY OF IMPORTANT EQUATIONS

EQUATION	COMMENTS
$\Delta l = \alpha l\, \Delta T$	Thermal expansion of a rod
$\dfrac{pV}{T} = \text{a constant}$	Pressure, volume, and temperature of a fixed amount of a gas
$\Delta U = \text{Work} + Q$	First law of thermodynamics
$Q = Cm\, \Delta T$	Heat needed to raise the temperature
$\text{Relative humidity} = \dfrac{\text{humidity}}{\text{sat. den.}} \times 100\%$	Definition of relative humidity
$\text{Efficiency} = \dfrac{\text{energy or work output}}{\text{energy or work input}} \times 100\%$	
$\text{Efficiency} = \dfrac{T_h - T_l}{T_h} \times 100\%$	Carnot efficiency

QUESTIONS

1. What are the three common temperature scales? What are the normal boiling and freezing points of water in each scale?

2. What is the significance of absolute zero?

3. What happens to the atoms and molecules in a substance as its temperature increases?

4. In the most general terms, why do the moon and the planet Mercury not have atmospheres?

5. Explain what a bimetallic strip is and how it functions.

6. A certain engine part made of iron expands 1 millimeter in length as the engine warms up. What would be the approximate change in length if the part were made of aluminum instead of iron?

7. What is unusual about water below the temperature of 4° C?

8. What are the two general ways to increase the internal energy of a substance? Describe an example of each.

9. Air is allowed to escape from an inflated tire. Is the temperature of the escaping air higher than, lower than, or equal to the temperature of the air inside the tire? Why?

10. Is it possible to compress air without causing its internal energy to increase? If so, how?

11. Describe the three methods of heat transfer. Which of these are occurring around you at this moment?

12. A potato will cook faster if a large nail is inserted into it. Why?

13. A coin and a piece of glass are both heated to 60° C. Which will feel warmer when you touch it?

14. A submerged heater is used in an aquarium to keep the water above room temperature. Should it be placed near the surface of the water or near the bottom to be most effective?

15. On a cool night with no wind, people facing a camp fire feel a breeze on their backs. Why?

16. The temperature of the air in a one-room building that is not insulated is kept at exactly 22° C year round. A person inside the building feels cooler when it is cold outside than when it is warm outside. Why?

17. When heating water on a stove, does it take longer for a full pan of water to reach the boiling point than for a pan that is half full?

18. A piece of aluminum and a piece of iron fall without air resistance from the top of a building and stick into the ground on impact. Will their temperatures change by the same amount? Explain.

19. The specific heat capacity of water is extremely high. If it were much lower, say one fifth as large, what effect would this have on processes like fire fighting and cooling automobile engines?

20. Why does the temperature of water not change while it is boiling?

21. Describe how changing the air pressure affects the temperature at which water boils.

22. What is saturation density? How does it change when the temperature increases?

23. What effect does heating the air in a room have on the relative humidity?

24. Explain what a heat engine does. Repeat for a heat mover.

25. In the winter the amount of heat energy a heat pump delivers to a house is greater than the electrical energy it uses. Does this violate the law of conservation of energy? Explain.

PROBLEMS

1. Your jet is arriving in London, and the pilot informs you that the temperature is 30°C. Should you put on your jacket? Use Figure 5.2 to determine the temperature in degrees Fahrenheit.

2. On a nice winter day at the South Pole, the temperature rises to −60°F. What is the approximate temperature in degrees Celsius?

3. An iron railroad rail is 700 ft long when the temperature is 30°C. What is its length when the temperature is −10°C?

4. A copper vat is 10 m long at room temperature (20°C). How much longer is it when it contains boiling water under 1 atm pressure?

5. A machinist wishes to insert a steel rod with a diameter of 5 mm into a hole with a diameter of 4.997 mm. By how much would the machinist have to lower the temperature of the rod to make it fit the hole?

6. A brick oven 2 m long (at 20°C) is designed to allow for an expansion of 5 mm without damage. To what temperature would the oven have to be heated to expand that much?

7. A gas is compressed inside a cylinder (see Figure 5.15). An average force of 50 N acts to move the piston 0.1 m. During the compression, 2 J of heat is conducted away from the gas. What is the change in internal energy of the gas?

8. Air in a balloon does 50 J of work while absorbing 70 J of heat. What is its change in internal energy?

9. How much heat is needed to raise the temperature of 5 kg of silver from 20°C to 960°C?

10. A bottle containing 3 kg of water at a temperature of 20°C is placed in a refrigerator where the temperature is kept at 3°C. How much heat is removed from the water to cool it to 3°C?

11. Figure 5.36 shows a graph of the temperature of ice as heat is added to it.
    a) How much heat does each kilogram of ice absorb from point a to point b?
    b) How much heat does each kilogram of water absorb from point c to point d?

12. Aluminum is melted during the recycling process.
    a) How much heat must be given to each kilogram of aluminum to bring it to its melting point, 660°C, from room temperature, 20°C?
    b) About how many cups of coffee could you make with this much heat (see Example 5.2)?

13. A 1,200-kg car going 25 m/s is brought to a stop using its brakes. Let's assume that a total of approximately 20 kg of iron in the brakes and wheels absorb the heat produced by the friction.
    a) What was the car's original kinetic energy?
    b) After the car has stopped, what is the change in temperature of the brakes and wheels?

14. A 0.02-kg lead bullet traveling 200 m/s strikes an armor plate and comes to a stop. If all of its energy is converted to heat that it absorbs, what is its temperature change?

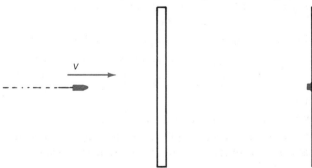

15. A 10-kg lead brick is dropped from the top of the World Trade Center towers and falls 1,350 ft (412 m) to the ground. Assuming all of its energy goes to heat it, what is its temperature increase?

16. Water flowing over the Lower Falls in Yellowstone National Park drops 94 m. If all of the water's energy goes to heat it, what is its temperature increase?

17. On a winter day, the air temperature is −15°C, and the humidity is 0.001 kg/m³.
  a) What is the relative humidity?
  b) When this air is brought inside a building, it is heated to 20°C. If the humidity isn't changed, what is the relative humidity inside the building?

18. On a summer day in Houston, the temperature is 35°C and the relative humidity is 77%.
  a) What is the humidity?
  b) To what temperature could the air be cooled before condensation would start taking place?

19. Inside a building, the temperature is 20°C, and the relative humidity is 40%. How much water vapor is in each cubic meter of air?

20. On a hot summer day in Washington, D.C., the temperature is 86°F, and the relative humidity is 70%. How much water vapor does each cubic meter of air contain?

21. An apartment has dimensions 10 m by 5 m by 3 m. The temperature is 25° C and, the relative humidity is 60%. What is the total mass of water vapor in the air in the apartment?

22. The total volume of a new house is 800 m³. Before the heat is turned on, the air temperature inside is 10° C, and the relative humidity is 50%. After the air is warmed to 20° C, how much water vapor must be added to the air to make the relative humidity 50%?

23. What is the Carnot efficiency of a heat engine operating between the temperatures of 300° C (573 K) and 100° C (373 K)?

24. What is the maximum efficiency that a heat engine could have when operating between the normal; boiling and freezing temperatures of water?

25. As a gasoline engine is running, an amount of gasoline containing 15,000 J of chemical potential energy is burned in 1 s. During that second, the engine does 3,000 J of work.
  a) What is the engine's efficiency?
  b) The burning gasoline has a temperature of about 4,000°F (2,500 K). The waste heat from the engine is released into the air at about 80°F (300 K). What is the Carnot efficiency of a heat engine operating between these two temperatures?

26. A proposed ocean thermal energy conversion (OTEC) system is a heat engine that would operate between warm water (25° C) at the ocean's surface and cooler water (5° C) 1,000 m below the surface. What is the maximum possible efficiency of the system?

CHALLENGES

1. A solid cube is completely submerged in a particular liquid and floats at a constant level. When the temperature of the liquid and the solid is raised, the liquid expands more than the solid. Will this make the solid rise upward, stay floating, or sink? Explain why.

2. Pyrex glassware is noted for its ability to withstand sudden temperature changes without breaking. Explain how its coefficient of thermal expansion contributes to this ability.

3. Is it possible to add heat to air without changing its temperature? Explain.

4. A heater is placed in water at 1°C.
  a) Sketch the convection circulation that would be produced in the water.
  b) Where in the water should the heater be placed? Should it be moved at some later point in time?

5. As air rises in the atmosphere, its temperature drops, even if no heat flows out of it.
  a) Based on what you learned in Sections 4.4 and 5.3, explain why this is so.
  b) Cumulus clouds form when rising air is cooled to the point where water droplets form because of condensation. Why are these clouds usually much higher above the ground in dry climates than in wet ones?

6. If air at 35°C and 77% relative humidity is cooled to 25°C, what mass of water would condense out of the air in a room that measures 5 by 4 by 3 m? (See Problem 18.)

7. The door on a refrigerator is left open. Assuming the refrigerator is in a closed room, will the air in the room eventually be cooled?

SUGGESTED READINGS

Cajori, Florian. *A History of Physics.* New York: Dover, 1962 (originally published by Macmillan in 1929). A good general source, it includes an account of the development of the Fahrenheit scale.

Fowler, John M. *Energy and the Environment* (2d ed). New York: McGraw-Hill, 1984. An analysis of the production and usage of energy in our society, including a chapter on solar energy.

Hamakawa, Yoshihiro. "Photovoltaic Power." *Scientific American* 256, no. 4(April 1987):87–92. Describes the current state of solar cell technology and the prospects for large, economically viable power plants.

Hunter, Lloyd, "The Art and Physics of Soaring." *Physics Today* 37, no. 4(April 1984):34–41. Includes a detailed look at soaring in thermals.

Segrè, Emilio. *From Falling Bodies To Radio Waves*. New York: W.H. Freeman and Co, 1984. Among other things, this book contains an account of the colorful life of Count Rumford.

Walker, Jearl. *The Flying Circus of Physics with Answers*. New York: Wiley, 1977. "a collection of problems and questions about physics in the real, everyday world." Chapter 3 addresses dozens of questions about heat and temperature.

OUTLINE

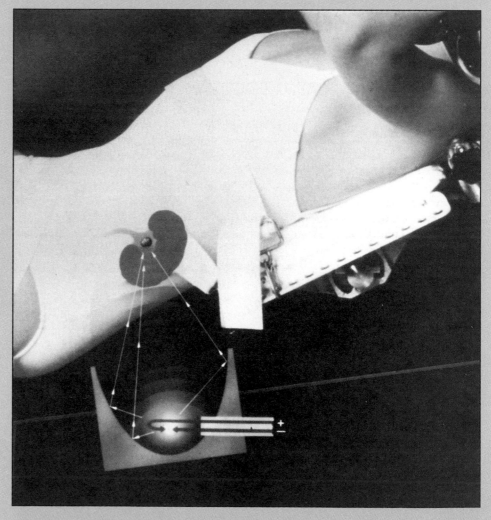

ESWL® being used on a kidney stone. The wave pulse is focused onto the stone because the stone and the spark producing the pulse are located at the foci of the ellipsoidal reflector.

WAVES AND SOUND

PROLOGUE: SOUND MEDICINE

Halley's comet, the planets, and the satellites that orbit the earth have something in common: the shape of the path each follows is an *ellipse*. As shown in Figure 2.42, this characteristic oval shape has two points associated with it that are called *foci* (plural of *focus*). The sun is at one focus of each planet's elliptical orbit, and the earth's center is at one focus of each satellite's orbit. Another property of the ellipse is that its shape can be used to focus light, sound, and other waves. Waves produced at one focus of a reflector shaped like an ellipse travel outward, reflect off the surface, and converge upon the other focus. If a room is shaped like an ellipse, a person standing at one focus can hear even faint sounds, such as whispering, produced by someone at the other focus. This is one type of "whispering chamber." (Legend has it that once a paranoid king arranged it so that his ministers conferred with each other at one focus of a whispering chamber while he stationed himself at the other focus.) Real whispering chambers can be found in some museums of science and in Statuary Hall in the US Capitol. A recently developed medical procedure uses this property of ellipses to save lives.

The formation of stones inside kidneys and gallbladders is a painful and often life-threatening affliction. (The mortality rate for those with kidney stone disease is 2% to 3%.) Surgical removal of kidney stones is effective but dangerous and expensive. *Extracorporeal shock-wave lithotripsy* (ESWL) is a nonsurgical procedure that uses focused sound-wave pulses to break up the stones. (The term extracorporeal is applied because no part of the apparatus is inside the body).

The first ESWL system was introduced in 1980. It uses a concave reflector in the shape of one half of an *ellipsoid*—a surface formed by rotating an ellipse about a line through its foci. The open end of the reflector is positioned near the patient so that the stone is at one focus. A powerful electric spark generated at the other focus produces a wave pulse—called a shock wave because the spark momentarily vaporizes the water nearby—that reflects off the ellipsoid and converges on the stone. This procedure requires that both the ellipsoid and the lower part of the patient's body be immersed in water. Because sound travels through human tissue with about the same speed that it travels through water, the wave pulse is largely unaffected as it enters the

body. The converging wave comes to a focus on the stone the same as if it were traveling in water.

In a typical application, 500 to 2,000 wave pulses are administered during a session lasting up to 1 hour. Each focused pulse lasts about a millionth of a second and can produce a sudden pressure of over 1,000 atmospheres on the stone. This battering breaks the stone up into smaller pieces that can pass out of the body naturally. The procedure is much safer than surgery and results in much less discomfort for the patient. Anesthesia is required, however, and care must be taken to limit the damage that can be done to the surrounding tissue. To ensure that the wave pulses hit the target, dual x-ray units are used to locate the stone and to position the ellipsoid.

The nature of waves and how they travel are two of the main topics of this chapter. As will be seen, waves are inherent in our everyday lives. Moreover they have become an integral health-promoting component in the medical sphere. Whether playing a guitar, listening to a radio, clocking the speed of a thrown baseball, or shattering kidney stones, the common denominator is a wave of some kind. The two senses used most often—sight and hearing—are highly developed wave-detection mechanisms. In the first part of this chapter we will study simple waves and examine some of their general properties. In the remainder of the chapter we will learn about sound—how it is produced, how it travels in matter, and how it is perceived by humans.

6.1   WAVES—TYPES AND PROPERTIES

Ripples moving over the surface of a still pond, sound traveling through the air, a pulse "bouncing" back and forth on a piano string, light from the sun illuminating and warming the earth—these are all waves (see Figure 6.1). We can see or feel the effects of some waves, such as water ripples and earthquake tremors (called seismic waves), as they pass. Others, such as sound and light, we sense directly. Technology has given us numerous devices that produce or detect

FIGURE 6.1
Like all waves, these water ripples involve oscillation.

FIGURE 6.2
A rope is a simple medium for waves.

waves that we cannot sense (radio, ultrasound, x-rays).

What are waves? Though many and diverse, they share some basic features. They all *involve vibration or oscillation* of some kind. Floating petals show the vibration of the water's surface as ripples move by. Our ears respond to the oscillation of air molecules and give us the perception of sound. Also, waves move and *carry energy* yet do not have mass. The sound from a loudspeaker can break a wineglass even though no matter moves from the speaker to the glass. We can define waves as follows:

> **WAVE** A traveling disturbance consisting of coordinated vibrations that transmit energy with no net movement of matter.

Sound, water ripples, and similar waves consist of vibrations of matter—air molecules or the water's surface, for example. The substance through which such waves travel is called the *medium* of the wave. Particles of the medium vibrate in a coordinated fashion to form the wave.

A rope stretched between two people is a nice medium for demonstrating a simple wave (see Figure 6.2). A flick of the wrist sends a wave down the rope; each short segment of the rope is pulled upward in turn by its neighboring segment. The forces *between* the parts of the medium are responsible for "passing along" the wave. This kind of wave is not unlike a row of dominoes knocking each other over, except the medium of a wave does not have to be "reset" after a wave goes by.

These waves—sound, water ripples, waves on a rope—*require* a material medium. They cannot exist in a vacuum. On the other hand, light, radio waves, microwaves, and x-rays can travel through a vacuum because they do not require a medium for their propagation. We will take a close look at these special waves—called "electromagnetic waves"—in Chapter 8.

Waves occur in a great variety of substances: in gases (sound), in liquids (water ripples), and in solids (seismic waves through rock). Some travel along a line (a wave on a rope), some across a surface (water ripples), and some throughout

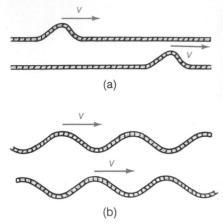

FIGURE 6.3
(a) Successive views of a wave pulse as it travels down a rope. (b) A continuous wave traveling down a rope.

space in three dimensions (sound). Many more examples could be listed for each. Clearly, waves are everywhere, and they are diverse.

A wave can be short and fleeting, called a *wave pulse,* or steady and repeating, called a *continuous wave.* The sound of a bursting balloon, a tsunami (large ocean wave generated by an earthquake), and the light from a camera flash are examples of wave pulses. The sound from a tuning fork and the light from the sun are continuous waves.[1] Figure 6.3 shows a wave pulse and a continuous wave on a long rope. You can see that a continuous wave is like a series or "train" of wave pulses, one after another.

If we take a close look at many different types of waves, we find that they can be classified according to the orientation of the wave oscillations. There are two wave types: *transverse* and *longitudinal.*

> TRANSVERSE WAVE   A wave in which the oscillations are perpendicular (transverse) to the direction the wave travels. Examples: waves on a rope, electromagnetic waves, some seismic waves.
>
> LONGITUDINAL WAVE   A wave in which the oscillations are along the direction the wave travels. Examples: sound in the air, some seismic waves.

Both types of waves can be produced on a Slinky—a short, fat spring that you may have seen "walk" down stair steps. With a Slinky stretched out on a flat, smooth tabletop, a transverse wave is produced by moving one end from side to side, perpendicular to the Slinky's length (Figure 6.4a). A longitudinal wave is produced by moving one end back and forth, first toward the other end then back (see Figure 6.4b). For each type of wave, one can produce either a wave pulse or a continuous wave.

A Slinky is not the only medium that can carry both transverse and longitudinal waves. Both kinds of waves can travel in any solid. Earthquakes and underground explosions produce both longitudinal and transverse seismic waves that travel through the earth. In liquids and gases, on the other hand, the absence of rigid bonds between the atoms or molecules (or both) means that only longitudinal waves can travel along them.

The *speed* of a wave is simply the rate of movement of the disturbance. (Do not confuse this with the speed of individual particles as they oscillate.) For a given type of wave, the speed is determined by the properties of the medium. In the waves that we have been discussing, the masses of the particles that oscillate and the forces that act between them affect the wave speed. As a wave travels down a Slinky, for example, each coil is accelerated back and forth by its neighbors. Basic mechanics tells us that the mass of each coil and the size of the force acting on it will determine how quickly it—and therefore the wave—moves. In general, weak forces or massive particles in a medium cause the wave speed to be low.

Often the speed of waves in a medium can be predicted by measuring some other properties of the medium. (After all, the factors that affect wave speed—masses of particles and the forces between them—also affect other properties of a substance.) For example, the speed of waves on a stretched rope, a Slinky, or a

---

[1]For now we can treat light as a simple wave. In Chapter 10, we will present the modern view of light as developed by Albert Einstein and others.

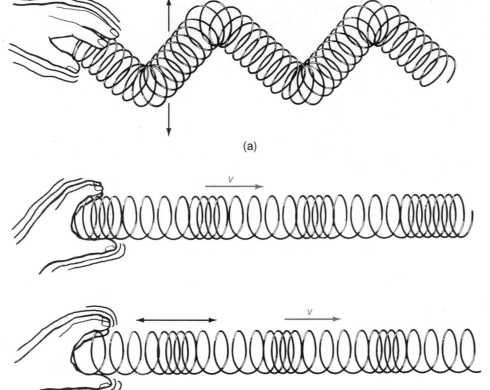

FIGURE 6.4
(a) A continuous transverse wave on a Slinky. In this figure each coil oscillates up and down as the wave travels to the right. (b) A continuous longitudinal wave on a Slinky. Each coil oscillates left and right as the wave travels to the right.

wire can be computed by using the force, F, which must be exerted to keep it stretched, and its *linear mass density,* ρ, which equals its mass m divided by its length l. In particular:

$$v = \sqrt{\frac{F}{\rho}} \quad \left(\text{wave on a rope or spring; } \rho = \frac{m}{l} \right)$$

Increasing this force, also called the "tension," will cause the waves to move faster. This is how stringed instruments like guitars and pianos are tuned. (More on this in Section 6.4.)

A student stretches a Slinky out on the floor to a length of 2 meters. The force needed to keep the Slinky stretched is measured and is found to be 1.2 newtons; the Slinky's mass is 0.3 kilograms. What is the speed of any wave sent down the Slinky by the student?

EXAMPLE 6.1

First we compute the Slinky's linear mass density.

$$\rho = \frac{m}{l} = \frac{0.3 \text{ kg}}{2 \text{ m}}$$

$$\rho = 0.15 \text{ kg/m}$$

The speed of waves on the Slinky is:

$$v = \sqrt{\frac{F}{\rho}} = \sqrt{\frac{1.2\,\text{N}}{0.15\,\text{kg/m}}}$$

$$v = \sqrt{8\,\text{m}^2/\text{s}^2} = 2.8\,\text{m/s}$$

The speed of sound in air or any other gas depends on the ratio of the pressure of the gas to the density of the gas. But for each gas, this ratio in turn depends only on the temperature. In particular, the speed of sound in a gas is proportional to the square root of the Kelvin temperature. For air:

$$v = 20.1 \times \sqrt{T} \quad \text{(speed of sound in air;} \quad v \text{ in m/s, } T \text{ in Kelvin)}$$

The speed of sound is lower at higher altitudes, not because the air is thinner but because it is colder.

EXAMPLE 6.2   What is the speed of sound in air at room temperature ($20°\,\text{C} = 68°\,\text{F}$)? The temperature in Kelvin is:

$$T = 273 + 20 = 293\,\text{K}$$

Therefore:

$$v = 20.1 \times \sqrt{T}$$

$$v = 20.1 \times \sqrt{293} = 20.1 \times 17.1$$

$$v = 344\,\text{m/s (770 mph)}$$

The numerical factor (20.1) in the equation is determined by the properties of the molecules that comprise air and therefore *applies to air only*. The speed of sound in any other gas will be different, and the corresponding equation for v will have a different numerical factor. Two examples:

Speed of sound in helium: $v = 58.8 \times \sqrt{T}$ (SI units)

Speed of sound in oxygen: $v = 19.1 \times \sqrt{T}$ (SI units)

For the remainder of this section we will take a look at some of the properties of a continuous wave. A convenient example is a transverse wave on a Slinky produced by moving one end smoothly side to side. Figure 6.5 shows a "snapshot" of such a wave. It shows the shape of the Slinky at some instant in time. Note that the wave has the same sinusoidal shape you've seen before (Figure 2.24).

The high points of the wave are called "peaks" or "crests," and the low points are called "valleys" or "troughs." The straight line through the middle represents the equilibrium configuration of the medium—its shape when there is no wave.

In addition to wave speed, there are three other important parameters of a continuous wave that can be measured. They are: *amplitude, wavelength,* and *frequency.* At any moment, the different particles of the medium are displaced from their equilibrium positions by different amounts. The maximum displacement is called the *amplitude* of the wave.

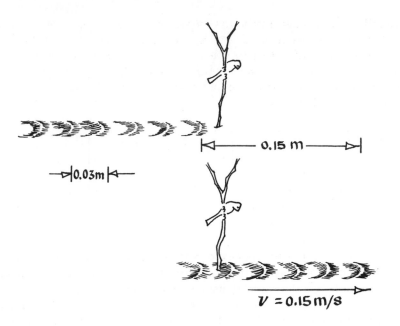

FIGURE 6.9
If five cycles of a wave pass by a point in 1 second, and the wavelength of the wave is 0.03 meter, then the wave is traveling 0.15 meters per second.

Wave speed = number of cycles per second × length of each cycle

But the two quantities on the right of the equal sign are just the frequency of the wave and the wavelength, respectively. Therefore:

$$v = f\lambda$$

The velocity of a continuous wave is equal to the frequency of the wave times the wavelength.

In many cases all waves that travel in a particular medium have the same speed. Wave pulses, low-frequency continuous waves, and high-frequency continuous waves all travel with the same speed. Sound is an important example of this; sound pulses, low-frequency sounds, and high-frequency sounds travel through the air with the same speed, 344 meters/second at room temperature. Similarly, light, radio waves, and microwaves travel with the same speed—3×10^8 meters/second—in a vacuum. According to the equation $v = f\lambda$, when the wave speed is the same for all waves, *higher* frequency waves must have proportionally *shorter* wavelengths. A 20-hertz sound wave has a wavelength of about 17 meters, while a 20,000-hertz sound wave has a wavelength of about 2 centimeters.

Before a concert, musicians in an orchestra tune their instruments to the note "A," which has a frequency of 440 hertz. What is the wavelength of this sound in air at room temperature? The speed of sound at this temperature is 344 m/s. So:

$$v = f\lambda$$

$$344 \text{ m/s} = 440 \text{ Hz} \times \lambda$$

$$\frac{344 \text{ m/s}}{440 \text{ Hz}} = \lambda$$

$$\lambda = 0.78 \text{ m} = 2.6 \text{ ft}$$

EXAMPLE 6.3

The wavelength of sound with a frequency of 220 Hz is twice as large— 1.56 m.

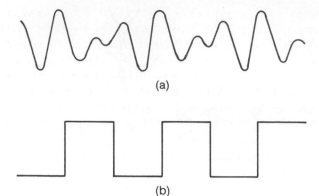

FIGURE 6.10
Two examples of complex waves. The lower wave, a "square wave," is used in many electronic devices.

Not all continuous waves have the simple sinusoidal shape shown in Figure 6.5. (In fact, waves with precisely that shape are relatively rare.) Any continuous wave that does not have a sinusoidal shape is called a *complex wave.* Figure 6.10 shows two examples. (Note that there are actually three different-sized peaks in each cycle of the upper wave.) The shape of a wave is called its *waveform.* The two complex waves in the figure have about the same wavelength and amplitude, but they have very different waveforms. The waveform is another feature that is needed when comparing complex waves. We will take a closer look at this in Section 6.6.

6.2   ASPECTS OF WAVE PROPAGATION

In this section we consider some of the things that waves do as they travel. For waves traveling along a surface or throughout space in three dimensions, it is convenient to use two different ways to represent the wave. We will call these the

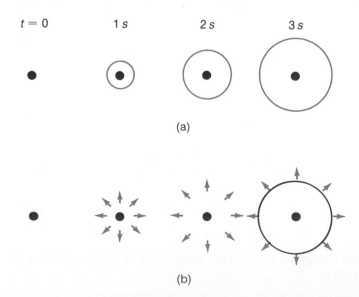

FIGURE 6.11
(a) A wavefront is used to show how a pulse spreads over water. (b) The same wave pulse represented with wave rays. The rays point in the direction the wave travels and are perpendicular to the wavefront.

wavefront model and the *ray model.* Figure 6.11 shows how each is used to illustrate a wave pulse on water as it travels from the point where it was produced. The wavefront is just a circle that shows the location of the peak of the wave pulse. A "ray" is a straight arrow that shows the direction a given segment of the wave is traveling. A laser beam or sunlight passing through a hole in a window shade are like individual rays of light that we can see. On the other hand, the rays of water ripples are not visible, but we do see the wavefronts.

For a continuous water wave, the wavefronts are concentric circles about the point of origin that represent individual peaks of the wave (see Figure 6.12). The largest circle shows the position of the first peak that was produced. Each successive wavefront is smaller because it came later and has not traveled as far. The distance between adjacent wavefronts is equal to the wavelength of the wave. Again, a continuous wave is like a series of wave pulses produced one after another. The rays for a continuous wave are lines radiating from the source of the wave. The wavefronts arriving at a point far from the source are nearly straight lines (far right in Figure 6.12). The corresponding rays are nearly parallel.

For a wave moving in three-dimensional space, the wavefronts are spherical shells around the wave source. The wavefront of a wave pulse expands like a balloon that is being inflated. For continuous three-dimensional waves, the wavefronts form a series of concentric spherical shells that expand like the circular wavefronts of a wave on a surface. A 440-hertz tuning fork produces 440 of these wavefronts each second. The surface of each wavefront expands outward with a speed of 344 meter/second (at room temperature). The rays for waves in three dimensions are also lines radiating from the wave source.

One inherent aspect of the propagation of waves on a surface or in three dimensions is that the amplitude of the wave necessarily decreases as the wave gets farther from the source. A certain amount of energy is expended to create a wave pulse or each cycle of a continuous wave. This energy is distributed evenly over the wavefront and determines the amplitude of the wave: the greater the amount of energy given to a wavefront, the larger the amplitude. As the wavefront moves out, it gets larger and so this energy is spread out more: it becomes less concentrated. This attenuation accounts for the decrease in loudness of sound as a noisy car moves away from you and for the decrease in brightness as you move away from a light bulb.

One can infer when the amplitude of a wave is changing by noting changes in the wavefronts or the rays. If the wavefronts are growing larger, the amplitude is getting smaller. The same thing is indicated when the rays are diverging (slanting away from each other).

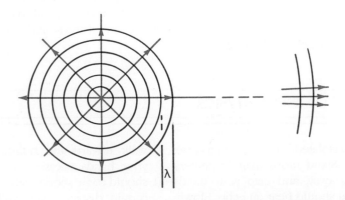

FIGURE 6.12
Wavefronts and rays for a continuous wave on a surface. Far away from the wave source the wavefronts are nearly straight, and the rays are nearly parallel.

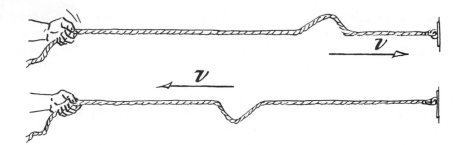

FIGURE 6.13
A wave pulse traveling on a rope is reflected at a fixed end. In this case the pulse is inverted.

At great distances from the source of a three-dimensional wave, the wavefronts become nearly flat and are called *plane waves*. The corresponding rays are parallel, and the wave's amplitude stays constant. The light and other radiation we receive from the sun come as plane waves because of the great distance between the earth and the sun.

With this background we will look at several phenomena associated with wave propagation.

Reflection

A wave is reflected whenever it reaches a boundary of its medium or encounters an abrupt change in the properties (density, temperature, and so on) of its medium. A wave pulse traveling on a rope is reflected when it reaches a fixed end (see Figure 6.13). It "bounces" off the end and travels back along the rope. Notice that the reflected pulse is inverted. When the end of the rope is attached to a very light (but strong) string instead, the reflected pulse is not inverted. This reflection occurs because of an abrupt change in the density of the medium: from high density (for the heavy rope) to low density (for the light string).

In a similar fashion, a wave on a surface or a wave in three dimensions is reflected when it encounters a boundary. The wave that "bounces back" is called the *reflected wave*. Rays are more commonly used to illustrate reflection because they show nicely how the direction of each part of the wave is changed. When a wave is reflected from a straight boundary (for surface waves) or a flat boundary (in three dimensions), the reflected wave appears to be expanding out from a point behind the boundary (Figure 6.14). This point is called the "image" of the original wave source. An echo is a nice example of this: sound that encounters a large flat surface, such as the face of a cliff, is reflected and sounds like it is coming from a point behind the cliff.

DO-IT-YOURSELF PHYSICS

Locate an isolated building with a large, flat side. Stand more than 15 meters (50 feet) away, and clap your hands once: you should hear an echo. Slowly walk toward the building, clapping your hands occasionally. The echoes should come sooner. When you get to a certain distance, you will no longer

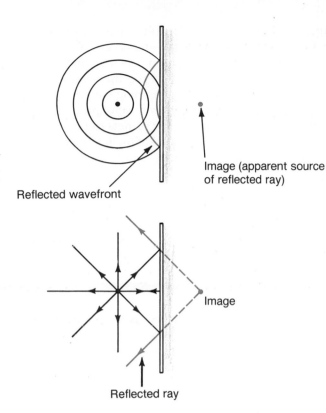

Image (apparent source of reflected ray)

Reflected wavefront

Image

Reflected ray

FIGURE 6.14
The reflection of water ripples off the side of a pool. Both models of the wave show that the reflected wave appears to diverge from a point behind the wall, called the "image."

hear the echo because it comes too soon after the direct sound. The ear can distinguish two successive sound pulses only if the second comes about 20 milliseconds or more after the first.

Our most common experience with reflection is that of light from a mirror. The image that you see in a mirror is a collection of reflected light rays originating from the different points on the object you see. (More on this in Section 9.2.)

Reflection from surfaces that are not flat (or straight) can cause useful things to happen to waves. Figure 6.15 shows a wave being reflected by a curved surface.

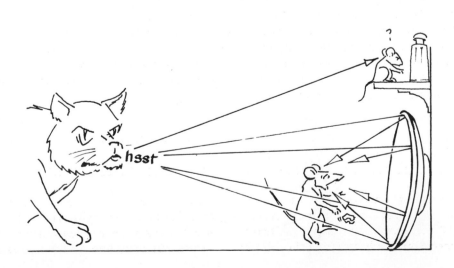

hsst

FIGURE 6.15
A sound wave reflecting off a concave surface. The reflected rays converge toward a point, indicating that the wave's amplitude increases.

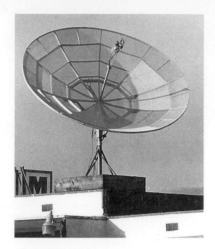

FIGURE 6.16
This parabolic "dish" focuses radio waves by reflection.

Note that the rays representing the reflected part of the. wave are converging toward each other. This means that the amplitude of the wave increases—the wave is being "focused." Parabolic microphones seen on the sidelines of televised football games use this principle to reinforce the sounds made on the playing field. Satellite receiving dishes do the same thing with radio waves (Figure 6.16).

A reflector in the shape of an ellipse has a useful property. (We saw in Section 2.9 that the orbits of satellites, comets, and planets can be ellipses.) An ellipse has two points in its interior called *foci* (plural of focus). If a wave is produced at one focus, it will converge on the other focus after reflecting off the ellipse. All rays originating from one focus reflect off the ellipse and pass through the other focus (Figure 6.17). A room shaped like an ellipse is called a *whispering chamber* because a person standing at one focus can hear faint sounds—even whispering— produced at the other focus. This property of the ellipse is also used in the medical treatment of kidney stones (as described in the Chapter 6 Prologue).

Doppler Effect

The *Doppler effect* is an apparent change in the frequency of a wave due to motion of the source of the wave or the receiver. Figure 6.18 shows the wavefronts emitted by a moving source—perhaps a bug moving over the surface of a pond or a train blowing its whistle. Each wavefront expands outward from the point where the source was when it emitted that wavefront. (The speed of a wave in a medium is constant and is not affected by any motion of the wave source.) In front of the moving source, the wavefronts are bunched together. This means that the *wavelength* is *decreased,* and therefore the *frequency* of the wave is *increased.* Behind the moving source the wavefronts are spread apart: the *wavelength* is *increased* and the *frequency* is *decreased.* In both places, the higher the speed of the wave source, the greater the frequency shift.

The frequency of sound that reaches a person in front of a moving train is higher than that perceived when the train is not moving. A person behind the moving train hears a lower frequency. As a train or a fast car moves by, you hear the sound shift from a higher frequency (pitch) to a lower frequency. (The change in the loudness of the sound, which you also hear, is *not* part of the Doppler effect: it involves a separate process.)

A similar shift in frequency of sound occurs if *you* are moving towards a stationary sound source (see Figure 6.19). This Doppler shift happens because the speed of the wave relative to you is higher than that when you are not moving. The wavefronts approach you with a speed equal to the wave speed *plus* your

FIGURE 6.17
The focusing property of an ellipse. Each ray of a wave produced at one focus reflects off the ellipse and passes through the other focus.

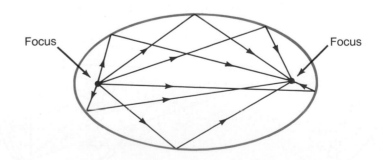

Focus   Focus

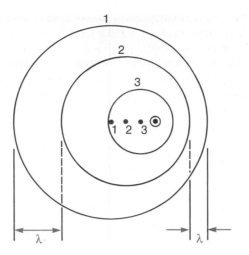

FIGURE 6.18
The Doppler effect with a moving source. A wave source moves to the right with constant speed. Each dot indicates the source's location when a wavefront was emitted. Wavefront 1 was emitted when the source was at position 1; similarly for 2 and 3. Ahead of the source (to the right), the wavelength of the wave is decreased. Behind the source (to the left), the wavelength is increased.

speed. Since the wavelength is not affected, the equation $v = f\lambda$ tells us that the frequency of the wave is increased in proportion to the speed of the wave relative to you. By the same reasoning, when one is *moving away* from the sound source, the frequency is *reduced.*[2]

The Doppler effect is routinely taken into account by astronomers. The frequencies of light emitted by stars in galaxies that are moving toward or away from the earth are shifted. If the speed of the star is known, the original frequencies of the light can be computed. If the frequencies are known instead, the speed of the star can be computed from the amount of the Doppler shift. Such information is essential for determining the motions of stars in our galaxy or of entire galaxies throughout the universe (see Physics Potpourri on the next page).

Echolocation is the process of using the reflection of waves off an object to determine its location. *Radar* and *sonar* are two examples of this. Basic echolocation uses reflection only: a wave is emitted from a point, reflected by an object of some kind, and detected on its return to the original point. The time between the emission of the wave and the detection of the reflected wave (the round-trip time) depends on the speed of the wave and the distance to the reflecting object. For example, if you shout at a cliff and hear the echo 1 second later, you know that the cliff is approximately 172 meters away. This is because the sound travels a total of 344 meters (172 meters each way) in 1 second (at room temperature).

---

[2]For the musically inclined: To shift the frequency of sound by a musical half step, the speed of the source of the sound (or the listener, if the source is at rest) must be about 20 meters/second (45 mph). For example, if the sound from a car's horn is the note "A" when the car is at rest, it will be heard as an "A sharp" when the car is approaching a listener at 20 meters/second and as an "A flat" when it is moving away at 20 meters/second. (The occupants of the car always hear an "A.") This value applies when the temperature is around 20° C. When it is colder, the required speed would be smaller. (Why?)

FIGURE 6.19
The Doppler effect with moving observers. The person in the car on the left hears a higher frequency than the pedestrian. The person in the car on the right hears a lower frequency.

We have mentioned that the Doppler effect can be used to determine the line-of-sight speeds for individual stars bright enough to allow accurate measurement of the shifts in frequency of the light emitted by them. The question of whether or not the speeds of approach or recession of large aggregates of stars, called "galaxies," could be measured from similar shifts in the combined light emitted by all the members of the galaxy was first investigated by Vesto Slipher beginning in about 1912. By 1925, Slipher had carefully studied the radiation emitted by 40 galaxies and found that he could indeed determine the speeds with which they were moving. What he discovered was quite surprising: a number of the galaxies had very high speeds, up to 5,700 kilometers/second (13 million miles per hour), and 38 out of the 40 showed frequency shifts that indicated they were moving away from us.

During the last 65 years, much has been accomplished in galactic astronomy, but the essentials of Slipher's results remain true. The vast majority of galaxies—and certainly all of those farther from us than about 3 million light-years—are moving away from us and are doing so with velocities that are, in many cases, a significant fractions of the speed of light. One of the more important additional facts to be discovered about galaxies during this period concerns the relationship between the recessional speed of a galaxy and its distance from us. Shortly after Slipher reported his results, Edwin Hubble noticed that galaxies with low recessional speeds were all relatively close to us, whereas those with high recessional speeds were more remote. By 1929 Hubble, in whose honor the Hubble Space Telescope was named, had amassed enough data to show a direct proportionality between the speeds

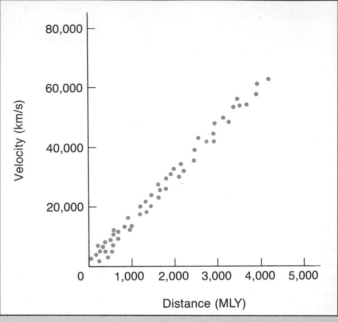

FIGURE 6.20
The Hubble law. Plotted here are the recessional velocities of galaxies versus their distances (1 light year = 9.46×10^{12} km).

with which galaxies are receding from us and their distances. This connection between speed and distance for all but the nearest galaxies has been verified by modern observations and is shown in Figure 6.20. It is now referred to as the *Hubble law*.

What is the significance of the Hubble law? Only that the universe is expanding! To see this we must recognize that galaxies are the "atoms" of the universe, and just as one speaks of an "expanding gas"

If it takes 2 seconds, the cliff is approximately 344 meters away, and so on (see Figure 6.22).

With sonar, a sound pulse is emitted from an underwater speaker, and any reflected sound is detected by an underwater microphone. The time between the transmission of the pulse and the reception of the reflected pulse is used to determine the distance to the reflecting object. Basic radar uses a similar process with microwaves that reflect off aircraft, raindrops, and other things.

Incorporating the Doppler effect in echolocation makes it possible to determine immediately the speed of an approaching or departing object. A moving

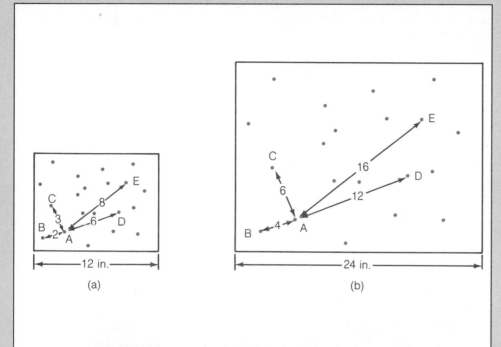

FIGURE 6.21
The "raisin cake" analogy for the expansion of the universe. As the cake rises (through the action of the yeast in the dough), it doubles in size in a time equal to, say, 1 hour. Consequently all the raisins move farther apart. Some characteristic distances from one of the raisins (chosen arbitrarily) to several others are shown. Since each distance doubles during an hour, each raisin must move away from the one selected as the origin at a speed proportional to its distance. This remains true no matter which raisin is chosen as the origin. The point here is that a uniform expansion leads to a Hubble-like relationship between velocity and distance. (From Exploration of the Universe *by* George Abell.*)*

when the atoms of the gas are moving away from one another, one talks about an "expanding universe" when the atoms of the universe, the galaxies, are observed to be systematically separating from each other. And this is precisely what the Hubble law is telling us. No matter in what direction we look, we find galaxies rushing away from us with speeds proportional to their distances: the farther away they are, the faster they are receding from us. The "raisin cake" analogy depicted in Figure 6.21 may be helpful to you in visualizing better what is happening here. The linear (straight line) correla-

tion between speed and distance is the trademark of an expanding system, and, in this case the system is the grandest one of all—the entire universe. The Hubble law (and its interpretation) have had a profound influence on *cosmology,* the study of the structure and evolution of the universe. At this stage, however, one should keep firmly in mind that were it not for the Doppler effect, this relation might never have been discovered.

object causes the reflected wave to be Doppler shifted. If the frequency of the reflected wave is *higher* than that of the original wave, the object is moving *toward* the source. If the frequency is *lower,* the object is moving *away.*

Doppler radar uses this combination of echolocation and the Doppler effect. The time between transmission and reception gives the distance to the object, while the amount of frequency shift is used to determine the speed. Law-enforcement officers use Doppler radar to check the speeds of vehicles, and Doppler radar is also used in baseball to clock the speed of a thrown ball (see Figure 6.23). Dust, raindrops, and other particles in air reflect microwaves, mak-

FIGURE 6.22
Simple echolocation. You can determine the distance to the cliff by timing the echo.

ing it possible to detect the rapidly swirling air in a tornado with Doppler radar. Another potentially life-saving application is the detection of wind shear—drastic changes in wind speed near storms that have caused low-flying aircraft to crash. (There are two more examples at the end of Section 6.3.)

FIGURE 6.23
Radar unit used to measure the speed of a baseball.

Bow Waves and Shock Waves

Figure 6.24 shows another series of wavefronts produced by a moving wave source. This time the speed of the wave source is *greater than* the wave speed. The wavefronts "pile up" in the forward direction and form a large-amplitude wave pulse called a *shock wave*. This is what causes the V-shaped bow waves produced by boats.

Aircraft flying faster than the speed of sound produce a similar shock wave. In this case the three-dimensional wavefronts form a conical shock wave, with the aircraft at the point. This conical wavefront moves with the aircraft and is heard as a *sonic boom* (a sound pulse) by persons on the ground.

Diffraction

Diffraction occurs as a wave moves through an opening in a barrier. Figure 6.25 shows wavefronts as they reach a gap in a wall. These might be ocean waves encountering a breakwater. The part of the wave that passes through the gap sends out wavefronts ahead and to the sides. The rays that represent this process show that the wave "bends" around the edges of the opening.

The extent to which the diffracted wave spreads out depends on the ratio of the size of the opening to the wavelength of the wave. When the opening is much larger than the wavelength, there is little diffraction: the wavefronts remain straight and do not spread out to the sides appreciably. This is what happens when light comes in through a window. The wavelength of light is less than a millionth of a meter, and consequently there is little diffraction. When the wavelength is roughly the same size as the opening, the diffracted wave spreads out much more (see Figure 6.26).

The sizes of windows and doors are well within the range of the wavelengths of sound waves. Sound produced in a room is diffracted through any open window or door. This is why we can hear sound coming through an open window even if the source isn't in our line of sight. The higher frequencies (shorter wavelengths) are not diffracted as well as the lower frequencies.

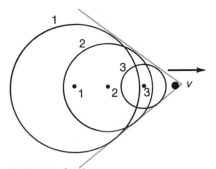

FIGURE 6.24
A shock wave is produced when the speed of a wave source exceeds the wave speed. Parts of the wavefronts combine along the two colored lines to form a V-shaped wavefront. As a boat moves through shallow water, we see the shock wave, called the "bow wave," but we do not see the circular wavefronts.

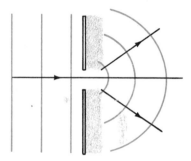

FIGURE 6.25
The diffraction of a wave as it passes through an opening in a barrier. The wavefronts spread out to the sides after passing through.

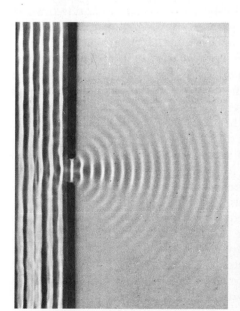
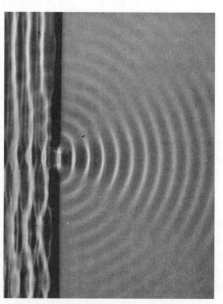

FIGURE 6.26
Diffraction of water waves passing through a narrow gap in a barrier. Short-wavelength waves (a) are not diffracted as much as long-wavelength waves (b) passing through an opening of the same size.

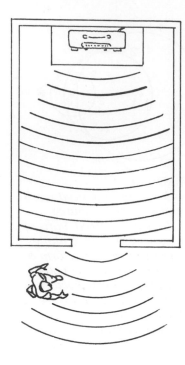

FIGURE 6.27

Here is a simple way to experience diffraction of sound. Stand near an open window or door of a building in which a continuous sound is being produced, such as recorded music (see Figure 6.27). Move back and forth past the opening, and notice that the sound is loudest when you are directly in front of the opening but that you can hear it when you are well off to one side. Carefully notice that not only is the diffracted sound (at the sides) not as loud, it has less treble: the higher frequencies are not diffracted as much.

Interference

Interference arises when two continuous waves, usually with the same amplitude and frequency, arrive at the same place. The sound from a stereo with the same steady tone coming from each speaker is an example of this situation. Another way to cause interference is to direct a continuous wave at a barrier with two openings in it. The two waves that emerge from the two openings will diffract (spread out), overlap each other, and undergo interference.

Let us consider the case of identical, continuous water waves produced by two small objects made to oscillate up and down in unison on the surface of the water. As these two waves travel outward, each point in the surrounding water moves up and down under the influence of both waves. If we move around in an arc about the wave sources, we find that at some places the water is moving up and down with a large amplitude. At other places the water is actually still—it is not oscillating at all (Figure 6.28).

To see why this characteristic pattern of large amplitude and zero amplitude motion arises, consider Figure 6.29 (a)—a sketch showing two water waves at one moment in time. The thicker lines represent peaks of the water waves, and the thinner lines represent the valleys. In Figure 6.29 (b), the dashed lines labeled "C" indicate the places where the two waves are "in phase"—the peak of one wave matches the peak of the other, and valley matches valley. The two waves reinforce each other, and the water oscillates with a large amplitude. This is called *constructive interference.* On the dashed lines labeled "D," the waves are "out of phase"—the peak of one wave matches the valley of the other, and the two waves cancel each other. (Whenever one wave makes the water go upward, the other makes it go downward, and vice versa. Therefore the net displacement is always zero.) This is called *destructive interference.* Figure 6.29 (c) shows the same waves a short time later, after the waves have traveled one half of a wavelength. The pattern of constructive and destructive interference is not altered as the waves travel outward. If the photograph shown in Figure 6.28 had been taken earlier or later, it would look the same.

Whether the two waves are in phase or out of phase depends on the relative distances they travel. To reach any point on the line "C_1" in Figure 6.29, the two waves travel the same distance and consequently arrive with peak matching peak and valley matching valley. Along the line "C_2," the wave from the source on the left must travel a distance equal to *one wavelength farther* than the wave from the source on the right. The reverse is true along the line on the left labeled "C." In general, there is *constructive interference at all points where one wave travels one, or two, or three, etc wavelengths farther than the other wave.*

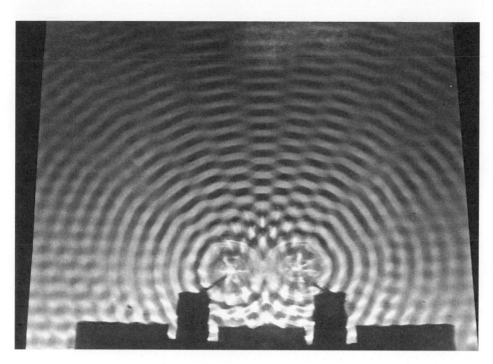

FIGURE 6.28
Interference pattern of water waves from two nearby sources. The thin lines of calm water indicate destructive interference. Between these lines are regions of large-amplitude waves, caused by constructive interference.

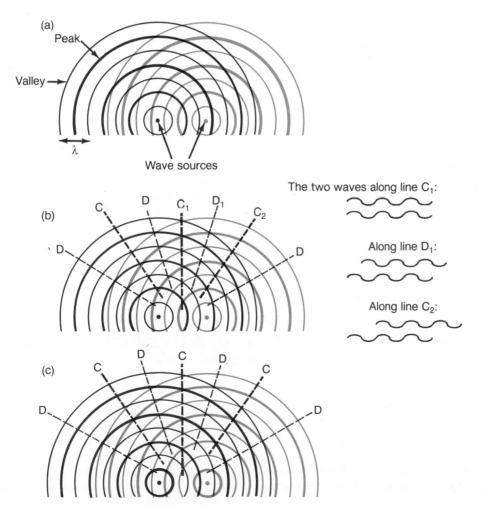

FIGURE 6.29
Interference of two waves.

On the other hand, along the line of destructive interference labeled "D_1," the wave from the source on the left has to travel *one half wavelength farther* than the wave from the source on the right. They arrive with peak matching valley and cancel each other. Along the far right line labeled "D," the wave from the source on the left has to travel $1\frac{1}{2}$ *wavelengths farther,* so the two waves again arrive out of phase. The reverse is true for the lines showing destructive interference on the left. In general, there is *destructive interference at all points where one wave travels one-half, or $1\frac{1}{2}$, or $2\frac{1}{2}$, etc wavelengths farther than the other wave.* At places in between constructive and destructive interference, the waves are not completely in phase or out of phase and partially reinforce or cancel each other.

Sound and other longitudinal waves can undergo interference in the same way. We can imagine Figure 6.29 representing sound waves with the peaks corresponding to compressions and the valleys corresponding to expansions. Along the lines of constructive interference one would hear a loud, steady sound. Along the lines of destructive interference one would hear no sound at all. In Chapter 9 we will look at the interference of light waves.

These are some of the more important phenomena associated with waves as they propagate. Later in this chapter and in Chapter 9 we will take a closer look at some of these and introduce others that are particularly important to light.

LEARNING CHECK

1. A _____ wave is one with the oscillations perpendicular to the direction of propagation.

2. Two sound waves traveling through the air have different frequencies. Consequently they must also have different:
   a) Amplitudes.   c) Wavelengths.
   b) Speeds.   d) All of the above.

3. As a sound wave or water ripple travels out from its source, its _____ decreases.

4. Which of the following affects the frequency of a wave?
   a) Reflection   c) Diffraction
   b) Doppler effect   d) All of the above

5. When two waves undergo _____ interference, the net amplitude is zero.

6.3   SOUND

Our most common experience with sound is in air, but it can travel in any solid, liquid, or gas. For example, when you speak, most of what you hear is sound that travels to your ears through the bones and other tissues in your head. That is why a recording of your voice does not sound the same to you as what you hear when you are talking. Table 6.1 lists the speed of sound in some common substances. The speed of sound in any substance depends on the masses of its constituent atoms or molecules and on the forces between them. The speed of sound is

TABLE 6.1 SPEED OF SOUND IN SOME COMMON SUBSTANCES		
	Speed*	
Substance	(m/s)	(mph)
Air		
At −20° C	320	715
At 20° C	344	770
At 40° C	356	795
Helium	1,006	2,250
Water	1,440	3,220
Human tissue	1,540	3,450
Aluminum	5,100	11,400
Granite	4,000	9,000
Iron and steel	5,200	11,600
Lead	1,200	2,700

*At room temperature (20°C) except as indicated.

generally higher in solids than in liquids and gases because the forces between the atoms and molecules in solids are very strong. Sound in gases and liquids is a longitudinal wave, whereas in solids it can be either longitudinal or transverse. In the rest of the chapter, we will concentrate on sound in air.

Sound is produced by anything that is vibrating and causing the air molecules next to it to vibrate. Figure 6.30 shows a representation of a sound wave being emitted by a vibrating tuning fork. The shading represents the air molecules that we, of course, cannot see. The wave looks very much like a longitudinal wave on a Slinky (Figure 6.8). These compressions and expansions travel at 344 meters/second (at room temperature).

The air pressure in each compression is higher than normal atmospheric pressure because the air molecules are squeezed closer together. Similarly, the pressure in each expansion is below atmospheric pressure. Beneath the sketch is a graph of the air pressure along the direction the wave is traveling. Note that it has the characteristic sinusoidal shape. So a sound wave can be represented by a series of pressure peaks and valleys—a pressure wave. It is more convenient to

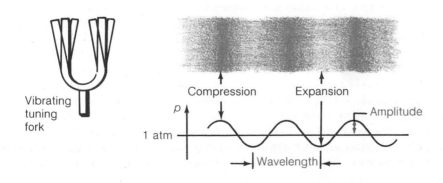

FIGURE 6.30
A representation of part of the sound wave emitted by a tuning fork. The air pressure is increased in each compression and reduced in each expansion, as shown in the graph. (From Physics in Everyday Life *by Richard Dittman and Glenn Schmieg.)*

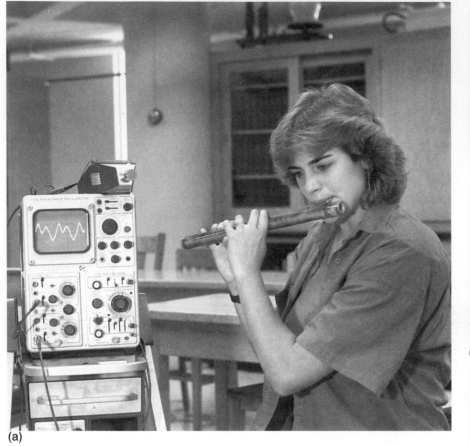

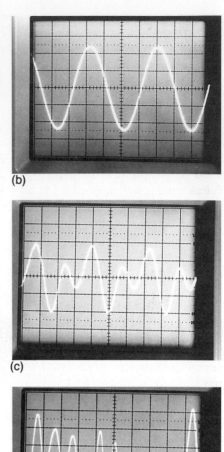

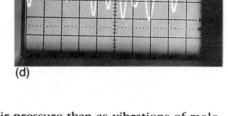

FIGURE 6.31
Examples of the three types of waveforms.
(a) Peggy displays a waveform on an oscilloscope. (b) Pure tone: sound
from a tuning fork. (c) Complex tone: a spoken "ooo" sound. (d) Noise:
sound of air rushing over microphone.

think of sound as regular fluctuations of air pressure than as vibrations of molecules, although it is both. The amplitude of a sound wave is just the maximum pressure change. For a very loud sound, this is only about 0.00002 atmosphere.

It is these pressure variations that our ears detect and convert into the sensation of sound. The eardrum is a flexible membrane that responds to pressure changes. (That is why your ears "pop" when you ride in a fast elevator.) The oscillating pressure of a sound wave forces the eardrum to vibrate in and out. A remarkable series of organs converts this oscillation of the eardrum into an electrical signal to the brain that is perceived as sound.

The *waveform* of a sound wave is the graph of the air-pressure fluctuations caused by the sound wave. The easiest way to display the waveform of sound is to connect a microphone to an oscilloscope, an electronic device often seen displaying heartbeats in television hospital shows. The oscilloscope shows a graph, of the pressure variations that are detected by the microphone (Figure 6.31 a).

We can classify sounds by their waveforms (see Figure 6.31 b,c,d). A *pure tone* is a sound with a sinusoidal waveform. A tuning fork produces a pure tone, as does a person carefully whistling a steady note. A *complex tone* is a complex sound wave. The waveform of a complex tone repeats itself but is not sinusoidal. Therefore any complex tone also has a definite wavelength and a definite frequency. Most steady musical notes are complex tones.

The third type of sound is called *noise*. Noise has a random waveform that does not repeat over and over. For this reason, noise does *not* have a definite wavelength or frequency. The sound of rushing air is a good example of noise. (In everyday speech, the term "noise" is often used to describe any unwanted sound, even if it is a pure tone or a complex tone.)

Sound with frequencies outside the range of 20 to 20,000 hertz cannot be heard by people. Inaudible sound with frequency less than 20 hertz is called "infrasound." High-amplitude infrasound can be felt, rather than heard, as periodic pressure pulses. The hearing ranges of elephants and whales extend into the infrasound region.

Ultrasound

Sound with frequencies higher than 20,000 hertz (20 kilohertz) is called *ultrasound*. The audible range of dogs extends into ultrasound frequencies; dogs can hear some very high-frequency sounds that humans cannot. Bats use ultrasonic frequencies in a highly sophisticated echolocation system that employs the Doppler effect (Figure 6.32). Each species uses a characteristic range of frequencies, anywhere from 16 to 150 kilohertz. The bat emits a short burst of sound that

FIGURE 6.32
Bats use ultrasonic echolocation to navigate in the dark.

reflects off surrounding objects like the ground, trees, and flying insects. The bat detects these echoes and uses the time it takes the sound to make the round trip to determine the distances to the objects. The shift in the frequency of sound reflected off moving objects is used by the bat to track down its dinner—flying insects. The bat even compensates for the Doppler shift in the frequency of the emitted sound caused by its own motion.

In this century many useful applications of ultrasound have been discovered: it is used in some burglar alarm systems, as well as to control rodents and insects, and to clean watches, jewelry, and intricate electronic and mechanical components. Some automatic-focus cameras use ultrasonic echolocation to determine the distance to the object being photographed. The widest application of ultrasonics is in medicine, where ultrasound is routinely used to form images of internal organs and unborn children (see Figure 6.33). High-frequency ultrasound, typically 3.5 million hertz, is sent into the body and is partially reflected as it encounters different types of tissue. These reflections are analyzed and used to form an image on a television monitor. Some sophisticated ultrasonic scanning devices also use the Doppler effect. The beating heart of an unborn child and the flow of blood in arteries can be monitored by detecting the frequency shift of the reflected ultrasound.

Another use of ultrasound in medicine is *ultrasonic lithotripsy*, a procedure that breaks up kidney stones that have migrated to the bladder. A large-amplitude 27,000-hertz sound wave travels through a steel tube inserted into the body and placed in contact with the stone. The ultrasound breaks the stone into small pieces, somewhat like a singer breaking a wineglass. A more recently developed

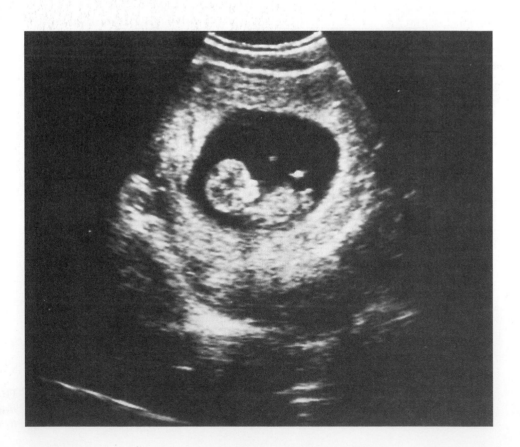

FIGURE 6.33
Ultrasonic image of an unborn child.

procedure called *extracorporeal shock-wave lithotripsy* (ESWL) uses a large-amplitude sound pulse that is focused on the stone using a specially shaped concave reflector. (See Chapter 6 Prologue.)

6.4 PRODUCTION OF SOUND

In the remaining sections of this chapter we will take a brief look at the three *P's* of acoustics: the production, propagation, and perception of sound.

The sounds that we hear range from simple pure tones like a steady whistle to complicated and random waveforms like those found on a noisy street corner. Most of the sound we hear is a combination of many sounds from different sources. The loudness usually fluctuates, as do the frequencies of the component sounds.

Sound is produced when vibration causes pressure variations in the air. Any flat plate, bar, or membrane that vibrates produces sound. The tuning fork shown in Figure 6.30 is a nice example. A dropped garbage can lid, a vibrating speaker cone, and a struck drumhead produce sound the same way. The tuning fork executes simple harmonic motion and produces a pure tone. The garbage can lid and the drumhead have more complicated motions and so produce complex tones or noise.

The various musical instruments represent some of the basic types of sound producers. Drums, triangles, xylophones, and other percussion instruments produce sound by direct vibration. Each is made to vibrate by a blow from a mallet or drumstick.

Guitars, violins, and pianos use vibrating strings to produce sound (Figure 6.34). By itself, a vibrating string produces only faint sound because it is too thin to compress and expand the air around it effectively. These instruments employ "soundboards" to increase the sound production. (The electric versions of guitars and violins pick up and amplify the string vibrations electronically.) One end of the string is attached to a wooden soundboard, which is made to vibrate by the string. The vibrating soundboard, in turn, produces the sound. Figure 6.35 shows

FIGURE 6.34
Guitarist Andrés Segovia makes the strings vibrate by plucking them. Most of the sound that we hear comes from the front plate of the guitar, which is made to vibrate by the strings.

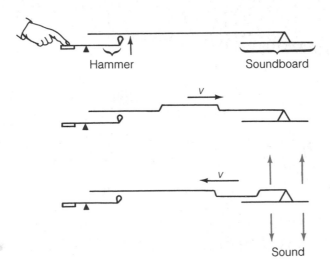

FIGURE 6.35
Sound production in a piano. The hammer creates a wave pulse that oscillates back and forth on the piano wire. (The amplitude is not to scale.) The pulse causes the soundboard to vibrate at the frequency of the pulse's oscillation.

a simplified diagram of the process used in pianos. When a note is played, a hammer strikes the piano wire and produces a wave pulse. The pulse travels back and forth on the string, being reflected each time at the ends. The soundboard receives a "kick" each time the pulse is reflected at that end. This makes the soundboard vibrate at a frequency equal to the frequency of the pulse's back-and-forth motion. The sound dies out because the pulse loses energy to the soundboard during each cycle.

The strings on guitars are plucked instead of struck, giving the pulses a different shape. This is part of the reason why the sound of a guitar is different from that of a piano. Violin strings are bowed, resulting in even move complicated wave pulses. In all three instruments, the frequency of the pulse's motion depends on the speed of waves on the string and on the length of the string. When a string is tuned by being tightened, the wave speed is increased. The pulse moves faster on the string and makes more "round trips" each second—the frequency of the sound is raised. Different notes are played on the same guitar or violin string by using a finger to hold down the string some distance from its fixed end. The pulse travels a shorter distance between reflections, makes more round trips per second, and produces a higher frequency sound.

A similar process is used in flutes, trumpets, and other wind instruments. Here it is a pressure pulse in the air inside a tube that moves back and forth (Figure 6.36). Initially a sound pulse is produced at one end of the tube by the musician. This pulse travels down the tube and is partially reflected and partially transmitted at the other end. The transmitted part spreads out into the air, becoming the sound that we hear. The reflected part returns to the mouthpiece end, where it is reflected again and reinforced by the musician. (The sound pulse is also inverted at each end: it goes from a compression to an expansion and vice versa.) The musician must supply pressure pulses at the same frequency that the pulse oscillates back and forth in the tube. This, of course, is the frequency of the sound that is produced.

Different notes are played by changing the length of the tube—by opening side holes in woodwinds and by using valves or slides in brasses. The speed of the pulses is determined by the temperature of the air. This is one reason why musicians "warm up" before a performance. The air inside the instrument is warmed by the musicians' breath and hands. Hence the frequencies of the notes are higher than when the air inside is cool.

FIGURE 6.36
Sound production in a flute. The musician causes a sound (pressure) pulse to oscillate back and forth in the tube. Each time the pulse reaches the right end, it sends out a compression—part of the sound that we hear. Opening a side hole shortens the path of the pulse, thereby increasing the frequency.

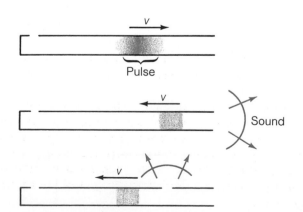

FIGURE 6.37
All singers produce sound by making their vocal cords vibrate. The position of the jaw, the size of the mouth opening, and other factors also influence the sound.

The human voice uses several types of sound production and modification mechanisms. Some consonant sounds like *"sss"* and *"fff"* are technically noise: they are hissing sounds produced by air rushing over the teeth and lips. The randomly swirling air produces sounds with random, changing frequencies. The vocal cords, located inside the Adam's apple in the throat, are the primary sound producers for singing and for spoken vowel sounds (see Figure 6.37). When air is blown through the vocal cords, they vibrate and produce pressure pulses (sound) much like a reed of a saxophone. This sound is modified by the shapes of the air cavities in the throat, mouth, and nasal region. Muscles in the throat are used to tighten and to loosen the vocal cords, thereby changing the pitch of the sound. Moving the tongue or jaws changes the shape of the mouth's air cavity and allows for different sounds to be produced. A sinus cold can change the sound of one's voice because swelling changes the configuration of the nasal cavity.

Perhaps you've heard someone who had inhaled helium speak. (This is *not* a recommended exercise. It is possible to suffocate because of lack of oxygen in the lungs.) The speed of sound in helium is nearly three times that in air (refer back to Table 6.1). This raises the frequencies of the sounds and gives the speaker a falsetto voice.

Sound waves carry energy, as do all waves. This means that the source of the sound must supply energy. Speaking loudly or playing an instrument for extended periods of time can tire you out for this reason. For continuous sounds, it is more relevant to consider the power of the source, since the energy must be supplied continuously. Most instruments, including the human voice, are very inefficient: typically only a small percentage of the energy output of the performer is converted into sound energy.

6.5   PROPAGATION OF SOUND

Once a sound has been produced, what factors affect the sound as it travels to our ears? The general aspects of wave propagation discussed in Section 6.2 of course apply to sound waves. Of these, reflection, diffraction, and the reduction of amplitude with distance from the sound source are most important in influencing the sound that actually reaches us.

The simplest situation is a single source of sound in an open space—such as a person talking in an empty field. The sound travels in three dimensions, and its amplitude decreases as the wavefronts expand. In particular, the amplitude is inversely proportional to the distance from the sound source.

$$\text{Amplitude} \propto \frac{1}{d}$$

When you move to twice as far away from a steady sound source, the amplitude of the sound is decreased by one half. The sound becomes quieter as you move away from the sound source.

Sound propagation is more complicated inside rooms and other enclosures. First, diffraction and reflection of sound allow you to hear sound from sources that you can't see because they are around a corner. We are so accustomed to this phenomenon that it doesn't seem mysterious. Second, even when the source is inside the room with you, most of the sound that you hear has been reflected one or more times off the walls, ceiling, floor, and any objects in the room. This has a large effect on the sound that we hear.

Figure 6.38 shows that sound emitted by a source in a room can reach your ears in countless ways. Consider a single sound pulse like a hand clap. In an open field, you would hear only a single, momentary sound as the pulse moves by you. A similar pulse produced in a room is heard repeatedly: you hear the sound that travels directly to your ears; then sound that reflects off the ceiling, floor, or a wall before reaching your ears; then sound that is reflected twice, three times, and so

FIGURE 6.38
Some possible pathways for the sound in a room to travel from the source to your ears. D represents the sound that reaches your ears directly. Rays 1 and 2 are reflected once. Ray 3 is reflected twice. In most rooms the sound can be reflected more than a dozen times and still be heard.

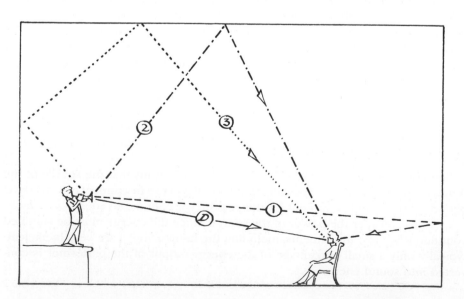

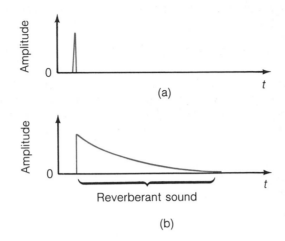

FIGURE 6.39
Comparison of the sound heard from a hand clap in (a) an open field and in (b) a room. A single pulse is heard in the field, whereas a continual, fading sound is heard in the room.

on. These reflected sound waves travel greater and greater distances before reaching your ears and so are heard successively later than the direct sound. This process of multiple, random reflections of sound in an enclosure is called *reverberation*. The single hand clap is heard as a continuous sound that fades quickly.

Figure 6.39 compares the sound from a hand clap as it is heard in an open field and in a room. Each graph shows the amplitude of the sound that is heard versus the time after the sound pulse is produced. The reverberation causes the sound to "linger" in the room. The indirect sound that one hears after the initial direct pulse is called the *reverberant sound*. The amount of time it takes for the reverberant sound to fade out depends on the size of the room and the materials that cover the walls, ceiling, and floor. Sound is never completely reflected by a surface: some percentage of the energy in an incoming wave is absorbed by the surface, leaving the reflected wave with a reduced amplitude. (Concrete absorbs only about 2% of the incident sound's energy, whereas carpeting and acoustical ceiling tile can absorb up to around 90%.) A room with a large amount of sound-absorbing materials in it will have little reverberation; after a few reflections the sound loses most of its energy and cannot be heard.

The *reverberation time* is used to compare the amount of reverberation in different rooms. It is the time it takes for the amplitude of the reverberant sound to decrease by a factor of 1,000. It varies from a small fraction of a second for small rooms with high sound absorption to several seconds for large, brick-walled gymnasiums and similar enclosures. (The Taj Mahal is made of solid marble, which absorbs very little sound. The reverberation time of its central dome is about 12 seconds.)

DO-IT-YOURSELF PHYSICS

You can make a rough measurement of the reverberation time of a large room, such as a lecture hall, with a stopwatch and another person to help you. Have the other person produce a single, loud hand clap, and start the stopwatch at that instant. Stop it the moment that you can no longer hear the reverberant sound.

FIGURE 6.40
Symphony Hall in Boston is very successful at using reverberation to enhance sound.

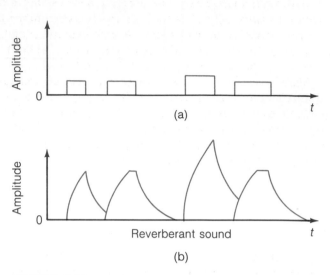

FIGURE 6.41
A series of steady sounds, such as spoken syllables, are heard (a) in an open field and (b) in a room. Reverberation in the room causes the sounds to merge. This tends to blend musical notes and makes speech more difficult to understand.

When a steady sound is produced in a room, such as a trumpet playing a long note in an auditorium, the sound that one hears is affected by reverberation in a number of ways:

1. The sound is louder than it would be if you had heard it at the same distance in an open field.

2. The sound "surrounds" you; it comes from all directions, not just straight from the source (Figure 6.38).

3. Beyond a short distance from the sound sourse, the loudness does not decrease as rapidly with distance as it would in an open field. That is why one can often hear as well near the back of an auditorium as in the middle.

4. Not only does the sound fade gradually when the source stops, the sound also "builds" when the source starts.

Moderate reverberation has an overall positive effect on the sound that we hear, particularly music (see Figure 6.40). However, excessive reverberation adversely affects the clarity of both speech and music. In essence, speech and music are a series of short, steady sounds interspersed with short moments of silence. Each note, word, or syllable is followed by a brief pause. If we again graph the amplitude of sound versus time, we can see the effect of reverberation (see Figure 6.41). In an open field one hears each syllable or note as a distinct, separate sound. In a room, the individual sounds begin to merge. As a new note is played or a new word is spoken, the reverberant sound from the preceding one can still be heard. The longer the reverberation time the more the sounds overlap each other and the harder it is to understand speech. (Racquetball courts have hard, smooth walls and very high reverberation times; that is why it is very difficult for players to converse unless they are close to each other.) It is recommended that the reverberation time of rooms used for oral presentations and lectures should be around 0.5 to 1 second. For concert halls it should be from 1 to 2 seconds, depending on the type of music being performed.

Many other factors besides reverberation time must be taken into account by architects and building designers. For example, a balcony must be high above the main floor and not extend out too far, or little reverberant sound will reach the seats below it. Also, sound is focused by concave walls and ceilings. Building an elliptical (egg-shaped) auditorium could result in the sound being concentrated in a small area of the room (Figure 6.17).

6.6   PERCEPTION OF SOUND

In this section we consider some of the aspects of sound perception: how the physical properties of sound waves are related to the mental impressions we have when we hear sound. We will be comparing psychological sensations, which can be quite subjective, to measurable physical quantities. (A similar situation: "hot" and "cold" are subjective perceptions that are related to temperature, which is a measurable physical quantity.) To make things simple, we will limit ourselves to only steady, continuous sounds. This frees us from having to include such effects as reverberation in a room; that is, how the sound builds up, how it decays, and so on.

The main categories that we use to subjectively describe sounds are *pitch, loudness,* and *tone quality.*

> The *pitch* of a sound is the perception of highness or lowness. The sound of a soprano voice has a high pitch and that of a bass voice has a low pitch. The pitch of a sound depends primarily on the *frequency* of the sound wave.
>
> The *loudness* of a sound is self-descriptive. We can distinguish among very quiet sounds (difficult to hear), very loud sounds (painful to the ears), and sounds with loudness somewhere in between. The loudness of a sound depends primarily on the *amplitude* of the sound wave.
>
> The *tone quality* of a sound is used to distinguish two different sounds even though they have the same pitch and loudness. A note played on a violin does not sound quite like the same note played on a flute. The tone quality of a sound (also referred to as the "timbre" or "tone color") depends primarily on the *waveform* of the sound wave.

Pitch

Pitch is perhaps the most accurately discriminated of the three categories, particularly by trained musicians. It depends almost completely on the frequency of the sound wave: the higher the frequency, the higher the pitch. (Noise does not have a definite pitch, since it does not have a definite frequency.)

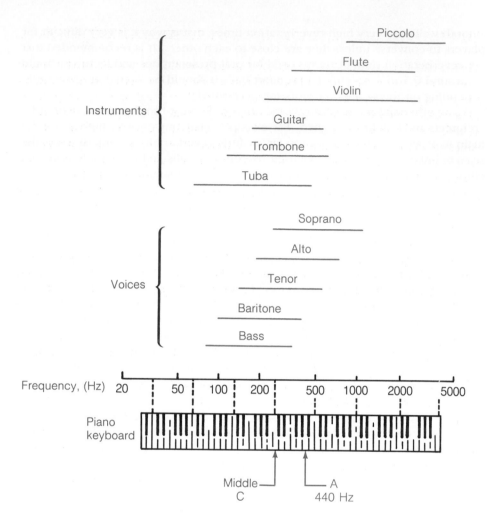

FIGURE 6.42
The frequencies of the notes on a piano are indicated and compared with the approximate frequency ranges of different singing voices and musical instruments. This frequency scale, and therefore the piano keyboard, is a logarithmic scale: the interval between 50 and 100 hertz is the same as the interval between 500 and 1,000 hertz. (From Physics in Everyday Life *by Richard Dittman and Glenn Schmieg.)*

Pitch is, of course, essential to nearly all music. There is a great deal of arithmetic in the musical scale; each note has a particular numerical frequency. Figure 6.42 shows the frequencies of the notes on the piano keyboard. There are seven octaves on the piano, each consisting of 12 different notes. Within each octave, the notes are designated A through G, plus five sharps and flats. Each note in a given octave has exactly *twice* the frequency of the corresponding note in the octave below. For example, the frequency of the lowest note on the piano, an A, is 27.5 hertz; the A in the next octave has frequency 55 hertz; the third A has 110 hertz, and so on. The frequency of middle C is 261.6 hertz.

Certain combinations of notes are pleasing to the ear, whereas others are not. It was nearly 2,500 years ago that Pythagoras indirectly discovered that two different notes are in harmony when their frequencies have a simple whole-number ratio. For example, a musical "fifth" is any pair of notes whose frequencies are in the ratio 3 to 2. Any E and the first A below it have this ratio of frequencies, as do any G and the first C below it.

Figure 6.42 also shows the approximate ranges of singing voices and some instruments. For normal speech, the ranges are approximately 70 to 200 hertz for men and 140 to 400 hertz for women. When whispering, you do not use your vocal cords, and you produce much higher frequency "hissing" sounds.

Loudness

The loudness of a sound is determined mainly by the amplitude of the sound wave. The greater the amplitude of the sound wave that reaches your eardrums, the greater the perceived loudness of the sound. The actual pressure amplitudes of normal sounds are extremely small: typically around one millionth of 1 atmosphere. This causes the eardrum to vibrate through a distance of around 100 times the diameter of a *single atom.* An extremely faint sound has an amplitude of less than one billionth of 1 atmosphere, and it makes the eardrum move *less* than the diameter of an atom. The ear is an amazingly sensitive device.

There is a specially defined physical quantity that depends on the amplitude of sound but is more convenient for relating amplitude to perceived loudness. This is the sound pressure level or simply the *sound level.*

The standard unit of sound level is the *decibel (dB).* The range of sounds that we are normally exposed to have sound levels from 0 decibel to about 120 decibels. Figure 6.43 shows some representative sound levels along with their relative perceived loudness. Sound level does not take into account the irritability of the sound. Your favorite music played at 100 decibels may not sound as loud as an annoying screech of fingernails on a blackboard with a sound level of 80 decibels.

The relationship between the amplitude of a steady sound and its sound level is based on factors of 10. A sound with 10 times the amplitude of another sound has a sound level that is 20 decibels higher. A 90-decibel sound has 10 times the amplitude of a 70-decibel sound.

The following five statements describe how the perceived loudness of a sound is related to the measured sound level. These are general trends that have been identified by researchers after testing large numbers of people. For a particular person, the actual numerical values of the sound levels can vary somewhat from those listed. Also some of the values given, particularly in number 1 and 3 below, depend on the frequencies of the sounds.

   1. The sound level of the quietest sound that can be heard under ideal conditions is 0 decibel. This is called the "threshold of hearing."

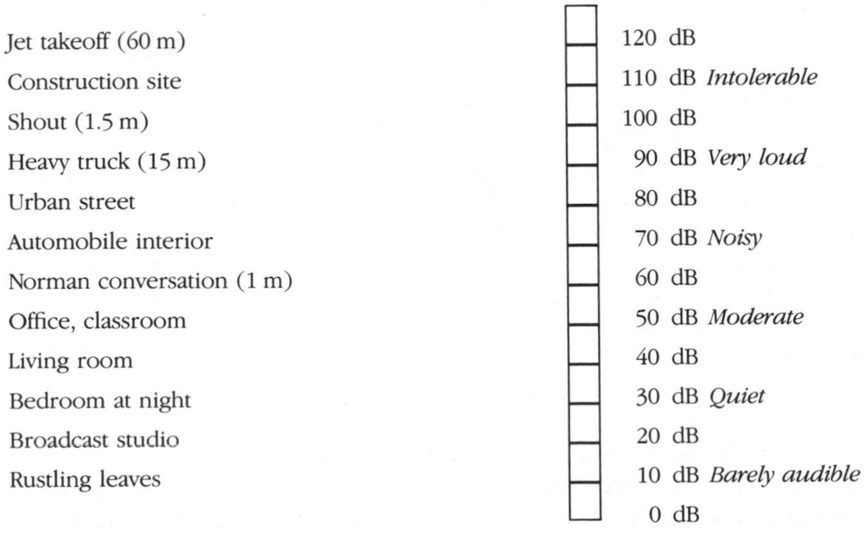

Jet takeoff (60 m)	120 dB	
Construction site	110 dB	*Intolerable*
Shout (1.5 m)	100 dB	
Heavy truck (15 m)	90 dB	*Very loud*
Urban street	80 dB	
Automobile interior	70 dB	*Noisy*
Norman conversation (1 m)	60 dB	
Office, classroom	50 dB	*Moderate*
Living room	40 dB	
Bedroom at night	30 dB	*Quiet*
Broadcast studio	20 dB	
Rustling leaves	10 dB	*Barely audible*
	0 dB	

FIGURE 6.43
The decibel scale with representative sounds that produce approximately each sound level. Sound levels above about 85 decibels pose a danger to hearing. Levels above 120 decibels cause ear pain and the potential for permanent hearing loss. (From The Physics Teacher, *Part 1 by Thomas D. Rossing.)*

2. A sound level of 120 decibels is called the "threshold of pain." Sound levels this high cause pain in the ears and can result in immediate damage to them.

3. The minimum increase in sound level that makes a sound noticeably louder is approximately 1 decibel. For example, if a 67-decibel sound is heard and after that a 68-decibel sound, we can just perceive that the second sound is louder. If the second sound had a sound level of 67.4 decibels, we could not notice a difference in loudness.

4. A sound is judged to be twice as loud as another if its sound level is about 10 decibels higher. A 44-decibel sound is about twice as loud as a 34-decibel sound. A 110-decibel sound is about twice as loud as a 100-decibel sound. This is a cumulative factor: a 110-decibel sound is about four times as loud as a 90-decibel sound, and so on.

5. If two sounds with equal sound levels are combined, the resulting sound level is about 3 decibels higher. If one lawn mower causes an 80-decibel sound level at a certain point nearby, starting up a second identical lawn mower next to the first will raise the sound level to about 83 decibels. It turns out that 10 similar sound sources are perceived to be twice as loud as a single source (see Figure 6.44).

The loudness of pure tones and, to a lesser degree, of complex tones, also depends on the frequency. This is because the ear is inherently less sensitive to low- and high-frequency sounds. The ear is most sensitive to sounds in the frequency range of 1,000 to 5,000 hertz. For example, a 50-hertz pure tone at 78 decibels, a 1,000-hertz pure tone at 60 decibels, and a 10,000-hertz tone at 72 decibels all sound equally loud. (At very high sound levels, 80 decibels and above, the ear's sensitivity does not vary as much with frequency as it does at lower sound levels.) One reason for this variation in sensitivity is that a considerable amount of low-frequency sound is produced inside our bodies by flowing blood and flexing muscles. The ear is less sensitive to low-frequency sounds, so these internal sounds do not "drown out" the external sounds that we need to hear.

Sound levels are measured with sound-level meters (Figure 6.45). Most sound-level meters are equipped with a special weighting circuit (called the "A" scale) that allows them to respond to sound much as the ear does; the response to low and high frequencies is diminished. The readings on the A scale are designated "dBA." When operating in the normal mode (called the "C" scale) a sound-level meter measures the sound level in decibels, treating all frequencies equally. When in the A scale mode, a sound-level meter responds like the human ear and therefore indicates the relative loudness of the sound.

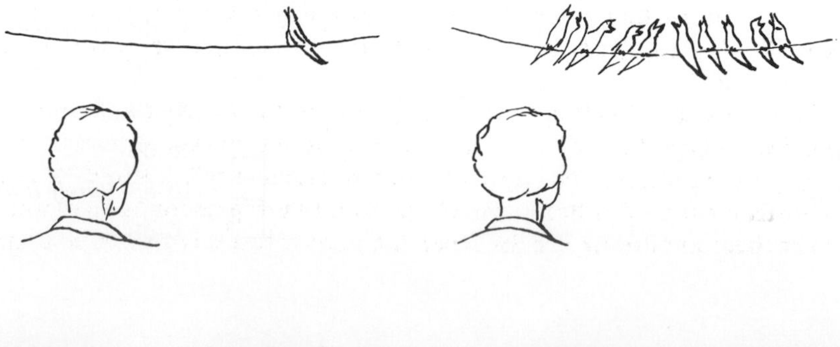

FIGURE 6.44
Ten equivalent sources sound about twice as loud as one.

Loud sounds can not only damage your hearing but can also affect the physiological and psychological balance of your body. Since the beginning of the human race, the sense of hearing has been used as a warning device: loud sounds often indicate the possibility of danger, and the body automatically reacts by becoming tense and apprehensive. Constant exposure to loud or annoying sounds puts the body under stress for long periods of time and consequently jeopardizes the physical and mental well-being of the individual.

The Occupational Safety and Health Administration (OSHA) has established standards designed to protect workers from excessive sound levels. Workers must be supplied with sound-protection devices such as earplugs if they are exposed to sound levels of 90 dBA or higher (Table 6.2). Some communities also have noise ordinances designed to reduce the sound levels of traffic and other activities.

FIGURE 6.45
A sound-level meter.

Tone Quality

The tone quality of a sound is not as easily described as loudness or pitch. Comparisons such as full versus empty, harsh versus soft, or rich versus dry are sometimes used. The tone quality of a sound is very important to our ability to identify what produced the sound. The sound of a flute is different from the sound of a clarinet, and we notice this even if they produce the same note at the same sound level. The tone quality of a person's voice helps us to identify the speaker.

The tone quality of a sound depends primarily on the waveform of the sound wave. If two sounds have different waveforms, we usually perceive different tone qualities. The simplest waveform is that of a pure tone: sinusoidal. Pure tones have a soft, pleasant tone quality (unless they are very loud or high pitched). Complex tones with waveforms that are nearly sinusoidal share the same characteristics. Unlike frequency and sound level, a waveform cannot be expressed as a single numerical factor.

What determines the waveform of a sound wave? The following mathematical principle gives us a way to comparatively analyze waveforms.

> Any complex waveform is equivalent to a combination of 2 or more sinusoidal waveforms with definite amplitudes. These component waveforms are called "harmonics." The frequencies of the harmonics are whole number multiples of the frequency of the complex waveform.

For our purposes this means that *any complex tone is equivalent to a combination of pure tones.* These pure tones, called the "harmonics," have frequencies that are equal to 1, 2, 3, . . . etc, multiplied by the frequency of the complex tone. For example, Figure 6.46 shows the waveform of a complex tone whose frequency is 100 hertz. It is equivalent to a combination of the three pure tones shown whose frequencies are 100, 200, and 300 hertz. This means that we could artificially produce this complex tone by carefully playing the individual pure tones simultaneously. This is one method that electronic music synthesizers use to create sounds.

The tone quality of a complex tone depends on the number of harmonics that are present and on their relative amplitudes. These two factors give us a quantitative way of comparing waveforms. A spectrum analyzer is a sophisticated electronic instrument that indicates which harmonics are present in a complex tone and what their amplitudes are. In general, complex tones with a large number of

TABLE 6.2	OSHA NOISE LIMITS
Sound Level	Daily Exposure
(dBA)	(hours)
90	8
92	6
95	4
97	3
100	2
102	1.5
105	1
110	0.5
115	0.25

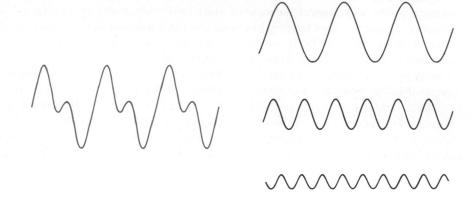

FIGURE 6.46
The complex tone waveform on the left is a combination of the three pure tone waveforms on the right. The pure tones (harmonics) have frequencies that are 1, 2, and 3 times the frequency of the complex tone.

harmonics have a rich tone quality. Notes played on a recorder or a flute contain only a couple of harmonics, and so they sound similar to pure tones. Violins and clarinets, on the other hand, have over a dozen harmonics in their notes and consequently have richer tone qualities.

The waveform of noise is a random "scribble" that doesn't repeat. This is because noises are composed of large numbers of frequencies that are not related to each other: they are not harmonics. "White noise" contains equal amounts of all frequencies of sound, just as white light contains equal amounts of all colors. The sound of rushing air approximates white noise.

The existence of higher frequency harmonics in musical notes explains why high-fidelity sound-reproduction equipment must respond to frequencies up to about 20,000 hertz. Even though the frequencies of musical notes are generally less than 4,000 hertz, the frequencies of the harmonics *do* go up to 20,000 hertz and higher. (Since our ears can't hear any harmonic with a frequency above 20,000 Hz, it isn't necessary to reproduce them.) To accurately reproduce a complex tone, each higher-frequency harmonic must be reproduced (see Figure 6.47).

The study of sound perception brings together the fields of physics, biology, and psychology. The mechanical properties of sound waves interact with the physiological mechanisms in the ear to produce a psychological perception. One of the challenges is to relate subjective descriptions of different sound perceptions to measurable aspects of the sound waves. The hearing apparatus itself is one of the most remarkable in all of nature. It responds to an amazing range of frequencies and amplitudes and still captures the beauty and subtlety of music. At

FIGURE 6.47
High-fidelity sound equipment must be able to reproduce high-frequency harmonics. This equalizer has 10 frequency bands that span the range from 20 hertz to 20 kilohertz.

the same time it is very fragile. We need to learn more about how our sense of hearing works and how it can be protected.

HISTORICAL NOTES

Waves

The struggle to understand the nature of waves and wave propagation was waged on two fronts: optics and acoustics. Optics is the formal study of light, and acoustics is the study of sound and vibration. Progress in both areas over the last four centuries or so has produced a nearly complete picture of wave phenomena. The two areas of investigation complemented each other nicely because some wave properties are more easily exhibited with light, whereas others are more easily illustrated with sound. For example, although diffraction of sound is a very common occurrence, the diffraction of light was "discovered" first by the Italian mathematician Francesco Maria Grimaldi (1618–1663). A beam of light passing through a tiny hole in a plate produces a projected circle of light on a screen that is larger than the size of the hole. This is because the light spreads out as it passes through the hole. Isaac Newton also experimented with light diffraction.

In a similar manner, the Doppler effect was first investigated with sound. Predicted for both light and sound by the Austrian physicist Christian Johann Doppler (1803–1853) in 1842, the Doppler effect is not easily observed with light because of its extreme speed. In 1845 a Dutch meteorologist named Christoph Heinrich Dietrich Buys-Ballot (1817–1890) illustrated the Doppler effect using trains to provide the motion for a sound source, trumpets. He also used the trumpets as a stationary source and put listeners on the moving train.

Sound

Acoustics is one of the oldest branches of physics and yet one of the newest. It is one of the oldest because the first recorded discovery in acoustics occurred 200 years *before* Aristotle; it is one of the newest because most of our present knowledge of sound has been developed only in the last 150 years. Why did acoustics progress more slowly than, say, mechanics and thermodynamics? One main reason is that fairly sophisticated equipment is needed for experimentation because of the relatively high speed and rapid fluctuations (high frequencies) of sound. The development of calculus in the 18th century allowed for greater theoretical understanding of general wave phenomena, but experimental verification and investigation were limited by the crude technology.

Another reason for the slow development of acoustics is that certain areas, most notably the design of musical instruments, defied the complete mathematical analysis made possible in mechanics by Newton's work. To this day there is no "magic equation" that can be used to design a flute, for example, based solely on mathematical principles. The exact length, location of tone holes, and sizes of

FIGURE 6.48
Marin Mersenne

FIGURE 6.49
Joseph Fourier

the tone holes have been determined as much by trial and error on the part of instrument makers as by mathematical analysis.

The earliest discoveries in acoustics were apparently made around 550 BC by Pythagoras, a Greek philosopher and mathematician whose name is immortalized by a theorem in geometry. Although Pythagoras left no record of his work, the accounts of his successors indicate that he used a stretched string to discover that two pitches sound good together (are in harmony) if the ratio of their frequencies is simple: 2 to 1, 3 to 2, and so on. A treatise written by Aristotle indicates that he had a good understanding of the nature of sound and sound production. A rare contribution to science by the Romans was made by the architect Vitruvius around 50 BC. In a treatise on the design of theaters, he discussed the effects of reverberation and echoes on sound clarity.

The next notable discoveries in acoustics were made by the father of experimental science, Galileo. He completed the analysis of the vibrations of strings and correctly determined the relationship between the frequency of vibrations and the string's length, mass density, and tension. Galileo also showed that the pitch of a sound is determined by its frequency. At about this same time, a French Franciscan friar named Marin Mersenne (1588–1648) made fairly good measurements of the speed of sound in air and of the actual frequencies of musical notes (Figure 6.48). Other experimenters measured the speed of sound quite accurately by measuring the time between the flash and the arrival of the sound from a distant gun.

In his *Principia,* Isaac Newton showed how the properties of air can be used to theoretically predict the speed of sound. However, his predicted value was markedly lower than the measured value because he did not take into account the fact that the air is rapidly heated and cooled by the compressions and expansions associated with a sound wave. Newton then "fudged" his analysis by introducing erroneous compensating factors that raised his theoretical value so as to agree with the measured one. Even the great master was not correct on all matters.

A major step toward understanding the nature of complex tones was made by the French mathematician Joseph Fourier (1768–1830; Figure 6.49). He discovered that any complex waveform is actually a composite of simple, sinusoidal waveforms (Section 6.6). Fourier did not apply this principle to sound. The German physicist Georg Ohm (1787–1854), famous for his work on the conduction of electricity, used Fourier's discovery to distinguish pure tones and complex tones and to show that complex musical tones are combinations of pure tones.

Perhaps the greatest contributions to the science of acoustics, particularly sound perception, were made by Hermann von Helmholtz (1821–1894; Figure 6.50). Helmholtz was a man of prodigious analytical talent, which he applied as a surgeon, physiologist, physicist, and mathematician. In addition to his work in acoustics, Helmholtz made important contributions to electricity and magnetism and to the principle of the conservation of energy. It has been said that as a doctor he began to study the eye and the ear but found that he needed to learn more physics. Then he found that he needed to study mathematics to learn physics. In this way he became a noted physiologist, a noted physicist, and a noted mathematician.

Helmholtz used his medical understanding of the ear to make several discoveries about sound perception. He distinguished the sensations of loudness, pitch, and tone quality in a sound. He showed that the tone quality of a complex tone depends on the harmonics that it contains. To detect these harmonics, he designed a set of "Helmholtz resonators"—hollow spheres tuned to particular frequencies that would amplify an individual pure tone in a complex tone. Most of his discoveries are beyond the level of this text. Helmholtz's book *Sensations of Tone* was perhaps *the* musical acoustics textbook for nearly a century. Only with the advent of modern electronic instrumentation in this century has it been possible to improve upon Helmholtz's findings about sound perception.

In recent years there has been a resurgence of interest in the field of musical acoustics. One example of this is the establishment of the *Institut de Recherche et Coordination Acoustique/Musique* in Paris in the late 1970s. Here musicians and scientists work together, using computers and other sophisticated electronic instruments to study and create musical sounds. Many physicists find musical acoustics to be a particularly gratifying field of research because it combines the art and beauty of music with the elegance and depth of physical science.

FIGURE 6.50
Hermann von Helmholtz

SUMMARY

Waves are everywhere. They can be classified as transverse or longitudinal according to the orientation of the wave oscillations. The speed of a wave depends on the properties of the medium through which it travels. For continuous periodic waves, the product of the frequency and the wavelength equal the wave speed.

Once waves are produced, they are often modified as they propagate. Reflection and diffraction cause waves to change their direction of motion when they encounter boundaries. Interference of two waves produces alternating regions of larger amplitude and zero-amplitude waves. The Doppler effect and the formation of shock waves are phenomena associated with moving wave sources. The former also occurs when the receiver of the waves is moving.

Although sound can refer to a broad range of mechanical waves in all types of matter, we often restrict the term to the longitudinal waves in air that we can hear. Sound waves in air are generally represented by the air-pressure fluctuations associated with the compressions and rarefactions in the sound wave. Sound with frequency too high to be heard by humans, above about 20,000 hertz, is called "ultrasound." Ultrasound is used for echolocation by bats and for internal imaging in medicine. The sounds that we can hear can be divided into pure tones, complex tones, and noise.

Sound is produced in many different ways, all resulting in rapid pressure fluctuations that travel as a wave.

Different musical instruments employ diverse and sometimes multiple sound-production mechanisms, including vibrating plates or membranes, vibrating strings, or vibrating columns of air in tubes.

Sound propagation inside rooms and other enclosures is dominated by repeated reflection, called reverberation. This causes individual sounds to linger after they are produced. The length of time that reverberant sound can be heard is approximately equal to the reverberation time—the time it takes for the amplitude of the fading sound to decrease by a factor of 1,000. Moderate reverberation causes a positive blending of successive sounds, such as musical notes, but can adversely affect the clarity of speech.

We use three main characteristics to classify a steady sound we perceive: pitch, loudness, and tone quality. The pitch of a sound depends mainly on the frequency of the sound wave. The loudness of a sound depends mainly on the amplitude (or sound level) of the sound but is also affected by the frequency; the ear is more sensitive to frequencies between 1,000 and 5,000 hertz. The sound level of a sound wave, measured in decibels, is related to the sound's amplitude. The tone quality of a sound depends on the waveform of the sound wave. The waveform of a complex tone depends on the number and amplitudes of the separate pure tones, called harmonics, that comprise it.

SUMMARY OF IMPORTANT EQUATIONS

EQUATION	COMMENTS
$\rho = \dfrac{m}{l}$	Linear mass density of rope, wire, string, Slinky, etc
$v = \sqrt{\dfrac{F}{\rho}}$	Speed of waves on a rope, wire, string, Slinky, etc
$v = 20.1 \times \sqrt{T}$	Speed of sound waves in air (similar for other gases)
$v = f\lambda$	Relates frequency, wavelength, and speed for continuous waves

QUESTIONS

1. Take a close look at the pulse traveling on the rope in Figure 6.2. It was produced by moving the hand quickly up, then down. Notice that the rope is blurred ahead of the peak of the pulse and behind it but that the peak itself is not. Explain why.

2. What is the difference between a longitudinal wave and a transverse wave? Give an example of each.

3. A popular distraction in large crowds at sporting events during the 1980s was the "wave." The people in one section would quickly stand up then sit down; the people in the neighboring sections would follow suit in succession, resulting in a visible pattern in the crowd that would travel around the stadium. Which of the two types of waves is this? Compared to the waves described in this chapter, how is the stadium wave different?

4. A long row of people are lined up behind one another at a service window. Joe E. Clumsy stumbles into the back of the person at the end and pushes hard enough to generate a wave in the people waiting. What type wave is produced?

5. The speed of an aircraft is sometimes expressed as a "Mach number": Mach 1 means that the speed is equal to the speed of sound. If you wish to determine what the speed of an aircraft is that is going Mach 2.2, for example, in terms of meters per second or miles per hour, what additional information do you need?

6. Based on information in Section 4.1 and in the Periodic Table of the Elements (Appendix B), why is it not surprising that sound travels faster in helium than in air?

7. If you were actually in a battle fought in space like the ones shown in science fiction movies, would you hear the explosions that occur? Why or why not?

8. Explain what the amplitude, frequency, and wavelength of a wave are.

9. When trying to hear a faint sound from something far away, we sometimes "cup" a hand behind an ear. Explain why this can help.

10. What useful thing happens to a wave when it encounters a concave reflecting surface?

11. When a wave passes through two nearby gaps in a barrier, interference will occur, provided that there is also diffraction. Why must there be diffraction?

12. A recording of a high-frequency pure tone is played through both speakers of a portable stereo placed in an open field. A person a few meters in front of the stereo walks slowly along an arc around it. How does the sound that is heard change as the person moves?

13. A person riding on a bus hears the sound from a horn on a car that is stopped. What is different about the sound when the bus is approaching the car compared to when the bus is moving away from the car?

14. Explain the process of echolocation. How is the Doppler effect sometimes incorporated?

15. In the past, ships often carried small cannons that were used when approaching shore in dense fog to estimate the distance to the hidden land. Explain how this might have been done.

16. While a shock wave is being generated by a moving wave source, is the Doppler effect also occurring? Explain.

17. Describe some of the things that would happen if the speed of sound in air suddenly dropped to, say 20 m/s. What would it be like living next to a freeway?

18. If a boat is producing a bow wave as it moves over the water, what must be true about its speed?

19. As a loud, low-frequency sound wave travels by a small balloon, the balloon's size is affected. Explain. (The effect is too small to be observed under ordinary circumstances.)

20. Describe the waveforms of pure tones, complex tones, and noise.

21. What is ultrasound, and what is it used for?

22. Describe how sound is produced in string instruments. Why does tightening a string change the frequency of the sound it makes?

23. A special room contains a mixture of oxygen and helium that is breathable. Two musicians play a guitar and a flute in the room. Does each instrument sound different than when played in normal air? Why or why not?

24. What is reverberation? How does reverberation affect how we hear sounds?

25. What are the three categories used to describe our mental perception of a sound? Upon what physical properties of sound waves does each depend?

26. A 100-Hz pure tone at a 70-dB sound level and a 1,000-Hz pure tone at the same sound level are heard separately. Do they sound equally loud? If not, which is louder, and why?

27. The highest musical note on the piano has a frequency of 4,186 Hz. Why would a tape of piano music sound terrible if played on a tape player that reproduces frequencies only up to 5,000 Hz?

PROBLEMS

1. Two children stretch a jump rope between them and send wave pulses back and forth on it. The rope is 3 m long, its mass is 0.5 kg, and the force exerted on it by the children is 40 N.
   a) What is the linear mass density of the rope?
   b) What is the speed of the waves on the rope?

2. The force stretching the D string on a certain guitar is 150 newtons. The string's linear mass density is 0.005 kg/m. What is the speed of waves on the string?

3. What is the speed of sound in air at the normal boiling temperature of water?

4. The coldest and hottest temperatures ever recorded in the United States are $-80°F$ (211 K) and $134°F$ (330 K), respectively. What is the speed of sound in air at each temperature?

5. A 4-Hz continuous wave travels on a Slinky. If the wavelength is 0.5 m, what is the speed of waves on the Slinky?

6. A 500-Hz sound travels through pure carbon dioxide. The wavelength of the sound is measured to be 0.54 m. What is the speed of sound in carbon dioxide?

7. A wave traveling 80 m/s has a wavelength of 3.2 m. What is the frequency of the wave?

8. What frequency of sound traveling in air at 20°C has a wavelength equal to 1.7 m, the height of an average person?

9. Verify the figures given in Section 6.1 for the wavelengths (in air) of the lowest and highest frequencies of sound that people can hear (20 and 20,000 Hz, respectively).

10. What is the wavelength of 3.5 million Hz ultrasound as it travels through human tissue?

11. The frequency of middle C on the piano is 261.6 Hz.
    a) What is the wavelength of sound with this frequency as it travels in air at room temperature?
    b) What is the wavelength of sound with this frequency in water?

12. A steel cable with total length 30 m and mass 100 kg is connected to two poles. The tension in the cable is 3,000 N, and the wind makes the cable vibrate with a frequency of 2 hertz. Calculate the wavelength of the resulting wave on the cable.

13. In a student laboratory exercise, the wavelength of a 40,000-Hz ultrasound wave is measured to be 0.868 cm. Find the air temperature.

14. A 1,720-Hz pure tone is played on a stereo in an open field. A person stands at a point that is 4 m from one of the speakers and 4.4 m from the other. Does the person hear the tone? Explain.

15. A person stands directly in front of two speakers that are emitting the same pure tone. The person moves to one side until no sound is heard. At that point the person is 7 m from one of the speakers and 7.2 meters from the other. What is the frequency of the tone being emitted?

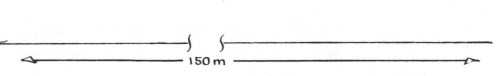

FIGURE 6.51
Problem 16.

|50 m

16. A baseball fan sitting in the "cheap seats" is 150 m from home plate (see Figure 6.51). How much time elapses between the instant the fan sees a batter hit the ball and the moment the fan hears the sound?

17. A geologist is camped 8,000 m (5 miles) from a volcano as it erupts.
   a) How much time elapses before the geologist hears the sound from the eruption?
   b) How much time does it take the seismic waves produced by the eruption to reach the geologist's camp, assuming the waves travel through granite as sound waves do?

18. A person stands at a point 300 m in front of the face of a sheer cliff. If the person shouts, how much time will elapse before an echo is heard?

19. A sound pulse emitted under water reflects off of a school of fish and is detected at the same place 0.01 s later. How far away are the fish?

20. The sound level in a quiet room is 45 dB. Approximately how many times louder is it when a TV is turned on and the sound level is 75 dB?

21. Approximately how many times louder is a 100-dB sound than a 60-dB sound?

22. What are the frequencies of the first four harmonics of middle C (261.6 Hz)?

23. The highest note on the piano has frequency 4,186 Hz.
   a) How many harmonics of that note can we hear?
   b) How many harmonics of the note one octave below it can we hear?

CHALLENGES

1. Redraw Figure 6.15 for a convex boundary (one with the curve in the opposite direction). What happens to the amplitude of the reflected wave?

2. Perform the Do-It-Yourself Physics experiment on page 260 and measure the distance to the wall when you could no longer distinguish the echo. Use this distance to determine the amount of time between the direct sound that you heard and the echo.

3. Jack and Jill go for a walk along an abandoned railroad track. Jack puts one ear next to a rail, while Jill, 200 m away, taps on the rail with a stone. How much sooner does Jack hear the sound through the steel rail than through the air?

4. A stationary siren produces a sound with frequency 500 Hz. A person hears the sound while riding in a car traveling 30 m/s toward the siren. Compute the frequency of the sound that is heard by using the relative speed of the listener and the sound wave. Assume the air temperature is 20° C.

5. An entrepreneur decides to invent and to market a device that will fool the Doppler radar units used to detect cars that are speeding. The device would be placed at the very front of the car and would detect the radar signal, determine its

frequency, and transmit back its own radar signal that would make the radar unit register a legal speed. What would the frequency of the "fake" signal have to be in comparison to the original? If a second unit were designed to be placed at the back end of the car, what would be different about the frequency it would have to use compared to that used by the unit at the front?

6. The guitar string described in problem 2 is approximately 0.6 m long. What is the frequency of oscillation of a pulse on the string? (Hint: The frequency equals 1 divided by the period, the amount of time it takes the pulse to go back and forth once.)

7. Use the relationship between amplitude and sound level described in Section 6.6 to determine how many times larger the amplitude of a 120-dB sound is compared to the amplitude of a 0-dB sound.

8. The frequency of the lowest note played on a flute at room temperature (20°C) is 262 Hz. What would be the frequency of the same note when the flute is played outside in the winter and the air temperature in the flute is 0°C?

FIGURE 6.52
Challenge 3.

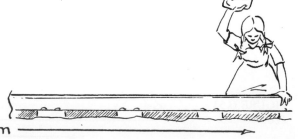

← 200 m →

SUGGESTED READINGS

Cajori, Florian. *A History of Physics*. New York: Dover, 1962 (originally published by Macmillan in 1929). Includes short sections on sound in several chapters.

Miller, Dayton Clarence. *Anecdotal History of the Science of Sound*. New York: Macmillan, 1935. Perhaps the best source on the history of acoustics up to 1935.

Neuweiler, Gerhard. "How Bats Detect Flying Insects." *Physics Today* 33, no. 8 (August 1980): 34–40.

The Physics of Music. Readings from *Scientific American*. San Francisco: Witt. Freeman and Co., 1978. A collection of articles on the singing voice, the piano, woodwinds, brasses, violins, and on three other topics.

Rossing, Thomas D. "Physics and Psychophysics of High-Fidelity Sound." *The Physics Teacher*. Part 1, 17, no. 9 (December 1979): 563; Part 2, 18, no. 4 (April 1980): 278; Part 3, 18, no. 6 (September 1980): 426; Part 4, 19, no. 5 (May 1981): 293. A series of four articles on basic acoustics, with applications to speakers, turntables, tape recording, etc.

Segrè, Emilio. *From Falling Bodies To Radio Waves*. New York: W.H. Freeman and Co, 1984. Chapter 5 has a nice segment on Hermann von Helmholtz and his work.

Walker, Jearl. *The Flying Circus of Physics with Answers*. New York: Wiley, 1977. "A collection of problems and questions about physics in the real, everyday world." Chapter 1 addresses dozens of questions about sound.

A magnet floats above a wafer of yttrium-based superconductor cooled in a bath of liquid nitrogen.

ELECTRICITY

PROLOGUE: A TRULY SUPER CONDUCTOR

This one image represents a remarkable event that stirred the world in 1987. It shows a tiny magnet, the kind that is used to make light-weight stereo headphones, hovering above a high-temperature ("high T_c") superconductor. First viewed in physics and chemistry laboratories around the world, it soon appeared at meetings of business leaders and politicians, in high school science laboratories, in science museum demonstrations, and on national television. It represents a revolutionary discovery, but it is not the physical principles that are new. Magnetic levitation can be accomplished with ordinary magnets repelling each other. Superconductors, early versions of the dark disc in the photo, have been around since 1911. Even the combination of a magnet levitated above a superconductor, one demonstration of what is called the *Meissner effect,* had been done before. What made this image so magical? It represented one of the most stunning, recent discoveries in physics, and yet it was so accessible, not hidden inside some million-dollar apparatus. It was there for the average person to see and to touch. High-school students even manufactured their own superconductors as science projects.

This demonstration captured the world's attention because the superconductor had to be cooled only to the temperature of liquid nitrogen (77 Kelvin or $-320°$ F). Superconductivity was first identified in mercury that was cooled to 4 Kelvin. Its electrical resistance was measured to be zero, meaning it was a perfect conductor of electricity. (That discovery was like finding a Teflon-style coating with zero friction. Such perfections are almost unknown in the real world.) Scientists tested thousands of substances and found that hundreds of them are superconductors, but only at very low temperatures. After nearly 75 years of searching, the world record had inched up to 23 Kelvin. The Nobel Prize winning theory that explained superconductivity predicted that the upper limit might be around 40 Kelvin. Then, during a 13-month period ending in February 1987, the record leaped to 90 Kelvin.

Many superconducting devices have been in use for decades (some are described in Section 7.3), but they are hampered by the fact that the low-temperature superconductors have to be kept cold with liquid helium—a very expensive and cumbersome operation. The new high-temperature superconductors developed in 1987 can be cooled with liquid nitrogen—about a thousand times cheaper to use than liquid helium. This sudden breakthrough in a well-tilled field of physics was amplified by the hope of economic and technological payoffs.

The immediate impact of this discovery was evidenced in many ways: it appeared almost daily in the news media, thousands of scientists dropped their projects to join the search for even higher T_c superconductors, fax machines became indispensable for communicating the latest information, thousands of patents were filed, a race began to see which nation or company could cash in on it, over 2,000 physicists at a normally sedate national meeting jammed an all-night session to hear the latest reports (the event was dubbed the "Woodstock of physics"). Perhaps the high point of this story occurred when the 1987 Nobel Prize in physics was awarded to J. George Bednorz and K. Alex Müller for their January 1986 discovery of this new class of superconductors. Usually many years or even decades pass before a discovery is so rewarded. Albert Einstein waited 16 years. The coinventors of the transistor were rewarded 8 years after their breakthrough. Earning the Nobel Prize in their first year of eligibility (it cannot be awarded in the same year as the discovery) is phenomenal.

Superconductivity and its uses, both now and in the future, are among the basic concepts of electricity discussed in this chapter. Probably no physical phenomenon has ever affected so many aspects of human society as electricity has in the last century. It has drastically changed the home and the workplace, revolutionized communication and transportation, and accelerated the advancement of science almost beyond measure. The first part of the chapter deals with *electrostatics*— effects associated with electrical charges that are not moving. In the remainder of the chapter we discuss the physics of moving charges—*electric currents*. We introduce the important quantities *voltage, resistance,* and *current* and show how they are involved in many devices around us and in living things. The last two sections deal with power and energy in electrical circuits and the two types of electric current—*AC* and *DC.*

7.1   ELECTRIC CHARGE

The word electricity comes from "electron," which is based on the Greek word for amber. Amber is a fossil resin that attracts bits of thread, paper, hair, and other things after it has been rubbed with fur. This phenomenon, known as the "amber effect," was documented by the ancient Greeks, but its source remained a mystery for more than 2 millenia (Figure 7.1). The results of numerous experiments, some conducted by the American statesman Benjamin Franklin, indicated that under certain circumstances, matter exhibited a "new" property. This property was eventually traced to the atom and is called *electric charge.*

> ELECTRIC CHARGE   An inherent physical property of certain subatomic particles that is responsible for electrical and magnetic phenomena. Charge is represented by q, and the unit of measure is the *coulomb,* C.

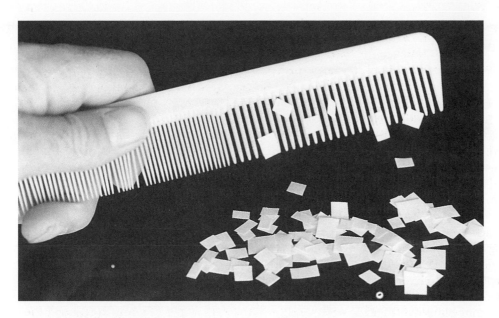

FIGURE 7.1
The amber effect. The comb attracts bits of paper because it was charged by rubbing.

Originally we stated that there are three fundamental things that physicists can quantify or measure: space, time, and properties of matter (Section 1.2). Until now mass has been the only fundamental property of matter that we have used. Electric charge is another basic property of matter, but it is intrinsically possessed only by electrons, protons, and certain other subatomic or "elementary" particles. Unlike mass, there are two different (and opposite) types of electric charge, appropriately named *positive* and *negative* charge. One coulomb of positive charge will "cancel" one coulomb of negative charge. In other words, the *net* electric charge would be zero ($q = -1$ C $+ 1$ C $= 0$ C).

Recall that every atom is composed of a nucleus surrounded by one or more electrons (Section 4.1). The nucleus itself is composed of two types of particles, protons and neutrons (Figure 7.2). Every electron has a charge of -1.6×10^{-19} C, and every proton has a charge of $+1.6 \times 10^{-19}$ C. (To have a total charge of -1 C, over 6 billion billion electrons are needed.) Neutrons are so named because they are neutral; they have no net electric charge. Normally an atom will have the same number of electrons as it has protons, which means the atom as a whole is neutral. This number is the atomic number, which determines the atom's identity. For example, an atom of the element helium has two protons in the nucleus and two

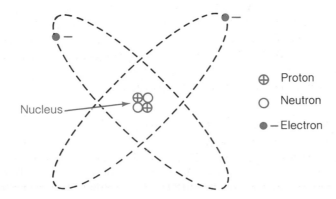

⊕ Proton
○ Neutron
●− Electron

FIGURE 7.2
Simplified model of an atom of the element helium. The nucleus contains two protons; that is why helium's atomic number is two. Also in the nucleus are uncharged neutrons, two in this case. Orbiting the nucleus are electrons. As long as the number of electrons equals the number of protons, the atom has no net charge.

PHYSICS POTPOURRI

Benjamin Franklin: "The Inventor of the Lightning Rod ... and the Republic."

The description above, reportedly given by the French novelist Honoré de Balzac, may well summarize what most of us think when the name Benjamin Franklin is mentioned. We remember Franklin (Figure 7.3) best for his practical inventions, like the lightning rod and the Franklin stove (which he called the "Pennsylvanian fireplace"), his activities as a printer (including publishing his own *Poor Richard's Almanac*), and his achievements as a diplomat and statesman during the founding of the American Republic. Few of us, however, including many physicists, have any appreciation of Franklin's contributions to pure or theoretical science or his stature in the scientific community of his day. To do justice to both of these topics would take many pages, so let us try to get just a little better understanding of Franklin the scientist by briefly examining some of his efforts in the field of electricity.

Franklin performed his electrical experiments during the late 1740s and early 1750s. In describing his results, he imagined that electricity (or "electrical fire") is a single fluid found in all objects that is capable of being circulated among objects. The flow of electrical fluid is initiated by rubbing (friction), a process in which one object, say B, acquires

FIGURE 7.3
Benjamin Franklin (1706–1790).

an excess of electrical fire, while another object, say A, receives a deficit. Franklin described this circumstance as follows: "Hence have arisen some new terms among us: we say, B ... is electrised positively; A negatively. Or rather, B is electrised plus; A, minus." Thus Franklin introduced the terms "positive" and "negative" to describe electricity, not to distinguish different types of charges but to describe the states of objects having a greater or lesser amount of electrical fire in proportion to their fair share.

electrons in orbit about the nucleus. The positive charge possessed by the two protons is exactly opposite that possessed by the two electrons, so the net charge is zero. Most of the substances that we normally encounter are electrically neutral simply because the total number of electrons in all of the atoms is equal to the total number of protons.

A variety of physical and chemical interactions can cause an atom to gain one or more electrons or to lose one or more of its electrons. In these cases the atom is said to be *ionized*. For example, if a helium atom gains one electron, it has three negative particles (electrons) and two positive particles (protons). This atom is said to be a *negative ion*, since it has a net negative charge (Figure 7.4). The value of its net charge is just the charge on the "extra" electron, $q = -1.6 \times 10^{-19}$ C. Similarly, if a neutral helium atom loses one electron, it becomes a *positive ion*, since it has two positive particles and only one negative particle. Its net charge is $+1.6 \times 10^{-19}$ C.

In many situations ions are formed on the surface of a substance by the action of friction. When a piece of amber, plastic, or hard rubber is rubbed with fur, negative ions are formed on its surface. The contact between the fur and the

Franklin's renown among the scientists of his day was based primarily on his electrical theory and its application to the analysis of phenomena associated with the Leyden jar (a precursor of modern charge-holding devices called *capacitors*) and lightning. Using kites and rods, Franklin demonstrated that lightning is nothing more than the discharge of electrical fluid on a grand scale.[1] His book on electricity was widely consulted by scientists during the latter half of the 18th century: it appeared in 10 editions in four languages, including French, German, and Italian. Franklin was awarded the Copley Gold Medal by the Royal Society of London in 1753 for his scientific research and in 1772 was elected a foreign associate of the Académie Royale des Sciences, Paris, only one of eight such members allowed in the society at any one time. Franklin's reputation as a scientist was independent of his stature as a printer and was established *before* his career as a statesman. Indeed, it has been argued that Franklin's success as a diplomat derived in no small way from his international fame as a scientist.

The same wit, wisdom, and forthrightness that Franklin displayed in his *Poor Richard* chronicles were present in his scientific writings and letters, as illustrated by the following excerpt to John Lining on 18 March 1755:

> The fact is singular. You require the reason; I do not know it. Perhaps you may discover it, and then you will be so good as to communicate it to me. I find a frank acknowledgement of one's ignorance is not only the easiest way to get rid of a difficulty, but the likeliest way to obtain information, and therefore I practise it: I think it an honest policy.

Perhaps all of us should seriously consider these words and not be afraid to admit our shortcomings. Perhaps, like Franklin, we should be more willing to "go fly a kite," in spite of possible ridicule, in an effort to learn more about our universe.

[1]*Do not* fly a kite during a thunderstorm! Franklin took precautions and was not hurt, but at least two scientists have been electrocuted while trying similar experiments.

---

material causes some of the electrons in the atoms of the fur to be transferred to some of the atoms on the surface of the solid. The fur acquires a net positive charge, since it has fewer electrons than protons. Similarly, the amber, plastic, or hard rubber acquires a net negative charge, since it has an excess of electrons (Figure 7.5). Combing your hair can charge the comb in the same way. Rubbing

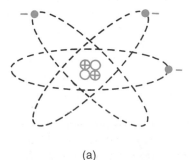

(a)

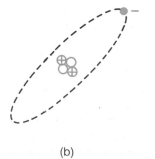

(b)

FIGURE 7.4
An atom is ionized when the number of electrons does not equal the number of protons. (a) a negative helium ion. (b) A positive helium ion.

FIGURE 7.5
The contact between fur and plastic when they are rubbed together causes some of the electrons in the fur to be transferred to the plastic. This leaves the plastic with a net negative charge and the fur with a net positive charge.

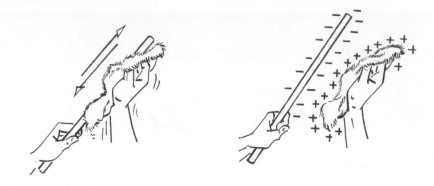

glass with silk causes the glass to acquire a net positive charge. Some of the electrons in the surface atoms of the glass are transferred to the silk, which becomes negatively charged.

Ion formation by friction is a complicated phenomenon that is not completely understood. It is affected by many factors such as which materials are used and the relative humidity.

7.2   ELECTRIC FORCE AND COULOMB'S LAW

The original amber effect illustrates that electric charges can exert forces. You may have noticed hair being pulled toward a charged comb or "static cling" between items of clothes removed from a dryer. These represent the most common situation—two objects with opposite charges attracting each other (Figure 7.6). The negatively charged comb exerts an attractive force on the positively charged hair. In addition, two objects with the same kind of charge (both positive or both negative) repel each other. When two similarly charged combs are suspended from threads, they push each other apart. Just remember this simple rule: *like charges repel, unlike charges attract.*

This force between charged objects is extremely important in the physical world, particularly at the atomic level. An atom can exist because the positively charged protons in the nucleus exert attractive forces on the negatively charged electrons. The electrical force on each electron keeps it in its orbit, much like the gravitational force exerted by the sun keeps the earth in its orbit. The forces

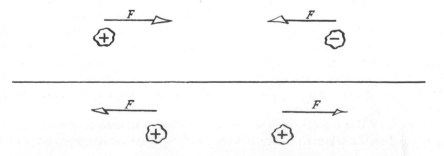

FIGURE 7.6
Objects with opposite electric charges exert attractive forces on one another. Objects with the same kind of charge repel each other.

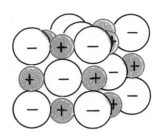

Sodium ion

Chlorine ion

FIGURE 7.7
Ordinary table salt consists of positive sodium ions and negative chlorine ions. The strong attractive forces between the oppositely charged ions hold them in a rigid crystalline array.

between the atoms in many compounds arise because opposite charges attract. For example, when salt is formed from the elements sodium and chlorine, each sodium atom gives up an electron to a chlorine atom. The resulting ions exert attractive forces on one another because they are oppositely charged (Figure 7.7).

The size of the force exerted by one charged body on another depends on the amount of charge on each object and on the distance between them. The exact relationship is known as Coulomb's law.

> COULOMB'S LAW   The force acting on each of two charged objects is directly proportional to the net charges on the objects and inversely proportional to the square of the distance between them.
>
> $$F \propto \frac{q_1 q_2}{d^2}$$
>
> The constant of proportionality in the SI system is 9×10^9 N-m^2/C^2. Therefore:
>
> $$F = \frac{9 \times 10^9 q_1 q_2}{d^2}$$
>
> (SI units—F in *newtons*, q_1 and q_2 in *coulombs*, and d in *meters*)

The force on q_1 is equal and opposite to the force on q_2, by Newton's third law of motion. Note that if both objects have the same kind of charge (both positive or both negative), the force F is positive. This indicates a repulsive force. If one charge is negative and the other positive, the force F is negative, indicating an attractive force. If the distance between two charged objects is doubled, the forces are reduced to one fourth their original values (Figure 7.8).

Notice that Coulomb's law has the same form as Newton's law of universal gravitation (Section 2.9). Perhaps this is not a surprise because both mass and charge are fundamental properties of the particles that comprise matter. We must remember, however, that the (gravitational) force between two bodies due to their masses is *always* an attractive force, while the (electrostatic) force between two bodies due to their electric charges can be attractive or repulsive, depending on whether they have opposite charges or not. Also, all matter has mass and so experiences and exerts gravitational forces, whereas the electrostatic force acts between objects only when there is a net charge on one or both of them. Generally, when two objects have electrical charges, the electrostatic force between them is much stronger than the gravitational force. For example, the electrostatic force between an electron and a proton is 10^{39} times as large as the gravitational force.

It is possible for a charged object to exert a force of attraction on a second object that has not net charge. This is what happens when a charged comb is used

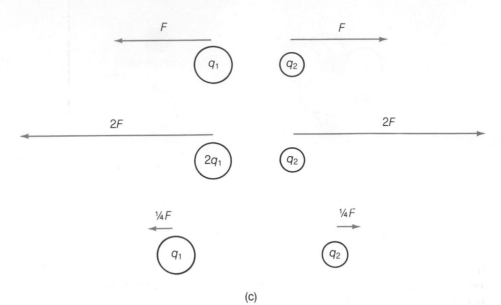

FIGURE 7.8
Coulomb's law. Doubling the charge on one object, (b), doubles the force on both. Doubling the distance between the charges, (c), reduces the forces to one fourth their original values.

to pick up bits of paper or thread. Here the negatively charged comb attracts the nuclei of the atoms and repels the electrons. The orbits of the electrons are distorted so that the electrons are, on the average, farther away from the charged comb than the nuclei (Figure 7.9). This results in a net attractive force because the repulsive force on the slightly more distant, negatively charged electrons is smaller than the attractive force on the closer, positively charged protons.

The electrostatic force is another example of "action at a distance." As with gravitation, the concept of a field is useful. In the space around any charged object, there is an *electric field.* This field is the "agent" of the electrostatic force: it will cause any charged object to experience a force. The electric field around a charged particle is represented by lines that indicate the direction of the force that the field would exert on a positive charge. Thus the electric *field lines* around a positively charged particle point radially outward, and the field lines around a negatively charged particle point radially inward (Figure 7.10).

The strength of an electric field at a point in space is equal to the size of the force that it would cause on a given charged object placed at that point, divided by the size of the charge on the object.

FIGURE 7.9
A charged comb exerts a net attractive force on a neutral piece of paper because the electrons are displaced slightly away from the negatively charged comb. (The distorted electron orbits are exaggerated in this drawing.) The attractive force on the closer nuclei is stronger than the repulsive force on the electrons.

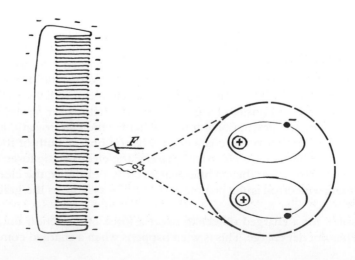

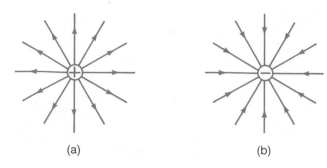

(a)   (b)

FIGURE 7.10
Electric field lines in the space around a positive charge (a) and around a negative charge (b). The arrows show the direction of the force that would act on a positive charge placed in either field.

$$\text{Electric field strength} = \frac{\text{force on a charged object}}{\text{charge on the object}}$$

Where the field is strong, a charged object will experience a large force. The strength of the electric field is indicated by the spacing of the field lines: where the lines are close together, the field is strong. The electric field around a charged particle is clearly weaker at greater distances from it.

Any time a positive charge is in an electric field, it experiences a force in the *same* direction as the field lines. A negative charge in an electric field feels a force in the *opposite* direction of the field lines (Figure 7.11). It is important to remember that it is the electric field that exerts the force on a charged object; in Chapter 8 we will show that electric fields can be produced without directly using electric charges.

Perhaps you have had the experience of walking across a carpeted floor and receiving a shock when you touch a metal doorknob. (This is more likely to happen in winter than in summer because the relative humidity is usually lower, and electrostatic charging takes place more readily.) The shock, due to charges flowing between you and the doorknob, may be accompanied by a visible spark. Air normally does not allow charges to flow through it. A spark occurs when there is an electric field strong enough to ionize atoms in the air. Freed electrons accelerate in a direction opposite to the direction of the electric field, and positive ions accelerate in the same direction as the field. The electrons and ions pick up speed and collide with other atoms and molecules, ionizing them or causing them to emit light. Lightning is produced in this same way on a much larger scale (Figure 7.12). In Chapter 10 we will discuss how atomic collisions cause the emission of light.

Compared to the number of electrical devices that use electric currents, there are few that make use of electrostatics. One very important example of the latter is the *electrostatic precipitator,* an air-pollution control device (Figure 7.13). Tiny particles of soot, ash, and dust are major components of the airborne emissions from fossil–fuel–burning power plants and from many industrial processing plants. Electrostatic precipitators can remove nearly all of these particles from the emissions. The flue gas containing the soot, ash, or dust particles (or a combination thereof) is passed through a series of positively charged metal plates and negatively charged wires (Figure 7.14). The strong electric field around the wires creates negative ions in the particles. These negatively charged particles are attracted by the positively charged plates and collect on them. Periodically the plates are shaken so the collected soot, ash, and dust slides down into a collection hopper. This "fly ash" must then be disposed of, but in some cases it has uses, for example, as a filler in concrete.

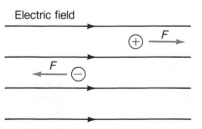

Electric field

FIGURE 7.11
In an electric field, the force on a positively charged body is parallel to the field, while the force on a negatively charged body is opposite the direction of the field.

FIGURE 7.12
Lightning and sparks are caused by very strong electric fields.

FIGURE 7.13
This electrostatic precipitator removes about 99% of the dust in the exhaust gases at a cement plant.

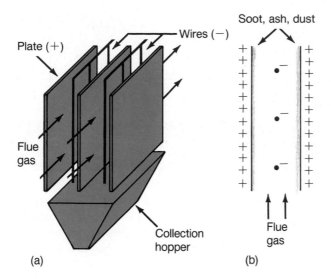

Plate (+)

Wires (−)

Flue gas

Collection hopper

(a)

Soot, ash, dust

Flue gas

(b)

FIGURE 7.14
Schematic of an electrostatic precipitator. (a) The emissions (flue gas) flow around negatively charged wires hanging between positively charged plates. (b) Top view of one channel. The particles are charged negatively by the strong electric field around the wires. Consequently they are attracted to and collect on the plates.

DO-IT-YOURSELF PHYSICS

(This experiment does not work well when the relative humidity is high.)

A number of electrostatic effects can be illustrated with a pair of Styrofoam cups, some thread, a piece of aluminum foil (a foil gum wrapper works), and small pieces of paper. Attach lengths of thread to the bottoms of the cups and hang one cup upside down from some fixed support (a cupboard handle works well). "Charge" the two cups by rubbing them vigorously in your hair.

1. Hold the end of the thread attached to the second cup and dangle the two cups side by side. Since the two cups acquired the same kind of charge, they repel each other. Notice that the force of repulsion, indicated by how far each thread is angled from vertical, increases when the distance between the cups is decreased.

2. Attraction can be shown by using a charged cup to pick up pieces of paper.

3. Suspend a wad of foil from a piece of thread and bring it near one of the charged cups. It will be attracted until it touches the cup and some of the charge flows into the foil. Then the foil and cup have the same kind of charge and will repel one another. Under ideal conditions (very low relative humidity), the foil will recoil instantly.

4. Turn on a water faucet just enough to have a very small stream trickling out. Bring a charged cup near the stream and notice the attraction of the water to the cup (Figure 7.15).

5. The operation of an electrostatic precipitator can be illustrated using a lit cigarette or burning incense. In a room where the air is still, bring a charged cup near a rising stream of smoke. The tiny particles composing the smoke are strongly attracted to the cup.

FIGURE 7.15

The cups can be "recharged" as needed. To "discharge" a cup, you can rub its outside on a metal water faucet.

7.3   ELECTRIC CURRENTS—SUPERCONDUCTIVITY

An electric current is simply a flow of charged particles. The cord on an electrical appliance is actually two separate metal wires covered with insulation. When the appliance is plugged in and operating, electrons inside the wires move back and forth. Inside a television picture tube, free electrons are accelerated from the back of the tube to the screen at the front. There is a near vacuum inside the picture tube, so the electrons can travel without colliding with gas molecules (Figure 7.16). When salt is dissolved in water, the sodium and chlorine ions

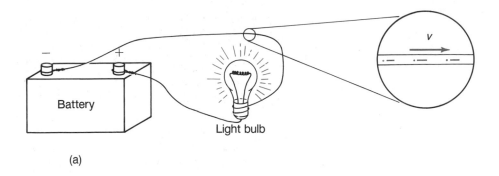

(a)

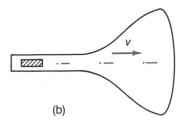

(b)

FIGURE 7.16
Examples of electric current. (a) Electrons flowing inside a metal wire. (b) Electrons moving through the near vacuum inside a television picture tube. (c) Positive ions and negative ions, from dissolved salt, flowing through water.

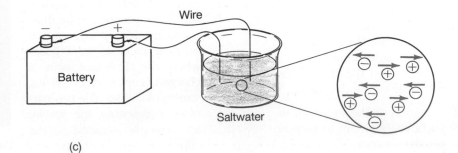

(c)

separate and can move about just like the water molecules. If an electric field is applied to the water, the positive sodium ions will flow one way (in the direction of the field), and the negative chlorine ions will flow the other way.

Regardless of the nature of the moving charges, the quantitative definition of electric current is as follows:

CURRENT   The rate of flow of electric charge. The amount of charge that flows by per second.

$$\text{Current} = \frac{\text{charge}}{\text{time}} \qquad I = \frac{q}{t}$$

The unit of current is the *ampere* (A or amp), which equals one coulomb per second. Current is measured with a device called an *ammeter*.

A current of 5 amperes in a wire means that 5 coulombs of charge flow through the wire each second. (Table 7.1 lists some representative currents.) Either positive charges or negative charges can comprise a current. The effect of a positive charge moving in one direction is the same as that of an equal negative charge moving in the opposite direction. Benjamin Franklin guessed that it was positive charges that flowed through metals. Even though it was determined later that it is negative electrons that flow, this convention was retained: a current is represented as a flow of positive charges, even if it is actually a flow of negative charges in the opposite direction. If positive ions are flowing to the right in a liquid, then the current is to the right. If negative charges (like electrons) are flowing to the right, the direction of the current is to the left. In Figure 7.16, the current is to the left in (a) and (b), and to the right in (c).

The ease with which charges move through different substances varies greatly. Any material that does not readily allow the flow of charges through it is called an electrical *insulator*. Substances like plastic, wood, rubber, air, and pure water are insulators because the electrons are tightly bound in the atoms, and electric fields are usually not strong enough to rip them free so they can move. An (electrical) *conductor* is any substance that readily allows charges to flow through it. Metals are very good conductors because some of the electrons are only loosely bound to atoms and so are free to "skip along" from one atom to the next when an electric field is present. In general, solids that are good conductors of heat are also good conductors of electricity. As mentioned before, liquids such as water are

TABLE 7.1 TYPICAL CURRENTS IN COMMON DEVICES	
Device	Current (A)
Calculator	0.0001
Spark plug	0.001
Clock radio	0.03
60-watt light bulb	0.5
Color television	2
1,200-watt hair dryer	10

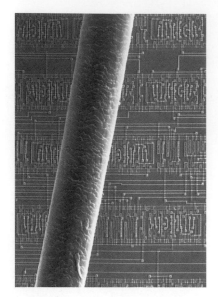

FIGURE 7.17
Greatly enlarged image of electronic circuits etched on a semiconductor chip. The human hair lying on the chip gives you an idea of the chip's small size.

conductors when they contain dissolved ions. Most drinking water has some natural minerals and salts dissolved in it and so conducts electricity. Solid insulators can become conductors when wet because of ions in the moisture.

Semiconductors are substances that fall in between the two extremes. The elements silicon and germanium, both semiconductors, are poor conductors of electricity in their pure states, but they can be modified chemically ("doped") to have very useful electrical properties. Transistors, solar cells, and numerous other electronic components are made out of such semiconductors. The electronic revolution in the second half of this century, including the development of inexpensive calculators, computers, sound-reproduction systems, and other devices, has come about because of semiconductor technology (Figure 7.17).

Resistance

In many situations it is desirable to limit the size of the current that flows through a conductor. For example, the current in a light bulb is controlled by using a very thin wire as a filament. Higher wattage bulbs use thicker filaments and therefore allow more current to flow. We can associate with such devices a quantity called *resistance.*

> **RESISTANCE**   A measure of the opposition to current flow. Resistance is represented by "R," and the unit is the *ohm,* Ω.

Resistance is a bit like friction: the former inhibits the flow of electric charge, and the latter inhibits relative motion between two substances. In metals, electrons in a current move among the atoms and in the process collide with them and give them energy. This causes the metal to gain internal energy and thereby raises its temperature. The consequence of resistance is the same as that of kinetic friction—heating. The larger the current through a particular device, the greater the heating.

In general, conductors will have low resistances, and insulators will have high resistances. The actual resistance of a particular piece of conducting material, a metal wire for example, depends on several factors:

Composition: The particular metal making up the wire affects the resistance. For example, an iron wire will have a higher resistance than an identical copper wire.

Length: The longer the wire is, the higher its resistance.

Diameter: The thinner the wire is, the higher its resistance.

Temperature: The higher the temperature of the wire, the higher its resistance.

Superconductivity

In 1911 the Dutch physicist Heike Kamerlingh Onnes made an important discovery while measuring the resistance of mercury at extremely low temperatures. He found that the resistance decreased steadily as the temperature was lowered, until at 4.2 K it suddenly dropped to zero. Electric current flowed through the mercury with *no* resistance. Onnes named this phenomenon *superconductivity* for good reason: mercury is a perfect conductor of electric current below what is called its *critical temperature* (referred to as "T_c") of 4.2 K. Subsequent research showed that hundreds of elements, compounds, and metal alloys become superconductors, but only at very low temperatures. As recently as 1985 the highest known T_c was 23 K for a mixture of the elements niobium and germanium.

Superconductivity seems too good to be true: electricity flowing through wires with *no* loss of energy to heating. Once a current is made to flow in a loop of superconducting wire it can flow *for years* with no battery or other source of energy. A great deal of the electrical energy that is wasted as heat in conducting wires could be saved if conventional conductors could be replaced with superconductors. But the superconducting state for a given material has limitations. Resistance returns if the temperature is raised above the superconductor's T_c, if the current in it becomes too large, or if it is placed in a magnetic field that is too strong.

In the 1960s practical superconductors were developed and are now widely used in science and medicine. Most of them are compounds of the element niobium. *Superconducting electromagnets,* the strongest magnets known, are used to study the effects of magnetic fields on matter and to direct high-speed, charged particles. The huge particle accelerator at the Fermi National Accelerator Laboratory near Chicago, shown in Figure 8.15, uses superconducting electromagnets to guide protons as they are accelerated to nearly the speed of light. An entire experimental passenger train has been built that is levitated by superconducting electromagnets. *Magnetic resonance imaging (MRI)* uses superconducting electromagnets to form incredibly detailed images of the body's interior. (Magnetism, electromagnets, and MRI are discussed in Chapter 8).

Another important development is the *superconducting quantum interference device (SQUID),* the most sensitive detector of tiny electrical voltages and very weak magnetic fields. In a technique call *magnetoencephalography,* SQUIDs are used to detect the minute magnetic fields produced by electrical activity inside the brain (Figure 7.18). SQUIDs are also used by physicists to search for exotic particles such as quarks and by geologists to detect subtle changes in the earth's magnetic field associated with oil and mineral deposits. The key component of SQUIDs, the *Josephson junction,* is used in the fastest oscilloscope on the market and one day may be used in place of transistors to make super-fast computers.

Widespread practical use of these superconductors is severely limited because they must be kept cold using liquified helium. Helium is very expensive and requires sophisticated refrigeration equipment to cool and to liquify. Once a superconducting device is cooled to the temperature of liquid helium, bulky insulation equipment is needed to limit the flow of heat into the helium and the superconductor. These factors combine to make the so-called "low T_c" superconductors unwieldy or uneconomical except in certain special applications where there are no alternatives.

But hope for wider use of superconductivity blossomed in 1987 when a new family of "high T_c" superconductors was developed with critical temperatures as high as 90 K, since then pushed upward to 125 K. (See Chapter 7 Prologue.) This was an astounding breakthrough because these materials can be made superconducting using liquid nitrogen (boiling point 77 K). This liquid is widely available, it is cheap compared to liquid helium, and it can be used with much less sophisticated insulation. However, the new high T_c superconductors are handicapped by a couple of unfortunate properties: they are brittle and consequently are not easily formed into wires, and they aren't very tolerant of strong magnetic fields or

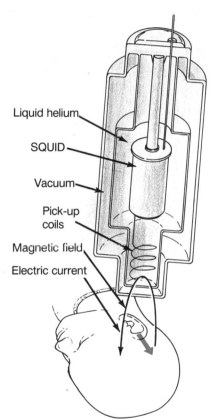

Liquid helium

SQUID

Vacuum

Pick-up coils

Magnetic field

Electric current

FIGURE 7.18
Magnetoencephalography. Superconducting quantum interference devices detect the weak magnetic fields generated by minute electric currents in the brain. (From "The New Superconductors Prospects For Applications," by Alan M. Wolsky, Robert A. Giese, and Edward J. Daniels. © 1989 by Scientific American, Inc. All rights reserved.)

large electric currents. If these problems can be overcome, a new revolution in superconducting technology will occur.

7.4   ELECTRIC CIRCUITS AND OHM'S LAW

An electric current will flow in a light bulb, a radio, or other such device only if an electric field is present to exert a force on the charges. A flashlight works because the batteries produce an electric field that forces electrons to flow through the light bulb. An *electric circuit* is any such system consisting of a battery or other electrical *power supply,* some electrical device like a light bulb, and wires or other conductors to carry the current to and from the device (see Figure 7.19). The power supply acts like a "charge pump": it forces charges to flow out of one terminal, go through the rest of the circuit, and flow into the other terminal. (Electrons typically move through a circuit quite slowly: about 1 millimeter per second.) In this respect an electric circuit is much like the cooling system in a car. The water pump forces coolant to flow through the engine, radiator, and the hoses connecting them.

Instead of using the concept of an electric field and the resulting force on the charges in a wire, it is more convenient to use the idea of energy and work to quantify the effect of a power supply in a circuit. In a flashlight, for instance, the batteries cause electrons to flow through the bulb's filament. Since a force acts on the electrons and causes them to move through a distance, *work is done* on the electrons *by the batteries.* In other words, the batteries give the electrons energy. This energy is converted into heat and light as the electrons go through the light bulb. This approach leads to the definition of electric *voltage.*

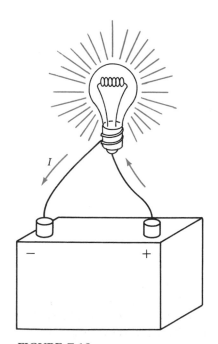

FIGURE 7.19
A simple electric circuit. The battery supplies the energy or "pressure" needed to move the charges through the circuit. Charge does not build up anywhere; for each coulomb of charge that leaves the positive terminal, 1 coulomb enters the negative terminal.

> VOLTAGE   The work that a charged particle can do divided by the size of the charge. The energy per unit charge given to charged particles by a power supply.
>
> $$V = \frac{\text{Work}}{q} \qquad V = \frac{E}{q}$$
>
> The unit of voltage is the *volt,* V, which is equal to *one joule per coulomb.* Voltage is measured with a device called a *voltmeter.*

A 9-volt battery gives 9 joules of energy to each coulomb of electric charge that it moves through a circuit. Each coulomb does 9 joules of work as it flows through the circuit. (Table 7.2 lists some typical voltages.)

If we return to the analogy of a battery as a charge pump, the voltage plays the role of pressure. A higher voltage causing charges to flow in a circuit is similar to a higher pressure causing a fluid to flow (Figure 7.20). Even when the circuit is disconnected from the power supply and there is no charge flow, the power supply still has a voltage. In this case, the electric charges have potential energy. (Voltage is also referred to as electric *potential.*)

The size of the current that flows through a conductor depends on its resistance and on the voltage causing the current. Ohm's law, named after its discoverer, Georg Simon Ohm, expresses the exact relationship.

TABLE 7.2 EXAMPLES OF COMMON VOLTAGES	
Description	Voltage (V)
Nerve impulse	0.1
D cell battery	1.5
Car battery	12
Wall socket	120 (varies)
Power plant generator (typical)	24,000
High-voltage transmission line (typical)	345,000

> **OHM'S LAW**   The current in a conductor is equal to the voltage applied to it divided by its resistance.
>
> $$I = \frac{V}{R} \quad \text{or} \quad V = IR$$

The units of measure are consistent in the two equations: if I is in amperes and R is in ohms, V will be in volts.

The higher the voltage for a given resistance, the larger the current. The larger the resistance for a given voltage, the smaller the current. By applying different-sized voltages to a given conductor, one can produce different-sized currents. A graph of the voltage versus the current will be a straight line whose slope is equal to the conductor's resistance (Figure 7.21). Reversing the polarity of the voltage (switching the "+" and "−" terminals) will cause the current to flow in the opposite direction.

A light bulb used in a 3-V flashlight has a resistance equal to 6 Ω. What is the current in the bulb when it is switched on?   **EXAMPLE 7.1**

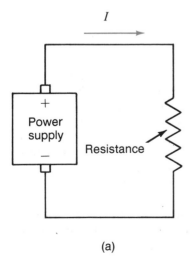

(a)

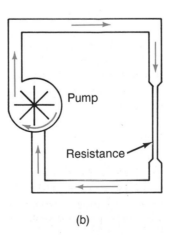

(b)

FIGURE 7.20
(a) The flow of charge in an electric circuit is much like (b) the flow of water through a closed path. The power supply corresponds to the water pump and the resistance corresponds to the narrow segment of pipe. The pressure on the output side of the pump is much like the voltage on the "+" terminal of the power supply. The electric current corresponds to the rate of flow of the water.

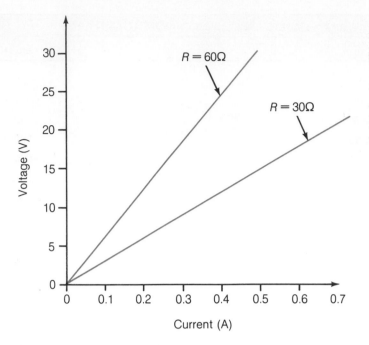

FIGURE 7.21
Graph of voltage versus current for two conductors with different resistances (R). *For each resistance, the voltage needed to produce a given current is proportional to the current.*

By Ohm's law:

$$I = \frac{V}{R} = \frac{3\,\text{V}}{6\,\Omega}$$

$$I = 0.5\,\text{A}$$

EXAMPLE 7.2   A small electric heater has a resistance of 15 Ω when the current in it is 2 A. What voltage is required to produce this current?

$$V = IR = 2\,\text{A} \times 15\,\Omega$$

$$V = 30\,\text{V}$$

Often the resistance of a conductor changes when the voltage changes. At higher voltages a larger current flows through the filament of a light bulb so its temperature is also higher. The resistance of the hotter filament is consequently greater (Figure 7.22). Some semiconductor devices, called diodes, are designed to have very low resistance when current flows through them in one direction but very high resistance when a voltage tries to produce a current in the other direction. Water with salt dissolved in it generally has lower resistance when higher voltages are applied to it: doubling the voltage will more than double the current. A graph of *V* versus *I* for ordinary tap water is less steep at higher voltages.

Many electrical devices are controlled by changing a resistance. The volume control on a radio or a television simply varies the resistance in a circuit. Turning up the volume reduces the resistance, so more current flows in the circuit, resulting in louder sound.

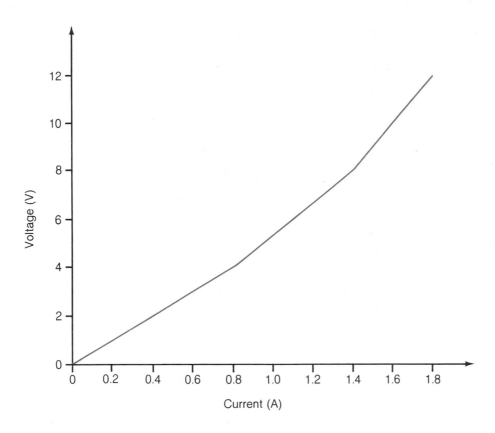

FIGURE 7.22
Graph of voltage versus current for a brake light bulb on a car. The graph curves upward becasuse the resistance is higher with larger currents. This is because the filament is hotter.

Series and Parallel Circuits

In many situations several electrical devices are connected to the same electrical power supply. A house may have a hundred different lights and appliances all connected to one cable entering the house. An automobile has dozens of devices connected to its battery. There are two basic ways in which more than one device can be connected to a single electrical power supply: by a series circuit and by a parallel circuit.

In a *series circuit* there is only one path for the charges to follow, so the same current flows in each device (see Figure 7.23). In such a circuit the voltage is divided among the devices: the voltage on the first device plus the voltage on the second device, and so on, equals the voltage of the power supply. For example, if three light bulbs with the same resistance are connected in series to a 12-volt battery, the voltage on each bulb is 4 volts. If the bulbs had different resistances, each one's "share" of the voltage would be proportional to its resistance.

Notice that the current in a series circuit is stopped if any of the devices breaks the circuit (path; Figure 7.24). A series circuit is not normally used with, say, a number of light bulbs because if one of them burns out, the current stops and all of the bulbs go out. A string of Christmas lights that flash at the same time uses a series circuit so that all the bulbs go on and off together.

In a *parallel circuit,* the current through the power supply is "shared" among the devices while each has the same voltage (Figure 7.25). The current flowing in the first device plus the current in the second device, etc, equals the current put out by the power supply. There is more than one path for the charges to follow—in this case, three. If one of the devices burns out or is removed, the

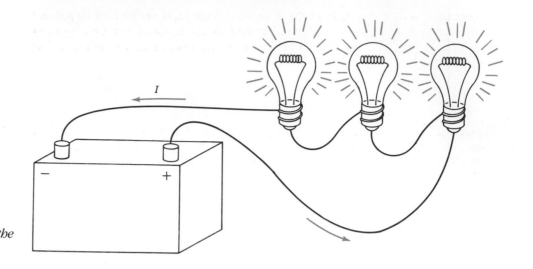

FIGURE 7.23
A simple series circuit. The current is the same in each of the bulbs.

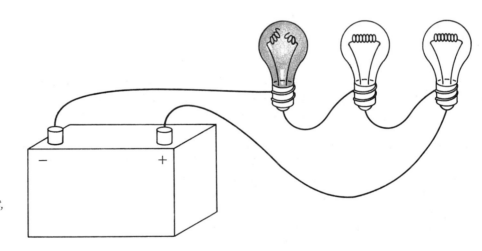

FIGURE 7.24
If one device in a series circuit fails, such as a bulb burning out, the current stops, and all of the devices go off.

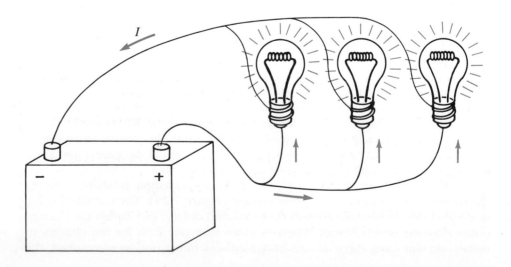

FIGURE 7.25
A simple parallel circuit. Each bulb has the same voltage (the voltage of the power supply). If one bulb burns out, current can still flow through the other bulbs.

others still function. The light bulbs in multiple-bulb light fixtures are in parallel so that if one bulb burns out, the others still function. Often the two types of circuits are combined: one switch may be in series with several light bulbs that are in parallel.

Three light bulbs are connected in a parallel circuit with a 12-V battery. The resistance of each bulb is 24 Ω. What is the current produced by the battery?   **EXAMPLE 7.3**
    The voltage on each bulb is 12 V. Therefore the current in each bulb is

$$I = \frac{V}{R} = \frac{12\,V}{24\,\Omega}$$

$$I = 0.5\,A$$

    The total current supplied by the battery equals the sum of the currents in the three bulbs.

$$I = 0.5\,A + 0.5\,A + 0.5\,A$$

$$I = 1.5\,A$$

    The concept of voltage is quite general and is not restricted to electric power supplies and electric circuits. Whenever there is an electric field in a region of space, a voltage exists because the field has the potential to do work on electric charges. The strength of an electric field can be expressed in terms of the voltage change per unit distance along the electric field lines. For example, air conducts electricity when the electric field is strong enough to ionize atoms in the air. The minimum electric field strength required for this to happen is between 30,000 and 75,000 V per inch, depending on the conditions. This means that if there is a spark one fourth of an inch long between your finger and a doorknob, the voltage that causes the spark is at least 7,500 V.

LEARNING CHECK

1. Like charges _____ ; unlike charges _____ .

2. A positive ion is formed when:
   a) A neutral atom gains an electron. c) A negative ion loses an electron.
   b) A neutral atom loses an electron. d) All of the above.

3. A charged particle creates an electric _____ in the space around it.

4. An electric current can consist of:
   a) Positive ions flowing in a liquid.   c) Electrons moving through a
   b) Negative ions flowing in a liquid.   wire.
                                           d) All of the above.

5. The size of the current flowing through a conductor equals the _____ divided by the _____ .

Electricity plays a vital role in living organisms. The normal function and reproduction of individual cells involve ions in complicated chemical processes. Photosynthesis, the process by which plants convert solar energy into chemical energy, involves a flow of electrons that produces a series of chemical reactions. The chemical energy stored in plants by photosynthesis is the ultimate energy source that sustains nearly the entire web of living organisms—including humans.

Most organisms, even single-celled protozoans, use electrical signals to transmit information. The nervous system in higher animals is a highly sophisticated network of different types of *neurons* (nerve cells). Sensory neurons provide input to the brain, associative neurons process information in the brain, and motor neurons carry signals from the brain to muscles and some glands. There are approximately 12 billion nerve cells in the human body, some up to a meter long.

A signal is transmitted along an individual neuron in the form of a small-voltage change between the inside and the outside of the cell. A neuron at rest has an excess of negative ions at its inner surface, resulting in a voltage of about −0.07V. A nerve signal involves a momentary flow of positive sodium ions into the neuron until the polarity is reversed and the voltage is about +0.03 V. That part of the neuron quickly returns to the resting state, but the small-voltage "pulse" travels along the neuron at up to 100 meters per second.

The specialized *sensory cells* associated with touch, taste, sight, and the other senses produce electrical signals when they are stimulated by the outside environment. These signals are passed on from neuron to neuron, through the spinal cord, and into various regions of the brain. Command signals from the brain are transmitted by *motor neurons* in the opposite direction to muscles, which contract after the electrical signal is received. The brain itself contains a constant swarm of neural signals as it processes information and formulates response signals. (As you read these words, think of the miracles taking place in your brain as the signals from the nerves in your eyes are transformed into a mental image, and from that, information is extracted.) An *electroencephalogram* (EEG) is a record of the periodic voltage changes associated with brain activity. EEG's are made by monitoring the minute voltages that appear on the scalp near different parts of the brain. EEGs show characteristic patterns for different levels of mental activity

7.5 POWER AND ENERGY IN ELECTRIC CURRENTS

Since a battery or other electrical power supply must continually put out energy to cause a current to flow, it is important to consider the *power output—the rate at which energy is delivered to the circuit.* The power is determined by the voltage of the power supply and the current that is flowing. Think of it this way: the power output is the amount of energy expended per unit amount of time. The power supply gives a certain amount of energy to each coulomb of charge that flows through the circuit. Consequently the energy output per unit time equals the *energy given to each coulomb of charge* multiplied by the *number of coulombs that flow through the circuit per unit time.*:

Energy per time = energy per coulomb × number of coulombs per time (fully alert wakefulness, asleep, at rest with the eyes closed, etc) and can indicate certain disorders such as epilepsy.

Magnetoencephalography is a newer, more sensitive way to monitor brain activity. It uses superconducting devices to detect the weak magnetic fields generated by the brain's electric currents (Figure 7.18).

The heart is an organ that pumps blood by coordinated muscle contraction followed by relaxation. Unlike other muscles, such as those in your arms, the heart does not rely on a signal from the brain to contract. A small group of cells in the heart produces electrical impulses spontaneously at a rate of around 72 per minute. (The actual frequency of the signal, and therefore the heart rate, is regulated by the brain and autonomic nervous system. For example, the heart rate is increased during exercise.) This impulse initiates a wave of contraction of heart muscle in the required coordinated fashion. An *electrocardiogram* (ECG, more commonly called EKG) is similar to an EEG in that it is a record of the voltages produced by various parts of the heart as it beats (Figure 7.26). EKGs are routinely used to detect irregularities in the heart's operation, such as a heart murmur.

FIGURE 7.26
An electrocardiogram is the tracing made by an electrocardiograph, an instrument for recording the changes of electrical potential during a heartbeat.

Often, with age, the intrinsic ability of the heart to control its beating diminishes. A common problem is that different parts of the heart start to beat at different rates because the controlling signal is too weak. A pacemaker can often be implanted near the heart to remedy this problem. This device supplies

These three quantities are just the power, voltage, and current, respectively. Consequently the power output of an electrical power supply is:

$$\text{Power} = \text{voltage} \times \text{current}$$

$$P = VI$$

The units work out correctly in this equation also: *joules per coulomb* (volts) multiplied by *coulombs per second* (amperes) equals *joules per second* (watts).

The power output of a battery is proportional to the current that it is supplying: the larger the current, the higher the power output.

In a previous example (7.1) we computed the current that flows in a flashlight bulb. What is the power output of the batteries?

EXAMPLE 7.4

a small electrical shock to the heart at regular intervals to reestablish coordinated heart contraction.

Because the body relies so heavily on small voltages for its normal function, it is extremely vulnerable to electricity from the outside. Interestingly, it is the *current* that actually flows through the body, not the voltage, that is most important in determining the physiological effect. A single shock of static electricity received from a doorknob may be caused by several thousand volts, but the effect is localized (a sharp pain in your finger) because only a small amount of charge flows through the interior of your body. The effect of a steady current on your body depends on the size of the current and on the part of the body through which it flows. Currents less than about 1 milliampere (mA) usually cannot be felt. A current of around 7 mA or above through your hand (for example) would block signals to your muscles, and you would not be able to move your fingers. A current of 100 mA or more through the heart can induce uncoordinated contractions, called *fibrillation,* leading to death. Steady currents above 300 mA through the heart are usually fatal.

You should realize that a current cannot flow through your body unless it is part of a complete circuit. A person hanging by the arms from a 240-volt power line will have no current flowing through his or her body. But touching the ground or a different wire would provide a complete circuit for the charge to flow through.

Sometimes an electrical current is purposely sent through the human body for medical treatment. A momentary current sent through a fibrillating heart can often restore a normal heartbeat. Electroshock therapy involves sending a current through the brain to treat certain mental disorders. Typically a current as high as several hundred milliamperes is applied for less than a second.

A variety of species of fish (about 500) have developed some interesting uses for electricity. These fish exploit living in an environment (water) that conducts electricity and have specially evolved cells that produce voltages between different parts of their bodies. The most well-known example is the electric eel: it can produce a formidable electric shock (350 to 650 volts with less than 1 amp) to stun prey or to ward off enemies (Figure 7.27). Many other fish naturally produce small, periodic electrical currents in the water around them. The electric eel and certain other species that inhabit murky

Recall that the two batteries produce 3 V and that the current in the light bulb is 0.5 A. The power output is:

$$P = VI = 3\,\text{V} \times 0.5\,\text{A}$$

$$P = 1.5\,\text{W}$$

The batteries supply 1.5 J of energy each second.

What happens to the energy delivered by an electrical power supply? In a light bulb less than 5% is converted into visible light, and the rest becomes heat energy. Even the visible light emitted by a light bulb is absorbed eventually by the surrounding matter and is transformed into heat. (Interior lighting is actually used to heat some buildings.) Electric motors in hair dryers, vacuum cleaners, and the like convert about 60% of their energy input into mechanical work or energy, while the remainder goes to heat energy. The mechanical energy generally is dissipated as heat through friction. In a similar way we can trace the energy

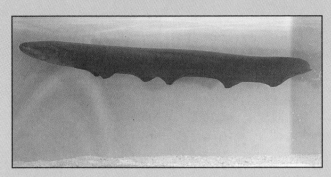

FIGURE 7.27
An electric eel can produce electricity for paralyzing prey and repelling enemies.

electric fields produced inside them by electromagnetic induction (see Chapter 8) as they swim through the earth's magnetic field. This they can use as an electromagnetic compass for navigation.

It was recently reported that the duckbilled platypus has electric field receptors in its bill. Researchers think that the platypus uses these to help locate shrimp and other underwater prey.

From the very "glue" that holds together atoms and molecules to the vehicle for thought and communication in the human body, electricity is essential to the existence of matter and life itself.

waters use such signals to communicate with each other. These fish are equipped with special organs that detect small electrical currents in the water. Some predators, such as the dogfish shark, use the signals emitted by their prey as an aid in hunting. Still other species of fish produce weak electric fields around themselves and can detect distortions in the field caused by the presence of other objects or fish. Some sharks apparently can detect small

conversions in other electrical devices, and the outcome is the same: the electrical energy eventually becomes heat energy.

Ordinary metal wire converts electrical energy into heat energy whenever there is a current flowing. You may have noticed when using a hair dryer that its cord becomes warm. This heating, called *ohmic heating,* occurs in any conductor that has resistance, even when the resistance is quite small. The huge cables used to conduct electricity from power plants to cities are heated by this effect. This represents a loss of usable energy.

The temperature that a current-carrying wire reaches due to ohmic heating depends on the size of the current and on the wire's resistance. Increasing the current in a given wire will raise its temperature. The resistances of heating elements in toasters and electric heaters are chosen so they glow red hot. The filament in an incandescent light bulb is made so thin that ohmic heating causes it to glow white hot (see Figure 7.28).

A sufficiently large current in any wire can cause it to become very hot—hot enough to melt any insulation around it or to ignite combustible materials nearby. Fuses and circuit breakers are put into electrical circuits as safety devices to

FIGURE 7.28
The current causes the thin filament wire to become white hot, while the larger wires connected to it stay much cooler.

FIGURE 7.29
Two 15-amp fuses used in automobiles. At some point, the current in the fuse on the left exceeded 15-amperes so it burned out.

prevent dangerous overheating of wires. If something goes wrong or if too many devices are plugged into the circuit and the current exceeds the recommended safe limit for the size of wire used, the fuse or circuit breaker will automatically "break" the circuit. (A fuse is simply a fine wire or piece of metal inside a glass or plastic case. When the current exceeds the fuse's design limit, the metal melts away, and the circuit is broken; Figure 7.29.) Designers of electrical circuits in cars, houses, and other buildings must choose wiring that is large enough to carry the currents needed without overheating. They must also include fuses or circuit breakers that will disconnect a circuit if it is overloaded.

Most electrical devices are rated by the power that they consume in watts. The equation $P = VI$ can be used to determine how much current flows through the device when it is operating.

EXAMPLE 7.5   An electric hair dryer is rated at 1,200 W when operating on 120 V. What is the current flowing through it?

$$P = VI$$

$$1{,}200 \, \text{W} = 120 \, \text{V} \times I$$

$$\frac{1{,}200 \, \text{W}}{120 \, \text{V}} = I$$

$$I = 10 \, \text{A}$$

The wires in the electric cord must be large enough to allow 10 A to flow through them without becoming dangerously hot.

The highest current that can flow in a particular wire without causing excessive heating depends on the size of the wire. This is one reason why electric utilities use high voltages in their electrical power supply systems. The electricity delivered to a city, subdivision, or individual house must be transmitted with wires. Since $P = VI$, using a large voltage makes it possible to transmit the same power with a smaller current. If low voltages were used, say 100 V instead of the more typical 345,000 V, much larger cables would have to be used to handle the larger currents.

Customers pay for the electricity supplied to them by electric companies on the basis of the amount of energy they use. An electric meter keeps track of the total energy used by monitoring the power (rate of energy use) and the amount of time each power level is maintained (see Figure 7.30). Recall the equation used to define power:

$$P = \frac{E}{t}$$

Therefore:

$$E = Pt$$

The amount of energy used is equal to the power times the time elapsed. If P is in watts and t is in seconds, E will be in joules.

If the hair dryer discussed in the previous example is used for 5 minutes, how much energy does it use?   **EXAMPLE 7.6**

The power is 1,200 W. To get E in joules, we must convert the 5 minutes into seconds.

$$t = 5 \text{ min} = 5 \times 60 \text{ s} = 300 \text{ s}$$

So:

$$E = Pt = 1{,}200 \text{ W} \times 300 \text{ s}$$

$$E = 360{,}000 \text{ J}$$

This is a large quantity of energy—about the same as the kinetic energy of a small car going 60 mph (Example 3.6). Another comparison: a 150-pound person would have to climb 1,770 feet (about 170 floors) to gain 360,000 joules of potential energy.

A typical household can consume several billion joules of electrical energy each month. For this reason a more appropriately sized unit of measure is used for electrical energy—the *kilowatt-hour* (kW-h). Energy in kilowatt-hours is computed by expressing power in kilowatts (kW) and time in hours. The conversion factor between joules and kilowatt-hours is

$$1 \text{ kW-h} = 1 \text{ kilowatt} \times 1 \text{ hour}$$

$$= 1{,}000 \text{ watts} \times 3{,}600 \text{ seconds}$$

$$1 \text{ kW-h} = 3{,}600{,}000 \text{ joules}$$

In the previous example, the energy used by the hair dryer equals 0.1 kilowatt-hour. The cost of electricity varies from region to region, but it is typically around 10 cents per kilowatt-hour. This means that it costs about 1 cent to run the hair dryer 5 minutes. (Would you climb 1,770 feet for 1 cent?)

Perhaps you have wondered why a regular 1.5-volt D-cell battery is actually larger than a 9-volt battery. The voltage of a battery really has nothing to do with its physical size. Different 1.5-volt batteries range from the size of a button (for wristwatches) to larger than a beer can. The size is more of an indication of the amount of electrical energy stored in the battery. A large battery can supply the

FIGURE 7.30
An electric meter registers the amount of energy consumed in kilowatt-hours.

same current (and the same power) for a longer time than a small battery with the same voltage.

7.6 AC AND DC

The electrical current supplied by a battery is actually different from the current supplied by a normal household wall socket. Batteries supply *direct current* (DC), and household outlets supply *alternating current* (AC). A DC power supply, such as a battery, causes a current to flow in a fixed direction in a circuit (Figure 7.31). The current flows out of the positive (+) terminal of the power supply, moves through the circuit, and flows into the negative (−) terminal of the battery. If the total resistance in the circuit doesn't change, the size of the current remains constant. A graph of the current *I* versus time is simply a horizontal line.

In an AC power supply, the polarity of the two output terminals switches back and forth—the voltage alternates. This causes the current in any circuit connected to the power supply to alternate as well. It flows counterclockwise, then clockwise, than back to counterclockwise, and so on. All the while the size of the current is increasing, then decreasing, and so forth. A graph of the current in an AC circuit shows this variation in the size and direction of the current. (When *I* goes below zero, it means that the direction has reversed; see Figure 7.32.)

We have seen this kind of oscillation before in Section 2.6 and throughout Chapter 6. One can even think of AC as a kind of "wave" causing the charges in a conductor to oscillate back and forth. Almost all public electric utilities in the United States supply 60-hertz AC. The voltage between the two slots in a wall outlet oscillates back and forth 60 times per second. (In Europe, the standard frequency of AC is 50 hertz.)

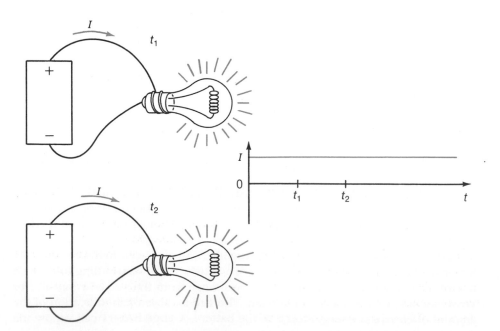

FIGURE 7.31
Direct current. The current flows in one direction and doesn't increase or decrease, as shown in the graph of current versus time.

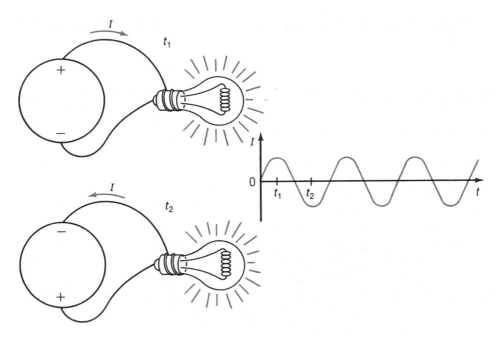

FIGURE 7.32
Alternating current. The direction of the current switches back and forth, and the size of the current varies continuously.

Both AC and DC have their advantages and disadvantages. Some electronic devices (such as light bulbs) can operate on AC or DC, while others require one or the other. Electric motors and generators must be designed to operate on or to produce either AC or DC. There are devices that can convert an AC voltage to a DC voltage, and vice versa. Batteries can produce direct current only. For this reason automobiles have DC electrical systems. (The alternator in an automobile generates AC, which is then converted into DC so as to be compatible with the battery.)

AC has one distinct advantage over DC: simple, highly efficient devices called *transformers* can "step up" or "step down" AC voltages. This makes it possible to generate AC at a power plant at some intermediate voltage, step it up to a very high voltage (typically over 300,000 volts) for economical transmission, and then step it down again to lower voltages for use in homes and industries. There is no counterpart of the transformer for DC. Another important use of AC is in electronic sound equipment. One example: if a 440-hertz tone is recorded on tape and then played back, the "signal" going to the speaker will be an alternating current with a frequency of 440 hertz. (We will discuss transformers and sound reproduction in Chapter 8.)

DO-IT-YOURSELF PHYSICS

Fluorescent lights emit light only when a current is flowing through them. When AC is used, the current flows for a short time, stops momentarily, then flows in the other direction. Because of this, a fluorescent light actually emits "bursts" of light 120 times each second (two bursts during each cycle, 60 cycles each second). This causes a "stroboscope" effect, which you can illustrate with a pen, pencil, or other thin object that is white or light colored. Position yourself so that the fluorescent light is behind or above you. Hold the pen in such a way that the light shines on it and there is a dark background behind it. Move the pen rapidly back and forth (sideways) with your hand. You should see faint lines parallel to the pen. These lines show the position of the pen when each burst of light illuminates it. The faster you move the pen, the farther apart the lines will be.

Incandescent light bulbs do not show the stroboscope effect nearly as well because they rely on a glowing filament to emit the light. In the short time that the alternating current goes to zero, the filament does not cool off enough to reduce its light output appreciably.

HISTORICAL NOTES

The first person to undertake a successful systematic analysis of electric as well as magnetic effects was William Gilbert (1540–1603), a contemporary of Galileo. Born into an English family of comfortable means, Gilbert studied medicine at Cambridge and was later appointed physician to Queen Elizabeth I. This position left him with sufficient time and financial resources to pursue his studies of electricity and magnetism (see Figure 7.33). Gilbert showed that the two types of effects are distinct, and he dispensed with a number of misconceptions about them. He carefully tested various substances to see which exhibited the amber effect. Some of the terminology that we use today, such as "pole," originated with Gilbert.

FIGURE 7.33
A 1903 painting of William Gilbert demonstrating electrostatics to Queen Elizabeth I.

FIGURE 7.34
Stephen Gray's demonstration that the human body conducts electricity.

After Gilbert published his findings in the book *De Magnete* in 1600, nothing new was discovered for some 60 years. Otto von Guericke, famous for his experiments with atmospheric pressure (see p. 191), constructed a huge electrostatic machine to illustrate the amber effect on a large scale. It consisted of a large, rotating sulfur ball that was charged by friction. It could attract feathers, bits of paper, and other things from considerable distances. He was also the first to record electrostatic repulsion. During the next 200 years, demonstrations of electrostatic effects with such machines became very popular as parlor amusements and as topics of public lectures. Benjamin Franklin's interest in electricity began after he viewed such a demonstration in Boston in 1746.

The findings of two other experimenters, the Englishman Stephen Gray (1666–1736) and the Frenchman Charles Dufay (1698–1739), are also noteworthy. Gray discovered, somewhat by accident, that the electrostatic charge flowed through some substances but not through others. Thus he discovered electrical conductors and insulators. Materials that are good at showing the amber effect turn out to be insulators. Conductors like iron cannot be charged in the usual way because the charge will simply flow into the person holding it. Conductors can be charged by keeping them out of contact with other conductors. Gray showed that the human body is itself a conductor by suspending a person by nonconducting strings from a framework (Figure 7.34).

Dufay studied electrostatic attraction and repulsion and concluded that there must be two different kinds of electric charge to account for them. The two types, which he called vitreous and resinous, obeyed the rule "like charges repel and unlike charges attract." Our current names for the two types of charges, positive and negative, come from Benjamin Franklin (Section 7.1).

Coulomb's law, the force law for static electric charges, seems to have been discovered independently by three different men: Charles Coulomb (1736–1806), John Robison (1739–1806), and Henry Cavendish (1731–1810). (Cavendish, who we encountered in Section 2.9, was a noted recluse who amassed a wealth of experimental findings that were not published until long after his death.) But the Frenchman Coulomb is credited with discovering the law. His experience as an engineer, including a 9-year stint in Martinique, provided him with the skills he needed to construct precision force balances.

The investigation of electric currents was triggered by a chance observation of an Italian biologist, Luigi Galvani (1737–1798). Frog legs that he was preparing

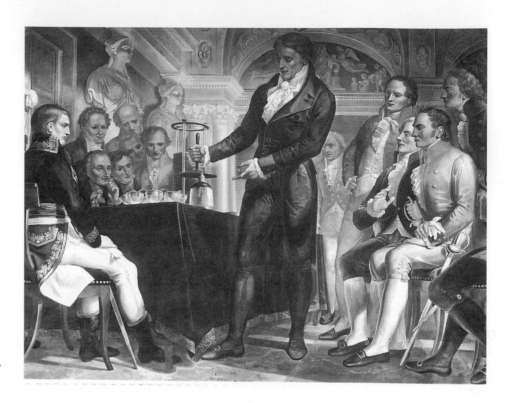

FIGURE 7.35
Allessandro Volta, inventor of the battery.

twitched when touched by a charged scalpel. A careful experimenter, Galvani undertook an extensive study of the phenomenon and thus made the first discoveries of the role of electricity in living systems. However, Galvani clung to the mistaken notion that the electricity in living organisms was somehow different from ordinary electricity.

Galvani's work came to the attention of the physicist Allessandro Volta (1745–1827) in northern Italy. Volta was already an accomplished experimenter in electricity: he was the first to devise a way to measure what was later named voltage. In the process of verifying and extending the discoveries of Galvani, Volta invented a type of battery (see Figure 7.35). This was a pivotal discovery: batteries were able to supply much larger currents than were possible with electrostatic generators. The invention brought immediate acclaim to Volta. In 1801 Napoleon viewed a demonstration of the device and was very impressed.

Experimenters throughout Europe quickly began building their own "Voltaic cells," and the understanding of moving electricity advanced rapidly. The foremost researcher in this area was a German schoolteacher named Georg Simon Ohm (1787–1854). Working in isolation from the more famous physicists of the time, Ohm introduced the important quantities of voltage, current, and resistance. He carefully studied the factors that affect the resistance of conductors and discovered the relationship that bears his name. Ohm's discoveries, unlike Volta's, were first met with skepticism and even contempt, perhaps because he was not a member of the traditional scientific community. It was only a few years before his death that Ohm was finally given the respect and recognition that he deserved.

The large currents that could be produced by batteries were a boon to science. (Some of the larger batteries built had power outputs of several thousand watts.)

It was soon discovered that sending a current through water decomposed it into hydrogen and oxygen. This process, known as *electrolysis,* was applied to many other substances and led to the discovery of several new elements. It also pointed out that electricity is involved in the very structure of matter.

SUMMARY

Electrons, protons, and certain other subatomic particles possess a physical property, called electric charge, that is the basic source of electrical and magnetic phenomena. Forces act between any objects that possess net electrical charge, positive or negative. The electrostatic force, expressed by Coulomb's law, is responsible for binding electrons to the nucleus in atoms, for the amber effect such as static cling, and for a number of other phenomena. Electric charges produce electric fields in the space around them. This field is the agent for the electrostatic force, just as the gravitational field is the agent of the gravitational attraction between objects.

Most useful applications of electricity involve electric currents. In most cases electrons are made to flow through metal wires by an electrical power supply, such as a battery. The flow of charge is analyzed using voltage, current, and resistance. Ohm's law states that the current in a circuit equals the voltage divided by the resistance. The power consumption in a circuit depends on the voltage and the current. The electrical energy needed to cause a current to flow through a resistance is converted into heat. This "ohmic heating" is exploited by incandescent light bulbs to produce light. Fuses and circuit breakers are used to automatically disconnect a circuit if the current is large enough to cause excessive ohmic heating.

At extremely low temperatures many materials become superconductors—they have zero resistance. Consequently no energy is lost to heating when electric currents flow through them. Superconductors now in use are limited to special-purpose scientific instruments, but the new high-temperature superconductors may bring about widespread applications in common devices.

There are two types of electric current: alternating (AC) and direct (DC). Since batteries produce DC, battery-powered devices generally employ DC. Transformers can be built to "step up" or to "step down" an AC voltage from one value to another. This makes AC particularly convenient for electrical supply networks, such as electric utilities.

SUMMARY OF IMPORTANT EQUATIONS

EQUATION	COMMENTS
$F = \dfrac{9 \times 10^9 \, q_1 q_2}{d^2}$	Coulomb's law
$I = \dfrac{q}{t}$	Definition of current
$V = \dfrac{E}{q}$	Definition of voltage
$I = \dfrac{V}{R}$	Ohm's law
$V = IR$	Ohm's law
$P = VI$	Electrical power consumption
$E = Pt$	Energy used during time t

QUESTIONS

1. All matter contains both positively and negatively charged particles. Why do most things have no net charge?

2. A particular solid is electrically charged after it is rubbed, but it is not known whether its charge is positive or negative. How could you determine which charge it has by using a piece of plastic and fur?

3. What is a positive ion? A negative ion?

4. What remains after a hydrogen atom is positively ionized?

5. Describe the similarities and the differences between the gravitational force between two objects and the electrostatic force between two charged objects.

6. a) A negatively charged iron ball (on the end of a plastic rod) exerts a strong attractive force on a penny even though the penny is neutral. How is that possible?
   b) The penny accelerates toward the ball, hits it, and then is immediately repelled. What causes this sudden change to a repulsive force?

7. What is an electric field? Sketch the shape of the electric field around a single proton.

8. At one moment during a storm the electric field between two clouds is directed toward the East. What is the direction of the force on any electron in this region? What is the direction of the force on a positive ion in this region?

9. Explain what an electrostatic precipitator is and how it works.

10. Materials can be classified into four categories based on the ease with which charges can flow through them. Give the names of these categories and describe each one.

11. A student using a sensitive meter that measures resistance finds that the resistance of a thin wire is changed slightly when it is picked up with a bare hand. What causes the change in the resistance, and does it increase or decrease?

12. If a new material is found that is a superconductor at all temperatures, what parts of some common electric devices would definitely *not* be made out of it?

13. Explain what current, resistance, and voltage are.

14. Describe Ohm's Law.

15. A power supply is connected to two bare wires that are inserted into a glass of saltwater. The resistance of the water *decreases* as the voltage is *increased*. Sketch a graph of the voltage versus the current in the water showing this type of behavior.

16. There are two basic schemes for connecting more than one electrical device in a circuit. Name, describe, and give the advantages of each.

17. Draw a sketch of an electric circuit that contains a switch and two light bulbs connected in such a way that if either bulb burns out the other remains on but if the switch is turned off both bulbs go out.

18. Two 9-volt batteries are connected in series in an electric circuit. Use the concept of energy to explain why this combination is equivalent to a single 18-volt battery. When connected in parallel, what are two 9-volt batteries equivalent to?

19. A simple electric circuit consists of a constant-voltage power supply and a variable resistor. What effect does reducing the resistance have on the current in the circuit and on the power output of the power supply?

20. What is the purpose of having fuses or circuit breakers in electrical circuits? How should they be connected in circuits so they will be effective?

21. A 20 amp fuse in a household electric circuit burns out. What catastrophe could occur if it is replaced by a 30 amp fuse?

22. Why is it economical to use extremely high voltages for the transmission of electrical power?

23. Explain what AC and DC are. Why is AC used by electric utilities? Why is DC used in flashlights?

24. If the electric utility company where you live suddenly changed the frequency of the AC to 20 Hz, what problems might this cause?

PROBLEMS

1. Five coulombs of charge flow through a car stereo every 20 s. What is the electric current? $I =$

2. A typical lightning stroke lasts 0.05 seconds, and involves a flow of 100 C. What is the current?

3. A current of 0.7 A goes through an electric motor for 1 min. How many coulombs of charge flow through it during that time?

4. A calculator draws a current of 0.0001 A for 5 min. How much charge flows through it?

5. A current of 12 A flows through an electric heater operating on 120 V. What is the heater's resistance?

6. A 120-V circuit in a house is equipped with a 20-A fuse that will "blow" if the current exceeds 20 A. What is the smallest resistance that can be plugged into the circuit without causing the fuse to blow?

7. The resistance of each brake light bulb on an automobile is 6.6 Ω. Use the fact that cars have 12-V electrical systems to compute the current that flows in each bulb.

8. The light bulb used in an overhead projector has a resistance of 24 Ω. What is the current through the bulb when it is operating on 120 V?

9. The resistance of the skin on a person's finger is typically about 20,000 Ω. How much voltage would be needed to cause a current of 0.001 A to flow through the finger?

10. A 150-Ω resistor is connected to a variable-voltage power supply.
    a) What voltage is necessary to cause a current of 0.3 A in the resistor?
    b) What current flows in the resistor when the voltage is 18 V?

11. Compute the power consumption of the electric heater in Problem 5.

12. All of the electrical outlets in a room are connected in a single parallel circuit (see Figure 7.36). The circuit is equipped with a 30-A fuse, and the voltage is 120 V.
    a) What is the maximum power that can be supplied by the outlets without blowing the fuse?
    b) How many 1,200-W appliances can be plugged into the sockets without blowing the fuse?

13. A car's headlight consumes 40 W when on low beam and 50 W when on high beam.
    a) Find the current that flows in each case (V = 12 V).
    b) Find the resistance in each case.

14. Find the current that flows in a 40-W bulb used in a house (V = 120 V). Compare this with the answer to the first part of Problem 13.

15. An electric clothes dryer is rated at 4,000 W. How much energy does it use in 40 min?

16. A clock consumes 2 W of electrical power. How much energy does it use each day?

17. Which costs more, running a 1,200-W hair dryer for 5 minutes or leaving a 60-W lamp on overnight (10 hours)?

18. A representative lightning strike is caused by a voltage of 200,000,000 V and consists of a current of 1,000 A that flows for a fraction of a second. Calculate the power.

19. A toaster operating on 120 V uses a current of 9 A.
    a) What is the toaster's power consumption?
    b) How much energy does it use in 1 min?

20. A certain electric motor draws a current of 10 A when connected to 120 V.
    a) What is motor's power consumption?
    b) How much energy does it use during 4 h of operation? Express the answer in joules and in kilowatt-hours.

21. The generator at a large power plant has an output of 1,000,000,000 W at 24,000 V.
    a) If it were a DC generator, what would be the current in it?
    b) What is its energy output each day—in joules and in kilowatt-hours?
    c) If this energy is sold at a price of 10 cents per kilowatt-hour, how much revenue does the power plant generate each day?

22. A light bulb is rated at 60 W when connected to 120 V.
    a) What current flows through the bulb in this case?
    b) What is the bulb's resistance?
    c) What would be the current in the bulb if it were connected to 60 V, assuming the resistance stays the same?
    d) What would be its power consumption in this case?

23. An electric car is being designed to have an average power output of 4,000 W for 2 hours before needing to be recharged. (Assume there is no wasted energy.)
    a) How much energy would be stored in the charged up batteries?
    b) The batteries operate on 30 V. What would the current be when they are operating at 4,000 W?
    c) To be able to recharge the batteries in 1 hour, how much power would have to be supplied to them?

24. The resistance of an electric heater is 10 Ω. How much energy does it use during 30 minutes of operation?

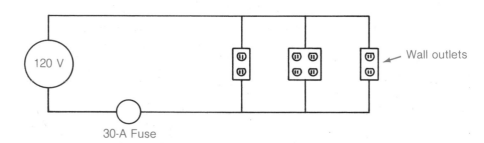

120 V

30-A Fuse

Wall outlets

FIGURE 7.36
Problem 12.

CHALLENGES

1. Compute the electric force acting between the electron and the proton in a hydrogen atom. The radius of the electron's orbit around the proton is about 5.3×10^{-11} m.

2. Use the result from Challenge 1 and the equation for centripetal force from Chapter 2 to compute the speed of the electron as it moves around the proton. The electron's mass is 9.1×10^{-31} kg.

3. Compute the number of electrons that flow through a wire each second when the current in the wire is 0.2 A.

4. Using your understanding of the nature of internal energy and temperature, explain why one might expect the resistance of a solid to increase if its temperature increases.

5. The current that flows through an incandescent light bulb immediately after it is turned on is higher than the current that flows moments later. Why?

6. An electrical device called a diode is designed to have very low resistance to current flow in one direction through it but very large resistance to current flow in the other direction. Sketch a graph of the voltage versus current for such a device.

7. Imagine a company offering a line of hair dryers that operate on different voltages: say, 12, 30, 60, and 120 V. If all are rated at 1,200 W, find the current that would flow in each as it operates. What would be different about the heating filament wires and the motors in the various hair dryers?

8. Perform the calculation referred to in the last sentence of Example 7.6.

9. Combine Ohm's Law and the equation for power consumption to derive the equation that gives the power in terms of current and resistance. Use the result to answer the following: A cable carrying electrical energy wastes 10 kilowatt-hours of energy each day because of ohmic heating. If the current in the cable is doubled but the cable's resistance remains the same, how much energy will it waste each day?

SUGGESTED READINGS

Cajori, Florian. *A History of Physics.* New York: Dover, 1962 (originally published by Macmillan in 1929). Includes sections on electricity in several chapters.

Kuffler, Stephen W., and Nicholls, John G. *From Neuron to Brain.* Sunderland, Mass: Sinauer Associates, Inc., 1976. "Directed to the reader who is curious about the workings of the nervous system but does not necessarily have a specialized background in biological sciences."

Loeb, Leonard B. "The Mechanism of Lightning." *Scientific American* 180, no. 2(February 1949): 22–27.

Roller, Duane, and Roller, Duane, H. D. "The Development of the Concept of Electric Charge: Electricity from the Greeks to Coulomb." In *Harvard Case Histories in Experimental Science,* edited by J. B. Conant, vol 2. Cambridge, Mass.: Harvard University Press, 1957.

Schecter, Bruce. *The Path of No Resistance: The Story of the Revolution in Superconductivity.* New York: Simon and Schuster, 1989. A captivating account written for nonscientists by a physicist.

Segrè, Emilio. *From Falling Bodies to Radio Waves.* New York: W. H. Freeman and Co, 1984. The first part of Chapter 4 is a good account of the lives and work of Gilbert, Franklin, Volta, and others.

Walker, Jearl. The Flying Circus of Physics with Answers. New York: Wiley, 1977. "A collection of problems and questions about physics in the real, everyday world." Chapter 6 addresses dozens of questions about electricity.

Williams, Earle R. "The Electrification of Thunderstorms." Scientific American 259, no. 5(November 1988). Describes models that have been developed to explain the cause of lightning.

Wolsky, Alan M.; Robert F. Giese; and Edward J. Daniels. "The New Superconductors: Propects for Applications." *Scientific American* 260, no. 2(February 1989): 60–69. Describes current and proposed uses of superconductors.

CHAPTER 8

OUTLINE

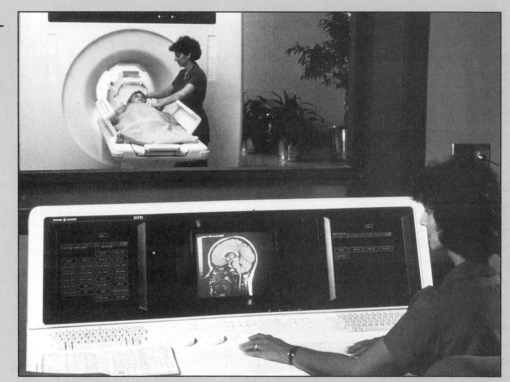

MRI unit in use. Note the patient inside the electromagnet and the computer display of the MRI image.

s cooled to the temperature of
ric currents. The patient is inserted
s the body. Auxiliary coils can
egions are highlighted. Selected
can be viewed individually.
lanted steel plates or pins, or
ntense magnetic field could
)
, MRI units can cost more than
) systems were installed worldwide
to 1987). The field holds great
ly if the new high-temperature

tromagnetic waves (such as radio
chapter. There is perhaps no better
than MRI.
p with electricity are the subjects
anent magnets and the earth's
onstrate how electric fields and
notion or change is involved. These
mmon electrical devices operate.
agnetic (EM) waves. The
f EM waves are the main topics of

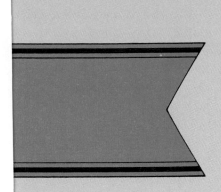

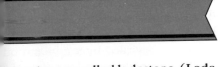

ccurring ore called lodestone. (Lode-
n ancient city in Asia Minor.) Small
ls are attracted by lodestones, similar
by charged plastic (Figure 8.1). The
that a piece of lodestone will orient
ead or floated in water on a piece of
ion because it allowed one to deter-
loudy weather. It was also one of the
the 19th century.
sizes and shapes out of special alloys
n lodestone. All magnets exhibit the
it is attracted to the north, and the
h. The north-seeking part of a magnet
ng part is its *south pole.* All magnets
pieces, each part will have its own
e magnet exerts a mutually attractive
The south poles of two magnets repel
8.2). Simply: *like poles repel, unlike*

agnets are said to be *ferromagnetic.*
them when they are near a magnet. If

superconducting electromagnets—large coi
liquid helium (4 K) and carrying large elec
into the coil so the magnetic field permeate
adjust the magnetic field so that particular
cross-sections of different parts of the body
(Incidentally, persons with pacemakers, im
unremoved shrapnel cannot use MRI. The
dislodge the metal and do internal damage

Because of their sophisticated technolog
$2 million each. In spite of this, about 1,30
in the first decade of their existence (1977
promise for further development, particula
superconductors can be used.

Magnetic fields, electromagnets, and elec
waves) are some of the main topics of this
example of these concepts currently in use

Magnetism and its useful interrelationsh
of this chapter. First the properties of perm
magnetic field are described. Next we dem
magnetic fields are intertwined whenever
concepts are used to explain how many co
They also suggest the existence of electror
properties and uses of the different types o
the latter half of the chapter.

8.1   MAGNETISM

Magnetism was first observed in a naturally
stones were prominent around Magnesia,
pieces of iron, nickel, and certain other met
to the way pieces of paper are attracted
Chinese were probably the first to discove
itself north and south if suspended by a th
wood. The *compass* revolutionized naviga
mine the direction of north even at sea in
few useful applications of magnetism up to

Now magnets are made into a variety of
that exhibit much stronger magnetism tha
same compass effect—one end or part of
opposite end or part is attracted to the sou
is called its *north pole,* and the south-seek
have both poles. If a magnet is broken int
north and south poles. The south pole of o
force on the north pole of a second magnet
each other, as do the north poles (Figure
poles attract.

Metals that are strongly attracted by m
Such materials have magnetism induced in

ELECTROMAGNETISM AND EM WAVES

PROLOGUE: MRI

An important part of the great advance of medical science in the last 100 years has been the development of noninvasive ways to "see" inside the human body. The 1800s ended with the application of newly discovered x-rays in medicine—still the most widely used imaging process. This trend continued throughout the 20th century as new scientific discoveries and advances in technology were applied to medical imaging. Of the resulting new imaging technologies that evolved, the one regarded as the most significant advance since the x-ray is *magnetic resonance imaging (MRI)*. (It was originally called *nuclear magnetic resonance [NMR]* imaging. The "nuclear" was dropped by the medical profession, most likely so that patients would not get a mistaken impression that nuclear radiation is involved.) The appeal of MRI lies both it what it *can do* and in what it *does not do*: it can show amazing detail, such as tumors not revealed by x-rays and other imaging processes, but it does not require the injection or ingestion of foreign "tracer" substances, and the body is not irradiated with ionizing radiation as with x-rays.

MRI uses a process that has been used in chemical analysis for quite some time. When hydrogen atoms are placed in a strong magnetic field, their nuclei, which are simply protons, align with the field. Like compass needles, the protons line up with the magnetic field because they are themselves tiny magnets. But the protons also precess—they wobble much like a spinning top or gyroscope—at a precise frequency that depends on the strength of the magnetic field. When a short burst of radio waves with that same frequency hits the protons, they are momentarily knocked out of alignment. In the process of spiraling back into position, the protons emit their own weak radio waves. These faint signals are detected and are then processed by a computer to form an image on a monitor.

MRI works well with living tissue because hydrogen atoms are the most numerous in our bodies; they are in water molecules (H_2O) as well as in thousands of different organic molecules. The signals emitted by the protons in different types of tissue and in regions of varying densities of hydrogen are not the same. These subtle differences lead to surprisingly high contrast in the images. The figure at left is just one example.

Magnetic resonance imaging requires a very strong magnetic field, up to 60,000 times as strong as the earth's field. Most MRI systems use superconducting electromagnets—large coils cooled to the temperature of liquid helium (4 K) and carrying large electric currents. The patient is inserted into the coil so the magnetic field permeates the body. Auxiliary coils can adjust the magnetic field so that particular regions are highlighted. Selected cross-sections of different parts of the body can be viewed individually. (Incidentally, persons with pacemakers, implanted steel plates or pins, or unremoved shrapnel cannot use MRI. The intense magnetic field could dislodge the metal and do internal damage.)

Because of their sophisticated technology, MRI units can cost more than $2 million each. In spite of this, about 1,300 systems were installed worldwide in the first decade of their existence (1977 to 1987). The field holds great promise for further development, particularly if the new high-temperature superconductors can be used.

Magnetic fields, electromagnets, and electromagnetic waves (such as radio waves) are some of the main topics of this chapter. There is perhaps no better example of these concepts currently in use than MRI.

Magnetism and its useful interrelationship with electricity are the subjects of this chapter. First the properties of permanent magnets and the earth's magnetic field are described. Next we demonstrate how electric fields and magnetic fields are intertwined whenever motion or change is involved. These concepts are used to explain how many common electrical devices operate. They also suggest the existence of electromagnetic (EM) waves. The properties and uses of the different types of EM waves are the main topics of the latter half of the chapter.

8.1   MAGNETISM

Magnetism was first observed in a naturally occurring ore called lodestone. (Lodestones were prominent around Magnesia, an ancient city in Asia Minor.) Small pieces of iron, nickel, and certain other metals are attracted by lodestones, similar to the way pieces of paper are attracted by charged plastic (Figure 8.1). The Chinese were probably the first to discover that a piece of lodestone will orient itself north and south if suspended by a thread or floated in water on a piece of wood. The *compass* revolutionized navigation because it allowed one to determine the direction of north even at sea in cloudy weather. It was also one of the few useful applications of magnetism up to the 19th century.

Now magnets are made into a variety of sizes and shapes out of special alloys that exhibit much stronger magnetism than lodestone. All magnets exhibit the same compass effect—one end or part of it is attracted to the north, and the opposite end or part is attracted to the south. The north-seeking part of a magnet is called its *north pole,* and the south-seeking part is its *south pole.* All magnets have both poles. If a magnet is broken into pieces, each part will have its own north and south poles. The south pole of one magnet exerts a mutually attractive force on the north pole of a second magnet. The south poles of two magnets repel each other, as do the north poles (Figure 8.2). Simply: *like poles repel, unlike poles attract.*

Metals that are strongly attracted by magnets are said to be *ferromagnetic.* Such materials have magnetism induced in them when they are near a magnet. If

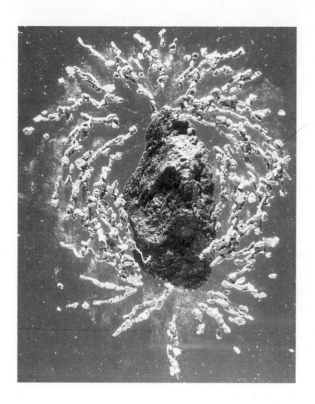

FIGURE 8.1
Lodestone, a natural magnet, attracting pieces of metallic ore.

a piece of iron is brought near the south pole of a magnet, the part of the iron nearest the magnet has a north pole induced in it, and the part farthest away has a south pole induced in it (Figure 8.3). Once the iron is removed from the vicinity of the magnet, it loses most of the induced magnetism. Some ferromagnetic metals actually retain the magnetism induced in them—they become *permanent*

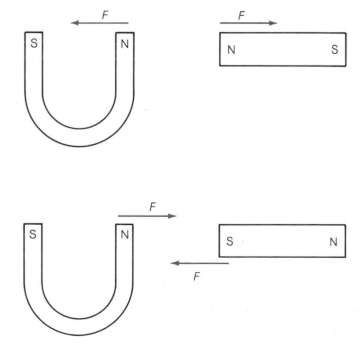

FIGURE 8.2
The poles of two magnets exert forces on each other. Like poles repel each other, and unlike poles attract each other.

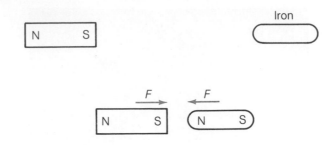

FIGURE 8.3
When a piece of ferromagnetic material (like iron) is brought near a magnet, it has magnetism induced in it. That is why it is attracted by the magnet.

magnets. Regular magnets and compass needles are made of such metals. Ferromagnetism is also the basis of magnetic tape recording. More on this later.

As with gravitation and electrostatics, it is useful to employ the concept of a field to represent the effect of a magnet on the space around it. A *magnetic field* is produced by a magnet and acts as the agent of the magnetic force. The poles of a second magnet experience forces when in the magnetic field: its north pole has a force in the same direction as the magnetic field, while its south pole has a force in the opposite direction. A compass can be thought of as a "magnetic field detector" because its needle will always align itself with a magnetic field (Figure 8.4). The shape of the magnetic field produced by a magnet can be "mapped" by noting the orientation of a compass at various places nearby. Magnetic *field lines* can be drawn to show the shape of the field, just as electric field lines are used to show the shape of an electric field (Figure 8.5). The direction of a field line at a particular place is the direction that the north pole of a compass needle will point.

Since magnets respond to magnetic fields, the fact that compass needles point north indicates that the earth itself has a magnetic field. The shape of the earth's field has been mapped very carefully over the course of many centuries because of the importance of compasses in navigation. The earth's magnetic field has the same general shape as the field around a short bar magnet with its poles tilted about 12° with respect to the axis of rotation (Figure 8.6). The direction of "true north" shown on maps is determined by the orientation of the earth's axis of rotation. (The axis is aligned quite closely with Polaris, the North Star.) Because of the tilt of the earth's "magnetic axis," *at most places on earth compasses do not point to true north.* For example, in the western two thirds of the United States, compasses point to the right (east) of true north, while in New England compasses point to the left (west) of true north. The actual difference, in degrees, between the direction of a compass and the direction of true north varies from location to location and is referred to as the *magnetic declination.* (In parts of Alaska, the magnetic declination is as high as 30° east.) This must be taken into account when navigating with a compass.[1]

Lodestones, the original magnets, are pieces of naturally occurring ferromagnetic ore. They have been weakly magnetized by the earth's magnetic field.

Another thing to note about the earth's magnetic field: the earth's *north* magnetic pole is at (near) its *south* geographic pole and vice versa. Why?

Magnetic field

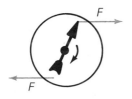

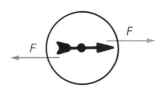

FIGURE 8.4
The forces on the poles of a compass placed in a magnetic field are in opposite directions. This causes the needle to turn until it is aligned with the field.

---

[1]The earth's magnetic poles actually move around: the south magnetic pole is not exactly where it was 20 years ago. At most places on earth, a compass does not point *exactly* in the same direction it did 20 years ago. Detailed maps used for navigation are periodically corrected to show any change in magnetic declination.

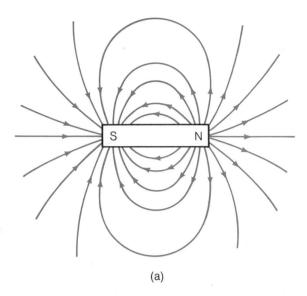

(a)

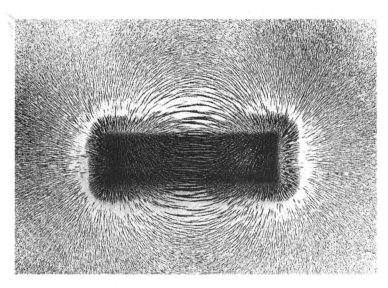

FIGURE 8.5
(a) Sketch of the magnetic field in the space around a bar magnet. Note that the field lines point toward the south pole and away from the north pole. (b) Photograph of iron filings around a magnet. Each tiny piece of iron becomes magnetized and aligns itself with the magnetic field.

1. The north pole of a magnet is attracted to the south pole of a second magnet.

2. The north pole of a compass needle points to the north.

Therefore it is the earth's south magnetic pole to which compasses point. This is not a physical contradiction: it is a result of naming the poles of a magnet after directions instead of, say, $+$ and $-$ or A and B.

Some organisms use the earth's magnetic field to aid navigation. Particles of ferromagnetic metal in the skulls of pigeons give them a built-in compass. (It has been suggested that whales are similarly equipped.) Certain species of bacteria contain tiny ferromagnetic particles that they use to orient themselves along magnetic field lines as they swim.

Superconductors, so named because of their ability to carry electrical current with zero resistance, react to magnetic fields in a rather startling fashion. When in

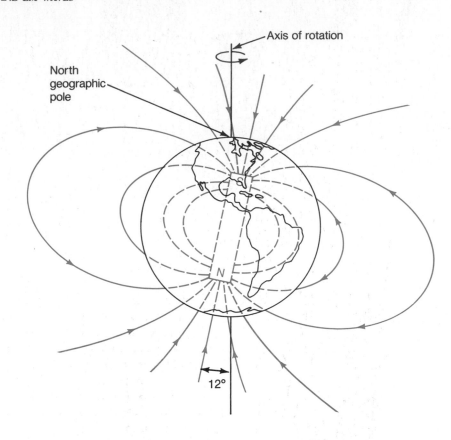

FIGURE 8.6
The earth's magnetic field. It is shaped as if there were a huge bar magnet deep inside the earth, tilted 12° relative to the earth's axis of rotation.

Refer back to superconductivity in Section 7.3.

the superconducting state, the material will expel any magnetic field from its interior. This phenomenon, known as the *Meissner effect*, is why strong magnets are levitated when placed over a superconductor (see photo at beginning of Chapter 7). When trying to determine if a material is in the superconducting state, it is easier to test for the presence of the Meissner effect than it is to see if the resistance is exactly zero.

You have probably noticed that magnetism and electrostatics are very similar: there are two kinds of poles and two kinds of charges. Like poles repel as do like charges. There are magnetic fields and electric fields. However, there are some important differences. Each kind of charge can exist separately, while magnetic poles always come in pairs. (Modern theory indicates the possible existence of a particular type of subatomic "elementary particle" that does have a single magnetic pole. As of this writing, such a "magnetic monopole" has not been found.) Furthermore, all conventional matter contains positive and negative charges (electrons and protons) and can exhibit electrostatic effects by being "charged." But, with the exception of ferromagnetic materials, most matter shows very little response to magnetic fields.

We should also point out that the electrostatic and magnetic effects described so far are completely independent. Magnets have no effect on pieces of charged plastic, for instance, and vice versa. This is the case as long as there is no motion of the objects or changes in the strengths of the electric and magnetic fields. As we will see in the following sections, a number of fascinating and useful interactions between electricity and magnetism take place when motion or change in field strength occurs.

8.2 INTERACTIONS BETWEEN ELECTRICITY AND MAGNETISM

We might summarize the basic concepts of electrostatics and magnetism as follows:

Electric charges produce electric fields in the space around them.

An electric field, regardless of its origin, causes a force on any charged object placed in it.

Magnets produce magnetic fields in the space around them.

A magnetic field, regardless of its origin, causes forces on the poles of any magnet placed in it.

These statements have been worded in this particular way because, as we will see, it is the electric and magnetic *fields* that are involved in the interplay between electricity and magnetism. We will first describe some basic observations of these interactions, then discuss some useful applications of them, and finally summarize them in the form of basic principles similar to those in the previous paragraph. We wish to stress here the ways in which electricity and magnetism interact, not discuss why these interactions occur.

> Observation 1: A moving electric charge produces a magnetic field in the space around it. An electric current produces a magnetic field around it.

A single charged particle creates a magnetic field only when it is *moving*. The magnetic field that is produced is in the shape of circles around the path of the charge (see Figure 8.7). For a steady (DC) current in a wire, which is basically a succession of moving charges, the field is steady, and its strength is proportional to the size of the current and inversely proportional to the distance from the wire. (The field is quite weak unless the current is large. A current of 10 A or more will produce a field strong enough to be detected with a compass; Figure 8.8.) Reversing the direction of the current in the wire will reverse the directions of the magnetic field lines.

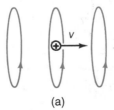

(a)

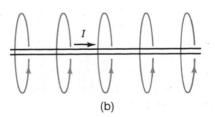

(b)

FIGURE 8.7
Magnetic field produced by (a) a moving charge and by (b) a wire carrying DC. The field lines are circles concentric with the path of the charges. (The power source for the current is not shown.)

FIGURE 8.8
(a) The compass needles align with the earth's magnetic field when no current is in the wire. (b) The compass needles show the circular shape of the magnetic field produced by a large current (15 amperes) flowing in the wire.

DO-IT-YOURSELF PHYSICS

As an automobile is being started, a huge current flows through the battery and starter motor—typically around 100 A. This produces strong magnetic fields around the cables that carry the current.

Open the hood of a car and locate the battery and the large cables that carry current from it to the starter. Close the hood and hold a compass over the hood just above where a cable is located. Have a friend start the car, and notice the deflection of the compass needle while the starter is engaged.

FIGURE 8.9
(a) The magnetic field produced by a current in a coil of wire. The field has the same shape as that produced by a bar magnet. When the direction of the current is reversed (b), the polarity of the magnetic field is also reversed.

Most applications of this phenomenon use coils—long wires wrapped in the shape of a cylinder, often around an iron core. (The magnetism induced in the iron greatly enhances the magnetic field of the coil.) The magnetic field of such a coil (when carrying a direct current) has the same shape as the field around a bar magnet. (See Figure 8.9 and compare it to Figure 8.5.) This device is an *electromagnet*; it behaves just like a permanent magnet as long as there is a current flowing. One end of the coil is a north pole, and the other is a south pole. Electromagnets have an advantage over permanent magnets in that the magne-

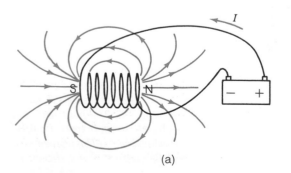

(a)

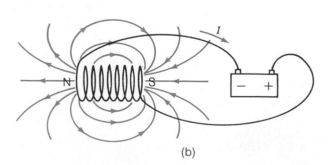

(b)

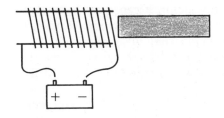 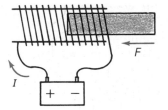

FIGURE 8.10
The iron rod is pulled into the coil (solenoid) when the current flows. One use of this phenomenon is in electronic door locks.

tism can be "turned off" simply by switching off the current. (Large electromagnets are routinely used to pick up scrap iron.) A coil whose length is much greater than its diameter is called a *solenoid.* If an iron rod is partially inserted into a solenoid with a hollow core, the rod will be pulled in when the current is switched on (see Figure 8.10). Some of the ways this is used in common devices include striking door bell chimes, opening valves to allow water to enter and to leave washing machines, withdrawing deadbolts in electric door locks, and engaging starter motors on car and truck engines.

Electromagnets are used to produce the strongest magnetic fields on earth. Two factors contribute to stronger fields: wrapping more coils around the cylinder and using a larger electric current. The former suggests the use of thinner wire so that more coils can fit into the same amount of space. But smaller wires necessitate smaller electric currents so the wire do not overheat and melt. This limitation is overcome in *superconducting electromagnets* (Figure 8.11). When

Refer back to superconductivity in Section 7.3.

FIGURE 8.11
The pattern of magnetic field lines surrounding a pair of powerful superconducting electromagnets is made visible by scattering iron rails on a piece of white plywood.

Aluminum

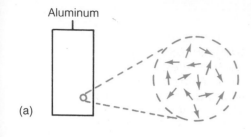

(a)

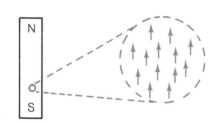

(b)

FIGURE 8.12
The arrows represent the magnetic fields of electrons in individual atoms. (a) The fields remain randomly oriented in nonferromagnetic materials. (b) Inside ferromagnetic material that is magnetized, the individual magnetic fields are aligned.

the wire used in an electromagnet is a superconductor, it can carry huge electrical currents with no ohmic heating because there is no resistance. Very small superconducting electromagnets can generate very strong magnetic fields while using much less electrical energy than a conventional electromagnet. (We describe several uses of superconducting electromagnets in Section 7.3 and the Chapter 8 Prologue.)

Superconducting electromagnets do have limitations though. The superconducting state is lost if the temperature, electric current, or magnetic field strength exceed certain values. The superconducting electromagnets now found in science laboratories throughout the world use a compound of niobium and tin that must be kept cold with liquid helium ($T = 4$ Kelvin). The added cost of the liquid helium system is offset by the high magnetic fields achieved and the great reduction in use of electric energy compared to conventional electromagnets. The new high-temperature superconductors hold the promise of much lower cost and a corresponding wider range of applications.

The polarity of an electromagnet is reversed if the direction of the current is reversed (see Figure 8.9b). An alternating current in a coil will produce a magnetic field that oscillates: it increases, decreases, and switches polarity with the same frequency as the current. Such an oscillating magnetic field will cause a nearby piece of iron to vibrate. The oscillating magnetic field of a coil with AC in it is used in many common devices, as we will see in the following sections.

Since electrons in atoms are charged particles in motion about the nucleus, they produce magnetic fields. Also, the electrons themselves have their own magnetic fields associated with their spin. In any unmagnetized material, the individual magnetic fields of the electrons are randomly oriented and cancel each other out (Figure 8.12). In ferromagnetic materials, these fields can be aligned with one another by an external magnetic field; the material then produces a net magnetic field. So we can conclude that moving electric charges are the causes of magnetic fields even in ordinary bar and horseshoe magnets.

This brings us back to a statement made at the beginning of Chapter 7: electric charges are the cause of both electrical and magnetic effects. We might regard electricity and magnetism as two different manifestations of the same thing—charge.

> Observation 2:   A magnetic field exerts a force on a moving electric charge.[2] Therefore a magnetic field exerts a force on a current-carrying wire.

A stationary electric charge is not affected by a magnetic field, but a moving charge is. Note that this second observation is a logical consequence of the first: anything that produces a magnetic field will itself be affected by other magnetic fields.

A curious characteristic of electromagnetic phenomena is that the *effects* are often *perpendicular* to the *causes*. The direction of the magnetic field from a current-carrying wire is perpendicular to the direction the current is flowing (Figure 8.7). Similarly, the force that a magnetic field exerts on a moving charge or on a current-carrying wire is perpendicular to both the direction of the magnetic field and the direction the charge is flowing. For example, if a horizontal

---

[2]If a charge's velocity or the direction of a current is parallel to the magnetic field or in the opposite direction, the magnetic field does not exert a force on it.

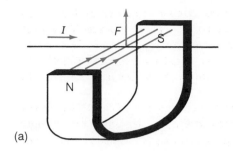

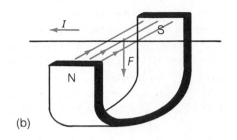

(a)

(b)

FIGURE 8.13
(a) The force on a current-carrying wire in a magnetic field. (b) When the direction of the current is reversed, the direction of the force is also reversed.

magnetic field is directed away from you and a wire is carrying a current to your right, the force on the wire is *upward* (Figure 8.13). If the direction of the current is reversed, the direction of the force is reversed (downward). An alternating current would cause the wire to experience a force that alternates up and down.

Electric motors, used in various things from hair dryers to elevators, exploit this electromagnetic interaction. The simplest type of electric motor consists of a coil of wire mounted so that it can rotate in the magnetic field of a horseshoe-shaped magnet (Figure 8.14). A direct current flows through the coil, the magnetic field causes forces on the sides of the coil, and it rotates. Once the coil has completed half of a rotation, a simple mechanism reverses the direction of the current. This reverses the force on the coil, causing it to rotate another half turn. This process is repeated, and the coil spins continuously. Motors designed to run on AC can exploit the fact that the direction of the current is automatically reversed 120 times each second.

The effect of magnetic fields on moving charged particles is used in several large-scale machines used in experimental physics. High-temperature plasmas cannot be kept in any conventional metal or glass container because it would melt. Since plasmas are composed of charged particles, magnetic fields can be used to contain them in what is known as a "magnetic bottle." This is one approach being used in the attempt to harness nuclear fusion as an energy source (Section 11.7).

In the absence of other forces, a charged particle moving perpendicularly to a magnetic field will travel in a circle: the force on the particle is always perpendicular to its velocity and therefore is a centripetal force. An electron, proton, or other charged particle can be forced to move in a circle and then can be gradually accelerated during each revolution. Various types of particle accelerators used in experimental atomic and nuclear physics operate on this principle. One such accelerator, the *Tevatron* at the *Fermi National Accelerator Laboratory (Fermilab)* near Batavia, Illinois, is 2 kilometers (1.2 miles) in diameter (Figure 8.15). The *LEP* accelerator near Geneva, Switzerland is newer and even larger than the Tevatron: with a diameter of 8.6 kilometers (5.3 miles), LEP is the largest scientific instrument built so far. Both Tevatron and LEP have positively charged particles circulating in one direction and negatively charge particles circulating in the opposite direction. The head-on collisions of these oppositely charged particles yield information about the fundamental forces and particles in nature.

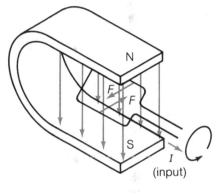

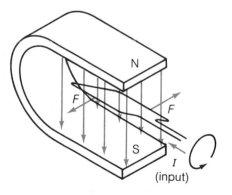

FIGURE 8.14
Simplified sketch of an electric motor. The loop of wire rotates because of the forces on its sides. Each time the loop becomes horizontal, the direction of the current is reversed, and the rotation continues.

> Observation 3: A moving magnet produces an electric field in the space around it. A magnet moving through a coil of wire will induce a current in the wire.

FIGURE 8.15
Fermi National Accelerator Laboratory, near Batavia, Illinois. Inside the huge ring, charged particles are accelerated to nearly the speed of light. Magnetic fields are used to keep the charges moving in a circle.

The electric field around a moving magnet is in the shape of circles around the path of the magnet. This circular electric field will force charges in a coil of wire to move in the same direction—as a current (Figure 8.16). (Note the similarity between Observations 1 and 3.) The process of inducing an electric current with a magnetic field is known as *electromagnetic induction*. All that is required is that the magnet and coil move relative to each other. If the coil moves and the magnet remains stationary, a current is induced. If the motion is steady in either case, the induced current is in one direction. If either the coil or the magnet oscillates back and forth, the current alternates with the same frequency—it is AC.

Electromagnetic induction is used in the most important device for the production of electricity: the generator. The simplest generator is basically an electric motor. When the coil is forced to rotate, it moves relative to the magnet, so a current is induced in it (see Figure 8.17). We might call this device a "two-way energy converter." When electrical energy is supplied to it, it is a motor, and it converts this electrical energy into mechanical energy of rotation. When it is mechanically turned (by hand cranking, by a fan belt on a car engine, or by a turbine in a power plant) it is a generator, and it converts mechanical energy into electrical energy. Only rarely is a single device used as both a motor and a generator. An efficiently designed generator does not make a very practical motor and vice versa.

FIGURE 8.16
(a) The electric field produced by a moving magnet is circular. (The magnetic field of the magnet is not shown.) (b) A magnet will induce a current in a coil as it passes through.

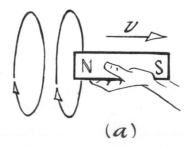

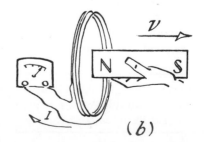

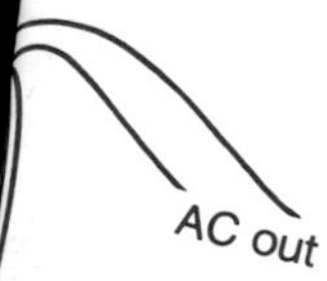

AC out

...t would not induce a current in
... DC.
...ifferent from the voltage of the
...as the same voltage induced in
... that the more turns there are
...e ratio of the number of turns
...d output voltages. In particular:

$$\frac{\text{...s in output coil}}{\text{...s in input coil}}$$

...in the input coil, the output
...e third as many turns in the
...input voltage. Thus the AC
...desired amount by simply
...coils.

...put with a 120-volt input.
...w many turns must there

...us
...wo

...field.

...empha-
...that the
...be used
...strength
...ion of the
...magnetic
...romagnetic

...p up or step
...ns of electro-
...ils of wire in
...d the "input"
...d the "output"
...roduces an *os-*
...have both coils
...he magnetic field
...erefore changing)
...oil. Note that a DC

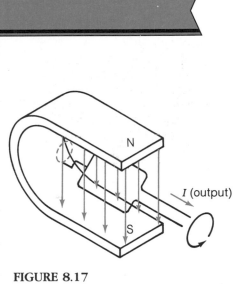

FIGURE 8.17
*Simplified sketch of a generator.
As the loop of wire rotates
relative to the magnetic field, a
current is induced in it.*

I (output)

N

S

...l distribution systems,
...nces. Most electrical
...oltages much smaller
...r household AC must
...intensity desk lamps
...y (Figure 8.20). The
...ated using a type of
...output coil is many

3
⊕

FIGURE 8.18
*The electric field at the point
changes as the charged particle
moves by. The arrows indicate
the magnitude and direction of
the electric field for three
different locations of the particle.
The magnetic field that is
induced at that point is directed
straight out of the paper.*

FIGURE 8.19
Simplified diagram of a transformer. The alternating magnetic field produced by the AC in the input coil induces an alternating current in the output coil. In this case, the output voltage would be higher than the input voltage.

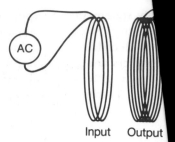

Input   Output

input would produce a *steady* magnetic field th
the output coil. Transformers do not work with
   Now, how can the voltage of the output be o
input? Each "loop" or "turn" of the output coil h
it. The voltages in all of the turns add together s
in the output coil, the higher the total voltage. Th
in the two coils determines the ratio of the input an

$$\frac{\text{Voltage of output}}{\text{Voltage of input}} = \frac{\text{number of turn}}{\text{number of turn}}$$

$$\frac{V_o}{V_i} = \frac{N_o}{N_i}$$

If there are twice as many turns in the output coil as
voltage will be twice the input voltage. If there are o
output coil, the output voltage will be one third the
voltage can be stepped up or stepped down by any
adjusting the ratio of the number of turns in the two

EXAMPLE 8.1   A transformer is being designed to have a 600-volt ou
If there are to be 800 turns of wire in the input coil, h
be in the output coil?

$$\frac{V_o}{V_i} = \frac{N_o}{N_i}$$

$$\frac{600 \text{ V}}{120 \text{ V}} = \frac{N_o}{800}$$

$$800 \times 5 = N_o$$

$$N_o = 4{,}000$$

   In addition to being used to change voltages in electrica
transformers are used in a wide variety of electrical appli
components used in radios, calculators, and the like require
than 120 volts. Such appliances designed to operate on regula
include transformers to reduce the voltage accordingly. High
also use transformers; that is what makes their bases so heav
spark used to ignite gasoline in automobile engines is gener
transformer called simply a "coil." The number of turns in the

These three observations are simply statements of experimental facts. They can be illustrated quite easily using a battery, wires, a compass, a large magnet, and a sensitive ammeter. The fact that electricity and magnetism interact only when there is motion (and then the effects are perpendicular to the causes) is somewhat startling when compared to, say, gravitation and electrostatics. Nonetheless, they are the key to our modern electrified society.

8.3   PRINCIPLES OF ELECTROMAGNETISM

The interactions between electricity and magnetism described in the previous section, along with several other similar observations, suggest the following two general statements. We might call these the *principles of electromagnetism*:

1. An electric current or a changing electric field induces a magnetic field.

2. A changing magnetic field induces an electric field.

These two statements summarize the previous observations and also emphasize the symmetry that exists. In both cases, a "changing" field means that the strength or the direction of the field is changing. The first principle can be used to explain the first observation: as a charge moves by a point in space, the strength of the electric field increases and then decreases. All the time the direction of the field is changing as well (Figure 8.18). The effect of this is to cause a magnetic field to be produced. Similarly, the second principle explains electromagnetic induction.

As mentioned in Section 7.6, a transformer is a device used to step up or step down AC voltages. It represents one of the most elegant applications of electromagnetism. In essence, a transformer consists of two separate coils of wire in close proximity. An AC voltage is applied to one of the coils, called the "input" or "primary" coil, and an AC voltage appears at the other coil, called the "output" or "secondary" coil (Figure 8.19). The AC in the primary coil produces an *oscillating magnetic field* through both coils. (Most transformers have both coils wrapped around a single ferromagnetic core so as to intensify the magnetic field and guide it from one coil to the other.) This oscillating (and therefore changing) magnetic field induces an alternating current in the output coil. Note that a DC

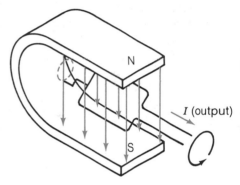

FIGURE 8.17
Simplified sketch of a generator. As the loop of wire rotates relative to the magnetic field, a current is induced in it.

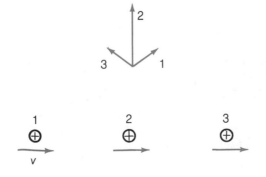

FIGURE 8.18
The electric field at the point changes as the charged particle moves by. The arrows indicate the magnitude and direction of the electric field for three different locations of the particle. The magnetic field that is induced at that point is directed straight out of the paper.

FIGURE 8.19
Simplified diagram of a transformer. The alternating magnetic field produced by the AC in the input coil induces an alternating current in the output coil. In this case, the output voltage would be higher than the input voltage.

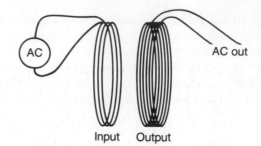

Input   Output

input would produce a *steady* magnetic field that would not induce a current in the output coil. Transformers do not work with DC.

Now, how can the voltage of the output be different from the voltage of the input? Each "loop" or "turn" of the output coil has the same voltage induced in it. The voltages in all of the turns add together so that the more turns there are in the output coil, the higher the total voltage. The ratio of the number of turns in the two coils determines the ratio of the input and output voltages. In particular:

$$\frac{\text{Voltage of output}}{\text{Voltage of input}} = \frac{\text{number of turns in output coil}}{\text{number of turns in input coil}}$$

$$\frac{V_o}{V_i} = \frac{N_o}{N_i}$$

If there are twice as many turns in the output coil as in the input coil, the output voltage will be twice the input voltage. If there are one third as many turns in the output coil, the output voltage will be one third the input voltage. Thus the AC voltage can be stepped up or stepped down by any desired amount by simply adjusting the ratio of the number of turns in the two coils.

EXAMPLE 8.1   A transformer is being designed to have a 600-volt output with a 120-volt input. If there are to be 800 turns of wire in the input coil, how many turns must there be in the output coil?

$$\frac{V_o}{V_i} = \frac{N_o}{N_i}$$

$$\frac{600\ \text{V}}{120\ \text{V}} = \frac{N_o}{800}$$

$$800 \times 5 = N_o$$

$$N_o = 4{,}000$$

In addition to being used to change voltages in electrical distribution systems, transformers are used in a wide variety of electrical appliances. Most electrical components used in radios, calculators, and the like require voltages much smaller than 120 volts. Such appliances designed to operate on regular household AC must include transformers to reduce the voltage accordingly. High-intensity desk lamps also use transformers; that is what makes their bases so heavy (Figure 8.20). The spark used to ignite gasoline in automobile engines is generated using a type of transformer called simply a "coil." The number of turns in the output coil is many

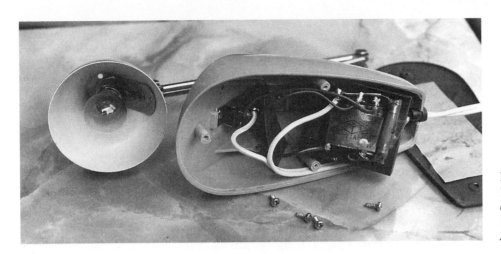

FIGURE 8.20
The transformer in this lamp converts 120 volts AC into 14 volts AC ("Hi") or 12 volts AC ("Lo").

times the number of turns in the input coil. A spark is produced by first sending a brief current into the input. A magnetic field is produced that quickly disappears. This induces a very high voltage (around 25,000 volts) in the output, which is conducted to the spark plugs to ignite the fuel.

8.4 APPLICATIONS TO SOUND REPRODUCTION

One hallmark of this century is that an overwhelming percentage of people have access to high-quality music. This has been made possible by the advent of affordable sound-reproduction equipment. The first Edison phonographs were strictly mechanical and did a fair job of reproducing sound. But it was the invention of electronic recording and playback machines that brought true high fidelity to sound reproduction. The chain that begins with sound in a recording studio and ends with the reproduced sound coming from a speaker in your home or car includes some basic components that employ electromagnetism.

The key to electronic sound recording and playback is to first translate the sound into an alternating current and then later to translate the AC back into sound. The first step requires a *microphone,* and the second step requires a *speaker.* Although there are several different types of microphones, we will take a look at what is called a *dynamic* microphone. It consists of a magnet surrounded by a coil of wire attached to a diaphragm (see Figure 8.21). The coil and diaphragm are free to oscillate relative to the stationary magnet. When sound waves reach the microphone, the pressure variations in the wave push the diaphragm back and forth, making it and the coil oscillate. Since the coil is moving relative to the magnet, an oscillating current is induced in it. The frequency of the AC in the coil is the same as the frequency of the diaphragm's oscillation, which is the same as the frequency of the original sound. That is all it takes. (This type of dynamic microphone is referred to as a "moving coil" microphone. The alternative is to attach a small magnet to the diaphragm and keep the coil stationary—a "moving magnet" microphone.)

Let us skip ahead now to when the sound is played back. The output of the tape player, radio, or whatever is an alternating current that has to be converted back

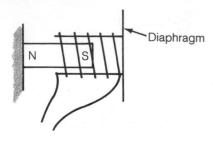

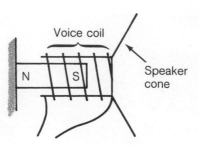

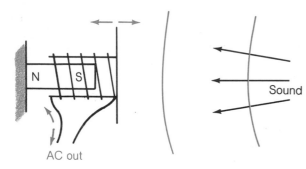

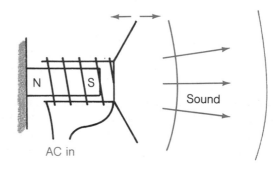

FIGURE 8.21
Simplified diagram of one type of microphone. Sound waves cause the diaphragm and coil to oscillate relative to the magnet. This induces AC in the coil.

FIGURE 8.22
Simplified sketch of a speaker. An alternating current in the voice coil causes the cone to be forced in and out, thereby producing sound.

into sound by a speaker. The basic speaker is quite similar to a dynamic microphone. In this case, the coil (called the "voice coil") is connected to a stiff paper cone instead of to a diaphragm (Figure 8.22). Recall from Section 8.1 that an alternating current in the voice coil and the presence of the magnet will cause the coil to experience an alternating force. The voice coil and the speaker cone oscillate with the same frequency as the AC input. The oscillating paper cone produces a longitudinal wave in the air—sound.

Microphones and speakers are classified as *transducers:* they convert mechanical oscillation due to sound into an alternating current (microphone), or they convert alternating current into mechanical oscillation and sound (speaker). They are almost identical; in fact, a microphone can be used as a speaker, and a speaker can be used as a microphone. But, as with motors and generators, each is best at doing what it is designed to do.

Most sound recording, from simple cassette recorders to sophisticated studio tape machines, is done on magnetic tape. The tape is a plastic film coated with a thin layer of fine ferromagnetic particles that retain magnetism. Sound is recorded on the tape using a recording head, a ring-shaped electromagnet with a very narrow gap (Figure 8.23). During recording, an AC signal (from a microphone, phonograph cartridge, or whatever) produces an alternating magnetic field in the gap of the recording head. As the tape is pulled past the gap, the particles in each part of the tape are magnetized according to the polarity of the head's magnetic field at the instant they are in the gap. The polarity of the particles changes from north-south to south-north and so on along the length of the tape.

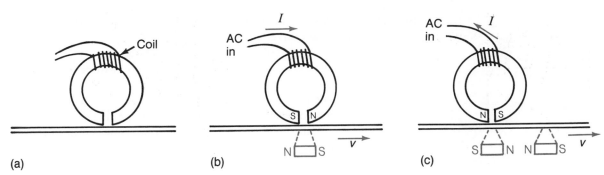

(a) (b) (c)

FIGURE 8.23
(a) Simplified sketch of a tape head. During recording, (b) and (c), AC in the coil induces alternating magnetism in the tape.

To play back the recording, the tape is pulled past a playback head, often the same head used for recording. The magnetic field of the particles in the tape oscillates back and forth and induces an oscillating magnetic field in the tape head (Figure 8.24). This oscillating magnetic field induces an oscillating current (AC) in the coil—electromagnetic induction again.

Magnetic recording is not limited to sound reproduction. Television video recorders (VCRs) record both sound and visual images on magnetic tape. Computers store information magnetically on tape, floppy disks, and hard disks (Figure 8.25).

The AC signals produced by microphones, compact disc players, and tape playback heads are quite weak. Amplifiers are used to increase the power of these signals before they are sent to speakers. Amplifiers also allow the listener to modify the sound by adjusting its loudness with the volume control and its tone quality with the bass and treble controls. The actual details of how an amplifier works are beyond the scope of this text. An amplifier can be regarded as simply a device that can increase the amplitude of an alternating current.

Digital Sound—CD and DAT

A revolution in sound reproduction occurred in the 1980s with the advent of digital sound reproduction. This method is used in *compact discs (CDs)* and in *digital audio tapes (DATs)*. In a process known as *analog-to-digital* conversion, the sound wave to be recorded is measured and stored as numbers. The actual voltage of the AC signal from a microphone is measured 44,100 times *each second* for the CD format and 48,000 times per second for the DATs (Figure 8.26). (Note that these frequencies are more than twice the highest frequency that people can hear.) In essence, the waveform of the sound is "chopped up" into tiny segments and is recorded as numerical values. These numbers are stored as binary numbers

Refer back to waveforms in Section 6.3.

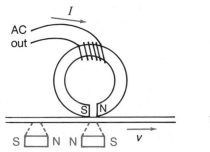

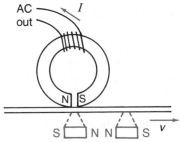

FIGURE 8.24
During playback, the alternating magnetism in the tape induces AC in the coil on the tape head.

FIGURE 8.25
The contents of this book are stored on just a few hundred feet of this magnetic tape.

using 0s and 1s, just as information is stored in computers. To playback the sound, a *digital-to-analog* conversion process reconstructs the sound wave by generating an AC signal whose voltage at each instant in time equals the numerical value that was originally recorded. After being "smoothed" with an electronic filter, the waveform is almost a perfect copy of the original.

A huge amount of data is associated with digital sound reproduction—over $2\frac{1}{2}$ million numbers for each minute of music. The optical compact disc stores this data in the form of microscopic pits in a spiral line some 3 miles long. A tiny laser focused on the pits reads them as 0s and 1s. The amount of information stored on a 70-minute CD is equivalent to more than a dozen full-length encyclopedias. (Some computers are equipped to read "CD-ROM"—compact discs on which huge amounts of information are stored.) The DAT stores the data magnetically in the same way that a video cassette recorder (VCR) does. DAT tapes are smaller than regular cassettes, but they can hold more information than a CD.

The superior quality of digital sound arises because the playback device looks only for numbers. It can ignore such things as imperfections in the disc or tape, the weak random magnetization in a tape that becomes tape hiss on cassettes, and the mechanical vibration of motors that we hear as a rumble on phonographs. A sophisticated error-correction system can even compensate for missing or garbled numbers. Because the pickup device in a CD player does not touch the disc, each CD can be played over and over without the slow deterioration that a needle in a phonograph groove causes. This combination of high fidelity and disc durability made the CD system an immediate hit with audiophiles.

This is just a glimpse of some of the factors involved in state-of-the-art high-fidelity sound reproduction. Perhaps we are all so accustomed to it that we cannot really appreciate just how much of a technological miracle it really is. The next time you listen to high-quality recorded music, remember that it is all possible because of the simple observations described in Section 8.1.

FIGURE 8.26
Digital sound reproduction. (a) To record the sound, the voltage of the waveform is measured 44,100 times each second (for CD's). The resulting numbers are stored on magnetic tape for later use. (b) During playback, the voltage of the output at each time is set equal to the numerical value that was stored originally. The reconstructed waveform is then "smoothed" using an electronic filter. The resulting waveform is an almost perfect reproduction of the original.

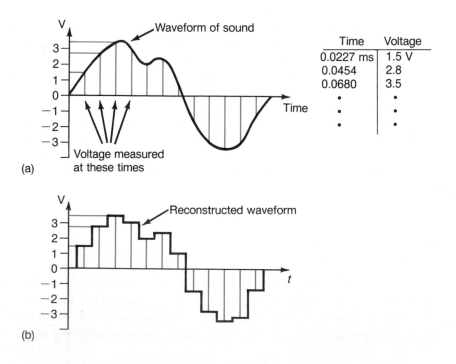

✍ | LEARNING CHECK

1. Magnetic _____ indicates how far away from true north a compass needle points.

2. A _____ electric charge produces a magnetic field around it.

3. A changing _____ produces an electric field.

4. Which of the following does *not* use electromagnetic induction?
   a) Transformer   c) Microphone
   b) Generator   d) Electromagnet

5. A coil of wire has an alternating current induced in it when it is in a magnetic field that is _____ .

6. The first step in recording sound electronically is to convert the sound wave into _____ .

8.5   ELECTROMAGNETIC WAVES

Eyes, radios, televisions, radar, x-rays, microwave ovens, heat lamps, tanning lamps . . . What do all of these things have in common? They all use *electromagnetic waves* (EM waves). As you can see, EM waves occupy prominent places both in our daily lives and in our society's technology. These waves are also involved in many natural processes and are essential to life itself. In the remainder of this chapter we will discuss the nature and general properties of electromagnetic waves and will look at some of their important roles in today's world.

As the name implies, EM waves involve both electricity and magnetism. The existence of these waves was first suggested by the 19th century physicist James Clerk Maxwell while he was analyzing the interactions between electricity and magnetism. Consider the two principles of electromagnetism stated in Section 8.3. Let us say that an oscillating electric field is produced at some place. The electric field switches back and forth in direction while its strength varies accordingly. This *oscillating electric field will induce an oscillating magnetic field* in the space around it. But the oscillating magnetic field will then induce an oscillating electric field. This will then induce an oscillating magnetic field and so on. This is an endless "loop": the principles of electromagnetism tell us that a continuous succession of oscillating magnetic and electric fields will be produced. These fields travel as a wave, an EM wave.

> ELECTROMAGNETIC WAVE   A transverse wave consisting of a traveling combination of oscillating electric and magnetic fields.

Electromagnetic waves are transverse waves because the oscillation of both of the fields is perpendicular to the direction the wave travels. Figure 8.27 shows a "snapshot" of an electromagnetic wave traveling to the right. (The three axes are

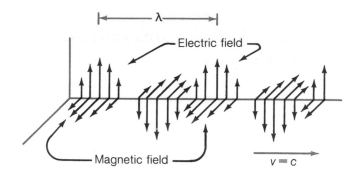

FIGURE 8.27
Representation of a portion of an electromagnetic wave. The electric field is always perpendicular to the magnetic field. The entire pattern moves to the right at the speed of light.

perpendicular to each other.) In this particular case, the electric field is vertical. As the wave travels by a given point in space, the electric field oscillates up and down, just like a floating petal oscillates on a water wave. The magnetic field at the point oscillates horizontally.

Figure 8.27 should remind you of the transverse waves we described in Chapter 6 (Figure 6.5, for example). Electromagnetic waves do differ from mechanical waves in two important ways. First, they are a combination of two waves in one: an electric field wave and a magnetic field wave. *These cannot exist separately.* Second, EM waves do not require a medium in which to travel. Electromagnetic waves can travel through a vacuum: the light from the sun does this. EM waves also travel through matter; light through air and glass and x-rays through your body are common examples.

Electromagnetic waves travel at an extremely high speed. Their speed in a vacuum, called the "speed of light" because it was first measured using light, is represented by the letter c. Its value is

$$c = 299,792,458 \text{ m/s} \quad \text{(speed of light)}$$

or

$$c = 3 \times 10^8 \text{ m/s} \quad \text{(approximately)}$$
$$= 300,000,000 \text{ m/s}$$
$$= 186,000 \text{ miles/second} \quad \text{(approximately)}$$

All of the parameters introduced for waves in Chapter 6 apply to EM waves. The wavelength can be readily identified in Figure 8.27. The amplitude is the maximum value of the electric field strength. The equation $v = f\lambda$ holds with v replaced by c. There is an extremely wide range of wavelengths of EM waves, from the size of a single proton, about 10^{-15} meters, to almost 4,000 kilometers for one type of radio wave. The corresponding frequencies of these extremes are about 10^{23} hertz and 76 hertz, respectively. Most EM waves used in practical applications have extremely high frequencies compared to sound.

EXAMPLE 8.2   An FM radio station broadcasts an EM wave with a frequency of 100 megahertz. What is the wavelength of the wave?

The prefix "mega" means one million. Therefore the frequency is 100 million hertz.

$$c = f\lambda$$

$$300,000,000 \text{ m/s} = 100,000,000 \text{ Hz} \times \lambda$$

$$\frac{300,000,000 \text{ m/s}}{100,000,000 \text{ Hz}} = \lambda$$

$$\lambda = 3 \text{ m}$$

Electromagnetic waves are named and classified according to frequency. In order of increasing frequency, the groups, or "bands," are radio waves, microwaves, infrared radiation, visible light, ultraviolet radiation, x-rays, and gamma rays (γ-rays). (Use of the word "radiation" instead of "waves" is not significant.) Figure 8.28 shows these groups along with frequency and wavelength scales. This is called the *electromagnetic wave spectrum.* Notice that the groups overlap. For example, a 10^{17} hertz EM wave could be ultraviolet radiation or an x-ray. In cases of overlap, the name applied to an EM wave depends on how it is produced.

We will discuss briefly the properties of each group in the electromagnetic wave spectrum—how they are produced, what their uses are, and how they can affect us. The great diversity of uses of EM waves arises from the variety of ways in which they can interact with different kinds of matter. All matter contains charged particles (electrons and protons), so it seems logical that EM waves can affect and be affected by matter. The oscillating electric field can cause AC currents in conductors; it can stimulate vibration of molecules, atoms, or individual electrons; or it can interact with the nuclei of atoms. Which sort of interaction occurs, if any, depends on the frequency (and wavelength) of the EM waves and on the properties of the matter through which it is traveling (its density, molecular and atomic structure, and so on).

In principle, any electromagnetic wave could be produced by simply forcing one or more charged particles to oscillate at that frequency. The oscillating field of the charges would initiate the EM wave. The "lower-frequency" EM waves (radio waves and microwaves) are produced this way: a transmitter generates an AC signal and sends it to an antenna. At higher frequencies this process becomes

FIGURE 8.28
The electromagnetic wave spectrum.

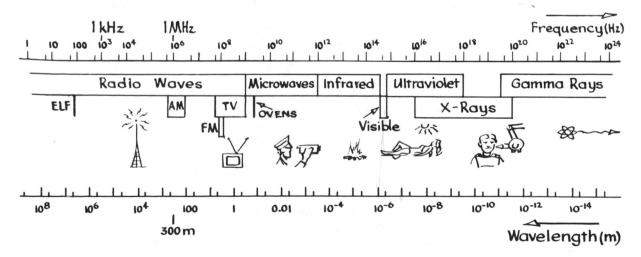

FIGURE 8.29
A radio is an electromagnetic wave detector that can select a single frequency radio wave.

increasingly difficult. Electromagnetic waves above the microwave band are produced by a variety of processes involving molecules, atoms, and nuclei. Note that charged particles are present in all of these.

We might point out one other factor to keep in mind: electromagnetic waves are a form of energy. Energy is needed to produce EM waves, and energy is gained by anything that absorbs EM waves. The transfer of heat by way of heat radiation is one example of this.

Radio Waves

Radio waves, the lowest frequency EM waves, extend from less than 100 hertz to about 10^9 Hz (1 billion hertz or 1,000 megahertz; Figure 8.29). Within this range are a number of frequency bands that have been given separate names: for example, ELF (extremely low frequency), VHF (very high frequency), and UHF (ultra high frequency). Most frequencies are given in kilohertz (kHz) or megahertz (MHz). Sometimes radio waves are classified by wavelength: long wave, medium wave, or short wave.

As mentioned earlier, radio waves are produced using AC with the appropriate frequency. Radio waves propagate well through the atmosphere, which makes them practical for communication. Lower frequency radio waves cannot penetrate the upper atmosphere, so higher frequencies are used for space and satellite communication. Only the very lowest frequencies can penetrate ocean water.

By far the main application of radio waves is in communication. The basic process involves broadcasting a certain frequency of radio wave with sound, video, or other information "encoded" in the wave. The radio wave is then picked up by a receiver, which recovers the information from it. Sometimes this is a one-way process (commercial AM and FM radio and television), but in most other applications it is two-way: each party can broadcast as well as receive. Narrow frequency bands are assigned for specific purposes. For example, frequencies from 88 to 108 megahertz (88 million hertz to 108 million hertz) are reserved for commercial FM radio. There are dozens of bands assigned to government and private communication.

Microwaves

The next band of EM waves, with frequencies higher than those of radio waves, is the microwave band. The frequencies extend from the upper limit of radio waves to the lower end of the infrared band, about 10^9 to 10^{12} hertz. The wavelengths range from about 0.3 m to 0.3 mm.

One use of microwaves is in communication. (One might argue that microwaves could be included in the radio band.) Early experiments with microwave communication led to the most important use of microwaves, *radar* (*radio detection and ranging*; Figure 8.30). It was discovered that microwaves are reflected by the metal in ships and aircraft. As we discussed in Section 6.2, radar is simply echolocation using microwaves. The time it takes microwaves to make a round trip from the transmitter to the reflecting object and back is used to determine the distance to the object. Radar systems are quite sophisticated: Doppler radar can determine the speed of an object moving toward or away from the transmitter by measuring the frequency shift of the reflected wave.

FIGURE 8.30
Radar was one of the earliest applications of microwaves.

Microwaves have gained wide acceptance as a way to cook food. The basic goal of cooking is to heat the food; in other words, increase the energies of the molecules in the food. Conventional ovens heat the air around the food and rely on conduction (in solids) or convection (in liquids) to transfer the heat throughout the food. Microwave ovens send microwaves (typically with $f = 2,450$ megahertz and $\lambda = 0.122$ meter) into the food. The microwaves penetrate the food and raise the energies of the molecules directly. Water molecules and some other molecules in food have a net positive charge on one side of the molecule and a net negative charge on the other side (Figure 8.31). The electric field of a microwave exerts forces on the two sides. These forces are in opposite directions and twist the molecule. Since the electric field is oscillating, the molecules are alternately twisted one way and then the other. This process increases the kinetic energy of the molecules and thereby raises the temperature of the food. Cooking with microwaves is very fast because energy is given directly to all of the molecules. It does not rely on the conduction of heat from the outside to the inside of the food—a much slower process.

FIGURE 8.31
(a) Simplified sketch of a water molecule showing the net charges on its sides. (b, c) The oscillating electric field of a microwave twists the molecule back and forth, giving it energy.

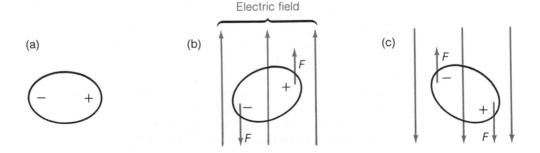

FIGURE 8.32
This remote control unit sends commands in the form of infrared radiation.

Infrared

Infrared radiation (IR; also called infrared light) occupies the region between microwaves and visible light in the electromagnetic wave spectrum. The frequencies are from about 10^{12} hertz to about 4×10^{14} hertz (400,000,000 megahertz). The wavelengths of IR range from approximately 0.3 millimeter to 0.00075 millimeter.

Infrared radiation is ordinarily the principal component of heat radiation (introduced in Section 5.4). Everything about you is both absorbing and emitting infrared radiation, as are you. The warmth you feel from a fire or heat lamp is due to your skin absorbing the IR. Infrared radiation is constantly emitted by atoms and molecules because of their thermal vibration. Absorption of IR by a cooler substance increases the vibration of the atoms and molecules, thus raising the temperature. We will take a closer look at heat radiation and its uses in Section 8.6.

Infrared radiation is commonly used in wireless remote control units for televisions, sound systems, and similar devices (Figure 8.32). The unit emits coded IR that is detected by the device. In this capacity, IR is used much like radio waves. Another use of IR is in lasers; some of the most powerful ones in use emit infrared light. (See Section 10.8.)

Visible Light

Visible light is a very narrow band of frequencies of EM waves that happen to be visible to human beings. Certain specialized cells in the eye, called rods and cones, are sensitive to EM waves in this band. They respond to visible light by transmitting electrical signals to the brain, where a mental image is formed. (The visible ranges of some animals such as hummingbirds and bees extend into the ultraviolet band. Some flowers that seem mundane to humans are quite attractive to these nectar eaters.)

Visible light is a component of the heat radiation emitted by very hot objects. About 44% of the sun's radiation is visible light: it glows white hot. Incandescent light bulbs produce visible light in the same way. Fluorescent and neon lights use excited atoms that emit visible light. In Chapter 10 we will discuss this process and describe how infrared, ultraviolet light, and even x-rays are emitted by excited atoms.

Within the narrow band of visible light, the *different frequencies* are perceived by people as *different colors*. The lowest frequencies of visible light, next to the infrared band, are perceived as the color red. The highest frequencies are perceived as violet. Table 8.1 shows the approximate frequencies and wavelengths of the six principal colors in the rainbow.

Note how narrow the frequency band is: the highest frequency is less than twice the lowest. By comparison, the range of frequencies of sound that can be heard is huge: the highest is 1,000 times the lowest.

Most colors that you see are combinations of many different frequencies. White represents the extreme: the white light coming from this paper consists of all frequencies of light combined equally. Rainbow formation involves reversing the process: white light is separated into its component colors. When no visible light reaches the eye, we perceive black.

Visible light is clearly the most important of all electromagnetic waves, at least to people. The entire next chapter is dedicated to optics, the study of light.

TABLE 8.1 APPROXIMATE FREQUENCIES AND WAVELENGTHS OF DIFFERENT COLORS

Color	Frequency Range ($\times 10^{14}$ Hz)	Wavelength Range ($\times 10^{-7}$ m)
Red	4.3–4.6	7.0–6.5
Orange	4.6–5.0	6.5–6.0
Yellow	5.0–5.5	6.0–5.5
Green	5.5–6.0	5.5–5.0
Blue	6.0–6.7	5.0–4.5
Violet	6.7–7.5	4.5–4.0

Ultraviolet Radiation

Ultraviolet radiation (UV), also called ultraviolet light, is a band of EM waves that begins just above the frequency of violet light and extends to the x-ray band. The frequency range is from about 7.5×10^{14} to 10^{18} hertz.

Ultraviolet light is also part of the heat radiation emitted by very hot objects. About 7% of the radiation from the sun is UV. This part of sunlight is responsible for suntans and sunburns. Ultraviolet radiation does not warm the skin as much as IR, but it does trigger a chemical process in the skin that results in tanning (Figure 8.33). Overexposure leads to sunburn as a short-term effect, and repeated overexposure during a person's lifetime increases the chances of developing skin cancer. In Section 8.7 we describe how the ozone layer protects us from excessive UV in sunlight.

A number of substances undergo "fluorescence" when irradiated with UV: they emit visible light. The inner surfaces of fluorescent lights are coated with such a

FIGURE 8.33
Tanning requires a source of ultraviolet light.

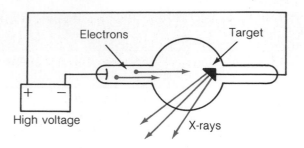

FIGURE 8.34
Simplified sketch of an x-ray tube. Electrons are acclerated to a very high speed by the high voltage. X-rays are emitted as the electrons enter the metal target.

substance. UV emitted by excited mercury atoms in the tube strikes the fluorescent coating, and visible light is produced. Some fluorescent materials appear to be colorless under normal light and can be used as a kind of invisible ink. They can be seen under a UV lamp but are invisible otherwise.

X-rays

The next higher frequency electromagnetic waves are x-rays. They extend from about 10^{16} to 10^{20} hertz. An important feature of x-rays is that their range of wavelengths (about 10^{-8} to 10^{-11} meters) includes the size of the spacing between atoms in solids. X-rays are partially reflected by the regular array of atoms in a crystal and so can be used to determine the actual arrangement of the atoms. X-rays also travel much greater distances through matter compared to UV, visible light, and other lower frequency EM waves.

X-rays are produced by smashing high-speed electrons into a "target" made of tungsten or some other metal (Figure 8.34). The electrons spontaneously emit x-rays as they are rapidly decelerated upon entering the metal. X-rays are also emitted by some of the atoms excited by the high-speed electrons.

Medical and dental "x-ray" photographs are made by sending x-rays through the body. Typically x-rays with frequencies between 3.6×10^{18} hertz and 12×10^{18} hertz are used. As x-rays pass through the body, the degree to which they are absorbed depends on the material through which they pass. Tissue containing elements with relatively large atomic numbers, like calcium ($Z = 20$), tend to absorb x-rays more effectively than those that contain predominantly light elements like carbon ($Z = 6$), oxygen ($Z = 8$), or hydrogen ($Z = 1$). (Lead, with atomic number 82, is a particularly good shield for blocking x-radiation.) Bones, which are rich in calcium, absorb x-rays better than soft tissue such as muscle or fat and hence show up more clearly on x-rays (Figure 8.35).

An x-ray, or, more properly, a *roentgenogram*, named after W.C. Roentgen (pronounced rent'gen), who discovered x-rays in 1895, is really an image of the x-ray shadows cast on film by various structures of the body. The greater the absorption of x-rays by the body tissue or structure the darker the shadow and the darker the image on the developed x-ray. (Originally x-ray images were called *skiagraphs* from the Greek meaning "shadow graphs.") Today most x-ray images are made using a special film sandwiched between two intensifying screens. The latter are just pieces of cardboard covered with crystals that absorb x-rays well and give off visible or UV light in response. The film is coated on both sides with a light-sensitive emulsion, and each side of the film produces a picture of the light emitted by the intensifying screen in contact with it. Because the intensifying screens are more efficient at producing x-ray images than a single piece of x-ray–sensitive film alone, the x-ray dosages required to give well-exposed images using

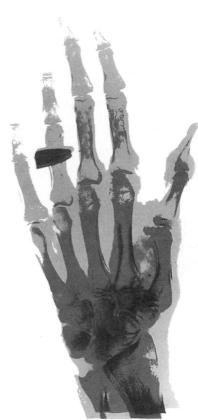

FIGURE 8.35
X-ray photograph of a human hand. In this negative image, areas that appear dark are those that strongly absorbed the incident x-radiation. Bones are much more efficient at absorbing x-rays because of their calcium content. Elements like gold and silver found in most jewelry are even better absorbers of x-rays because of their higher atomic numbers.

The years just prior to the beginning of the 20th century can be justly described as "revolutionary" in the history of modern physical science. Beginning in 1893 with Sir William Crookes' experiments with cathode ray tubes (the precursors of our modern television picture tubes) and continuing through 1895 with Roentgen's discovery of x-rays, the world of physics was turned topsy-turvy. Virtually every physicist in Europe was conducting experiments on these new phenomena. The period from 1895 to 1905 was one of intense activity, intense excitement, and intense rivalry—both between individuals and nations—in physics.

One of the laborers in the x-ray vineyards was René Blondlot, a professor of physics at the University of Nancy in France. Blondlot had already established a solid reputation as a physicist when in 1903, in connection with his research on the polarization of x-rays, he reported the discovery of a new type of radiation, which he christened "N-rays" after the place of their discovery, Nancy. Using first a small spark and later a low-intensity gas flame, whose increase in brightness gave witness to the presence of the N-rays, Blondlot found that these rays were emitted from a variety of sources: x-ray tubes, annular gas burners (but not ordinary Bunsen burners), sheets of iron and silver heated to glowing, and the sun. Originally thought to be similar to infrared radiation, N-rays possessed a number of unique qualities: they passed through a 4 millimeter thick platinum plate but not a 3 centimeter chunk of rock salt; they passed through dry but not wet cigarette paper; they could be produced by objects strained by compression (as a walking stick bent by the hands) or by hardening (as in a steel file). The marvels, mysteries, and means of manufacturing and manifesting these rays were described by Blondlot in great detail in 26 articles and a book (see Figure 8.36).

Once the discovery of N-rays was announced, experimenters around the world rushed to reproduce and to extend Blondlot's work. The results

FIGURE 8.36
Title page from Blondlot's book on N-rays.

these screens are about one tenth those needed without them. This is an important consideration, since medical x-rays are the largest source of artificially produced radiation in the United States, comprising about 36% of the total radiation received per year by the average resident in this country. Protecting the public against unnecessary exposure to damaging radiation used in diagnostic radiology is one of the greatest challenges to health and radiological physicists. Little wonder that such specialists recommend, when possible, the use of x-ray images were mixed. Some workers reported success—notably Augustin Charpentier, professor of biophysics at the University of Nancy (who found that rabbits and frogs gave off the rays)—but most reported failure, including such renowned physicists as Rayleigh, Langevin, and Rubens, a German physicist who pioneered studies in infrared radiation. Suspicion about the reality of N-rays began to grow, reaching the high point in the summer of 1904 when a group of concerned scientists decided unofficially to send an envoy to Blondlot's laboratory in Nancy to investigate the activities there firsthand. The spy was Professor Robert L. Wood of Johns Hopkins University, a well-known expert in optical phenomena and debunker of numerous spiritualist scams.

Upon arriving in Nancy, Wood was treated to the complete gamut of N-ray phenomena by none other than Blondlot himself. And having witnessed the demonstrations, Wood parted on good terms with his host and published his report. In it he recounted how, when asked to hold a steel file (a well-known emitter of N-rays) near Blondlot's forehead so as to enhance the latter's ability to see a dimly lit clock face, he instead substituted a piece of wood (one of the few objects known not to emit N-rays), with no adverse effects on the results of the experiment as reported by Blondlot. Wood himself reported no improvement in the clock's visibility when the file was placed near his line of vision. If this were not enough, Wood noted that during a critical experiment in which the spectrum of N-rays was to be produced by diffracting them through an aluminum prism, Wood pocketed the prism in the dark with no apparent alteration of the successful outcome of the demonstration as described by Blondlot. After Wood's exposé, the issue of N-rays was dead.

To this day explanations of the N-ray affair are unsatisfying. Probably the best that can be said is that problems associated with the subjective observation of low-intensity sources with varying energy output, coupled with the failure to perform a well-controlled experiment and the desire to achieve personal prestige and to foster national pride, led to spurious results. What does seem clear is that the story of N-rays is not a tale of deliberate fraud or deception on the part of Blondlot or his associates. Moreover, it was not a hoax. It was, quite simply, a mistake. In the words of Professor Josef Bolfa, when interviewed on the subject, *"C'est une erreur."*

Perhaps the point to take from all of this is that physics is a human endeavor carried out by human beings, and, as in all areas of human activity, mistakes are made. Physicists are not infallible. Moreover, the story reminds us of the tortuous path scientific advancement often takes, now surging forward along a straight, smooth course; now detouring along winding, rutted byways; and sometimes, as with N-rays, ending up in a narrow alley marked "No Exit." Fortunately the number of scientific dead ends has been small, the progress of science great, and the prospect for continued advancement bright.

produced with intensifying screens, despite the fact that they are somewhat blurrier (that is, less well defined) than those made directly on film.

X-rays and gamma rays can be harmful because they are *ionizing radiation*—radiation that produces ions as it passes through matter. Such radiation can "kick" electrons out of atoms, thus leaving a trail of freed electrons and positive ions. This process can break chemical bonds between atoms in molecules, thereby altering or breaking up the molecule. Living cells rely on very large, sophisticated molecules for their normal functioning and reproduction. Disruption of such molecules by ionizing radiation can kill the cell or cause it to mutate, perhaps into a cancer cell. The human body can (and does) routinely replace dead cells, but massive doses of x-rays or other ionizing radiation can overwhelm this process and cause illness, cancer, or death. We will discuss this more in Section 11.3.

Gamma Rays

The highest frequency electromagnetic waves are gamma rays (γ-rays). The frequency range is from about 3×10^{19} hertz to beyond 10^{23} hertz. The wavelength of higher frequency gamma rays is about the same distance as the diameter of individual nuclei. Gamma rays are emitted in a number of nuclear processes: radioactive decay, nuclear fission, and nuclear fusion, to name a few. We will study these processes in detail in Chapter 11.

This concludes our brief look at the electromagnetic wave spectrum. Even though the various types of waves are produced in different ways and have diverse uses, the only real difference in the waves themselves is their frequency and, therefore, their wavelength.

8.6 BLACKBODY RADIATION

Every object emits electromagnetic radiation because of the thermal motion of its atoms and molecules. We have already seen how this *heat radiation* is one method of transferring heat. Without radiation from the sun the earth would be a frozen rock. In this section we take a closer look at heat radiation and consider some of its uses.

The actual nature of the heat radiation emitted by a given object—the range of frequencies or wavelengths of EM waves present and their intensities—depends on the temperature of the object and on the characteristics of its surface (its color, for example). A hypothetical object that is perfectly black—one that absorbs all EM waves that strike it—would actually be the best at emitting heat radiation. Referred to as a *blackbody,* it would emit radiant energy at a faster rate than any other object at the same temperature, and the intensities of all of the wavelengths of EM waves emitted could be predicted quite accurately (Figure 8.37). The heat radiation emitted by a such an object is referred to as *blackbody radiation* (BBR).

Blackbody radiation, which is an idealized representation of heat radiation, has been analyzed very thoroughly and is well understood. The actual heat radiation emitted by real objects is usually not too different from BBR, so we can use it as a model of heat radiation.

The heat radiation emitted by any object (such as your own body, the sun, a blackbody) is simply a broad band of electromagnetic waves. Within this band, some wavelengths are emitted more strongly than others. By this we mean that the intensity of the different wavelengths, the amount of energy released per square meter per second, varies with wavelength. For example, the heat radiation from the sun contains more energy in each wavelength of visible light than in each wavelength of IR light. The intensity of the visible wavelengths is higher than that of the IR wavelengths.

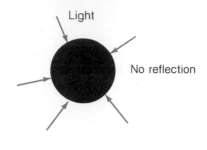

Light

No reflection

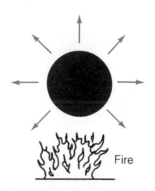

Fire

FIGURE 8.37
An ideal blackbody would absorb all EM radiation that strikes it. It would also radiate more heat radiation than any other object with the same size and the same temperature.

A graph showing the intensity of each wavelength of radiation emitted by a blackbody is called a *blackbody radiation curve* (Figure 8.38). The size and shape of the graph change with the object's temperature. The graph for a real object (not a blackbody) would be similar.

The total amount of radiation emitted per second by an object obviously depends on how large it is. A 100-watt light bulb is brighter (it emits more light) than a 10-watt light bulb because its filament is larger. Aside from this factor, it is the object's *temperature* that has the greatest influence on the amount and types of radiation emitted. There are three distinct aspects of heat radiation that are affected by the object's temperature.

1. The amount of each type of radiation (microwaves, IR etc) emitted increases with temperature.

An incandescent light bulb fitted with a dimmer control illustrates this nicely. Dimming the light causes the filament temperature to decrease. This reduces the amount of visible light that is emitted: the bulb's brightness is decreased. The amount of infrared is also reduced.

2. The total amount of radiant energy emitted per unit area per unit time increases rapidly with any increase in temperature. For a particular blackbody, the total radiant energy emitted per second (power) is proportional to the temperature in Kelvin raised to the fourth power.

$$P \propto T^4 \quad (T \text{ in Kelvin})$$

Doubling the Kelvin temperature of an object will cause it to emit *16* times as much radiant energy each second. The human body at 310 Kelvin (98.6°F) radiates about 25% more power than it would at room temperature, 293 Kelvin (68°F). If one could see infrared, humans would appear to glow more brightly than their cooler surroundings.

FIGURE 8.38
Typical blackbody radiation curve. It indicates the amount of energy emitted at each different wavelength of the EM wave spectrum.

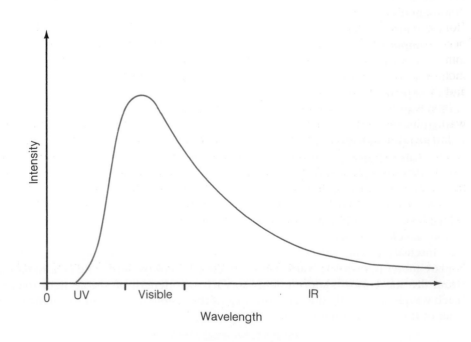

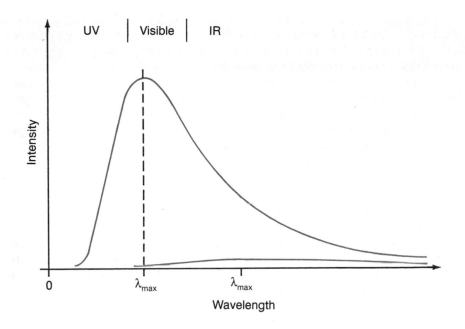

UV | Visible | IR

Intensity

0   λ_{max} λ_{max}

Wavelength

FIGURE 8.39
Blackbody radiation curves for the sun (upper) and a light bulb (lower).

3. At higher temperatures, more of the power is emitted at successively shorter wavelengths (higher frequencies) of electromagnetic radiation. For a blackbody, the wavelength that is given the maximum power (the peak of the blackbody radiation curve) is inversely proportional to its temperature.

$$\lambda_{max} = \frac{0.0029}{T} \qquad (\lambda_{max} \text{ in meters, } T \text{ in Kelvin})$$

Objects cooler than about 700 Kelvin (about 800°F) emit mostly IR with smaller amounts of microwaves, radio waves, and visible light. There is not enough of the latter to be detected by the human eye. Above this temperature, objects emit enough visible light to glow. They appear red hot because more red (longer wavelength) radiation is emitted than any other visible wavelength. The peak wavelength is still in the infrared region. (There are several factors that influence the minimum temperature needed to cause an object to glow. These include the size and color of the object, the brightness of the background light, and the acuity of the observer's eyesight.)

The filament of an incandescent light bulb can be as hot as 3,000 K. The peak wavelength of its heat radiation curve is in the infrared region, not far from the visible band. The visible light emitted is a bit stronger in the longer wavelengths, and so it has a slightly reddish tint to it. At 6,000 Kelvin the sun's surface emits heat radiation that peaks in the visible band (the wavelength of the peak is one half that of the light bulb's). It appears to be white hot (Figure 8.39). When comparing the curves, keep in mind that the sun's is higher *not* because the sun is bigger than a light bulb but because it is hotter.

Assuming that the sun is a blackbody with a temperature of 6,000 K, at what wavelength does it radiate the most energy?

EXAMPLE 8.3

$$\lambda_{max} = \frac{0.0029}{T}$$

One of the most important characteristics of a star is its surface temperature. Stars are classified as young or old, heavy or light, and so on, partly on the basis of their surface temperatures. Surface temperature controls the rate at which energy is released from each square meter of the star's emitting area. But how can we measure the temperature of a star? The majority of stars are so very far away that to consider sending an astronaut armed with a thermometer to do the job is quickly seen as wishful thinking: the journey would simply take too long, and the thermometer would probably melt at the high temperatures of most stars. So, just how do we deduce stellar temperatures?

One way is to use the fact that stars, while not true blackbodies, are reasonable approximations to them in terms of many of their radiative properties. Consequently we can use the blackbody radiation laws to determine the *color temperature* of the star.

Consider the graphs of the radiation curves of two stars—each assumed to radiate like a blackbody—as a function of wavelength shown in Figure 8.40. Star 1 is imagined to be the hotter star, so that it emits more energy at each and every wavelength than Star 2: it also emits more of its energy at short wavelengths (high frequencies) than does Star 2. These facts are just consequences of the radiation laws for blackbody emitters.

Suppose now that we chose to measure the brightness of both stars through two glass filters, one a deep blue, the other yellow. Such measurements allow us to sample the stars' intensities at two different wavelengths (colors). From the graphs you

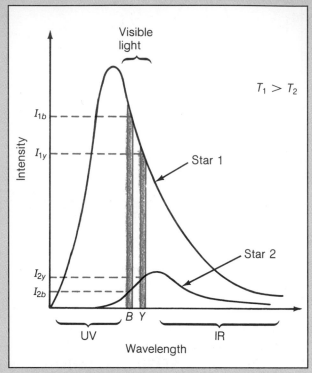

FIGURE 8.40
Blackbody intensity curves for Stars 1 and 2, with T_1 greater than T_2. The intensity for each star, as measured through two filters, B and Y, is shown.

can see that the hotter star (Star 1) emits more blue light than yellow light, whereas the converse it true for the cooler star (Star 2). Thus the *intensity difference, I(yellow) − I(blue)*, is negative for Star 1, but positive for Star 2.

$$\lambda_{max} = \frac{0.0029}{6,000 \text{ K}}$$

$$\lambda_{max} = 4.8 \times 10^{-7} \text{ m}$$

Table 8.1 shows this to be in the blue part of the visible band. The sun appears white, not blue, because all visible wavelengths are emitted with nearly the same brightness.

Some stars are hot enough to appear bluish (Sirius and Vega are two examples). The peaks of their radiation curves are in the UV, and so they emit more of

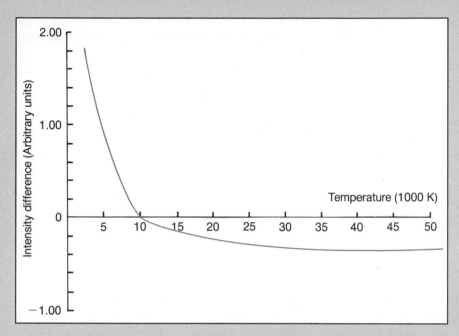

FIGURE 8.41
Graph of the intensity difference,
I(yellow) − I(blue), versus star
temperature.

The intensity differences just introduced can be directly related to the temperature of the star. Since there is one and only one blackbody curve for each temperature, there is, in principle, a single unique temperature associated with each intensity difference $I_y - I_b$. If we plot these associated values, the result is that shown in Figure 8.41. As is evident, the general trends found above for Stars 1 and 2 (that is, the more negative the intensity difference, the hotter the star) are confirmed. Thus by examining starlight through filters and using the blackbody radiation laws, a star's temperature may be taken over a long distance.

In practice, many things complicate this picture, not the least of which is that stars are only approximate blackbodies. In addition, the amount of radiation received from these stars is not precisely what is emitted because of such variables as absorption in the intervening interstellar medium, absorption by the earth's atmosphere, by the telescope optics, and by the filters themselves! In spite of these problems, measurements of the color temperatures of thousands of stars have been successfully made and provide us with valuable information about their natures.

the shorter wavelength visible light (blue) than the longer wavelengths (see Physics Potpourri, above).

The temperature dependence of blackbody radiation is responsible for a number of interesting phenomena, and it is used in some ingenious ways. The following are a few of them.

Temperature Measurement

The temperature of an object can be determined by examining the radiation that it emits. This is particularly useful when very high temperatures are involved, as in a furnace, because nothing has to come into contact with hot matter. Special devices called pyrometers measure the amount and types of radiation emitted and use the rules mentioned above to determine the temperature. This process is used to measure the temperature of the sun and other stars.

Detection of Warm Objects

Most things on earth have temperatures that cause them to emit primarily infrared light. Anything that can detect IR can use this fact to locate warmer-than-average objects, since they will emit more IR. Rattlesnakes and certain other snakes use IR to hunt mice and other warm-blooded animals at night. These snakes have sensitive organs that detect the higher intensity infrared emitted by objects warmer than their surroundings.

DO-IT-YOURSELF PHYSICS

Stars having positive $I_y - I_b$ values emit most of their energy at long wavelengths in the red region of the spectrum. Stars with negative $I_y - I_b$ values radiate strongly at short wavelengths, in the blue. On a clear night in winter (away from city lights), locate the constellation of Orion using the chart in Figure 8.42. Identify the bright stars Rigel and Betelgeuse. Notice the differences in their colors. Betelgeuse is a star whose surface temperature is only about 3,200 Kelvin (quite low by stellar standards but hot enough to melt iron), and it should appear distinctly reddish. Rigel is a much hotter star ($T = 10,000$ Kelvin) and should appear bluish white.

Can you find other stars with distinctive colors? What are their surface temperatures compared to those of Rigel and Betelgeuse? What are the colors of the majority of the *brightest appearing* stars? Most of the stars near the sun have rather low surface temperatures and should appear red orange. If this statement is at odds with your answer to the previous question, can you offer an explanation?

Infrared-sensitive photographic film, video cameras, and other detection devices have many practical uses. For instance, infrared photographs, called thermograms, can show where heat is escaping from a poorly insulated house and can detect ohmic heating caused by a short circuit in an electrical substation (Figure 8.43). They can locate warmer areas on the human body that might be caused by tumors. The military uses IR detectors to locate soldiers at night. Heat-seeking missiles automatically steer themselves toward the hot exhaust of aircraft.

8.7   EM WAVES AND THE EARTH'S ATMOSPHERE

A variety of substances in the atmosphere surrounding the earth interact with EM waves in important ways. Some of these interactions are crucial to the existence

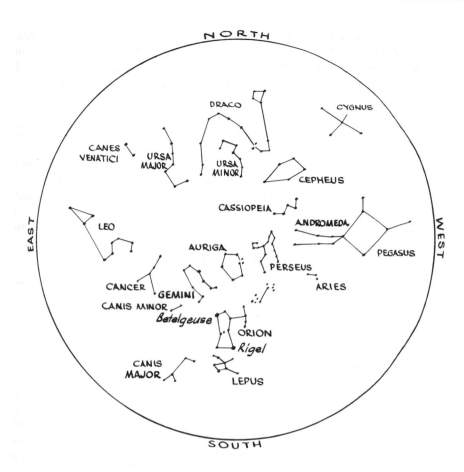

FIGURE 8.42
Winter constellations.

of life on this planet, another helps us communicate, and some add to the beauty that characterizes life on earth. The visible phenomena, rainbows, for example, are described in the chapter dedicated to visible light, Chapter 9. Some of the others are discussed here.

Ozone Layer

The sunlight that keeps the earth warm and provides the energy for plants to grow also contains ultraviolet radiation that is harmful to living things. But life has evolved on this planet because the atmosphere has protected it from this UV. In the region between about 20 kilometers and 40 kilometers above the earth, known as the *ozone layer,* there is a comparatively high concentration of *ozone* (O_3), the form of oxygen with 3 atoms in each molecule. This ozone absorbs most of the harmful UV in sunlight. The ozone layer has been a shield protecting living things on earth.

In 1974 it was reported that *chlorofluorocarbons* (CFCs), chemical compounds such as Freon used in refrigerators, air conditioners, and as an aerosol propellant in spray cans, could be depleting the ozone layer. CFCs that are released into the air drift upward to the ozone layer and chemically break up the ozone molecules with alarming efficiency. Because of this CFCs were banned for use as aerosol propellants in the United States. Then in 1985 it was discovered that a "hole" developed in the ozone layer over Antarctica during the later part

FIGURE 8.43
Thermograms—infrared photographs—show regions that are warmer or cooler than the surroundings.

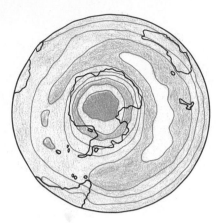

FIGURE 8.44
Map showing the ozone hole over Antarctica (outlined in the center) in October, 1985. The darkest area is where there was one-half the normal amount of ozone. (Based on data from the Nimbus 7 satellite.)

of each year. The concentration of ozone in a huge section of the atmosphere was reduced by about one half (Figure 8.44). A review of old satellite measurements revealed that this hole had developed in previous years as well and that the one in 1982 was twice the area of the United States. The discovery of the ozone hole startled atmospheric scientists because they had not anticipated such an occurrence. The exact cause of the ozone hole was not known, but this made it apparent that the ozone layer is indeed fragile.

The effects of a change in the ozone concentration could be dramatic. If the ozone concentration is reduced worldwide even moderately, the incidence of skin cancer could rise. Many plants, including important food crops, are adversely affected by increased UV, so global food production could be threatened. In 1987, nations from around the world drafted the *Montreal Protocol on Substances that Deplete the Ozone Layer.* Ratified by more than 50 nations, the Montreal Protocol outlines a reduction in the production and consumption of CFCs worldwide during the 1990s. Such global cooperation is crucial for attacking problems such as ozone depletion.

Greenhouse Effect

The *greenhouse effect* is so named because it is partly responsible for keeping greenhouses warm in cold weather. Glass and certain other materials allow visible light to pass through them but absorb or reflect the longer wavelength infrared radiation. A glass wall or roof on a building allows visible light to enter and to warm the interior. As the temperature of things inside increases, they emit more IR. Without the glass this IR would escape from the enclosure and carry away the added heat. But the glass blocks the IR, so the added heat energy is trapped and the enclosure is warmed. (The glass not only reduces heat loss by radiation, it also eliminates convection: the heated air that rises cannot leave and take heat energy with it.) A car parked in the sun with the windows up is much warmer than the outside air because of this heating. Windows on the sunny side of a building help to keep it warm in the same way.

The greenhouse effect occurs naturally in the earth's atmosphere. Water vapor, carbon dioxide (CO_2), and other gases in the air act somewhat like glass in that they allow visible light from the sun to pass through to the earth's surface but absorb part of the infrared radiation emitted by the warmed surface (Figure 8.45a). The atmosphere is heated by the IR that is absorbed; it is about 35°C warmer than it would be without this effect.

Since the middle of the 19th century, the carbon dioxide content of the atmosphere has risen 25% (Figure 8.45b). Evidence points to human activity as the likely cause. During the last century huge quantities of fossil fuels—coal, oil, and natural gas—have been burned. This has released vast amounts of carbon dioxide into the air. At the same time, forests have been cut down for building materials and to clear areas for farming and human occupation. This contributes to the problem because trees and other plants take carbon dioxide out of the air. Methane gas, released into the air by some animals and by rice as it grows, and the ozone-threatening chlorofluorocarbons are also "greenhouse gases" like water vapor and CO_2. The concentrations of these gases is also increasing. The result of this is the possibility of *global warming*—the entire atmosphere heating up. (Data suggest that the average worldwide temperature did rise about 0.5°C during the 100 years preceding 1989.)

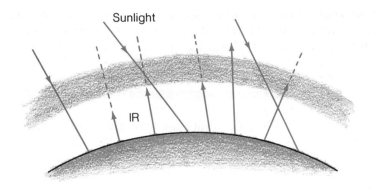

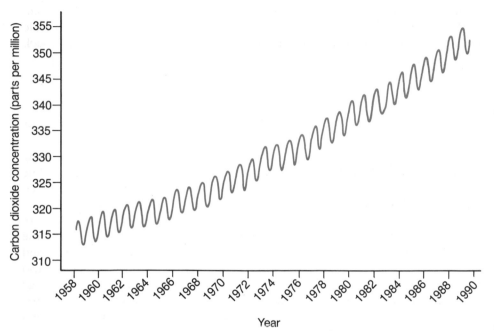

FIGURE 8.45
(a) The earth's atmosphere produces a greenhouse effect. Increased carbon dioxide content in the air could cause too much heating. (b) Graph of CO_2 content in the air versus time, measured at the Mauna Loa observatory atop a mountain in Hawaii. The concentration varies throughout each year because plants in the northern hemisphere take in more CO_2 in the summer when they are most active. (From "The Great Climate Debate," by Robert M. While, July 1990. © 1990 by *Scientific American,* Inc. All rights reserved.)

Anyone who has lived though a harsh winter might think that the warming of the atmosphere is not such a bad idea. The main concern is that we might be starting a runaway greenhouse effect that could lead to global disaster. The temperature of the earth's atmosphere could rise enough to trigger changes in the weather patterns. Once verdant areas could become deserts and vice versa. The polar ice caps could start melting, thereby raising the level of the oceans and flooding heavily populated coastal areas around the world. (A runaway greenhouse effect did occur naturally on the planet Venus, where the surface temperature is now 460°C. Conditions on Venus prevented excessive atmospheric carbon dioxide from being trapped in carbonate rocks or dissolved in oceans, as is the case on earth.)

The atmospheric greenhouse effect is such a complex phenomenon that it is nearly impossible to predict exactly what will happen. Different scenarios have been proposed: as the earth heats up, huge amounts of CO_2 dissolved in the oceans could be released and could exacerbate the global warming. Increased evaporation of water would raise the water vapor content to the atmosphere and possibly cause more heating. Or the higher humidity could lead to more cloud cover, which might cool the earth as less sunlight reaches its surface. About all we can be sure of is that we are altering the life-sustaining blanket in which we live even though we cannot predict what the effects will be.

The Ionosphere

Did you ever wonder why you can pick up AM radio stations that are several hundred miles away but usually cannot receive FM radio or television more than 50 miles away? About 50 to 90 kilometers above the earth's surface, in a region of the atmosphere known as the *ionosphere,* there is a relatively high density of ions and free electrons. Radio waves from transmitters travel upward and into the ionosphere. Higher frequency radio waves, such as those used in FM radio and television, pass through the ionosphere and out into space (Figure 8.46 b). Lower frequency radio waves, such as the 500 to 1,500 kilohertz waves used in AM radio, actually reflect off the ionosphere and return to earth (Figure 8.46 a). The range for high-frequency radio waves is limited to "line-of-sight" reception. The curvature of the earth eventually blocks the signal from the transmitting tower. Low-frequency radio waves can skip off the ionosphere and thereby travel farther around the planet. The radio waves used to communicate with spacecraft must have high frequency to pass through the ionosphere.

The ionosphere is also the home of the *aurorae*—the northern lights and the southern lights. Charged particles from the sun excite atoms and molecules in the ionosphere, causing them to emit light. (More on this in Chapters 9 and 10.)

Astronomy

Stars, galaxies, and other objects in space emit all types of electromagnetic radiation. Astronomers were originally limited to studying just visible light through optical telescopes. Now they examine the entire spectrum of EM radiation to gain a more complete understanding about the universe. The atmosphere is a hin-

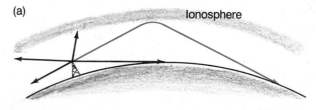
(a)

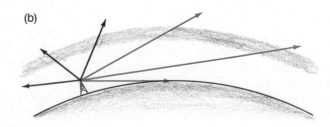
(b)

FIGURE 8.46
(a) Low-frequency radio waves are bent back to the earth's surface by the ionosphere. This allows commercial AM radio to be transmitted hundreds of miles. (b) High-frequency radio waves pass through the ionosphere. Because of this, transmission of commercial FM and television is limited to about 50 miles.

(a)

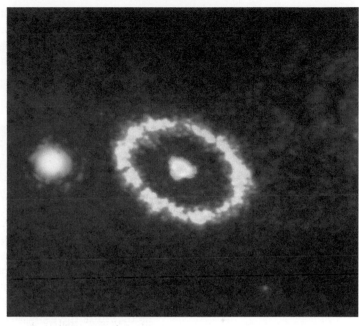

(b)

FIGURE 8.47
(a) The Hubble Space Telescope during deployment from the space shuttle Discovery. (b) Hubble Space Telescope photograph of a ring around the remnants of supernova 1987A. (The star on the left is not associated with the supernova.) See page 519 for more on 1987A.

drance for some of these investigations. Ultraviolet light from stars and galaxies is absorbed by ozone and other gases. Infrared light is also absorbed, mainly by water vapor. Lower frequency radio waves are absorbed in the ionosphere. Even visible light is affected by the random swirling of the air, which causes stars to twinkle and degrades the images formed in large telescopes. Microwaves and higher frequency radio waves are about the only EM waves from space that are not affected by the atmosphere.

Astronomers have taken to space to overcome the atmosphere's interference. The *Hubble Space Telescope,* the most sophisticated and expensive satellite ever built, was placed in orbit around the earth in 1990 (Figure 8.47). Since it is well above the atmosphere, the images it forms and sends to earth are free from the atmosphere's distortion. The Space Telescope promises to revolutionize our knowledge about very distant objects in the universe. The *International Ultraviolet Explorer (IUE),* launched in 1978, and the *Infrared Astronomical Satellite (IRAS),* launched in 1983, carried telescopes specifically designed to study the UV and IR emitted by celestial bodies. The wealth of data sent back to earth by these orbiting observatories is still being analyzed. Many other instruments, some sensitive to x-rays and γ-rays, have been placed in orbit or carried aloft with astronauts to operate them. Space exploration is giving astronomers access to the entire electromagnetic wave spectrum.

HISTORICAL NOTES

Refer back to Gilbert in Chapter 7, Historical Notes.

The first recorded investigation of magnetism was carried out by William Gilbert at the same time he was studying electricity. In the course of his work, Gilbert built a sphere out of lodestone to model the effects of the earth on a compass needle. With this *terrella* (little earth) he accounted for a number of phenomena, such as magnetic declination, that had been observed by navigators as they traveled around the globe. One of these was an effect observed by Columbus on his famous voyage: the declination angle changes with longitude.

As recently as 170 years ago, electricity and magnetism seemed to be similar but independent phenomena. Reports that compasses were affected by lightning storms suggested that some kind of interaction between the two existed, but nothing concrete was observed until 1820. In that year Hans Christian Oersted (1777–1851), a physics professor at the University of Copenhagen, announced that he had observed a compass needle being deflected by an electrical current in a nearby wire (Figure 8.48). Oersted had begun experimenting with electricity after he heard of Volta's invention of the battery. His initial attempts to observe the effect failed because, understandably, he was not expecting the deflection to be perpendicular to the current-carrying wire.

Within months, news of the discovery had spread throughout the European scientific community. The exact relationship between an electric current and the strength and configuration of the magnetic field that it produced were quickly deduced. Most of this work was done by the French physicist André Marie Ampère (1775–1836). The son of a prosperous merchant in a village near Lyons, Ampère showed signs of genius at an early age. His path to greatness was sidetracked when his father was executed during the French Revolution. It took Ampère a year to recover from the shock and to continue his education.

Before his studies of electromagnetism, Ampère had gained fame for his work in mathematics and chemistry. His talents and interests spread to many areas: during his career he held professorships in mathematics, philosophy, astronomy, and physics. (It seems he was also a classic absent-minded professor.)

Ampère's greatest work was triggered by Oersted's discovery; within a week of hearing about it, Ampère had thoroughly investigated it and prepared his own paper on the topic. Ampère showed that the magnetic field of a current could be concentrated by bending the wire into a loop or wrapping it up as a solenoid. He correctly suggested that the magnetism in permanent magnets is caused by tiny currents in the molecules. He made the second important observation of electromagnetism: a magnetic field exerts a force on a current-carrying wire. Ampère's most famous finding is that two current-carrying wires exert forces on each other (Figure 8.49). The force is due to magnetism even though no permanent

FIGURE 8.48
Hans Christian Oersted.

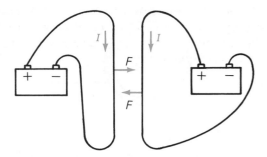

FIGURE 8.49
Two parallel current-carrying wires exert forces on each other. The magnetic field produced by the wire on the left exerts a force on the wire on the right and vice versa.

magnets are involved. This simple setup illustrates elegantly how magnetism is an inextricable part of moving electric charges.

An American physicist, Joseph Henry (1797–1878), is notable for his research on electromagnets. He improved their efficiency to the point of making an electromagnet capable of lifting 50 times its own weight when connected to a small battery. Henry also had the idea of using electromagnets for long-distance communication: the telegraph. He freely mentioned this idea to others, who later patented the process and became wealthy because of it.

The final major discoveries in electromagnetism were made by two of the greatest names in physics, Michael Faraday and James Clerk Maxwell. These men present interesting contrasts in their backgrounds and in the way in which they approached physics. Faraday was born into a poor family and was, for the most part, self-educated. He became the greatest experimenter in physics of the time. Maxwell came from a well-to-do family and was educated in Cambridge. His approach was highly theoretical: he incorporated sophisticated mathematics in his explanations of electromagnetism.

Faraday (1791–1867), son of a blacksmith, grew up near London. His apprenticeship as a bookbinder afforded him the opportunity to read extensively about the great discoveries in electricity. His break came when he was able to attend some lectures given by the great chemist Sir Humphry Davy (1778–1829). Davy gained fame for his use of electrolysis to discover sodium, potassium, and several other elements. Faraday took painstaking notes of the lectures, carefully bound them, and showed them to Davy. Davy was sufficiently impressed to hire Faraday and to launch him on his historical career. As Davy's assistant, Faraday came into contact with some of the greatest scientists of the time. He soon developed into a first-rate chemist and made some noteworthy discoveries.

Faraday first experimented with electricity in 1821, but his greatest work was done in the 1830s. Like many other experimenters, he assumed that magnetism could be used to produce electricity, since the reverse was true. His initial attempts failed because he used steady magnetic fields. But on 29 August 1831, using a device similar to a transformer, Faraday noted that a current was momentarily induced in one coil as the other was connected to or disconnected from a battery (Figure 8.50). Electromagnetic induction occurred in the coil only if the magnetic field in it was changing.

In the following months Faraday developed a keen understanding of electromagnetic induction. He constructed primitive electric motors, generators, and transformers. In addition to these practical inventions, he made important contributions to the theoretical understanding of electricity and magnetism. He introduced the concept of "lines of force" (field lines) to model what were later called electric and magnetic fields. Faraday's successors marveled at his ability to understand sophisticated phenomena without the aid of higher mathematics.

FIGURE 8.50
Apparatus used by Faraday to discover electromagnetic induction. A current is induced in the right coil only as the direct current in the left coil is either switched on or switched off. The compass is used to detect the induced current.

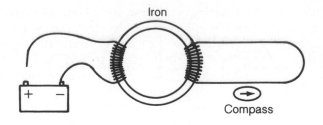

Most of his discoveries are beyond the level of this text. Faraday just missed predicting the existence of electromagnetic waves.

Faraday's scientific career was interrupted by a mental breakdown in 1840—perhaps caused by mercury poisoning. Five years later he resumed his work. Faraday was renowned as a great lecturer throughout his career. He filled lecture halls with excited patrons who paid to hear him explain the beauty of science (Figure 8.51). Eventually his mental problems, particularly loss of memory, forced him to resign. He left a monumental diary of several thousand pages detailing his experimental findings.

One of Faraday's greatest admirers was the young James Clerk Maxwell (1831–1879); (Figure 8.52). Maxwell grew up on his family's estate in rural Scotland. He quickly showed his intellectual ability in school. At the age of 15 years he had a mathematical paper published. At Cambridge he did some interesting work on the perception of color and proved mathematically that the rings around the planet Saturn could not be rigid solids. The latter established his prowess as a mathematical physicist.

Maxwell's work in electricity and magnetism began in 1855, about the time he accepted a physics professorship in Aberdeen, Scotland. He synthesized the findings of Faraday and others and provided a concrete mathematical framework for

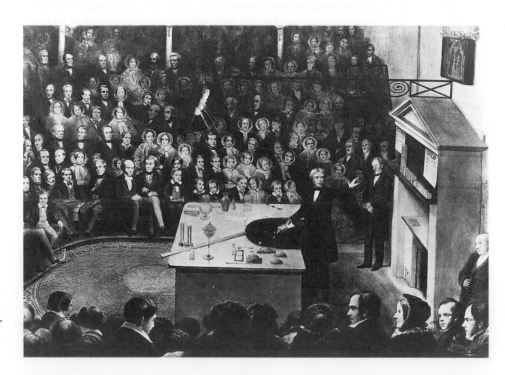

FIGURE 8.51
Michael Faraday delivering a public lecture on electricity and magnetism at the Royal Institution, London, in 1856. Faraday devoted a great deal of effort to his lectures and drew large crowds.

describing electrical and magnetic phenomena. He solidified Faraday's concept of fields, which was later applied to gravity as well as electricity and magnetism, and introduced an important clarification of the idea of current. He is most famous for a set of equations, called *Maxwell's equations,* that summarize the basics of electromagnetism. These equations are on the same level of importance in physics as Newton's laws.

Maxwell saw that his mathematical description of electricity and magnetism indicated that traveling waves of electric and magnetic fields could exist. That in itself was not particularly startling, but when he computed the speed of these waves, it turned out to be nearly equal to the measured speed of light. The conclusion was inescapable: light was one type of "electromagnetic wave." Not only had Maxwell integrated electricity and magnetism, he had brought light into the same realm.

Maxwell went on to make important discoveries in thermodynamics. He did not gain the level of recognition during his life that he deserved partly because his highly mathematical writings were rather inaccessible. In addition to his theoretical discoveries, Maxwell left his mark on the way physics is done. Like Newton nearly 200 years earlier, he showed that advanced mathematics plays a vital role in physics.

Some nine years after Maxwell's death, his prediction of electromagnetic waves was proved correct. The German physicist Heinrich Hertz (1857–1894), a former assistant to Hermann von Helmholtz, produced radio waves and showed that they had properties similar to light. In the century that has followed, our technological society has filled the skies with electromagnetic waves.

FIGURE 8.52
James Clerk Maxwell in 1855, while at Cambridge.

SUMMARY

Magnetism is a phenomenon most commonly illustrated with permanent magnets. Every magnet has a north pole and a south pole, so named because the two parts of the magnet are naturally attracted to the north and to the south, respectively. When two magnets are near each other, the like poles repel and the unlike poles attract. The magnetic field produced by a magnet causes a force on the poles of any other magnet. A compass is simply a small magnet that is free to rotate when in the presence of a magnetic field. It can be used to determine the direction of the magnetic field at any point in space. The earth has its own magnetic field that causes compasses to point north. The earth's magnetic poles do not coincide with its geographic poles, and so compasses do not point exactly north at most places on earth.

Many phenomena depend on the interactions between electricity and magnetism. These interactions occur only when there is some kind of change taking place, such as motion of charges or a magnet. The basic electromagnetic interactions can be stated in the form of three simple observations: moving charges produce magnetic fields, magnetic fields exert forces on moving charges, and moving magnets induce currents in coils of wire (electromagnetic induction). These processes are exploited in a variety of useful devices, from electromagnets to phonograph cartridges and speakers. In fact, nearly all electrical devices rely directly or indirectly on these interactions.

The principles of electromagnetism express the fundamental relationship between electricity and magnetism: "an electric current or a changing electric field induces a magnetic field" and "a changing magnetic field induces an electric field." These two statements summarize the three observations and predict the existence of electromagnetic waves—traveling combinations of oscillating electric and magnetic fields. These waves can be classified according to frequency. From low to high, the EM wave bands are radio waves, microwaves, infrared radiation, visible light, ultraviolet radiation, x-rays, and gamma rays. The different waves are involved in a diverse number of natural processes and technological applications.

Heat radiation is a broad band of electromagnetic waves emitted by objects because of the thermal motion of atoms and molecules. The amount of radiation emitted and the intensities of different wavelengths depend on the object's temperature. Even a small rise in temperature causes a significant increase in the amount of radiant energy emitted. This makes it possible to locate warmer objects even in the dark.

The earth's atmosphere interacts with UV and IR radiation in ways that are crucial to life on earth. The ozone layer absorbs most of the harmful UV present in sunlight. But the compounds known as CFCs drift upward and reduce the concentration of ozone. Carbon dioxide, water vapor, methane gas, and CFCs contribute to a greenhouse effect in the earth's atmosphere. They allow sunlight to pass through to the earth's surface and to warm it but then absorb much of the IR emitted by the heated surface. The result is a warming of the atmosphere that may increase because of heightened concentrations of these gases. The ionosphere, a region in the upper atmosphere containing ions and free electrons, reflects lower frequency radio waves back to the earth. This greatly increases the range of radio communication for these lower frequencies. Astronomers have overcome the absorption of different bands of EM waves by the atmosphere by placing telescopes and other instruments in space.

SUMMARY OF IMPORTANT EQUATIONS

EQUATION	COMMENTS
$\dfrac{V_o}{V_i} = \dfrac{N_o}{N_i}$	Input and output voltages of a transformer
$c = f\lambda$	Relates wavelength and frequency of EM wave
$\lambda_{max} = \dfrac{0.0029}{T}$	Peak wavelength of BBR curve

QUESTIONS

1. Draw a sketch showing the shape of the magnetic field around a bar magnet.

2. Explain why the rule concerning the force between poles on magnets leads to the conclusion that the north pole of the earth's magnetic field is at (near) the earth's south geographic pole.

3. Describe the three ways that one can show the basic interactions between electricity and magnetism.

4. Explain what *superconducting electromagnets* are. What advantages do they have over conventional electromagnets? What disadvantages do they have?

5. Name five different *basic* devices that use at least one of the electromagnetic interactions.

6. In many cases the effect of an electromagnetic interaction is perpendicular to its cause. Describe two different examples that illustrate this.

7. To test whether a material is a superconductor, a scientist decides to make a ring out of the material and then to see if a current will flow around in the ring with no energy input.
   a) Explain how a magnet could be used to initiate the current.

   b) At some later time, how could the scientist check to see if the current is still flowing in the ring but without touching the ring?

8. A coil of wire has a large alternating current flowing in it. A piece of aluminum or copper placed near the coil becomes warm even if it does not touch the coil. Explain why.

9. In the particle accelerator shown in Figure 8.15, what are the possible directions of the magnetic field that keep the particles traveling around the circular path?

10. Explain how each of the following devices uses electromagnetic interactions: speaker, electric motor, microphone, and transformer.

11. What would a speaker do just as a low-voltage battery is connected to it?

12. What is *analog-to-digital conversion,* and how is it used in sound reproduction?

13. Explain how electromagnetic waves are a natural outcome of the principles of electromagnetism.

14. An electromagnetic wave travels in a region of space occupied only by a free electron. Describe the resulting motion of the electron.

15. List the main types of electromagnetic waves in order of increasing frequency. Give at least one useful application for each type of wave.

16. Alternating current with a frequency of 1 million Hz flows in a wire. What in particular could be detected traveling outward from the wire?

17. What are the main uses of microwaves? Explain how each process works.

18. Aircraft equipped with powerful radar units are forbidden from turning them on when on the ground near people. Explain why this is.

19. Which type of EM wave is your body emitting the most?

20. What is different about our perceptions of the different frequencies within the visible light band of the electromagnetic wave spectrum?

21. A heat lamp is designed to keep food and other things warm. Would it also make a good tanning lamp? Why or why not?

22. How are x-rays produced?

23. Why are x-rays more strongly absorbed by bones than by muscles and other tissues?

24. What is "blackbody radiation"? How does the radiation emitted by a blackbody change as its temperature increases?

25. Explain how infrared light can be used to detect some types of animals in the dark. Can you think of a situation in which this would not work?

26. What is different about a star that appears reddish compared to one that appears bluish?

27. What effect does the ozone layer have on the EM waves from the sun? What is currently threatening the ozone layer?

28. Describe the greenhouse effect that is occurring in the earth's atmosphere.

29. How does the ionosphere affect the range of radio communications?

30. Explain why two wires, each with a current flowing in it, exert forces on each other even when they are not touching each other.

PROBLEMS

1. A radio is designed to operate on a 9-V battery or on household current. It contains a transformer that reduces 120 V AC to 9 V AC. What is the ratio of the number of turns in the output coil to the number of turns in the input coil?

2. The generator at a power plant produces AC at 24,000 V. A transformer steps this up to 345,000 V for transmission over power lines. If there are 2,000 turns of wire in the input coil of the transformer, how many turns must there be in the output coil?

3. Compute the wavelength of the carrier wave of your favorite radio station.

4. What is the wavelength of the 76-Hz ELF wave?

5. Compute the frequency of an EM wave with a wavelength of 1 in. (0.0254 m).

6. The wavelength of an electromagnetic wave is measured to be 600 m.
   a) What is the frequency of the wave?
   b) What type of EM wave is it?

7. Determine the range of wavelengths in the UV radiation band.

8. A piece of iron is heated with a torch to a temperature of 900 K. How much more energy does it emit as blackbody radiation at 900 K than it does at room temperature, 300 K?

9. The filament of a light bulb goes from a temperature of about 300 K up to about 3,000 K when it is turned on. How many times more radiant energy does it emit when it is on than when it is off?

10. What is the wavelength of the peak of the blackbody radiation curve for the human body ($T = 310$ K)? What type of EM wave is this?

11. What wavelength EM wave would be emitted most strongly by matter at the temperature of the core of a nuclear explosion, about 10,000,000 K? What type of wave is this?

12. What is the frequency of the EM wave emitted most strongly by a glowing element on a stove with temperature 1,500 K?

13. The blackbody radiation emitted from a furnace peaks at a wavelength of 1.2×10^{-6} (0.0000012) meters. What is the temperature inside the furnace?

14. What is the lowest temperature that will cause a blackbody to emit radiation that peaks in the infrared?

CHALLENGES

1. The earth's magnetic field lines are not parallel to its surface except in certain places: they actually "dip" downward at some angle to the ground. (An ordinary compass does not show this.)
   a) At what places on earth would the "magnetic dip" be the greatest?
   b) Where on the earth would the dip be zero?

2. A solenoid connected to a 60-Hz alternating current source will produce an oscillating magnetic field, as we have seen. If a permanent magnet is inserted into the solenoid, it will oscillate, but *not* with the same frequency that an unmagnetized piece of iron would. Why? (This is why the "hum" or "buzz" from electrical devices is sometimes 60 Hz sound and sometimes 120 Hz sound.)

3. The "right-hand rule" is a way to determine the direction of the magnetic field produced by moving charges. Imagine wrapping your right hand around the path of the charges so that the positive charges (or the current) flow from the "little finger" side of your fist to the thumb side. Then your fingers circle the path in the same direction as the magnetic field lines. Use this rule to verify the directions of the magnetic fields shown in Figures 8.7 and 8.8. How would you use the rule to find the direction of the magnetic field lines around a moving negative charge?

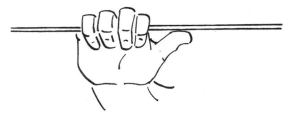

4. Sketch a diagram of an atom with one electron orbiting the nucleus. Use the right-hand rule (see Challenge 3) to determine the direction of the magnetic field produced by the electron.

5. If a coil of wire is connected to a very sensitive ammeter and then waved about in the air, a current will be induced in it even if there are no magnets around. Why?

6. Even though the output voltage of a transformer can be much larger than the input voltage, the power output is nearly the same as the power input. (There is some energy loss due to ohmic heating.) Use this to determine the relationship between the input and output currents and the number of turns in the input and output coils.

7. The highest frequency sound that can be recorded by a tape recorder depends on the size of the gap in the recording head. Why? Would a wider or a narrower gap be capable of recording higher frequencies?

8. Describe what happens to an electric charge as an electromagnetic wave passes through the region around it. Explain why the charge will produce another electromagnetic wave. (This process occurs in the ionosphere.)

SUGGESTED READINGS

Brodeur, Paul. "Annals of Chemistry (The Ozone Layer)." *The New Yorker* June 9, 1986, p 70. An account of the history and politics of the chlorofluorocarbon problem.

Cajori, Florian. *A History of Physics.* New York: Dover, 1962 (originally published by Macmillan in 1929). The chapter on the 19th Century includes a long section on electromagnetism.

Johnson, Kenneth W., and Walker, Willard C. *The Science of Hifidelity.* Dubuque, Ia: Kendall/Hunt, 1977. A very good book on the physics and acoustics of sound reproduction. It includes a separate chapter (11) on electromagnetism.

Lemonick, Michael D. "The Ultimate Quest." *TIME* April 16, 1990. Popular press description of the largest particle accelerators, including Tevatron and LEP, and what they are used for.

Rossing, Thomas D. "The Compact Disc Digital Audio System." *The Physics Teacher* 25, no. 9 (December 1987):556–562. A close look at CD technology.

Schneider, Stephen H. "The Greenhouse Effect: Science and Policy." *Science* 243, no. 4892 (February 10, 1989):771–781. A close look at the predictions and implications of global warming.

Segrè, Emilio. *From Falling Bodies to Radio Waves.* New York: W. H. Freeman and Co;, 1984. The latter two thirds of Chapter 4 is a good account of the lives and work of Ampère, Faraday, Maxwell, and others.

Sochurek, Howard. "Medicine's New Vision." *National Geographic* January 1987, pp. 2–41. Describes magnetic resonance imaging and other exotic medical imaging devices. Lots of color photos of images from inside the human body.

Stolarski, Richard S. "The Antarctic Ozone Hole." *Scientific American* 258 no. 1 (January 1988):30–36. Describes the ozone hole and its possible causes.

Walker, Jearl. *The Flying Circus of Physics with Answers*. New York: Wiley, 1977. "A collection of problems and questions about physics in the real, everyday world." Chapters 6 and 7 contain items of particular relevance to our discussions of EM waves.

OUTLINE

OPTICS

PROLOGUE: MORE THAN SKIN DEEP

This remarkable photograph was used on the cover of the first edition of this textbook. Why was it chosen? First and foremost, because it is beautiful. This long-exposure, nighttime photo shows the reddish hues of the *aurora borealis* (the northern lights) behind snow-covered Mt. Rainier near Seattle. This image is recreated and inverted by reflection off the tranquil surface of a lake. The dark zone separating the two is a forest, revealed by the silhouettes of the treetops against the white snow. This single photograph illustrates the stunning beauty that can be found on this planet.

The processes that create this beauty involve, you guessed it, physics! Just as the splendor of the ceiling in the Sistine Chapel only increases as one learns more about how Michelangelo produced this masterpiece, the beauty in this photograph can be enhanced by viewing it from a scientific perspective. You may discover that this photo is indeed "worth a thousand words," if for no other reason than that it tells the careful observer much about nature and the earth itself.

The white streaks passing through the aurora borealis were left by stars in the constellation *Ursa Major,* more commonly known as the Big Dipper. During the several minutes that the camera's shutter was open, the earth's rotation made the stars appear to be in motion. (They move through the sky just as the sun does during the day). Because Ursa Major is near Polaris, the North Star, the photo must have been taken at a point somewhere south of the mountain. Notice that the lake is obviously not frozen, yet the mountain is covered with snow and ice. This tells us that the lake is at a much lower elevation than the mountain and that the photo was probably not taken during the winter.

And what of the aurora itself? This awesome display arises from a veritable symphony of physical processes. It begins at the sun, where ionized atoms and freed electrons are flung out into space by the inferno. A journey of several days carries them to the earth, where the charged particles are guided by the magnetic field surrounding the planet. The lightweight electrons spiral around the magnetic field lines and follow them toward the north and south poles. Once these energetic electrons enter the atmosphere, they collide with oxygen atoms and nitrogen molecules and give them energy. Many of these atoms and molecules give up their energy by emitting the light that we see.[1] (More on this process in Chapter 10.)

Both the photograph itself and the mental image formed in the brain originate with optical devices that are actually quite similar: the camera and the eye. Light enters the camera, is bent as it passes through the camera lens, and is focused onto the film. There a chemical process occurs that forms a negative of the image. This is used to make the photograph. The eye is a complex instrument that uses physics and biochemistry. The light enters the eye and is focused on the retina at the back of the eye much like in the camera. The different colors of light stimulate special cells on the retina. The corresponding electrical signals sent to the brain by these cells is processed by the brain to form the colored mental image.

The beauty captured in this one photograph originates in physical processes that can be understood using the concepts of electromagnetism discussed in Chapter 8, the concepts of reflection and image formation discussed in this chapter, and the concepts of atomic physics presented in Chapter 10. This chapter is devoted to optics, the study of light and phenomena related to it. Light is used by people to gain information, to navigate, to detect possible dangers, and to appreciate beauty. We will describe the basic ways that light interacts with matter and look at how these give rise to things as useful as lenses and as beautiful as rainbows.

9.1   LIGHT WAVES

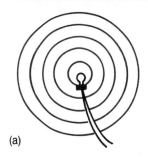

(a)

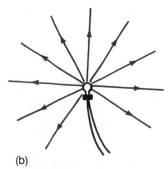

(b)

FIGURE 9.1
The light from a light bulb represented by (a) wavefronts and by (b) light rays.

Light generally refers to the narrow band of electromagnetic (EM) waves that can be seen by human beings. These are transverse waves (like all EM waves) with frequencies from about 4.3×10^{14} hertz to 7.5×10^{14} hertz. The corresponding wavelengths are so small that we will find it useful to express them in *nanometers* (*nm*). One nanometer is one billionth of a meter.

$$1 \text{ nanometer} = 10^{-9} \text{ meter} = 0.000000001 \text{ meter}$$

$$1 \text{ nm} = 10^{-9} \text{ m}$$

The wavelengths of visible light (in a vacuum or in air) range from about 700 nanometers for low-frequency red light to about 400 nanometers for high-frequency violet light. Two important points from Section 8.5 that you should keep in mind are (1) different frequencies of light are perceived as different colors and (2) white light is a combination of all colors.

The various properties of waves described in the first part of Chapter 6 of course apply to light. (You may find it useful to review Sections 6.1 and 6.2 at this time.) As with sound and water ripples, we will use both wavefronts and rays to represent light waves. Recall that a wavefront shows the location in space of one particular cycle of the wave. For a light bulb, the wavefronts are spherical shells (not unlike balloons) expanding outward at the speed of light (see Figure 9.1). A

---

[1]The mechanism that creates the aurora is much like that used in a common television. In a television picture tube, fast-moving electrons are guided along their paths to the screen by magnetic fields. The image we see on the screen is comprised of light emitted by atoms that are given energy when these electrons collide with them.

light ray is a line drawn in space representing a "pencil" of light that is part of a larger beam. Rays are represented as arrows and indicate the direction the light is traveling. A laser beam can often be thought of as a single light ray. The light from a light bulb can be represented by light rays radiating outward in all directions. (Be careful not to confuse these light rays with the electric and magnetic field lines discussed in the previous chapters.)

Some of the general characteristics of wave propagation, like reflection, are readily observed with light waves. But other phenomena are more rare in common experience because of two factors:

1. The speed of light is extremely high (3×10^8 meters/second).

2. The wavelengths of light are extremely short.

For example, we must turn to distant galaxies moving away from us at high speeds to see the Doppler effect with light (see the *Physics Potpourri* "The Hubble Law: Expanding Our Horizons" on page 264). As we saw in Section 6.2, the Doppler effect with sound is, on the other hand, quite common.

In these first two sections we will describe some of the phenomena that can occur when light encounters matter. The remaining sections of the chapter deal with important things that occur after light has traveled inside transparent material.

Reflection

Reflection of light waves is extremely common: except for the sun, light bulbs, and other light sources, everything we see is reflecting light to our eyes. There are two types of reflection: specular and diffuse.

Refer back to reflection in Section 6.2.

Specular reflection is the familiar type that we see in a mirror or in the surface of a calm pool of water. A mirror is a very smooth, shiny surface, usually made by coating glass with a thin layer of aluminum or silver. Specular reflection simply changes the direction the light wave is traveling (Figure 9.2). By changing the

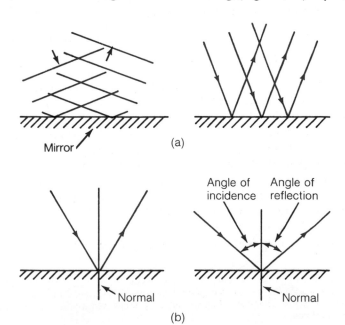

FIGURE 9.2
(a) Specular reflection using wavefronts and light rays. (b) Specular reflection of a single light ray with different angles of incidence.

angle of the incident (incoming) light ray and observing the reflected ray, we see that the light behaves somewhat like a billiard ball bouncing off a cushion on a pool table.

Figure 9.2b shows an imaginary line drawn perpendicular to the mirror and touching it at the point where the incident ray strikes it. This line is called the *normal.* The angle between the incident ray and the normal is called the *angle of incidence,* and the angle between the reflected ray and the normal is called the *angle of reflection.* Our simple observations indicate that these angles are always equal. The following law, first described in a book entitled *Catoptrics,* thought to have been written by Euclid in the third century B.C., states this formally:

> **LAW OF REFLECTION**   The angle of incidence equals the angle of reflection.

So specular reflection of light is much like sound echoing off a cliff, as described in Chapter 6.

The other type of reflection, *diffuse reflection,* occurs when light strikes a surface that is not smooth and polished, like the bottom of an aluminum pan. The light rays reflect off the random bumps and nicks in the surface and scatter in all directions (see Figure 9.3). The law of reflection still applies, but the rays encounter segments of the irregular surface oriented at different angles and therefore leave the surface with different directions. That is why you can shine a flashlight on the aluminum and see the reflected light from different angles around the pan. With specular reflection off a mirror, you could see the reflected light from only one direction.

Except for smooth, shiny surfaces like mirrors, every object we see is reflecting light diffusely. This diffuse reflection of ambient light results in light radiating outward from each point on a surface. You can see every point on your hand as you turn it in front of your face because each point on your skin is reflecting light in all directions.

Things can have color because light actually penetrates into the material and is partially reflected and partially absorbed along its way into and out of the material. The reflected light that leaves the surface will have color if pigments in the material absorb some frequencies (colors) more efficiently than others. A white surface, like this paper, reflects all frequencies of light equally. If you shine just red light on it, it will appear red. With just blue light, it will appear blue. A colored surface, like that of a red fire extinguisher, "removes" some frequencies

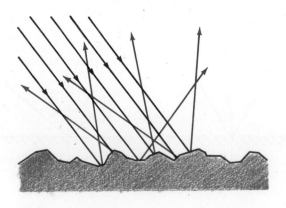

FIGURE 9.3
Diffuse reflection of light off a rough surface.

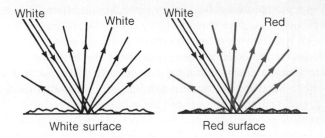

White White White Red

White surface Red surface

FIGURE 9.4
A white surface reflects all colors. A red surface reflects red but absorbs blue, green, and other colors in white light.

of the light. A red surface reflects the lower frequency light (red) most effectively and absorbs much of the rest (see Figure 9.4). If you shine red light on it, it will appear red. With blue or any other single color, it will appear black: very little of the light will be reflected.

Diffraction

As with all waves, diffraction of light as it passes through a hole or slit is observable only when the width of the opening is not too much larger than the wavelength of the light. This means that light doesn't spread out after passing through a window as nearly as much as sound does, but diffraction is observed when a very narrow slit (about the width of a human hair) is used (Figure 9.5). The narrower the slit, the more the light spreads out.

Interference

In Section 6.2 we described how waves can undergo interference. Recall that when two identical waves arrive at the same place, they add together. If the two waves are "in phase"—peak matches peak—the resulting amplitude is doubled. This is called *constructive interference*. At any point where the two waves are

FIGURE 9.5
Diffraction of light. (a) Light passing through a narrow slit spreads out, as shown on the screen. (b) Photograph of laser light projected onto a screen after passing through a slit 0.008 centimeters wide. The screen was 10 meters from the slit.

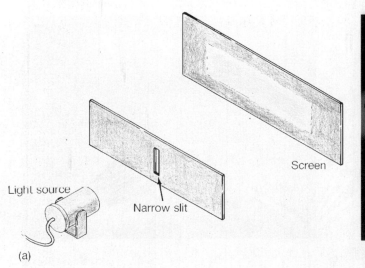

Screen

Light source

Narrow slit

(a)

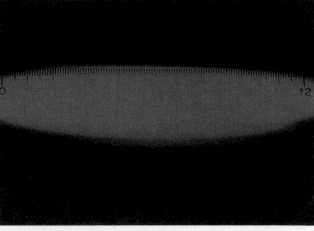

Refer back to interference in Section 6.2.

"out of phase"—peak matches valley—they cancel each other. This is *destructive interference.*

Interference of light waves is an important phenomenon for a couple of reasons. First, in experiments conducted around 1800, the British physician Thomas Young used interference to prove that light is indeed a wave. Second, interference is routinely used to measure the wavelength of light. We will consider two types of interference: *two-slit* interference and *thin-film* interference.

When a light wave passes through two narrow slits that are close together, the two waves emerging from the slits diffract outward and overlap. If the light consists of a single frequency (color), a screen placed behind the slits where the two light waves overlap will show a pattern of bright spots alternating with dark spots (Figure 9.6). At each bright spot the two waves from the slits are in phase and undergo constructive interference. Conversely, at each dark spot the two waves are out of phase and undergo destructive interference—they cancel out each other. At the center of this *interference pattern* there is a bright spot because the two waves travel exactly the same distance in getting there so they are in phase. At the first bright spot to the left of center the wave from the slit on the right has to travel a distance exactly equal to one wavelength farther than the wave from the left. This puts them in phase again. Similarly, at each successive bright spot on the left side the wave from the right slit has to travel 2, 3, 4, etc, wavelengths farther than the wave from the left slit. At each bright spot on the right side of the pattern it is the wave from the left slit that has to travel a whole number of wavelengths farther.

At the first dark spot to the left of the center of the pattern, the wave from the right slit travels one half a wavelength farther than the wave from the left slit. The two waves are out of phase and interfere destructively. At the next dark spot on the left, the additional distance is $1\frac{1}{2}$ wavelengths, then $2\frac{1}{2}$ wavelengths at the next, and so on. True constructive and destructive interference actually occurs only at the centers of the bright and dark spots. At points in between, the waves are between exactly in phase and exactly out of phase, so they partially reinforce or partially cancel each other.

The distance between two adjacent bright or dark spots is determined by the distance between the two slits, the distance between the screen and the slits, and the wavelength of the light. Since the first two can be measured easily, their values

FIGURE 9.6

Two slit interference. (a) Light passing through two narrow slits forms an interference pattern on the screen. The bright spots occur at places where the light waves from the two slits arrive in phase. (b) Photograph of interference pattern formed by laser light passing through two narrow slits 0.025 centimeters apart. The screen was 10 meters from the slits.

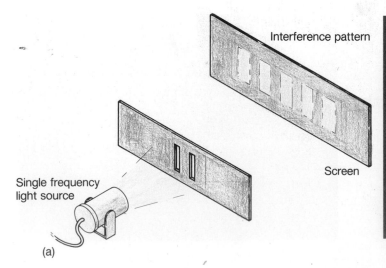

Interference pattern

Screen

Single frequency light source

(a)

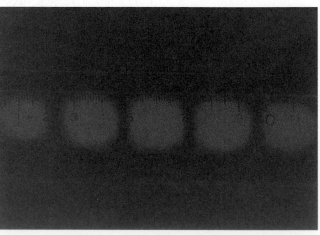

can be used to compute the wavelength of the light.

The swirling colors you see in oil or gasoline on wet pavement are caused by *thin-film interference.* Part of the light striking a thin film of oil reflects off it, and part of it passes through to be reflected off the water (Figure 9.7). The light wave that passes through the film before being reflected travels a greater distance than the wave that reflects off the oil. If the two waves emerge in step, there is constructive interference. If they emerge out of step, there is destructive interference.

The wavelength of the light, the thickness of the film, and the angle at which the light strikes the film combine to determine whether the interference is constructive, destructive, or in between. With single-color (one wavelength) light one would see bright spots and dark spots at various places on the film. With white light one sees different colors at different places on the film. At some places the film thickness and angle of incidence will cause constructive interference for the wavelength of red light, at other places for the wavelength of green light, and so on.

Interference in thin films in hummingbird and peacock feathers is the cause of their iridescent colors. Soap bubbles are also colored by interference of light reflecting off the two surfaces of the soap film (Figure 9.8)

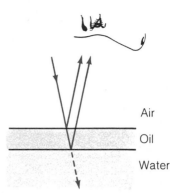

FIGURE 9.7
Interference of light striking a thin film of oil. The light reflecting off the upper surface of the film undergoes interference with the light that passes through to, and is reflected by, the lower surface. Part of the light continues on through the lower surface.

Polarization

The fact that light could undergo diffraction and interference convinced Young and other scientists of his time that light is a wave. The other model of light, which suggested that light is a stream of tiny particles, could not account for these distinctively wavelike phenomena. Polarization reveals that light is a *transverse* wave rather than a longitudinal wave.

A rope secured at one end can be used to demonstrate polarization. If you pull the free end tight and move it up and down, a wave travels on the rope that is *vertically polarized.* Each part of the rope moves in a vertical plane (Figure 9.9). In a similar manner, moving the free end horizontally produces a *horizontally polarized* wave on the rope. Moving the free end at any other angle with the vertical will also produce a polarized wave. Polarization is only possible with transverse waves.

FIGURE 9.8
Examples of colors generated by thin-film interference.

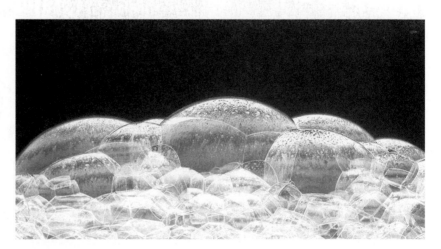

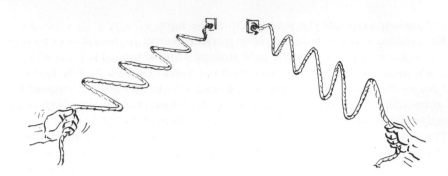

FIGURE 9.9
Waves on a rope polarized (a) horizontally and (b) vertically.

Light, being a transverse wave, can be polarized. A *Polaroid filter,* like the lenses of Polaroid sunglasses, absorbs light passing through it unless the light is polarized in a particular direction. This direction is coincident with the *transmission axis* of the filter. Light polarized in this direction passes through the Polaroid unaffected, light polarized perpendicular to this direction is blocked, and light polarized in some direction in between is partially absorbed (Figure 9.10).

The light that we get directly from the sun and from light fixtures is a mixture of light waves polarized in all different directions. The light is said to be "natural" or "unpolarized" because it has no preferred direction of polarization. When natural light encounters a Polaroid filter, it emerges polarized along the transmission axis (see Figure 9.11). The filter allows only those parts of the light wave that are polarized in this direction to pass through; the rest is absorbed.

Now if this light encounters a second Polaroid filter, the amount of light that emerges will depend on the orientation of the transmission axis of the second filter. If the axis of the second filter is aligned with that of the first, the light will pass through. If the axis of the second is perpendicular to that of the first, all of the light will be blocked by the second filter. This is referred to as "crossed Polaroids" (Figure 9.12). When the angle between the transmission axes of the two filters is other than zero or 90°, some of the light will pass through, with the intensity becoming progressively less for angles closer to 90°.

Polaroid sunglasses are very useful because light is polarized to some extent when it reflects off a surface, like water, asphalt, or the paint on the hood of a car. In particular, the reflected sunlight is partially polarized horizontally (Figure

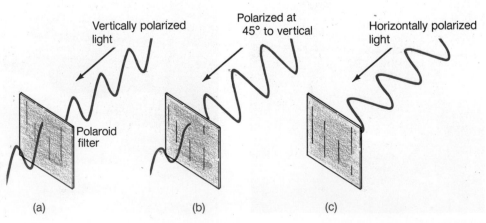

FIGURE 9.10
Light polarized in different directions encountering a Polaroid filter with its transmission axis vertical. The lines are not actually visible in a Polaroid filter but are drawn in here to show the direction of the filter's transmission axis.

9.13). This reflected light, called glare, is usually quite bright and annoying. Sunglasses using Polaroid lenses with their transmission axes vertical will block most of this reflected light, which makes it easier to see the surface itself.

Liquid crystal displays (LCDs) used in calculators, digital watches, miniature televisions and video games also use polarization. The liquid crystal is sandwiched between crossed Polaroids, and this assembly is placed in front of a mirror. Without the liquid crystal present, the display would be dark: light passing through the first Polaroid is polarized vertically and would be totally absorbed when it reached the second Polaroid. No light would reach the mirror, so none would be reflected back from the display. The specially chosen liquid crystal material between the Polaroids is arranged so that in its normal state its molecules change the polarization of the light passing through it. The polarization is changed from vertical to horizontal for light passing through from the front and from horizontal to vertical for light passing through from the rear. The light that was vertically polarized by the first Polaroid is made horizontally polarized by the liquid crystal, passes through the second Polaroid, is reflected back through the display, and emerges polarized vertically (see Figure 9.14a).

To make images on the display, segments of it are darkened. Here is where the liquid crystal property is used. An electric field is switched on in the parts of the display to be darkened. This causes the molecules of the liquid crystal to rotate and to become aligned in one direction. As a result they no longer change the polarization of light passing through. Light going through this segment is absorbed by the second Polaroid, and that part of the display is dark (see Figure 9.14b). (By the way, the transmission axis of the first Polaroid is oriented vertically so that you can see the display while wearing Polaroid sunglasses. If you rotate an LCD and look at it while wearing Polaroid sunglasses, at one point all of it will be dark.)

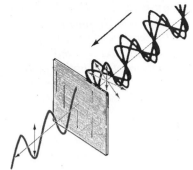

FIGURE 9.11
Unpolarized light being vertically polarized by a Polaroid filter. The star-shaped cluster of arrows represents a combination of light waves with all different directions of polarization. The filter blocks the horizontally polarized components in the light.

DO-IT-YOURSELF PHYSICS

Go to the sunglasses display in a department store and choose two polarized pairs. Arrange them so that light from a fixture passes through the right lens of one and the left lens of the other before reaching your eye. Rotate one pair and notice how the light gets dimmer until it is completely extinguished when the two sunglasses are perpendicular (ie, the Polaroids are crossed; Figure 9.15).

Look through one pair at a liquid crystal display on a watch or calculator. Rotate the LCD or the sunglasses and notice the display turn dark.

Find a place where bright light is reflected off the top of a glass display case. Look through a pair of sunglasses

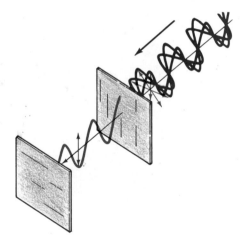

FIGURE 9.12
Crossed Polaroids. Light emerging from the first Polaroid is polarized vertically. The second Polaroid, with its transmission axis horizontal, blocks the light (see Figure 9.15).

FIGURE 9.13
Light reflected off water is partially polarized horizontally. Polaroid sunglasses with their transmission axes vertical block this glare.

FIGURE 9.14
Expanded view of part of a liquid crystal display. (a) The liquid crystal changes the polarization of the light so that it can make a round trip through the crossed Polaroids. (b) The voltage neutralizes this part of the liquid crystal so the light is blocked by the second Polaroid. This part of the display is darkened.

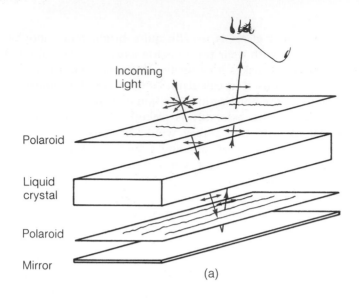

Incoming Light

Polaroid

Liquid crystal

Polaroid

Mirror

(a)

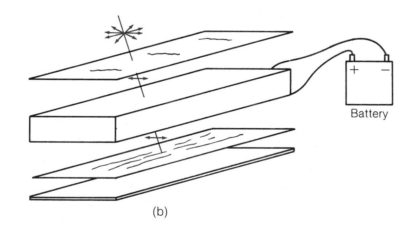

Battery

(b)

FIGURE 9.15
Polarized sunglasses completely block out light when their lenses are crossed perpendicularly.

at the glare, and rotate them 90° to the side. Notice that the glare is reduced the most when the sunglasses are in the normal position, and much less so when they are rotated 90°.

9.2   MIRRORS: PLANE AND NOT SO SIMPLE

Most mirrors that we use are *plane* mirrors: they are flat and almost perfect reflectors of light. When we use a mirror to "see ourselves," light that is diffusely reflected off our clothes and face strikes the mirror and undergoes specular reflection. Some of the rays leaving the mirror are going in the proper direction

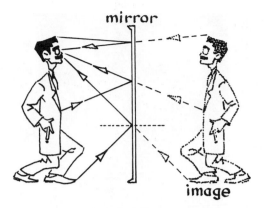

to enter our eyes and to give us an image or ourselves. The image appears to be on the other side of the mirror. (We are so accustomed to this that we don't think about it. But imagine your reaction if you had never seen a mirror image before.) Figure 9.16 shows a person viewing his image in a plane mirror. Instead of showing every light ray traveling outward from every point on the person, we show selected rays that happen to enter the person's eyes. The dashed lines from the image show the apparent paths taken by the rays when traced back to the image.

"One-Way Mirror"

A "one-way mirror" is made by partially coating glass so that it reflects some of the light and allows the rest to pass through. This is called a *half-silvered mirror* (Figure 9.17). When used as a window or wall between two rooms, it will function as a one-way mirror if one of the rooms is brightly lit and the other is dim. It will appear to be an ordinary mirror to anyone in the bright room, but it will appear to be a window to anyone in the dim room. This is because in the bright room the light reflected off the half-silvered mirror is much more intense than the light that passes through from the other room. In the dim room it is the transmitted light from the bright room that dominates (see Figure 9.18).

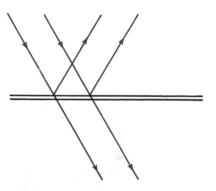

FIGURE 9.17
Light striking a half-silvered mirror. Part of the light is reflected, and part passes through.

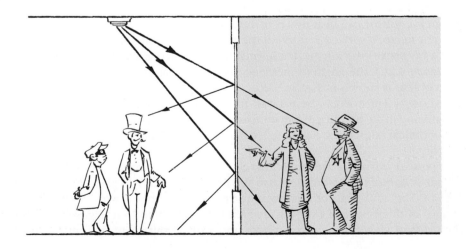

FIGURE 9.18
A half-silvered mirror used as a "one-way mirror" between two rooms. In the room on the left, a person sees mostly reflected light (a mirror). In the room on the right, a person sees mostly transmitted light (a window).

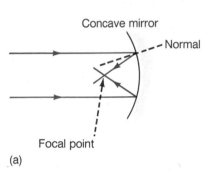

Concave mirror

Normal

Focal point

(a)

FIGURE 9.19

(a) In a well-lit room, the half-silvered mirror appears to be an ordinary mirror. (b) In a dark room on the other side, the half-silvered mirror appears to be a window.

A person in the dim room can see what is happening in the bright room without being seen by anyone in the bright room (see Figure 9.19) This device is often used in interview and interrogation rooms and as a means of observing customers in stores and gambling casinos. Note that if a bright light is turned on in the dimmer room, the one-way effect is destroyed. Ordinary window glass is a crude one-way mirror because it does reflect some of the light that strikes it. At night one can see into a brightly lit room through a window, but anyone in the room has difficulty seeing out.

Curved Mirrors

As we saw in Section 6.2, reflectors—mirrors in this case—that are curved have useful properties. Parallel light rays that reflect off a *concave mirror*—a mirror that is curved inward—are focused at a point called the *focal point* (Figure 9.20a). The energy in the light is concentrated at that point. Sunlight focused by a concave mirror can heat things to very high temperatures (Figure 9.20b). Even when a mirror's surface is curved, the law of reflection still holds at each point that a ray strikes the mirror. If a normal line is drawn at each point (as was done in Figure 9.2b), the angle of incidence equals the angle of reflection. One such normal line is shown in Figure 9.20a.

A concave mirror can be used to form images that are enlarged—magnified. Magnifying make-up mirrors are concave mirrors, as are the large mirrors used in astronomical telescopes. Figure 9.21b shows a magnified image seen in a concave mirror.

A *convex mirror* is one that is curved outward. The image formed by a convex mirror is *reduced*—it is smaller than the image formed by a plane mirror (Figure 9.21c). The advantage of a convex mirror is that it has a wide field of view—images of things spread over a wide area can be viewed in it. Figure 9.22 shows the fields of view for a convex mirror and a plane mirror of the same size. One

FIGURE 9.20

(a) Parallel rays reflecting off a concave mirror converge on a point called the focal point. (The normal for the upper ray shows that the law of reflection applies.) (b) Sunlight focussed by a concave mirror can generate high temperatures. This piece of wood was in flames a few seconds after being placed at the mirror's focal point.

(a)

(b)

(c)

FIGURE 9.21
*(a) Plane mirror image.
(b) Enlarged image in a concave mirror.
(c) Reduced image in a convex mirror.*

glance at a well-placed convex mirror on a bike path allows quick surveillance of a large area (Figure 9.23). Right-side rearview mirrors on cars and auxiliary left-side mirrors on trucks are convex so the driver can view a large space behind the vehicle. Care must be taken when using such a mirror because the reduced image makes any object appear to be farther away than it actually is.

DO-IT-YOURSELF PHYSICS

A common shiny metal spoon can serve as either a concave mirror or a convex mirror, although its shape is usually far from ideal. The front side of the spoon is a concave mirror. Place your finger close to the spoon, and look at its image in the spoon. If your finger is close enough, the image should be upright and magnified. Now hold the same side of the spoon a comfortable distance in front of your face. What is different about the image of your face that you see? The back side of a spoon is a convex mirror. Note the reduced size of the image of anything viewed in this side of the spoon as well as the wide field of view.

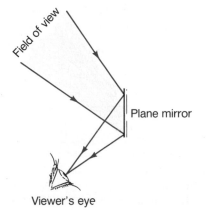

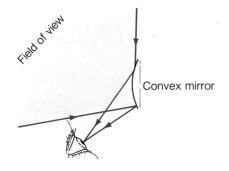

FIGURE 9.22
The field of view of a convex mirror is much larger than that of a plane mirror.

FIGURE 9.23
This convex mirror on a bike path at the University of Minnesota allows quick surveillance of a large area that would normally be hidden from view.

Astronomical Telescope Mirrors

The largest telescopes used by astronomers to examine stars, galaxies, and other celestial objects make use of curved mirrors. Figure 9.24 shows a typical design for such telescopes. Light from the distance source enters the telescope and reflects off a large concave mirror called the *primary mirror.* The reflected rays converge onto a much smaller convex mirror called the *secondary mirror.* The rays are reflected back toward the primary mirror, pass through a hole in its center, and converge to form an image at the focal point F. The primary mirror is the key component of the telescope.

Telescope mirrors have as their basic functions the gathering of light and the concentration of that light to a point. The ability of a mirror to collect light increases with its surface area. Thus to acquire enough radiation to study adequately faint objects, astronomers have sought to build instruments with larger and larger apertures ("openings").

The quality of the images produced by telescopes is greatly affected by the shapes of the mirrors. The easiest curved mirror to make is one whose surface has the shape of part of a sphere. But such a *spherical mirror* is not perfect for the task of focusing light rays. Figure 9.25a shows that parallel light rays reflecting off a spherical mirror are not all focused at the same point. An image formed using such a mirror will be somewhat blurred. This phenomenon is called *spherical aberration.* (We will see in Section 9.4 that the same thing happens with lenses.)

As the name implies, spherical aberration is a defect associated with spherical surfaces. A concave mirror in the shape of a *parabola* does not have this aberration. (You may recall that we saw the parabola in Sections 1.5 and 2.6.) A *parabolic mirror* will concentrate all the rays coming from a distant source at the same point (Figure 9.25b). Thus the ideal surface for a telescope mirror (or for that matter, reflectors in auto headlamps and household flashlight) is one shaped like a parabola. Fabricating very large mirrors, some as big as 8 meters (26 feet) in diameter, with the precise parabolic shape is an enormous technical challenge. One revolutionary technique recently developed exploits the fact that the surface of a liquid rotating at a steady rate has the required parabolic shape. (See Physics Potpourri, "Spin-casting" on page 397.)

Shortly after the Hubble Space Telescope (see Figure 8.47) was placed in orbit, scientists discovered that its primary mirror is afflicted with a type of spherical aberration. At the edge of the 2.4-meter diameter mirror, its surface is 0.002 millimeters off of what it is supposed to be. This seemingly minuscule error in the mirror's shape drastically reduces the telescope's ability to form sharp images. It is hoped that a corrective lens can be built that will allow the Space Telescope to perform as designed. This lens would be fitted onto a camera that is scheduled to be installed by astronauts on a space shuttle mission.

FIGURE 9.24
This is the basic design of a typical large astronomical telescope. The large, concave primary mirror and the small, convex secondary mirror combine to focus incoming light at the focal point F.

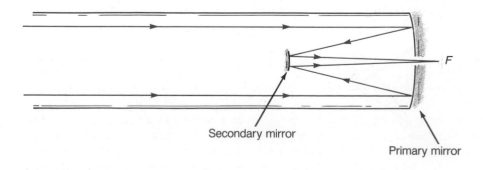

Secondary mirror

Primary mirror

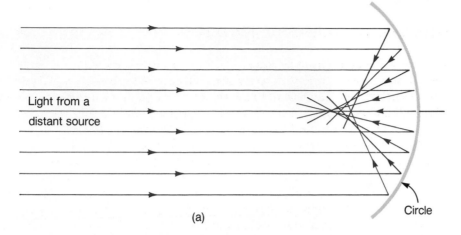

Light from a
distant source

Circle

(a)

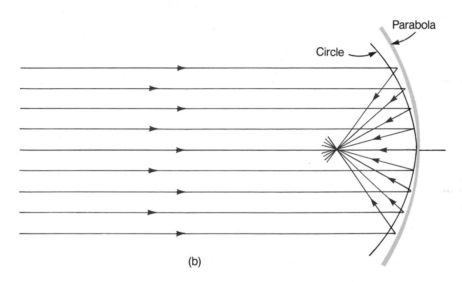

Parabola

Circle

(b)

FIGURE 9.25
(a) A spherical mirror does not reflect all incoming rays from a distant source to the same point. Rays located well off the symmetry axis of the mirror are brought to a focus closer to the mirror than those rays nearer the axis. This effect is called spherical aberration. *(b) A parabolic mirror focuses all the light from a distant source to a single point. It is the ideal shape for a telescope mirror, but it is generally harder to achieve without the use of special techniques.*

Some recent innovations in telescope design deserve mentioning. The *New Technology telescope (NTT)* in Chile has a 3.6-meter mirror that employs "active optics." Sensors and computer-controlled actuators are used to analyze and to adjust the actual shape of the mirror automatically to attain the best image quality while observations are underway. The *Multiple Mirror telescope* on Mount Hopkins in Arizona combines six 1.8-meter telescopes in a cluster that is equivalent to a 4.5-meter telescope. A similar approach is the use of segmented mirrors like those of the *Keck telescope,* now under construction in Hawaii. This instrument will employ 36 hexagonal mirrors, each 2 meters across, working together to achieve the light gathering power of one 10.5-meter telescope (Figure 9.26).

DO-IT-YOURSELF PHYSICS

If you have a phonograph turntable, you can make your own spinning parabolic mirror. Place a lightweight soup bowl—dark in color if possible—on

the turntable so that its center matches that of the platter. (You may have to remove the short spindle from the center or place a couple of useless records

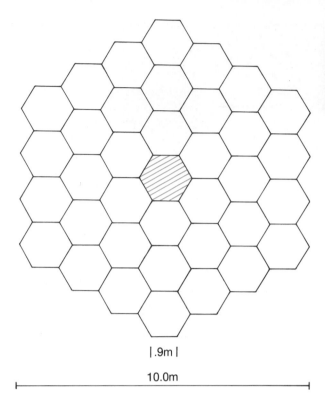

|.9m|

10.0m

FIGURE 9.26
(a) A mosaic of 36 hexagonal mirrors 2-meters in diameter will be assembled to form a single mirror with a 10-meter diameter in the Keck Telescope, now under construction in Hawaii. (b) The dome shown here which will house the telescope was completed in autumn of 1988. It rests on the rim of an extinct volcano nearly 14,000 feet above sea level.

on first so that the spindle doesn't hold up the center of the bowl.) Carefully pour cooking oil into the bowl until it is about half full (water works, but it sloshes a lot). Turn on the turntable and use the $33\frac{1}{3}$ revolutions per minute speed. If there is enough light in the room, place your hand about a foot above the bowl and to one side. Place your head above the bowl and off to the other side, and look for the dim magnified image of your hand in the oil's surface. If the room is not well lit, replace your hand with a dim lamp or a low-wattage frosted light bulb (Figure 9.27). (Be **very careful** when using a lamp, since its magnified image can be very bright.) Move your hand or the lamp up and down to show larger and smaller images. Turn the speed up to 45 revolutions per minute. What is different? Watch the image as you turn off the turntable and it slows to a stop.

FIGURE 9.27
Magnified image in a rotating bowl of oil.

9.3 REFRACTION

In this section we describe what happens when light interacts with glass and other transparent material. Imagine a light ray from some source in air arriving at the surface of a second transparent substance, like glass. The boundary between the air and glass is called the *interface*. As mentioned earlier, some of the incident light is reflected back into the air, but the rest of it is transmitted across the interface into the glass (see Figure 9.30). The law of reflection gives us the direction of the reflected light ray, but what about the transmitted light ray?

The light that passes into the glass is *refracted;* the transmitted ray is bent into a different direction than the incident ray. (If the incident ray is perpendicular to the interface, there is no bending.) This bending is caused by the fact that light travels slower in glass than in air. We again draw in the normal, a line perpendicular to the interface. The angle between the transmitted ray and the normal, called the *angle of refraction,* is smaller than the angle of incidence. We can also have the reverse process: a light ray traveling in glass reaches the interface and is transmitted into air (Figure 9.31). In this case the angle of refraction is larger than the angle of incidence.

The following law summarizes these observations:

> LAW OF REFRACTION   A light ray is bent toward the normal when it enters a transparent medium in which light travels slower. It is bent away from the normal when it enters a medium in which light travels faster.

Notice that the symmetry in Figures 9.30 and 9.31. This is one example of the *principle of reversibility:* the path of a light ray through a refracting surface is reversible. The path that a ray takes when going from air into glass is the same path it would take if it turned completely around and went from glass into air. In all of the figures in this and the following sections, the arrows on the light rays could be reversed and the paths would be the same.

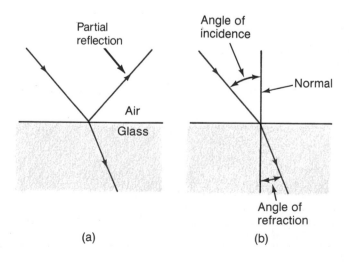

(a)

(b)

FIGURE 9.30
(a) Refraction of light as it enters glass, with partial reflection shown. (b) The normal and angles of incidence and refraction. Note that the ray is bent towards the normal.

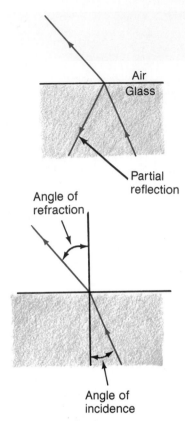

Air
Glass

Partial reflection

Angle of refraction

Angle of incidence

FIGURE 9.31
Refraction of light as it goes from glass into air. The path of the refracted ray is just the reverse of the path shown in Figure 9.30.

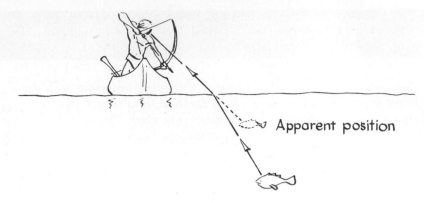

Apparent position

FIGURE 9.32
Because of refraction, the apparent position of a fish under water is not its real location.

Refraction of light affects how you see things that are in a different medium, such as under water. Figure 9.32 shows one ray of light traveling from a fish to a viewer's eye. The bending of the ray as it enters the air causes the fish to appear to be at a different position than it actually is.

The speed of light in any transparent material is less than the speed of light, c, in a vacuum. For example: $c = 3 \times 10^8$ meters/second, while the speed of light in water is 2.25×10^8 meters/second, and the speed of light in diamond is only 1.24×10^8 meters/second. Table 9.1 gives the speed of yellow light in selected transparent media. (Critical Angle and the variation of speed with color will be discussed later.)

At this point, you might have two questions in mind. Why is the speed of light lower in transparent media such as water and glass, and why does this cause

TABLE 9.1   SPEED OF (YELLOW) LIGHT AND CRITICAL ANGLES FOR SELECTED MATERIALS

Phase	Substance	Speed	Critical Angle*
Gases at 0°C and 1 atm	Carbon dioxide (CO_2)	2.9966×10^8 m/s	
	Air	2.9970	
	Hydrogen	2.9975	
	Helium	2.9978	
Liquids at 20°C	Benzene	1.997×10^8 m/s	41.8°
	Ethyl alcohol	2.203	47.3°
	Water	2.249	48.6°
Solids at 20°C	Diamond	1.239×10^8 m/s	24.4°
	Glass (dense flint)	1.81	37.0°
	Glass (light flint)	1.90	39.3°
	Glass (crown)	1.97	41.1°
	Salt (NaCl)	2.00	41.8°
	Fused silica	2.05	43.2°
	Ice	2.29	49.8°

*Critical angles are for air as the second medium.

refraction? To answer the first question we must remember that light is an electromagnetic wave—a traveling combination of oscillating electric and magnetic fields. As light enters glass, the oscillating electric field in the wave causes the electrons in the atoms of the glass to oscillate at the same frequency. These oscillating electrons emit EM waves (light) that travel outward to neighboring atoms. In essence, the light is absorbed, then reemitted over and over as it travels through the glass. (The light wave that emerges from the glass is not the same wave that entered it.) This process reduces the speed of the wave.

As to the second question, we can see why a change in the speed of light causes bending by carefully following wavefronts as they cross an interface. A good analogy to this is a marching band, lined up in rows, crossing (at an angle) the boundary between dry ground and a muddy area (see Figure 9.33). Each person in the band is slowed as his or her feet slip and sink in the mud. To remain aligned and properly spaced, each person has to turn a bit, and the whole band ends up marching in a slightly different direction. The slowing of the wavefronts of light entering glass produces the same kind of change in the direction of propagation of the light. Note that the reduction in speed also makes the wavelength *shorter.* This has to be the case because the frequency of the wave remains the same (why?) and we must have $v = f\lambda$.

The mathematical form of the law of refraction was apparently discovered independently around 1620 by the Dutch physicist Willebrord Snell and the French mathematician René Descartes. It gives the exact value of the angle of refraction when the angle of incidence and the speeds of light in the two media are known. Since it involves trigonometry, we will not present it here.

The graph in Figure 9.34 shows the relationship between the two angles for an air-glass boundary. (A similar graph could be constructed for different pairs of media.) The angle between the normal and the light ray in air is plotted on the vertical axis and the corresponding angle after the ray has entered the glass on the horizontal axis. Notice that the angle in glass is smaller than the angle in air. For a ray going from air into the glass, this means that the angle of refraction is smaller than the angle of incidence, as shown in Figure 9.30. For the largest possible angle of incidence, 90°, the angle of refraction is about 43°.

FIGURE 9.33
(a) Wavefront representation of light refracting as it enters glass. The change in direction occurs because the light travels slower in the glass. (b) For comparison, rows of a marching band alter their direction as they enter muddy ground.

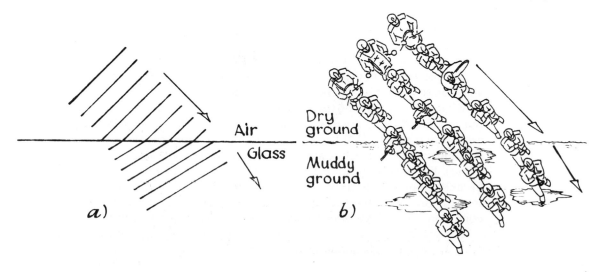

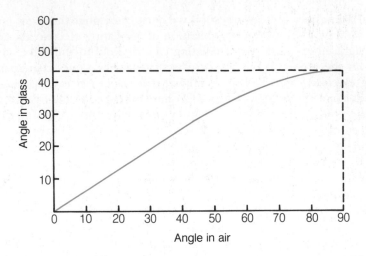

FIGURE 9.34
Graph showing the angles between a light ray and the normal when it passes through an air-glass interface. For a ray passing from air into glass, the angle of incidence is the "angle in air." When the ray passes from glass into air, the angle of incidence is the "angle in glass."

EXAMPLE 9.1

Figure 9.37 depicts a light ray going from air into glass with an angle of incidence of 60°. Find the angle of refraction.

We use Figure 9.34 with the angle in air equal to 60°. Imagine a vertical line drawn upward from 60°. From the point where that line intersects the graph, imagine a horizontal line drawn to the scale on the left. This line indicates the angle in glass is about 35°. Therefore the angle of refraction is 35°.

Total Internal Reflection

Consider now the following experiment shown schematically in Figure 9.35. Rays of light originating in glass strike the boundary (separating it from the surrounding air) at ever-increasing angles. In this case, since the speed of light in glass is smaller than the speed of light in air, we see that the ray is bent *away* from the normal. In other words, the angle of refraction is *larger* than the angle of incidence, as shown in Figure 9.35. Moreover, the angle of refraction increases as the angle of incidence does.

Because of reversibility we can use the same graph, Figure 9.34, to find the angle of refraction when the angle of incidence is known. In this situation the angle of incidence is the "angle in glass," while the angle of refraction is the "angle in air." For example, when the angle of incidence is 20°, we locate 20° on the vertical scale, move horizontally to the curve, then drop down to the axis where we see that the angle or refraction would be about 30°. This is shown in Figure 9.35c.

What distinguishes this case from the previous one, in which the incident ray was in air, is the existence of a *critical angle* of incidence for which the angle of refraction reaches 90°. When the angle of incidence equals this critical angle, the transmitted ray travels out *along* the interface between the two media (Figure 9.35e). For angles of incidence equal to or greater than the critical angle, the formerly transmitted ray is bent back into the incident medium and does not travel appreciably into the lower index medium. When this happens we have a condition called *total internal reflection*. This is shown in Figure 9.35e and f.

The graph in Figure 9.34 indicates that the critical angle for a glass-air interface, the angle in glass that makes the angle in air equal to 90°, is about 43°. In

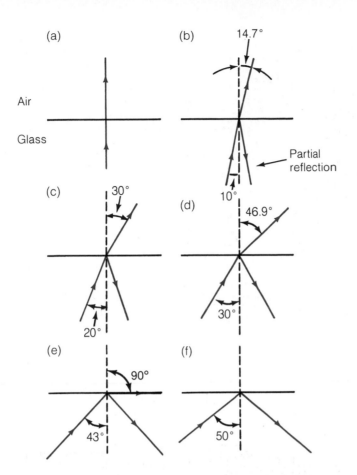

FIGURE 9.35
A light ray traveling in glass strikes a glass-air interface with different angles of incidence. The angle of incidence in (e) is the critical angle, *for which the angle of refraction is 90°. Total internal reflection (e and f) occurs when the angle of incidence is equal to or greater than the critical angle. No light enters the air.*

general, different media have different critical angles. Table 9.1 includes the values of the critical angle for light rays entering air from different transparent media. Notice that the critical angle is larger for materials in which the speed of light is lower. Diamond, with the lowest speed of light in the list, has a critical angle of less than 25°—little more than half that of glass.

A homeowner wishes to mount a floodlight on a wall of a swimming pool under water so as to provide the maximum illumination of the surface of the pool for use at night (see Figure 9.36). At what angle with respect to the wall should the light be pointed?

EXAMPLE 9.2

To illuminate the surface of the water, the refracted ray at the water-air interface should just skim the water's surface. This means that the angle of refraction must be 90°, so the incident angle must be the critical angle.

Table 9.1 shows that the critical angle for water is about 49°. The homeowner should direct the floodlight upwards so that the beam makes an angle of roughly 49° with the vertical side wall of the pool.

But what happens to light that passes completely through a sheet of glass, like a windowpane? Each light ray is bent toward the normal when it enters the glass and then is bent away from the normal when it reenters the air on the other side.

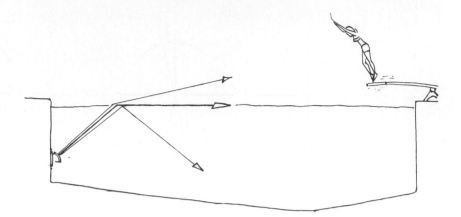

FIGURE 9.36

No matter what the original angle of incidence is, the two bends exactly offset each other, and the ray emerges from the glass traveling parallel to its original path (see Figure 9.37). Since all rays passing through a glass plate emerge parallel to their original paths, things look the same when seen through a window as when seen directly.

LEARNING CHECK

1. The two types of reflection are _____ and _____ .

2. Two otherwise identical light waves arriving at the same point with peak matching valley undergo

continued on next page

FIGURE 9.37
The light ray is refracted twice as it passes through the block of glass. Its final path is parallel to its initial path if the two surfaces of the glass are parallel. (Note the partial reflections.)

PHYSICS POTPOURRI

Reflections: Fiber Optics

Much of the charm and attraction of diamonds derives from their sparkle—and this is a direct consequence of the phenomenon of total internal reflection (TIR). Because of diamond's low speed of light, its critical angle of incidence is small (about 25°) compared to most other substances. Consequently most light rays entering a diamond undergo one or more TIRs within the gemstone before emerging. The light is briefly trapped inside the diamond, giving it a brilliant, fiery appearance.

Diamonds are not the only items that owe their beauty to TIR. Anyone who has seen colorful, decorative displays of thin glass fibers with light seemingly bursting from their ends has witnessed another example of TIR (Figure 9.38). Light entering the end of one of the fibers does so with a variety of angles of incidence. The rays are refracted toward the curved walls as they pass into the fiber. Rays that strike the walls with angles of incidence greater than the critical angle will be totally internally reflected. This process is then repeated over and over again many thousands of times before the ray finally emerges at the opposite end of the fiber (Figure 9.39). Light-conducting fibers are available with diameters of a few microns (10^{-6} meters) up to about 0.6 centimeters. The smaller filaments are

very flexible, and small bundles of them can be threaded into tiny passages within the body to permit physicians to examine the conditions of arteries, intestines, and the like. The larger diameter rods are usually referred to as "light pipes."

One recent and well-publicized use of the TIR phenomenon is in the development of fiber optic communication systems in which infrared radiation

FIGURE 9.38
Optical fiber display. Light from a lamp in the base of the display travels through the optical fibers and radiates from the ends.

a) Constructive interference.   c) Polarization.
b) Destructive interference.   d) All of the above.

3. Light is totally absorbed when sent through two Polaroid filters if their transmission axes are _____ .

4. Upon entering a medium in which the speed of light is 1.5×10^8 meters/second from a medium in which the speed of light is 2×10^8 meters/second, an oblique light ray will be
   a) bent away from,   d) remain undeviated with
   b) bent toward,   respect to
   c) travel along,

   the normal to the interface between the two media .

5. The critical angle is that angle of incidence (in the medium with the higher index of refraction) for which the angle of refraction equals _____ .

FIGURE 9.39
Multiple internal reflections within an optic fiber.

is "piped" from one location to another. In a typical application, a light-guide cable one-half inch in diameter containing 144 fibers is capable of carrying nearly 50,000 two-way conversations (Figure 9.40). To even come close to this capacity with conventional pairs of copper wires would require a cable more than 10 times larger in diameter. Given our growing dependence on rapid, reliable communication systems, fiber cables with high capacity and small size are likely to become quite important because they permit better use to be made of available underground duct space and may postpone the need to excavate new ducts—an expensive proposition in large cities. Commercially produced fibers currently available are so transparent that infrared signals can be propagated up to 14 kilometers in the fibers and still be detected without the need for amplification. Indeed, if seawater were as transparent, we would have little difficulty seeing to the bottom of the deepest ocean on earth!

The process of TIR thus contributes to making our environment more beautiful and more stimulating. Just think, in the not-too-distant future, it may be possible to select a gift for that special someone from a jewelry store across town using a videophone based upon the same physical principle that is responsible for the sparkle in the very diamonds being offered for sale. Something to reflect upon!

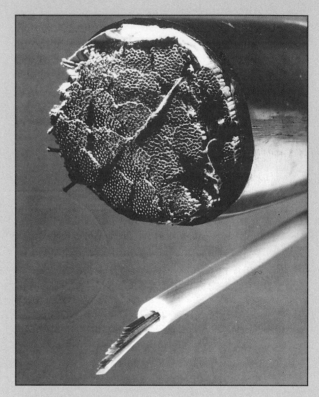

FIGURE 9.40
The fiber optic light-guide cable (below) can carry over three times as many telephone conversations as the 1,200-pair wire cable (above).

9.4   LENSES, IMAGES, AND THE EYE

When light passes through a window, it is useful to have the final direction of ray propagation remain unchanged from its initial direction. In other situations controlled deviation of light rays from their initial directions is very desirable. The enlarged (magnified) images we see with telescopes and microscopes are produced in this way. But how are such deviations brought about? The key is the use of curved boundaries (interfaces) rather than flat (planar) ones.

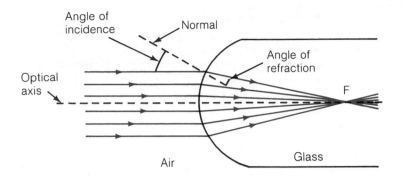

FIGURE 9.41
Refraction at a convex spherical surface showing the convergence of light rays. F' is the focal point, that is, the point at which the light rays are concentrated after having passed through the surface.

Suppose we grind a block of glass so that one end takes the shape of a segment of a sphere as shown in cross section in Figure 9.41. Let parallel rays strike the spherical surface at various points above and below the line of symmetry (called the *optical axis*) of the system. If one applies the law of refraction at each point to determine the angle of refraction of each ray, the results shown in Figure 9.41 are found. In particular, rays traveling along the optical axis emerge from the interface undeviated. Rays entering the glass at points successively above or below the optical axis are deviated ever more strongly toward the optical axis. The result is to cause the initially parallel bundle of rays to gradually *converge* together—to become focused—into a small region behind the interface. This point is called the *focal point* and is labeled F in the figure.

Figure 9.42 shows the behavior of parallel rays refracted across a spherical interface, which, instead of bowing outward, curves inward. In this case the emergent rays *diverge* outward as though they had originated from a point F' to the left of the interface. The ability to either bring together or spread apart light rays is the basic characteristic of devices called *lenses,* be they camera lenses, telescope lenses, or the lenses in human eyes.

Real lenses have two refracting surfaces instead of one, as in the previous examples, but their effect on parallel light rays is the same. A *converging lens* causes parallel light rays to converge to a point, called the focal point of the lens (Figure 9.43). The distance from the lens to the focal point is called the *focal length* of the lens. A more sharply curved lens has a shorter focal length. Conversely, if a tiny source of light is placed at the focal point, the rays that pass through the converging lens will emerge parallel to each other. This is the principle of reversibility again.

A *diverging lens* causes parallel light rays to diverge after passing through it. These emergent rays appear to be radiating from a point on the other side of the lens. This point is called the focal point of the diverging lens (Figure 9.44). The distance from the lens to the focal point is again called the focal length, but for

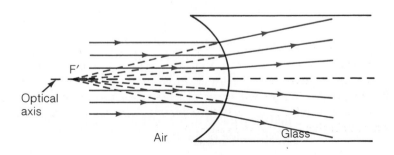

FIGURE 9.42
Refraction at a concave spherical surface showing the divergence of a beam of parallel light. F' is the focal point, that is, the point from which the rays appear to diverge after having passed through the surface.

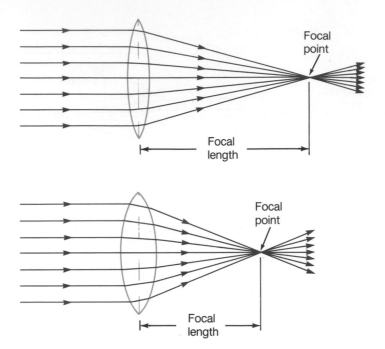

FIGURE 9.43
Converging lenses focusing parallel light rays at their focal points. A more sharply curved lens has a shorter focal length.

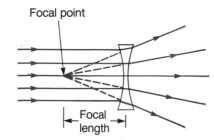

FIGURE 9.44
Parallel light rays diverging after passing through a diverging lens. The rays appear to radiate from the focal point, to the left of the lens.

a diverging lens it is given as a negative number, − 15 centimeters for example. If we reverse the process and send rays converging toward the focal point into the lens, they emerge parallel.

For both types of lenses there are two focal points, one on each side. Clearly, if parallel light rays enter a converging lens from the right side in Figure 9.43, they will converge to the focal point to the left of the lens.

Common lenses have surfaces that are in the shape of either a segment of a sphere or a flat plane. Whether a lens is diverging or converging can be determined quite easily: if it is *thicker* at the center than at the edges it is a converging lens; if it is *thinner* at the center it is a diverging lens (Figure 9.45).

Image Formation

The main use of lenses is to form images of things. In this section we present the basics of image formation when a converging lens is used. Our eyes, most cameras (both still and video), slide projectors, movie projectors, and overhead projectors all form images this way. Figure 9.46 illustrates how light radiating from an arrow, called the *object,* forms an *image* on the other side of the lens. One way of setting up this example would be to point a flashlight at the arrow so that light would reflect off the arrow and pass through the lens. The image would be projected onto a piece of white paper placed at the proper location to the right of the lens.

Although each point on the object has countless light rays spreading out from it in all directions, it is easier to first consider only three particular rays from a single point—the arrow's tip. These rays, shown in Figure 9.46, are called the *principal rays.*

> **1.** The ray that is initially parallel to the optical axis passes through the focal point (F) on the other side of the lens.

Converging   Diverging

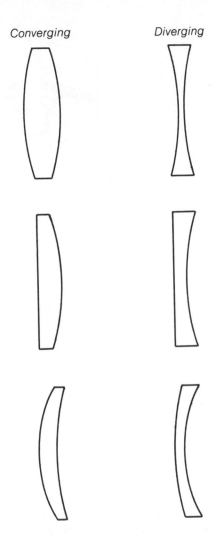

FIGURE 9.45
Examples of different types of lenses.

2. The ray that passes through the focal point (F′) on the same side of the lens as the object emerges parallel to the optical axis.

3. The ray that goes exactly through the center of the lens is undeviated because the two interfaces it encounters are parallel.

(Note that the image is *not* at the focal point of the lens. Only parallel incident light rays converge to this point.)

We could draw principal rays from each point on the object, and they would converge to the corresponding point on the image. This kind of image formation

FIGURE 9.46
Arrangement of a simple converging lens showing the object and image positions as well as the focal points F and F′, and focal lengths. Here s and p are on opposite sides of the lens and are both considered positive. The three principal rays from the arrow's tip are shown.

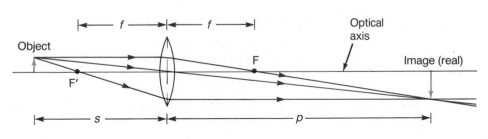

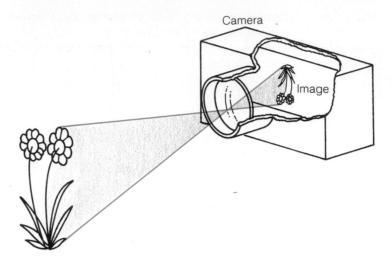

FIGURE 9.47
Image formation in a camera.
The third principal ray from a
point at the top and from a
point at the bottom of the object
are shown.

occurs when you take a photograph (Figure 9.47) or view a slide on a projection screen. In the latter case, light radiating from each point on the slide converges to a point on the image on the screen. Note that the image is inverted (upside down). That's why you load slides in upside down if you want their images to be right side up.

The distance between the object and the lens is called the *object distance,* represented by *s,* and the distance between the image and the lens is called the *image distance, p.* By convention, *s* is positive when the object is to the left of the lens, and *p* is positive when the image is to the right of the lens. If we place the object at a different point on the optical axis, the image would also be formed at a different point. In other words, if *s* is changed, *p* changes. Using a lens with a different focal length would also cause *p* to change. For example, the image would be closer to the lens if the focal length were shorter.

The following equation, known as the *lens formula,* relates the image distance, *p,* to the focal length, *f,* and the object distance, *s.*

$$p = \frac{sf}{s - f} \quad \text{(lens formula)}$$

The following example shows how the lens formula can be used.

EXAMPLE 9.3   In a slide projector a slide is positioned 0.102 meters from a converging lens that has a focal length of 0.1 meter. At what distance from the lens must the screen be placed so that the image of the slide will be in focus?

The screen needs to be placed a distance *p* from the lens, where *p* is the image distance for the given focal length and object distance. So:

$$p = \frac{sf}{s - f} = \frac{0.102 \text{ m} \times 0.1 \text{ m}}{0.102 \text{ m} - 0.1 \text{ m}}$$

$$= \frac{0.0102 \text{ m}^2}{0.002 \text{ m}}$$

$$p = 5.1 \text{ m}$$

If the slide-to-lens distance is increased to 0.105 meters, the distance to the screen (p) would have to be reduced to 2.1 meter.

If the lens is replaced by one that has a shorter focal length, the distance to the screen would have to be reduced as well.

The images formed in the manner just described are called *real images.* Such images can be projected onto a screen. Our eyes see the image on the screen because the light striking the screen undergoes diffuse reflection. A simple magnifying glass is a converging lens, but the image that it forms under normal use is not a real image—it can't be projected onto a screen. We see the image by looking *into* the lens, just as we see a mirror image by looking *into* the mirror. This type of image is called a *virtual image.* Figure 9.48 shows how the image is formed in a magnifying glass. In this case the object is between the focal point F′ and the lens so the object distance s is *less than* the focal length F of the lens. Note that the image is enlarged and that it is upright. It is also on the same side of the lens as the object, which means that p is negative. This situation is very much like the image formation with a concave mirror (Figure 9.21b).

A converging lens with focal length 10 centimeters is used as a magnifying glass. When the object is a page of fine print 8 centimeters from the lens, where is the image?

EXAMPLE 9.4

$$p = \frac{sf}{s-f} = \frac{8 \text{ cm} \times 10 \text{ cm}}{8 \text{ cm} - 10 \text{ cm}}$$

$$p = \frac{80 \text{ cm}^2}{-2 \text{ cm}}$$

$$p = -40 \text{ cm}$$

The negative value for p indicates that the image is on the same side of the lens as the object. Therefore it is a virtual image and must be viewed by looking through the lens.

A virtual image is also formed when you look at an object through a diverging lens (Figure 9.49). In this case the image is smaller than the object, as it is with a convex mirror. You can illustrate this by borrowing the eyeglasses of someone who is nearsighted (distant objects are out of focus) and looking through one lens as you would look through a magnifying glass.

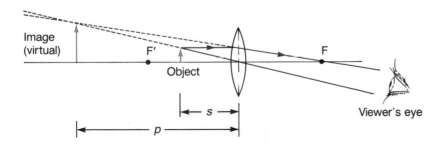

Image (virtual)

F′

Object

F

Viewer's eye

s

p

FIGURE 9.48
Image formation when the object is between the focal point and the lens. A virtual image of the arrow is formed, seen by looking through the lens toward the object.

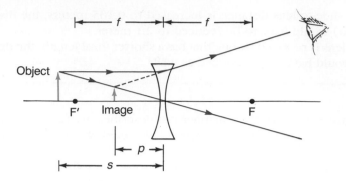

FIGURE 9.49
*Image formation with a
diverging lens. The image is
virtual so it must be viewed by
looking through the lens.*

Magnification

In both Figures 9.47 and 9.48 the size of the image is not the same as the size of
the object. This is one of the most useful properties of lenses: they can be used
to produce images that are enlarged (larger than the original object) or reduced
(smaller than the original object). In either case the *magnification, M,* of a
particular configuration is just the height of the image divided by the height of the
object.

$$M = \frac{\text{image height}}{\text{object height}}$$

If the image is twice the height of the object, the magnification is 2. If the image
is upright, the magnification is positive. If the image is inverted, the magnification
is negative (because the image height is negative).

The magnification that one gets with a particular lens changes if the object
distance is changed. Because of the simple geometry, the magnification also
equals *minus* the image distance divided by the object distance.

$$M = \frac{-p}{s}$$

From this we can conclude:

If p is positive (image is to the right of the lens and real), M is negative, and
the image is inverted (Figures 9.46 and 9.47).

If p is negative (image is to the left of the lens and virtual), M is positive, and
the image is upright (see Figures 9.48, 9.49, and 9.50).

EXAMPLE 9.5   Compute the magnification for the slide projector in Example 9.3 and for the
magnifying glass in Example 9.4.

In the first case in Example 9.3, $s = 0.102$ meters and $p = 5.1$ meters.
Therefore:

$$M = \frac{-p}{s} = \frac{-5.1 \text{ m}}{0.102 \text{ m}}$$

$$M = -50$$

FIGURE 9.50
The image seen through the magnifying glass is magnified, upright, and virtual.

The image is 50 times taller and wider than the object, and it is inverted (since M is negative). A slide that is 35 millimeters tall has an image on the screen that is 1,750 millimeters (1.75 meters) tall.

When s is 0.105 meters, p is 2.1 meters, and the magnification is -20.

In Example 9.4, $s = 8$ centimeters and $p = -40$ centimeters. Consequently:

$$M = \frac{-p}{s} = \frac{-(-40 \text{ cm})}{8 \text{ cm}}$$

$$M = +5$$

The image of the print seen in the magnifying is five times larger than the original, and it is upright (since M is positive).

Telescopes and microscopes can be constructed by using two or more lenses together. Figure 9.51 shows a simple telescope consisting of two converging lenses. The real image formed by lens 1 becomes the object for lens 2. The light that could be projected onto a screen to form the image for lens 1 simply passes on into lens 2. In essence, lens 2 acts as a magnifying glass and forms a virtual image of the object. In this telescope the image is magnified but inverted. Using a third lens in the design can result in an image that is upright.

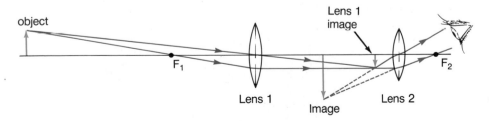

FIGURE 9.51
Simple telescope. The image formed by lens 1 is the object for lens 2.

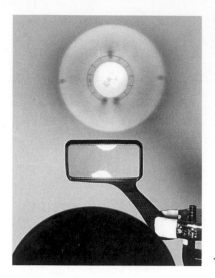

FIGURE 9.52

DO-IT-YOURSELF PHYSICS

You can check out the results of the examples preceding for yourself with a simple magnifying lens and a small electric light. A high-intensity desk lamp works well. Stand as far as you can away from the source and project a clear (real) image of it on a sheet of clean white paper. You should be able to see the lamp clearly and not just as a blur of light (see Figure 9.52). The image distance in this case approximates *f*. (This is a "quick-and-dirty" technique for determining the focal length of any converging lens!) Now move the lens toward the source, adjusting the position of the paper to produce a clear image. The distance of the paper from the lens at which a clear image is seen will increase. As *s* gets smaller, a clear image of the lamp filament can be pro-

jected, but only for ever-greater screen distances, *p*. Finally, when $s < f$, only a blur on the farthest wall will be seen where the now diverging rays intersect it. There is no real image to project. To see the image, which is now virtual, look directly through the lens at the lamp. Be sure to turn down the intensity so as to be able to comfortably view it! The image you see should be upright and magnified. This, of course, is just the principle behind common hand-held magnifiers and the more complex microscope.

Repeat the experiment, but this time mark the boundaries of the image as you gradually move the lens toward the source. Notice that for $s > 2f$, the image is reduced, while for $f < s < 2f$, the image is magnified, getting successively larger as *s* approaches *f*.

The Human Eye

One might well argue that the eye is the most sophisticated of our sense organs. But up to the point when the light rays are absorbed and the signal to the brain is formed, the eye is a relatively simple optical instrument.

Figure 9.53 shows a simplified cross section of the human eye. Light enters from the left and is projected onto the retina at the rear of the eyeball. The iris, the part of the eye that is colored, controls the amount of light that enters the eye.

FIGURE 9.53
The human eye. Light passes through the opening in the iris, the pupil, and forms an image on the retina. The cornea and the lens act as a single converging lens.

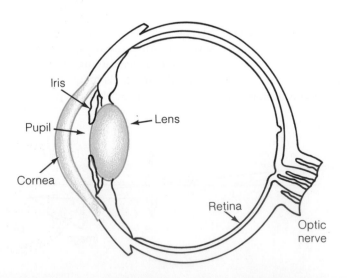

Its circular opening, the pupil, is large in dim light and small in bright light. As light enters the cornea, it converges because of the cornea's convex shape (Figure 9.41 shows this effect). The eye's *auxiliary lens,* simply called the *lens,* makes the light converge even more. This lens is used by the eye to ensure that the image is in focus on the retina. The net effect of the cornea and the lens is the same as if the eye were equipped with a single converging lens, so the image formation process is the same.

For an object to be seen clearly, the image must be in focus on the retina. In the eye, the image distance is always the same—the diameter of the eyeball. We are able to focus on objects near and far (with small and large object distances) because the focal length of the lens can be varied by changing its shape. When the eye is focused on a distant object (farther than, say, 5 meters away) the lens is thin and has a long focal length (Figure 9.54). For a near object, special muscles make the lens thicker. This shortens the focal length of the lens so that the image of the near object is focused on the retina. Unlike cameras and other optical devices, the eye has a constant value for p, the image distance, but it accommodates different values for s, the object distance, by changing its focal length, f.

The two most common types of poor eyesight, nearsightedness and farsightedness, are the result of improper focusing. *Nearsightedness,* or myopia, occurs when close objects are in focus but distant objects are not. The light rays from a distant object are brought into focus before they reach the retina. When the rays do reach the retina, they are out of focus (Figure 9.55). The problem is remedied by placing a properly chosen *diverging lens* in front of the eye. This lens makes the light rays diverge slightly before they enter the eye. This moves the point where the rays meet back to the retina.

Farsightedness, or hyperopia, is just the opposite: distant objects are in focus but near objects are not. The cornea and the lens do not make the rays from a near object converge enough. The rays reach the retina before they meet, and the image is out of focus (Figure 9.56). The remedy for this condition is a properly chosen *converging lens* placed in front of the eye. This lens makes the light rays

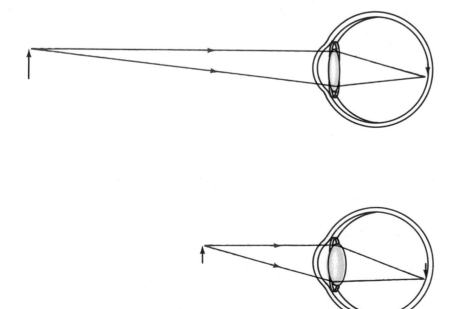

FIGURE 9.54
The eye can form images on the retina of either distant or near objects by changing the thickness (and therefore the focal length) of the lens.

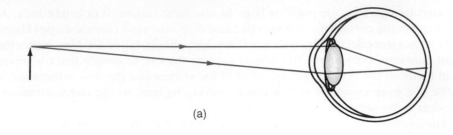

(a)

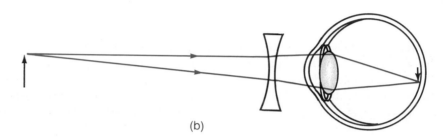

(b)

FIGURE 9.55
(a) A nearsighted eye causes rays from a distant object to converge too quickly. (b) A diverging lens corrects the problem.

converge slightly before they enter the eye, thereby bringing the light rays into focus on the retina.

Aberrations

The preceding discussion of image formation applies strictly to ideal lenses and to rays that enter the lenses at relatively small angles of incidence near the optical

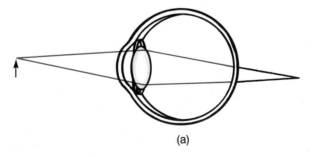

(a)

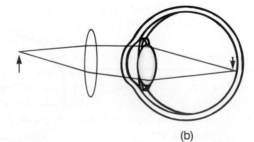

(b)

FIGURE 9.56
(a) A farsighted eye does not cause light rays from a near object to converge enough. (b) A converging lens corrects the problem.

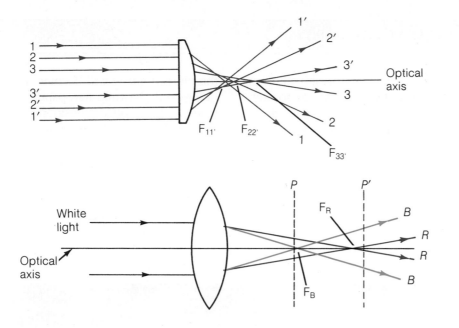

FIGURE 9.57
Spherical aberration for a convex lens illuminated by a beam of parallel light. Rays 1 and 1' are brought to a focus at $F_{11'}$, while rays 2 and 2', and 3 and 3' are focused at $F_{22'}$ and $F_{33'}$ respectively.

FIGURE 9.58
Diagram showing the positions of the focal points in various colors for a simple lens. The focal points for blue, green, and red are successively farther from the lens. A screen placed at point P will show an image whose outside edges are tinged red-orange, while a screen placed at point P' will reveal an image whose outside edges are bluish.

axis. In practice, rays enter real lenses with a variety of incident angles at points often far from the optical axis. If one applies the law of refraction and follows such rays through a lens, one finds that parallel rays do not all meet (for a converging lens) at a single point (Figure 9.57). As a result, images of real objects from which the rays emanate are not sharp and clear but blurred. In such cases we say that the lens system possesses *spherical aberrations,* that is, deviations from ideal image formation produced by the spherical shape of the lens surface. Lens aberrations of this type can be corrected, but this process is complicated and often necessitates the use of several simple lenses in combination.

One type of aberration shared by all simple lenses even when used under ideal conditions is *chromatic aberration.* A lens affected by chromatic aberration, when illuminated with white light, produces a sequence of more or less overlapping images, varying in size and color. If the lens is focused in the yellow-green portion of the EM spectrum where the eye is most sensitive, then all the other colored images are superimposed and out of focus, giving rise to a whitish blur or fuzzy overlay (see Figure 9.58). For a converging lens, the blue images would form closer to the lens than the yellow-green images, while the reddish images would be brought to a focus farther from the lens than the yellow-green ones.

The cause of chromatic aberration has its roots in the phenomenon of *dispersion,* to be discussed in the next section. The cure for this problem, originally thought to be insoluble by none other than Newton himself, was found around 1733 by C. M. Hall and later (in 1758) developed and patented by John Dolland, a London optician. It involves using two different types of glass mounted in close proximity. Figure 9.59 shows a common configuration called a Fraunhofer cemented achromat (meaning "not colored"). The first lens is made of crown glass, the second is made of dense flint glass (see Table 9.1). These materials are chosen because they have nearly the same dispersion. To the extent to which this is true, the excess convergence exhibited by the first lens at bluish wavelengths is compensated for by the excess divergence produced by the second lens at these same wavelengths. Similar effects occur at the other wavelengths in the visible spectrum, permitting cemented doublets of this type to correct more than 90% of the chromatic aberration found in simple lenses.

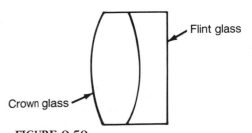

FIGURE 9.59
A Fraunhofer cemented achromatic doublet lens.

A thick magnifying glass and a burning candle can be used to demonstrate chromatic aberration. When illuminated with light from the candle flame, the lens can be made to cast a real image surrounded by a halo on a white piece of paper. If the paper is moved closer to the lens, the outside edge of the blurred image will become tinged orange-red. Moving the paper away from the lens, past the position of best focus, will give a blurred image with a perimeter that is bluish. Try examining the source directly by looking through the lens. The coloration will be far more apparent (Figure 9.58).

9.5   DISPERSION AND COLOR

Most of us at one time or another have seen decorative glass pendulums hanging in windows through which sunlight streamed. If so, you probably noticed patches of bright, rainbow-hued light playing about the room as the pendulums slowly turned in response to air currents. Did you ever wonder how such beauty was produced? Sir Isaac Newton did, and he performed several experiments in an attempt to determine the answer to this question. He concluded that sunlight—white light—was a mixture of all the colors of the rainbow and that upon being refracted through transparent substances like glass, it could be *dispersed,* or separated, into its constituent wavelengths (colors). The process of refraction was seen to be color dependent!

This color dependence of refraction comes about because the speed of light in any medium is slightly different for each color. In glass, diamond, ice, and most other common transparent materials, *shorter wavelengths* of light *travel slightly slower* than longer wavelengths. Violet travels a little slower than blue, blue travels slower than green, and so on. In the case of common glass, the speed of violet is about 1.95×10^8 meters/second, while the speed of red light is about 1.97×10^8 meters/second. The speeds of the other colors lie between these two values. This slight difference in the speeds of different colors causes dispersion. In diamond, the difference in speeds of violet and red light is comparatively larger—about 2%, compared to about 1% for glass—so the dispersion is greater. That is why one sees such brilliant colors in diamonds. The difference in speeds of violet and red light in water is comparatively smaller than it is in glass.

Up until this time we have treated the phenomenon of refraction as though it and we were colorblind, that is, we have ignored the fact that the speed of light in a medium depends on wavelength (or frequency). (Alternatively you can think of our previous study of refraction as having been done using light of only a single color or wavelength, something known as *monochromatic* light.) What effect does this have on how violet light rays are refracted at an air-glass interface relative to how red rays are refracted?

Consider Figure 9.60, showing an incoming ray of light that we will assume is a mixture of only blue and red wavelengths. Because the speeds of both blue light

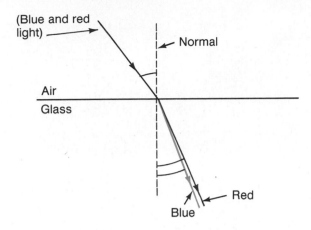

FIGURE 9.60
Dispersion at an air-glass interface produces a separation of red and violet rays. The angle of refraction of the blue ray is smaller than the angle of refraction of the red ray.

and red light are lower in glass than they are in air, we expect that both rays will be bent toward the normal based on our analysis in Section 9.3. But we know that the speed of blue light in glass is lower than the speed of red light in the same medium, so that the blue light will be bent slightly *more* toward the normal than the red light. In Figure 9.60, this results in the angle of refraction for blue light being a bit smaller than the angle of refraction for red light. Thus although both rays are bent toward the normal upon passing into the denser medium, the blue ray is refracted more strongly and emerges from the interface along a different path than the red ray. The colors have been dispersed, or separated, as a result of the refraction process because of the wavelength dependence of the speed of light.

If the incoming beam is now allowed to contain the remaining colors between red and blue, the emergent rays for each will fall between the limits set by the red and blue rays. What is produced is a *spectrum*—the different colors spread over a range of angles. It is the process of dispersion, acting to sort out the different colors, that causes the chromatic aberration in simple lenses described at the end of the previous section.

The difference between the angles of refraction for the red and violet rays above amounts to less than 0.5°. This may not sound like much, but it is some 30 times the minimum angular separation between rays that the human eye can detect under bright conditions and therefore would certainly be noticeable. If additional air-glass surfaces are introduced, more refractions may occur, and the angular spread in the emerging rays may be increased. The light is said to be more highly dispersed in this case.

A *prism* is a common device used to disperse light and form a spectrum (Figure 9.61). They were well known and highly prized by the Chinese (as indicated in missionary reports dating from the early 1600s) for their ability to generate color. Today they are highly valued by scientists for much the same reason. For example, one can analyze the kinds of radiation emitted by a source of light by dispersing its light into a spectrum and measuring the intensity (amount) of radiation coming off in the various wavelengths (colors). If the source radiates like a blackbody (see Section 8.6), this information might be used to determine the temperature of the source. As we shall see in Chapter 10, it is also possible to determine the chemical composition of a source by examining its spectrum. Figure 9.62 shows two common configurations of dispersing prisms used in research and the paths a single-color light ray follows through each.

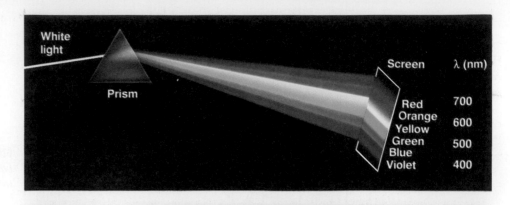

FIGURE 9.61
Dispersion of white light into a spectrum of colors as it passes through a 60° prism. Each color is bent a different amount because the speed is slightly different for each wavelength.

9.6 ATMOSPHERIC OPTICS: RAINBOWS AND HALOS

"My heart leaps up when I behold a rainbow in the sky." This is how the poet Wordsworth described his reaction to the phenomena of rainbows, and it is probably not too bad a description of how many of us feel upon seeing a dazzling, colored arc stretching across the sky. Rainbows are both beautiful and puzzling.

FIGURE 9.62
Two common forms of dispersing prisms used in laboratory work, with their angular deviations shown for monochromatic light. (a) The Pellin-Broca prism. (b) The Abbe prism. Used as shown, these prisms are very convenient because a light source and viewing system can be set up at a fixed angle—90° in (a)—and then the prisms rotated slightly to look at a particular wavelength (color).

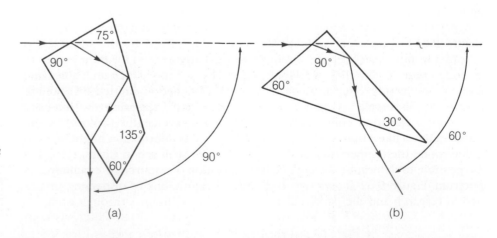

How do the elements of water and sunlight combine to produce such spectacles? Armed with the information of Sections 9.3 and 9.5, we are in a position to find out.

Before doing so, however, we need to point out some rainbow basics. First, rainbows consist of arcs of colored light (spectra) stretching across the sky, with the red part of the spectrum lying on the outside of the bow and the blue-violet part lying on the inside. Second, rainbows are always seen against a background of water droplets with the sun typically at our backs. These two basic characteristics of rainbows are what we seek to understand.

Imagine a beam of light from the sun striking a raindrop. For simplicity we will assume that raindrops are spherical, although real falling raindrops are shaped more like the squashed circular pillows that decorate sofas. (They are definitely *not* shaped like teardrops!) If we apply the law of refraction at each surface, concentrating only on those rays that make a U-turn in the drop and return in the general direction of the sun (so as to be consistent with the second rainbow basic above), we find the result shown in Figure 9.63. The ray striking the drop at its center (ray 1) returns directly back along its incident direction and defines the *axis* of the drop. Rays entering above the axis exit below the axis and vice versa.

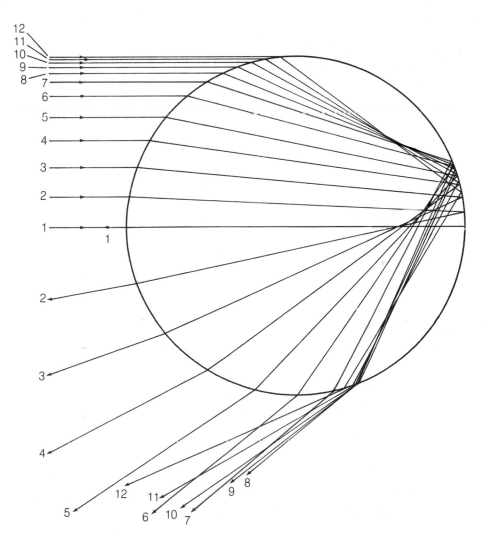

FIGURE 9.63
Paths of light rays through a water drop. Ray 7 is the Descartes ray.

The farther above the axis the ray enters, the greater its emergent angle *up to a point* defined by ray 7. This ray is called the *Descartes ray,* after René Descartes, who first suggested this explanation for rainbows in 1637.

For rays entering above the Descartes ray, the exit angles are *less* than that of the Descartes ray. Thus rays entering the drop on either side of the Descartes ray emerge at about the same angle as the Descartes ray itself, leading to a concentration of rays leaving the droplet at a maximum angle corresponding to that of the Descartes ray. This angle is about 41° for rays 6 through 10 in Figure 9.63.

It is this concentration of reflected and refracted sunlight at exit angles near 41° that produces rainbows. The Descartes model predicts that rainbows should consist of circles of light of angular radii equal to 41°, centered on a point opposite the sun in the sky—the *antisolar point.* Thus to see a rainbow we need to look for these concentrated rays in a direction about 41° from the "straight back" direction with the sun behind us (Figure 9.64). Notice, if the sun is above the horizon, the antisolar point will be below the horizon along the direction of your shadow. In this case the rainbow circle intersects the horizon, and we see only an arc of the circle. For earthbound observers the best rainbow apparitions occur when the sun is on the horizon, for then we see half of the rainbow circle. If the sun is higher in the sky than about 41° above the horizon, then no rainbow can be seen from the ground because the antisolar point lies 41° or more below the horizon, and the rainbow circle never reaches above the horizon. This is why observers throughout most of the continental United States rarely see rainbows at noon. When viewed from an aircraft, a rainbow can form a complete circle.

So far we have been able to understand several aspects of the shape and location of rainbows but not their colors. To do so, we must include the phenomenon of dispersion. Recall from Section 9.5 that blue light is more strongly deviated in passing through transparent media than is red light. This means that the maximum emergent angle from the raindrop for blue light will be smaller than the maximum emergent angle for red light (Figure 9.65). Therefore, the blue light is concentrated at slightly smaller angles than is the red light. Calculations show that blue-violet light is concentrated in a circle of angular radius of about 40°, while red light is concentrated at an angle of about 42°. The other "colors of the rainbow" fall in between. A more detailed model, including dispersion, then predicts that real rainbows should consist of bands of color in the sky a total of

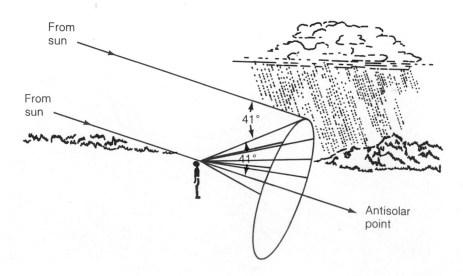

FIGURE 9.64
From Descartes' construction, the rainbow is predicted to be a circle of angular radius 41°, centered on the antisolar point.

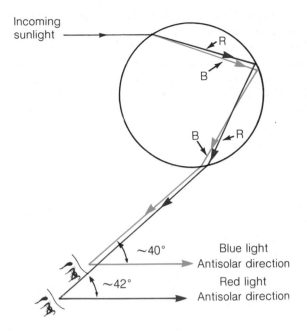

Incoming
sunlight

R

B

B

R

~40°

Blue light
Antisolar direction

~42°

Red light
Antisolar direction

FIGURE 9.65
*Dispersion of sunlight by a
spherical raindrop.*

some 2° or so wide, with blue-violet colors on the inside and red-orange colors on the outside. And this is precisely what is seen.

The application of the laws of reflection and refraction (including dispersion) to falling raindrops gives us an explanation for what is called the "primary" rainbow. The primary results from *one* internal reflection of the rays in the drop at the rear surface. Higher order rainbows may be produced by rays executing two or more internal reflections before leaving the droplet. Some of you no doubt have seen "secondary" rainbows lying outside the primaries along arcs of circles having angular radii of approximately 51° (Figure 9.66). The ordering of the

FIGURE 9.66
*Primary and secondary
rainbows over the Very Large
Array (VLA) near Socorro, New
Mexico.*

FIGURE 9.67
Schematic diagram showing the production of a secondary rainbow from two internal reflections in a raindrop. The path shown is for a typical ray of yellow light. The ray emerges at an angle of about 51° with respect to the antisolar direction. Red rays emerge with slightly smaller angles, while blue rays emerge with somewhat larger angles.

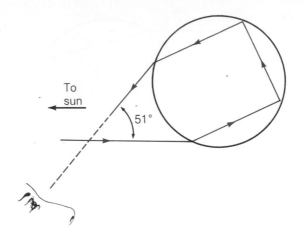

colors of these bows is reversed from that of the primaries. All of these properties are explicable in terms of the laws of geometrical optics with which we have become familiar (Figure 9.67).

Halos, circular arcs of light, often with reddish inner edges, surrounding the sun or full moon might be considered winter's answer to rainbows. When the temperature in the upper atmosphere drops below freezing, ice crystals form. At high elevations in the temperate regions of the earth, one common shape exhibited by such crystals is that of a hexagon, similar to a short, unsharpened pencil (Figure 9.68a). When seen in cross section, such crystals may be considered as pieces of 60° prisms and deviate light in a manner similar to them (see Figure 9.68b).

As in the case of rays entering raindrops, if one traces the paths of rays entering such a crystal at various incident angles, one finds that there is a concentration of exiting rays with deviation angles near 22°. Thus when light from the sun or the moon enters a cloud of such ice crystals having all possible orientations, the emergent rays tend to be clustered into circular arcs having angular radii of 22° centered on the source of illumination.

To see a ray of light forming part of a halo, we should look in a direction 22° away from the sun or moon (see Figure 9.69). When doing so you may notice that

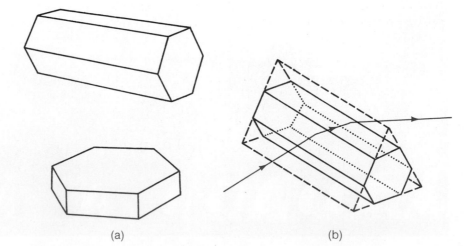

FIGURE 9.68
(a) Two simple ice-crystal forms: top, a columnar or pencil crystal; bottom, a plate crystal. (b) A light ray passing through a pencil crystal is refracted as if it were passing through a 60° prism.

(a)   (b)

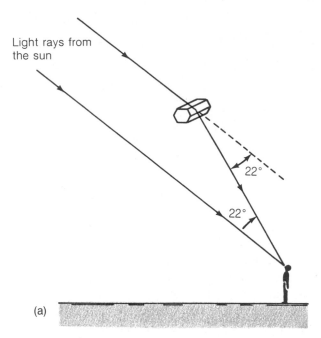

FIGURE 9.69
*(a) To see sunlight that is deviated by an angle of 22°
to form part of a halo, one looks at an angle 22°
away from the sun. (b) Actual halo around the sun.*

the inner edge of the halo circle is tinted red. This is again the result of dispersion. At each refraction the blue component of sunlight (or moonlight, which is merely reflected sunlight) is more strongly refracted than is the red component. Consequently the angle of concentration for the blue light is somewhat greater than it is for red light, and the latter piles up preferentially at the inner edge of the halo, as indicated in Figure 9.70.

What we have described is the well-known 22° halo. There are also 46° halos, which result from light entering one face of the pencil crystal and leaving through one end. These halos are much fainter than the 22° halos and are much harder to see—partly because they occupy such a large portion of the sky, having angular diameters of more than 90°! And these are but two of many, many phenomena

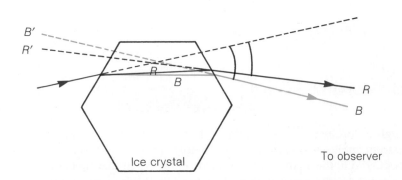

FIGURE 9.70
*Dispersion of sunlight by ice
crystals produces halos with
reddish inner edges. Because the
deviation of blue rays is larger
than that of red rays, the blue
rays appear to originate along
the line B'B, farther away from
the sun than the red rays, which
appear to come from the
direction R'R. Thus the inner
edge of the halo, that is, the part
nearest the sun, is tinged reddish.*

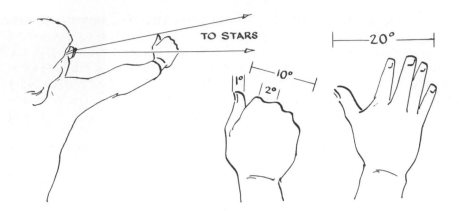

FIGURE 9.71
Using the human hand to measure angles roughly. (a) Extend the hand at arm's length and sight along it. (b) The thumb subtends about 1°, two knuckles about 2°, and the fist about 10°. (c) The hand subtends about 20°.

associated with ice-crystal reflection and refraction. Such magnificence surrounds us daily if only we allow our eyes to be open to it. A knowledge of physics can help us to appreciate these natural wonders more deeply.

DO-IT-YOURSELF PHYSICS

The next time you are washing your car or watering your garden, set the hose nozzle to give a fine mist of water droplets, position yourself with your back to the sun, and turn on the water. Observe your own personal rainbow by looking about 41° from the direction in which your shadow is cast. You can immediately check to see that the ordering of the colors is as it should be, given our model. In addition, you can check the dimensions of the primary bow by using the approximate angular measures shown in Figure 9.71. These angle-

measuring techniques or tricks have long been used by students of astronomy to measure the angular separations of celestial objects in the sky. They can be used to confirm the existence of a true 22° halo and to distinguish it from another commonly observed atmospheric phenomena seen mostly surrounding the moon, *coronae*. The latter have angular extents of only a few degrees, appear predominantly reddish brown in color, and are caused by completely different physical processes having to do with diffraction and not refraction.

HISTORICAL NOTES

Although a considerable body of knowledge about the way light propagates in transparent media had been amassed before 1600, the developments in optics that occurred in the 17th century quickly eclipsed all that had been done during the previous 1,500 years. The first decade of the 17th century saw the invention of the microscope and the refracting telescope, the latter being effectively used by Galileo to discover craters on the moon, the phases of Venus, and four of the moons of Jupiter (Figure 9.72). By 1611, Johannes Kepler (1571–1630) had discovered total internal reflection. In or about 1621, the law of refraction was found by Snell, a professor of mathematics at the University of Leiden, after many years of experimentation. Unfortunately Snell's results were not publicized until years after his death, and by that time Descartes had already succeeded in getting his version of the law of refraction into print.

René Descartes (1596–1650), "father of modern philosophy" and famous for his statement *Cogito, ergo sum* ("I think, therefore I am"), was also an accomplished mathematician and physicist (Figure 9.73). He founded the study of analytical geometry and was the first to develop the use of the hypothetical model as a tool of research. For example, his model for the propagation of light may be compared to that of a tennis ball moving uniformly. The law of reflection can be deduced by imagining an elastic collision of the ball with a stationary, impenetrable surface and applying the principle of conservation of linear momentum. It is interesting that Descartes was one of the first to recognize the momentum of a particle as an important physical quantity. These results seem to have been obtained in about 1626 and appeared in 1637 at the beginning of Descartes' *La Dioptrique.*

The laws of reflection and refraction reappear in another of Descartes' work entitled *Meteores.* Here he presents a mathematical explanation of primary and secondary rainbows, accounting for their angular dimensions and locations but not for their colors. It remained for Isaac Newton to produce the correct explanation for color in rainbows.

At the end of Chapter 2, several of the accomplishments of Sir Isaac Newton in the field of optics were mentioned, including his invention of the reflecting telescope and his explanation of the phenomenon of dispersion. These two items are very closely connected, the former being a direct consequence of the latter. In about 1666 Newton performed an *experimentum crucis* (a "critical experiment") in which, by a clever arrangement of two prisms, he was able to demonstrate that light of a single color undergoes no further dispersion upon being refracted through a prism (Figure 9.74). From this and similar experiments, Newton concluded, "Light itself is a heterogeneous mixture of differently refrangible rays," asserting that there is an exact correspondence between color and the

FIGURE 9.72
Galileo using the refracting telescope to demonstrate the true, imperfect geography of the moon.

FIGURE 9.73
René Descartes at the court of Queen Christine of Sweden.

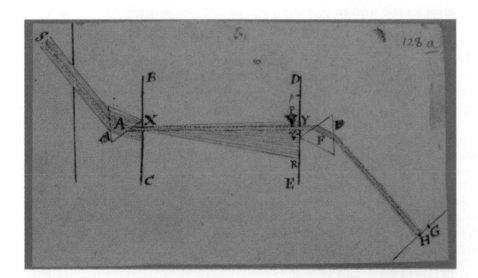

FIGURE 9.74
Newtons' original drawing of his experiment with two prisms. The slit Y allowed only one color in the spectrum from prism A to go through prism F. The light leaving the second prism was just the single color that entered it, not a complete spectrum. This showed that a prism does not somehow "add" color to light.

FIGURE 9.75
Thomas Young. His work with interference proved that light is a wave.

"degree of refrangibility," the least refrangible rays being "disposed to exhibit a red colour."

Newton's prismatic experiments seem to have convinced him of two things: first, that "light"—white light or sunlight—was not a "body" unto itself but an aggregate of corpuscles (tiny particles). "Each colour is caused by uniformly moving globuli," he wrote. "The uniform motion which gives the sensation of one colour is different from the motion which gives the sensation of any other colour." In particular, Newton believed the particles producing the sensation of blue light moved faster in glass than those associated with red light and that it was for this reason that blue rays were more strongly refracted. Newton's corpuscular theory of light permitted him to account for many of the properties of light, notably its straight-line propagation. But Newton recognized that certain phenomena, for example, diffraction and interference, could best be explained by treating light as a wave. Thus even before the beginning of the 18th century we see an apparent duality (particle versus wave) in the nature of light emerging. We will have more to say on this matter in Chapter 10.

The second thing of which Newton became convinced as a result of his color experiments was the impossibility of producing refracting telescopes (ie, ones using only combinations of lenses) that were free from chromatic aberrations. It was for this reason that Newton turned his attention to the development of reflecting telescopes.

Newton's discoveries about light and color were published in 1704 in his *Opticks*. Among the many investigations reported upon is one dealing with rainbows. In it Newton extended Descartes' model, explicitly taking into account dispersion, and calculated the size of the rainbow arcs for each of the different colors in the primary and secondary rainbows. This book ends with a series of "Queries"—questions about the nature of light and related phenomena in need of further explanation and study. It has been argued that these questions formed the most important part of the book insofar as they strongly influenced the course of research on light for the next two centuries.

Newton's particle model of light stood for less than a century before it was dealt a nearly mortal blow by Thomas Young (1773–1829; Figure 9.75). Young was born into an English family of Quakers and was quickly seen to be a genius. By the age of 14 years he knew several languages and had written an autobiography in Latin. Although Young studied medicine and became a physician, his intellectual pursuits spanned an incredible variety of disciplines, from physics to hieroglyphics (he worked on the Rosetta Stone) to insurance.

Young championed the wave model of light and showed that it could account for interference and diffraction of light. In his most famous experiment, Young passed light through two pinholes and produced an interference pattern similar to the one in Figure 9.6. From this pattern he was able to measure the wavelengths of different colors of light. Young was also the first to suggest that light could be polarized and was therefore a transverse wave. Young's success, along with that of other physicists in the early 1800s, caused the wave model of light to replace Newton's particle model—no small feat considering Newton's great stature. But as we shall see in Section 10.2, the particle model of light was reincarnated a century later.

We have said that Newton erred in his belief that achromatic refracting telescopes could never be built. Interestingly enough, the man who popularized the use of such instruments began as a staunch defender of Newton's viewpoint. John Dolland (1706–1761), a weaver by trade and a self-taught mathematician and optician, originally adhered to the Newtonian view on chromatic aberration.

After forming a partnership with his son in 1752 and establishing an optical instrument manufacturing business, however, he soon became convinced otherwise. In June 1758, in a paper read before the Royal Society of London, he described the achromatic doublet, which revolutionized lens making.

For some 50 years after Dolland's announcement, London opticians held first place in the field of achromatic lens fabrication. By about 1810, however, German optical specialists under the leadership of Joseph Fraunhofer (1787–1826) succeeded in wresting this position from their English confreres. Fraunhofer sought to apply optical theory to produce lens combinations having a minimum of aberrations. To do so, he first perfected methods of making large, flaw-free glass blanks that could be shaped into superior lenses and then proceeded to determine with unprecedented precision the optical constants and necessary speeds of light for such components. In this way Fraunhofer was able to design achromatic lenses using the principles of geometrical optics instead of the trial-and-error methods practiced by the London lens makers. It was while using sunlight to try to measure the characteristics of some glass samples that Fraunhofer noticed that the sun's spectrum was crossed by numerous fine dark lines. The significance of such spectral lines in terms of what they can tell us about the composition, temperature, density, and so on of the sun's atmosphere was not appreciated until long after Fraunhofer's death. Nevertheless the dark line spectrum of the sun is often referred to as the "Fraunhofer spectrum" in recognition of Fraunhofer's contributions to optical science.

FIGURE 9.76
Karl Friedrich Gauss.

Lens makers also profited from the work of another scientific giant born in the late 18th century, Karl Friedrich Gauss (1777–1855; Figure 9.76). Gauss was a child prodigy whose accomplishments in mathematics, physics, and astronomy mark him as one of the greatest intellects of all time. At the age of 19 years, Gauss proved that it was possible to construct a geometrical figure having 17 equal sides (a "17-gon") with only a straight edge and a compass, the first advance in matters of this type in more than 2,000 years. Among Gauss's many contributions to physics, we mention only his work in magnetism and, of course, optics. In collaboration with another German physicist, Wilhelm Weber, Gauss established an observatory dedicated to the study and mapping of terrestrial magnetism. He also developed a self-consistent set of absolute units (distance, mass, and time) that could be used to measure nonmechanical quantities like those encountered in his electromagnetic investigations.

In the field of optics, Gauss invented the helioscope, an instrument used in surveying and triangulation work, but only *after* he fully developed the necessary optical theory. In 1841 he published *Dioptrische Untersuchungen,* in which he analyzed the path of light through a system of lenses. In this work, in addition to determining how image distance, object distance, and the focal length of a lens are related to one another, he showed that any system of lenses is equivalent to a properly chosen single lens. This was Gauss's last significant contribution to science, and one of his biographers has called it his greatest work.

The advancement in optics by no means came to an end in the middle of the last century. Beginning shortly after World War II, applied optics began to flourish as the mathematical methods of communication theory and high-speed digital computers were brought into the discipline. Optical technology continues to advance, bringing improvements in our life-styles and in our understanding of the universe. But if rapid progress has recently occurred, it has happened because modern scientists have, in the words of Newton, "stood on the shoulders of giants," including Descartes, Gauss, and Newton himself.

SUMMARY

Light is a band of electromagnetic waves, with wavelengths between 400 and 700 nanometers visible to humans. Reflection of light can be specular, as with a mirror, or diffuse, as when the light reflects off this paper. The law of reflection states that the angle of incidence equals the angle of reflection. Interference occurs when two light waves travel different paths and then combine, as with light passed through two narrow slits. If the two waves are in phase, constructive interference (reinforcement) occurs. If they are out of phase, destructive interference (cancellation) occurs. Since light is a transverse wave, it can be polarized. Polaroid sunglasses and liquid crystal displays take advantage of this fact.

Mirrors come in a variety of useful forms: plane mirrors, the ordinary ones mounted on walls; half-silvered mirrors, used as one-way mirrors if proper lighting exists; concave mirrors, used to magnify images or concentrate light; and convex mirrors, used as "wide-angle" mirrors. Large astronomical telescopes use concave (parabolic) mirrors.

Refraction occurs when light passes from one transparent medium into another. The law of refraction reveals that the angle of refraction is smaller than the angle of incidence if the speed of light in the medium in which the refracted ray moves is smaller than that of the medium in which the ray originates. Conversely, if the speed of light in the substance in which the refracted ray travels is smaller than that of the substance in which the ray originates, then the angle of refraction will be greater than the angle of incidence. In the latter case it is possible to define a critical angle of incidence for which the angle of refraction is 90°. For angles of incidence greater than this critical angle, the incident rays undergo total internal reflection and remain trapped in the incident medium. The phenomenon of total internal reflection is largely responsible for the sparkle of diamonds and the operation of light pipes.

Reflection and refraction from curved boundaries can bring about a convergence or divergence of light rays. Lenses are optical devices that can be used to control the paths of light rays and to form images. The position of an image can be determined if the location of the object and the focal length of the lens are known. The nature of the image can be found by determining the negative ratio of the image to object distances—a quantity that is equal to the magnification.

Dispersion is the process whereby the individual wavelengths (colors) comprising a beam of light are separated at the boundary between two transparent media. Dispersion occurs because the speed of light is different for different wavelengths. Blue light moves more slowly through most material than does red light and is more strongly deviated toward the normal after passing into such material from air than is red light. Dispersion is responsible for chromatic aberration in simple lenses and can be used to determine the temperature, chemistry, and other properties of luminous sources.

The laws of reflection and refraction, including dispersion, may be applied to a multitude of physical systems. One particular application is in the area of atmospheric or meteorological optics, where these laws permit us to understand the strikingly beautiful phenomena of rainbows and halos.

SUMMARY OF IMPORTANT EQUATIONS

EQUATION	COMMENTS
$p = \dfrac{sf}{s - f}$	Lens formula
$M = \dfrac{\text{image height}}{\text{object height}}$	Magnification of a lens system
$M = \dfrac{-p}{s}$	Magnification using image distance and object distance

Questions

1. Why are the Doppler effect and diffraction not as commonly observed with light as they are with sound?

2. Describe specular reflection and diffuse reflection.

3. The law of reflection establishes a definite relationship between the angle of incidence of a light ray striking the boundary between two media and its angle of reflection. Describe this relationship.

4. Describe how light passing through two narrow slits produces an interference pattern.

5. A person looking straight down on a film of oil on water sees the color red. How thick could the film be to cause this? How thick could it be at another place where violet is seen? (Some useful information is given at the beginning of Section 9.1.)

6. If the wavelength of visible light were around 10 cm instead of 500 nm and we could still see it, what effect would this have on diffraction and interference?

7. An interference pattern is formed by sending red light through a pair of narrow slits. If blue light is then used, the spacing of the bright spots (where constructive interference takes place) won't be the same. How will it be different? Why?

8. What is polarized light? How do Polaroid sunglasses exploit polarization?

9. Describe how you could use two large, circular Polaroid filters in front of a circular window as a kind of window shade.

10. Just before the sun sets a driver encounters sunlight reflecting off the side of a building. Will Polaroid sunglasses stop this glare?

11. Compared to a person's height, what is the minimum length (top to bottom) of a mirror that will allow the person to see a complete image from head to toe?

12. What is different about an image formed (of a nearby object) with a convex mirror compared to an image formed with a concave mirror? What are the advantages of each type of mirror?

13. What is the exact shape of concave mirrors used in telescopes? What is one way to cast a mirror with this shape?

14. Describe how the path of a ray is deviated as it passes (at an angle) from one medium into a second medium in which the speed of light is lower. Contrast this with the case when the speed of light in the second medium is higher.

15. How would Figure 9.30 be different if the glass were replaced by water or by diamond?

16. The speed of light in a certain kind of glass is exactly the same as the speed of light in benzene—a liquid. Describe what happens when light passes from benzene into this glass, and vice versa.

17. A piece of glass is immersed in water. If a light ray enters the glass from the water with an angle of incidence greater than zero, in which direction is the ray bent?

18. What is total internal reflection, and how is it related to the critical angle?

19. Explain why images seen through flat, smooth, uniform, plate-glass windows are undistorted.

20. For transparent solids, distinguish between effects of surfaces that curve inward and those that curve outward on the paths of a parallel bundle of light rays incident on each.

21. Distinguish converging lenses from diverging lenses, and give examples of each type.

22. Of the three converging lenses shown in Figure 9.45, which would you expect to have the shortest focal length?

23. What are the three principal rays?

24. Contrast real images with virtual images in as many ways as you can.

25. Indicate whether each of the following is a real image or a virtual image.
   a) Image on the retina in a person's eye.
   b) Image one sees in a rearview mirror.
   c) Image one sees on a movie screen.
   d) Image one sees through eyeglasses or contact lenses.

26. How is magnification of a lens related to the object distance? How is it related to the image distance? What is the significance of the sign of the magnification? What is the significance of its magnitude (size)?

27. Estimate the values of the magnification in Figures 9.46, 9.48, and 9.49.

28. How is the eye able to form focused images of objects that are different distances away?

29. When a person is nearsighted, what happens in the eye when the person is looking at something far away? What is used to correct this problem?

30. What is chromatic aberration? How can it be cured?

31. Describe the phenomenon of dispersion, and explain how it leads to the production of a spectrum.

32. Two light waves that have wavelengths 700 and 400 nm enter a block of glass (from air) with the same angle of incidence. Which has the larger angle of refraction? Why? Would the answer be different if the light waves were going from glass into air?

33. What is a prism? Why are such devices useful to scientists?

34. Would a prism made of diamond be better than one made of glass? Why or why not?

35. Describe Descartes' model of rainbows. Specifically, discuss how the model accounts for the size, shape, location, and color ordering of primary rainbows.

36. Suppose an explosion at a glass factory caused it to "rain" tiny spheres made of glass. Would the resulting rainbow be different than the normal one? If so, how might it be different and why?

37. Compare the primary and secondary rainbows as regards their angular size, color ordering, and number of internal reflections that occur in the rain droplets.

38. How is a 22° halo formed? Describe a measurement technique you might use to distinguish a genuine 22° halo from some other "halolike" phenomenon seen surrounding the sun or the moon.

PROBLEMS

1. A light ray traveling in air strikes the surface of a slab of glass at an angle of incidence of 60°. Part of the light is reflected, and part is refracted. Find the angles the reflected and refracted rays make with respect to the normal to the air-glass interface.

2. A ray of yellow light crosses the boundary between glass and air, going from the glass into air. If the angle of incidence is 20°, what is the angle of refraction?

3. A camera is equipped with a lens with a focal length of 30 cm. When an object 2 m (200 cm) away is being photographed, how far from the film should the lens be placed?

4. When viewed through a magnifying glass, a stamp that is 2-cm wide appears upright and 6 cm wide. What is the magnification?

5. A person looks at a statue that is 2 m tall. The image on the person's retina is inverted and 0.005 m high. What is the magnification?

6. What is the magnification in Problem 3?

7. A small object is placed to the left of a convex lens and on its optical axis. The object is 30 cm from the lens, whose focal length is 10 cm. Determine the location of the image formed by the lens. Describe the image.

8. If the object in Problem 7 is moved toward the lens to a position 8 cm away, what will the image position be? Describe the nature of this new image.

CHALLENGES

1. When white light undergoes interference by passing through narrow slits, dispersion occurs. Why? What is the ordering of the various colors, starting from the middle of the pattern?

2. In Section 9.5 we described how the speed of light varies with wavelength (or frequency) for transparent solids. But the speed of light is also a function of temperature and pressure. This dependence is most marked for gases and is instrumental in producing such things as mirages and atmospheric refraction, the latter phenomenon being the displacement of an astronomical object (like the sun or another star) from its true position because of the passage of its light through the atmosphere. Because the earth's atmosphere is a gaseous mixture and easily compressed, its density is highest near the earth's surface (where the weight of the overlying layers exerts a strong compressional force) and gradually declines with altitude. (Refer back to the discussion of Section 4.4 and Figure 4.27). Thus the speed of light in the atmosphere is lowest near the surface and gradually gets higher, approaching c as one goes farther and farther into space. Using this fact and the law of refraction, sketch the path a light ray from the sun would follow upon entering the earth's atmosphere, and

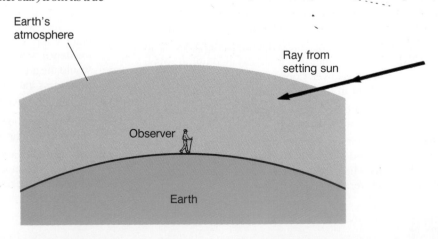

FIGURE 9.77
Challenge 2 (Drawing not to scale.)

predict the apparent position of the sun relative to its true position. What does this tell you about the actual location of the sun's disk relative to your local horizon when you see it apparently setting brilliantly in the west in the evening (Figure 9.77)

3. Would the critical angle for a *glass-water* interface be less than, equal to, or greater than the critical angle for a *glass-air* interface? Why?

4. The form of the lens formula used most commonly in physics is:

$$\frac{1}{f} = \frac{1}{s} + \frac{1}{p}$$

  a) Use this to derive the equation that gives *s* in terms of *p*.
  b) In a camera equipped with a 50-mm focal length lens, the maximum distance that the lens can be from the film is 60 mm. What is the closest an object can be to the camera if its photograph is to be in focus? What is the magnification?
  c) An extension tube is added between the lens and the camera body so that the lens can be positioned 100 mm from film. How close can the object be? What is the magnification?

5. A 35-mm camera is used to photograph a kitten. The camera is equipped with a standard lens that has a focal length of 50 mm. The camera is focused by moving the lens closer to or farther away from the film at the back of the camera.
  a) First the kitten is photographed when it is 350 mm (13.8 in) in front of the camera lens. Use the lens formula to compute the distance that the lens must be from the film for the image to be in focus.
  b) Compute the magnification and the height of the image, assuming that the kitten is 100 mm tall.
  c) The kitten is then photographed from a distance of 3,500 mm (11.5 ft). Compute the lens-to-film distance, the magnification, and the image height.

6. A camera is equipped with a telephoto lens that has a focal length of 200 mm. Repeat part (c) of Challenge 5 for this situation.

7. Why can't you go to the end of a rainbow? What happens to the rainbow when you walk toward one of its ends?

SUGGESTED READINGS

Akasofu, Syun-Ichi. "The Dynamic Aurora." *Scientific American* 260, no. 5(May 1989): 90–97. A detailed look at how the northern lights and the southern lights are produced.

"Atmospheric Phenomena." Introduction by Lynch, David K., Readings from *Scientific American,* Freeman, 1984. Included are articles on rainbows and halos, along with several on light phenomena not covered in this chapter.

Cajori, Florian. *A History of Physics.* New York: Dover, 1962 (originally published by Macmillan in 1929). Contains sections on light in several chapters.

Crane, Richard H. "Liquid Crystal Displays: Watches, Calculators and (Soon) Cars." *The Physics Teacher* 21, no. 7(October 1983): 467–469.

Greenler, Robert. *Rainbows, Halos, and Glories.* New York: Cambridge University Press, 1980. "This book describes beautiful things that can be seen in the sky, things that can be seen without special equipment or special location, things that can be seen by anyone who sees."

Katzir, Abraham. "Optical Fibers in Medicine." *Scientific American* 260, no.5(May 1989): 120–125. Describes how optical fibers are used for imaging, diagnosis, and therapy in medicine.

Koretz, Jane F., and George Handelman. "How the Human Eye Focuses." *Scientific American* 259, no.1(July 1988): 92–99. Well illustrated article about how the eye changes the shape of the lens to focus.

Rudd, M. Eugene. "The Rainbow and the Achromatic Telescope: Two Case Studies." *The Physics Teacher* 26, no.2(February 1988): 82–89. Describes the attempts to explain the cause of the rainbow and the efforts to overcome chromatic aberration in telescope lenses.

Segrè, Emilio. *From Falling Bodies To Radio Waves.* New York: W. H. Freeman and Co., 1984. Chapter 3 is dedicated to light.

Walker, Jearl. *The Flying Circus of Physics with Answers.* New York: Wiley, 1977. "A collection of problems and questions about physics in the real, everyday world." Chapter 5 deals primarily with light.

OUTLINE

High-speed IBM *laser printer, capable of 215 pages a minute in preparing bank statements, premium notices, and other high-volume documents.*

ATOMIC PHYSICS

PROLOGUE: PLOWSHARES, NOT SWORDS

"It is still a matter of wonder how the Martians are able to slay men so swiftly and so silently. Many think that in some way they are able to generate an intense heat in a chamber of practically absolute non-conductivity. This intense heat they project in a parallel beam against any object they choose by means of a polished parabolic mirror of unknown composition—much as the parabolic mirror of a lighthouse projects a beam of light. But no one has absolutely proved these details. However it is done, it is certain that a beam of heat is the essence of the matter. Heat, and invisible, instead of visible, light."

H. G. Wells

These words appeared in 1898 in the classic science fiction novel *War of the Worlds.* This description of the invading Martians' "heat ray" seems to presage the development of modern, high-intensity, infrared lasers. (Wells also coined the term "atomic bomb" some 35 years before the real thing was built.) Not actually invented until the late 1950s, the laser has been a favorite weapon of characters in science fiction novels, films, and television shows. Billions of dollars have been spent trying to develop real laser weapons. In fact, that is a principal goal of a recent project, the Strategic Defense Initiative, begun in the 1980s. But in a century in which countless technological advances have been used as weapons of war, the laser stands as a surprising disappointment. Currently lasers are used to help aim weapons and to do other auxiliary tasks but not directly to kill or to destroy.

The laser seems to appeal to people's imaginations, perhaps because of the sound of its name or because of the mental image of a "sword of light." Adding to its allure is the fact that we can trace its roots back to a paper published by the great physicist Albert Einstein in 1917. In this work Einstein included a reference to the physical process—later named stimulated emission—that is the basis for lasers. Perhaps better than any other device, the laser symbolizes the successful quest to understand the structure of the atom and the nature of light.

The laser has become an immensely useful tool. You probably use one every day, at least indirectly. Each time you buy something at a supermarket or a department store, chances are a laser scanner reads the price. Much of what you read (including these words) is first printed by a laser printer connected to a computer or, in the case of some newspapers, by a printing plate produced with the help of a laser. And when you listen to music on the radio, quite likely it comes from a compact disc that is "read" by a tiny laser. One has to believe that if Einstein were alive today, the famous pacifist and dedicated amateur violinist would be most pleased to see a product of modern physics involved in the reproduction of music. The laser, a device that seemed destined to be a tool of violence, has become a useful tool for commerce, publishing, medicine, entertainment, and much more.

The laser brings together the two main topics of this chapter: modern atomic physics and the quantum nature of light. The first part of the chapter describes how efforts to understand three physical processes led to the concept of photons and the Bohr model of the atom. This is followed by a description of the wave nature of atomic particles and the emergence of the revolutionary new physics known as quantum mechanics. The remainder of the chapter shows how the quantum mechanical model of the atom successfully explains the production of atomic spectra and x-rays and the operation of lasers.

10.1   THE QUANTUM HYPOTHESIS

By the end of the 19th century, physicists were quite satisfied with the great progress that had been made in the study of physics. Many questions that had been puzzling scientists for thousands of years had been answered. Advances such as Newton's great treatise on mechanics and the clarification of electromagnetism by Maxwell gave physicists cause to celebrate the deep understanding of the physical world that they had acquired. (In fact, some physicists worried that the field might be dying; they feared that *all* of their questions might soon be answered.) But some problems defied solution, and advancements in experimental equipment and techniques led to the uncovering of new ones. The implications arising from these discoveries were so revolutionary that shortly after the turn of the century many of these same physicists were bewildered and wondered just how much they really did know.

In the first three sections of the chapter we will discuss three of these problems and will describe how they led to a reinterpretation of the very nature of light and the way it is emitted. They are (1) *blackbody radiation*, (2) the *photoelectric effect*, and (3) *atomic spectra*. Physicists knew a great deal about these phenomena, but the fundamental understanding of their *causes* had eluded them. It was much like the period before Newton: astronomers knew a lot about the shapes of the orbits of the moon and the planets, but they didn't know what caused these particular shapes. Newton's mechanics and his law of universal gravitation provided the answer. In the same fashion, scientists some two centuries later were seeking the theoretical basis for these three phenomena.

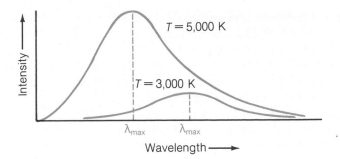

FIGURE 10.1
*Graph of intensity versus
wavelength for the
electromagnetic (EM) waves
emitted by blackbodies at two
different temperatures.*

Blackbody Radiation

Everything around you, as well as your own body, is constantly emitting electromagnetic (EM) radiation. At normal room temperature it is mostly infrared (IR) radiation that is emitted. Very hot bodies that are actually glowing, such as the sun and the heating elements in a toaster, emit visible light and ultraviolet (UV) radiation as well as copious amounts of infrared. Bodies that are very cold, on the other hand, emit very weak, long wavelength infrared and microwaves. This radiation emitted by bodies was studied carefully by scientists in the 19th century, and its properties were well known but not understood. It was determined that a perfectly "black" body, one that would absorb all light and other EM radiation incident upon it, would also be a perfect emitter of EM radiation. Such a body is called a *blackbody*, and the EM radiation emitted by it is called *blackbody radiation (BBR)*.

The characteristics of the BBR emitted by a given blackbody at a particular temperature can be illustrated with a graph. Imagine examining each wavelength of radiation in turn, from very short wavelength ultraviolet to much longer wavelength microwaves, and measuring the energy that is emitted each second (the power) from the blackbody. The graph of the resulting data, plotted as the intensity of the radiation versus wavelength, is called a *blackbody radiation curve*. Figure 10.1 shows two BBR curves for blackbodies at two different temperatures. These graphs illustrate two important ways that BBR changes when the temperature of the body is increased:

*This is an overview of the
characteristics of blackbody
radiation. For more details,
including applications, refer
back to Section 8.6.*

1. More *energy* is emitted per second *at each wavelength* of EM radiation. (The graph is raised.)

2. The *wavelength* at which the *most energy* is emitted per second (in other words, the *peak* of the BBR curve) *shifts to smaller values.* (That is why toaster elements glow red hot while extremely hot stars glow blue hot.)

Why does a blackbody emit different amounts of EM radiation at different wavelengths in precisely this way? Using the principles of electromagnetism, we can easily see why EM waves are emitted. Atoms and molecules are continually oscillating, and they contain charged particles—electrons and protons. We saw in Chapter 8 that this kind of system will produce EM waves. But the exact mechanism involved, one that would account for the three features above, was a mystery.

The clue to solving the puzzle was discovered by the German physicist Max Planck (Figure 10.2) in the year 1900. First, by trial and error, Planck determined

FIGURE 10.2
*Max Planck (1858–1947),
whose work started what can be
called the quantum revolution.*

a mathematical equation that fit the shape of the BBR curves. (This is a common first step in theoretical physics. The mathematical shapes of the planetary orbits— ellipses —were known about a century before Newton explained *why* they were ellipses.) This in itself did little to increase the understanding of the fundamental process. But Planck also developed a model that would account for his equation.

Planck proposed that an oscillating atom in a blackbody can have only certain fixed values of energy. It can have zero energy or a particular energy *E*, or 2, 3, 4, 5, and so on times the energy *E*. In other words, the energy of each atomic oscillator is *quantized*. The energy *E* is called the fundamental *quantum* of energy for the oscillator. The allowed values of energy for the atom are just integral multiples of this:

$$\text{Allowed energy} = 0, \text{ or } E, \text{ or } 2E, \text{ or } 3E, \text{ etc.}$$

An ordinary oscillator, such as a mass hanging from a spring or a child on a swing, can have a continuous range of energies, not just certain values.

We can illustrate the difference between quantized and continuous energy values by comparing a stairway and a ramp. A cat lying on a stairway has quantized potential energy: it can only be on one of the steps, and each step corresponds to a particular *PE*. A cat lying on a ramp does not have quantized potential energy. It can be anywhere on the ramp; its height above the ground can have any value within a certain range, and therefore its *PE* is one of a continuous range of values (Figure 10.3).

The concept of energy quantization for the oscillating atoms was revolutionary. There seemed to be no logical reason for it. But it worked. Planck's quantized atomic oscillators could emit light only in bursts as they went from a higher energy level to a lower one. He showed that light emitted in this fashion by a blackbody resulted in the correct blackbody radiation curves. Moreover, he determined that the basic quantum of energy was proportional to the oscillator's frequency. In particular:

$$E = hf \qquad h = 6.63 \times 10^{-34} \, \text{J} - \text{s}$$

The constant *h* is called *Planck's constant*. The allowed energies are then:

$$\text{Allowed energy} = 0, \text{ or } hf, \text{ or } 2hf, \text{ or } 3hf, \text{ etc.}$$

Planck himself was unsure of the implication of his model. He regarded it mainly as a helpful gimmick that gave him the correct result. But it turned out to be the first of many scientific discoveries about how things are quantized at the atomic level.

10.2   THE PHOTOELECTRIC EFFECT AND PHOTONS

The photoelectric effect was accidentally discovered by Heinrich Hertz during his experiments with electromagnetic waves. Hertz was producing EM wave pulses by generating a spark between two conductors. The EM wave would travel out in all directions and induce a spark between two metal knobs used to detect the wave. Hertz noticed that when these knobs were illuminated with ultraviolet light, the sparks were much stronger. The UV was somehow increasing the current in the spark. This is one example of the photoelectric effect.

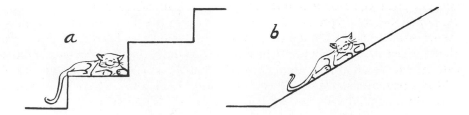

FIGURE 10.3
(a) A "quantized" cat. Its potential energy is restricted to certain values, one for each step. (b) On the ramp the cat can be anywhere, so its potential energy is not quantized.

The phenomenon was quickly investigated, and a great deal was learned. The photoelectric effect is exhibited by metals when exposed to x-rays, UV light, or, for some metals including sodium and potassium, high-frequency visible light. Somehow the EM waves give energy to electrons in the metal, and the electrons are ejected from the surface (Figure 10.4). (It was these freed electrons that enhanced the spark in Hertz's apparatus.) In light of the nature of EM waves, we shouldn't be too surprised that this sort of thing can happen. But again, some of the characteristics of the photoelectric effect indicated that the fundamental process was not understood. The biggest puzzle was the relationship between the speed or energy of the electrons and the incident light. It might seem reasonable to suppose that brighter light would cause the electrons to gain more energy and be ejected from the surface with higher speed. It was found that brighter light caused *more* electrons to be ejected each second, but it did not increase their energies. Even more of a surprise was the finding that only the frequency (color) of the light affected the electron energies. Higher frequency light ejected electrons with higher energy. Even extremely dim light with high enough frequency would immediately cause electrons to be ejected.

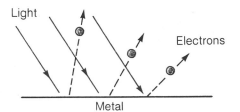

FIGURE 10.4
The photoelectric effect. Light striking a metal surface causes electrons to be ejected.

The explanation of the photoelectric effect was supplied in 1905 by Albert Einstein (Figure 10.5). Planck had suggested that light is emitted in discrete "bursts" or "bundles." Einstein took this idea one step further and proposed that the light itself remains in bundles or "packets" and is absorbed in this form. The electrons in metal can only absorb light energy by absorbing one of these discrete quanta of radiation. The amount of energy in each quantum of light depends on the frequency. In particular:

$$E = hf \quad \text{(energy of a quantum of EM radiation)}$$

This is the same equation for the energy of Planck's quantized atomic oscillators. Einstein suggested that light and other electromagnetic waves are quantized, just like the energy of oscillating atoms.

This idea, that the energy of an EM wave is quantized, allows us to picture the wave as being made up of individual particles now called *photons*. Each photon carries a quantum of energy, $E = hf$, and propagates at the speed of light. The total energy in an EM wave is just the sum of the energies of all of the photons in the wave. Notice how different this picture is from the wave picture of light developed in Chapter 8. Both pictures are correct and present mutually complementary aspects of EM waves. Under certain circumstances, such as those involving refraction or interference of light, the wave nature of light is manifested. In other circumstances, including those involving the emission or absorption of light, the particle aspects of light are demonstrated. These two different sides of light are rather like the front and back sides of a person (Figure 10.6). In certain instances we see only the front side of an individual. We recognize the individual by facial structure, eye color, hair color, and so on. In other circumstances we see only the back side of the person. Here again we may recognize the person, but

FIGURE 10.5
Albert Einstein (1879–1955) enjoying his favorite pastime—sailing.

(a)   (b)

FIGURE 10.6
(a) Recognition of an individual as she approaches is often made using such qualities as facial structure, eye color, hair color and style, and so on. (b) Recognition of the same person as she moves away from us can often be made using different qualities, such as body structure, gait, and so on.

now by virtue of body structure and gait. In each case we see different aspects of the same person, but we are still able to recognize the person, although for the most part the clues leading to recognition are not the same in the two instances. The situation with light is analogous. Different experiments reveal different aspects of what we recognize to be the same type of EM radiation.

With Einstein's proposal, the observed aspects of the photoelectric effect fell into place. Higher frequency light ejects the electrons with more energy because each photon has more energy to give. Brighter light simply means that more photons strike the metal each second. This results in more electrons being ejected each second, but it does not increase the energy of each electron. Einstein's explanation of the photoelectric effect earned him the 1921 Nobel Prize in physics. The new understanding of the nature of light and the way it interacts with matter profoundly altered the course of 20th century physics.

Just how much energy does a typical photon have? Not very much. The energy of a photon of visible light is only about 3×10^{-19} joules. On this scale it is convenient to use a much smaller unit of energy, called the *electron volt* (*eV*). One electron volt is the potential energy of each electron in a 1-V battery. Since voltage is energy per charge, the charge on an electron multiplied by voltage equals energy. So:

$$\text{one electron volt} = 1.6 \times 10^{-19}\,C \times 1 \text{ volt}$$

$$= 1.6 \times 10^{-19}\,C \times 1 \text{ joule/coulomb}$$

$$1\,eV = 1.6 \times 10^{-19}\,J$$

The energy of visible photons is on the order of 2 eV, a much easier number to deal with. In terms of the electron volt, the value of Planck's constant is:

$$h = 6.63 \times 10^{-34}\,J-s = 4.136 \times 10^{-15}\,eV/Hz$$

Figure 10.7 shows the photon energies corresponding to different types of EM waves.

Compare the energies associated with a quantum of each of the following EM waves.

EXAMPLE 10.1

$$\text{red light:} \quad f = 4.3 \times 10^{14}\,\text{Hz}$$

$$\text{blue light:} \quad f = 6.3 \times 10^{14}\,\text{Hz}$$

$$\text{x-ray:} \quad f = 5 \times 10^{18}\,\text{Hz}$$

For each one we use Planck's original equation for a quantum of energy.

$$E = hf$$

$$= 4.136 \times 10^{-15}\,\text{eV/Hz} \times 4.3 \times 10^{14}\,\text{Hz}$$

$$E = 1.78\,\text{eV} \quad \text{(red light)}$$

Using the same equation for blue light and x-rays, we get:

$$E = 2.61\,\text{eV} \quad \text{(blue light)}$$

$$E = 20,700\,\text{eV} \quad \text{(x-ray)}$$

Notice how much greater the energy of a quantum of x-radiation is compared to that of either red or blue light. Little wonder then that high doses of x-rays can be harmful to the human body.

Before we go on to the third puzzle that faced scientists at the turn of the century, let's look at some offshoots of the photoelectric effect. This phenomenon is the key to "interfacing" light with electricity. Just as the principles of electromagnetism make it possible to convert motion into electrical energy and vice versa, the ability of electrons to absorb the energy in photons makes it possible to detect, measure, and extract energy from light. Figure 10.8 shows a schematic of a device that can detect light. When no light strikes the metal, no current flows

FIGURE 10.7
Photon energies in the electromagnetic wave spectrum. Visible light photons range from approximately 1.7 to 3.2 eV (red light to blue light).

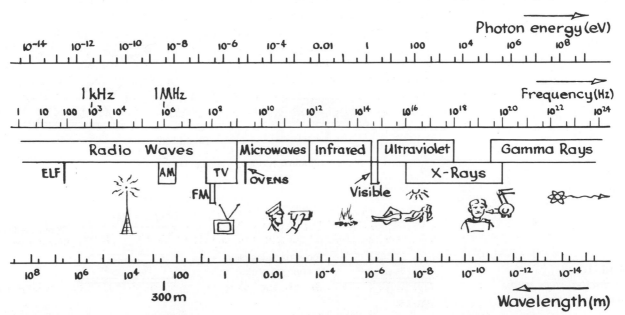

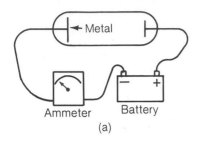

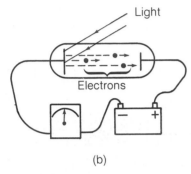

FIGURE 10.8
Schematic of a light detector. (a) No current flows in the circuit because there is nothing to carry charge through the tube. (b) Light releases electrons from the metal, allowing a current to flow. The size of the current indicates the brightness of the light.

in the circuit because there is nothing to carry the charge through the tube. When light strikes the metal, electrons are ejected and attracted to the positive terminal of the tube. The result is a current flowing in the circuit. This sort of detector could be used to automatically count people entering a building. The detector is placed on one side of the door and a light source is placed on the other side, pointed toward the detector. When something blocks the light going to the detector, the current stops. A counter connected to an ammeter in the circuit then automatically tallies one count. A similar setup could be used to automatically open and close a door.

Photocopying (Xerox) machines and laser printers use this interplay between electricity and light in a process known as *electrophotography*. The key part of the operation is a special *photoconductive* surface. It is normally an insulating material but it becomes a conductor when exposed to light. Electrons bound to atoms are freed when they absorb photons in the incident light.

The process of forming an image begins when the photoconductive surface is charged electrostatically. The surface retains the charge until light strikes it, and the freed electrons allow the charge to flow off of the surface. A mirror image of the material to be printed is formed on the photoconductive surface using light. In the case of a photocopying machine, a bright light shines on the original, and the reflected light strikes the charged surface (Figure 10.9). White areas on the original reflect most of this light onto the corresponding areas of the photoconductive surface, which are consequently discharged by the large number of incident photons. Dark parts of the original—printed letters, for example—reflect very little light; consequently the corresponding regions on the photoconductive surface retain their electrical charge. Then fine particles of *toner*, somewhat like a solid form of ink, are brought near the surface. The toner particles are attracted to the charged regions and collect on them, while the discharged areas remain clear. A blank piece of paper, also electrically charged, is placed in contact with it. The toner particles are attracted to the paper and collect on it. The final image is "fused" on the paper by melting the toner particles into the paper.

In a laser printer, a laser under computer control illuminates and discharges those parts of the photoconductive surface that will not be dark in the image to

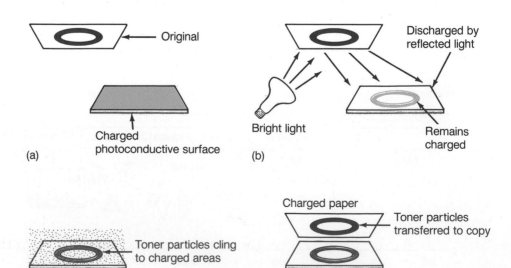

FIGURE 10.9
The principal steps in the photocopying process.

be printed. The rest of the process parallels the steps in photocopiers.

A variety of other specialized "photosensitive" materials, many of them semiconductors, have been devised for diverse uses. Light meters in cameras, the light-sensing elements in video cameras that convert optical images into electrical signals, scanning elements in fax machines, sensitive photodetectors used by astronomers to measure the faint light from distant stars and galaxies, and solar cells used to convert energy in sunlight into electricity (Figure 10.10) are just some of the devices that rely on extracting the energy in photons.

10.3 ATOMIC SPECTRA

The third problem that defied explanation at the beginning of the 20th century was the spectra produced by the various chemical elements. Suppose we use a prism to examine the light produced by the heated filament of an ordinary incandescent bulb. The light from the bulb will be dispersed upon passing through the prism, as described in Chapter 9, and a *spectrum* will be produced (Figure 9.61). The spectrum will appear as a continuous band of the colors of the rainbow, one color smoothly blending into the next. Such a spectrum is called a *continuous spectrum* and is characteristic of the radiation emitted by a hot, luminous solid.

Imagine now a sample of hydrogen gas confined in a narrow glass tube and induced to emit light, as by heating (Figure 10.11). If we examine the light from the luminous gas with a prism, we do not see a continuous spectrum. Instead we see an *emission line spectrum* consisting of a few, isolated, discrete lines of color. In this case the source is not emitting radiation at all wavelengths (colors) but only at certain selected wavelengths. Moreover, if a different type of gas (like helium) is investigated, one finds that the resultant line spectrum is different. *Each type of gas has its own unique set of spectral lines* (Figure 10.12).

The fact that luminous, vaporized samples of material produce line spectra when their light is dispersed (as by a prism) was discovered in the 1850s. Once chemists recognized that each element possessed its own special spectrum, it became possible for them, on a routine basis, to determine the compositions of substances in the laboratory by examining their spectra. Robert Bunsen (of Bunsen burner fame) and Gustav Kirchhoff, two German scientists, were pioneers in

FIGURE 10.10
Author Don Bord (left) and students on top of Emory Peak in Big Bend National Park, Texas. The solar cells absorb photons in sunlight to generate electrical energy that powers a radio transmitter for park personnel.

FIGURE 10.11
The spectrum of a hot gas consists of several discrete colors (three in this example).

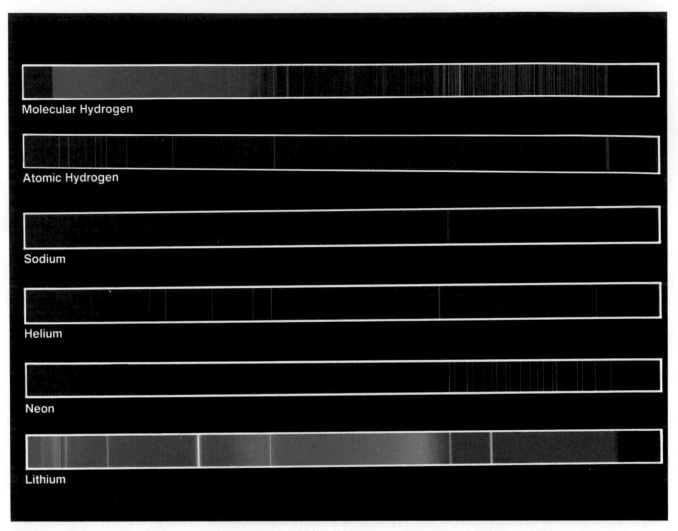

FIGURE 10.12
Emission spectra of selected elements.

this new field of *spectroscopy,* the study of spectra, during the last half of the 19th century. (See Physics Potpourri "Cosmic Chemistry" on page 445.) The problem that remained, of course, was to understand why luminous gases produce line spectra and not continuous spectra and how it is that each element has its own unique spectral "fingerprint" by which it can be identified.

Spectroscopy has grown to be one of the most useful tools for chemical analysis in fields as diverse as law enforcement and astronomy. Suspected poisons or substances found at a crime scene can be identified by comparing their line spectra to those in a catalog of known elements and compounds. Astronomers can determine what chemicals exist in the atmospheres of stars by examining spectra of the starlight with the aid of telescopes.

The explanation of atomic spectra, as it gradually evolved, inspired one of the most crucial periods of advancement in physics ever. In the remainder of this chapter we describe how a picture of the structure of the atom was created and refined to account for atomic spectra. This effort spawned a "new" physics for dealing with matter on the scale of atoms—quantum mechanics.

PHYSICS POTPOURRI

Cosmic Chemistry: "... To Dream of Such a Thing."

In 1844 the French philosopher Auguste Comte published the following view of the prospects of the science of astronomy:

> *The stars are only accessible to us by a distant visual exploration. This inevitable restriction therefore not only prevents us from speculating about life on all these great bodies, but also forbids the superior inorganic speculations relative to their chemical or even their physical natures.*

Like many other sweeping statements, Comte's was soon proved false. Indeed, within 20 years two German scientists would pioneer techniques that would allow the composition of substances to be discovered from an examination of the light they emit (see Figures 10.13 and 10.14).

Beginning in 1859, Robert Bunsen (1811–1899) and Gustav Kirchhoff (1824–1887) undertook a se-

ries of experiments at the University of Heidelberg that laid the foundations for what is now known as the field of spectroscopy. A key ingredient to their success was the burner developed by Bunsen a few years earlier. The Bunsen burner, which produces a flame with a very high temperature but low luminosity (brightness), is ideal for examining the distinctive colors imparted by chemical salts to flames containing them. Bunsen had originally used filters to distinguish the various hues emitted by different substances, but Kirchoff suggested that a much surer distinction between different chemicals might be obtained by examining the spectra of the colored flames. Together they designed and constructed the first laboratory spectroscope using a 60° hollow prism filled with carbon disulfide and carefully examined the light emitted by various hot gases. What they found was that each element produced its own special set of bright colored lines, the line patterns being as unique as a person's fingerprints. The importance of this observation was

FIGURE 10.13
Chemist Robert Bunsen, inventor of the Bunsen burner and codeveloper of the science of spectroscopy.

FIGURE 10.14
Physicist Gustav Kirchoff, who worked with Bunsen, left his mark in many areas of physics.

not lost on Bunsen and Kirchoff. "We can base a method of qualitative analysis on these lines," Bunsen wrote, "that greatly broadens the field of chemical research and leads to the solution of problems previously beyond our grasp."

One problem thought to be beyond the grasp of scientists in the mid-19th century (as evidenced by Comte's remark) was the determination of the chemical compositions of the stars. Bunsen and Kirchoff's efforts to do just that are thought to have originated from a fire in the city of Mannheim, some 10 miles west of Heidelberg. Bunsen and Kirchoff apparently observed the fire from their laboratory window in the evening and turned their spectroscope toward the flames. There they were able to detect barium and strontium. Some time later, while reflecting on this event, Bunsen remarked that if he and Kirchoff could analyze a fire in Mannheim, might they not do the same for the sun? "But," he added, "people would think we were mad to dream of such a thing."

But dreams sometimes have a way of coming true, and such was the case here. The realization of the dream began with Kirchoff's demonstration that a conspicuous dark line (labeled with the letter D by Fraunhofer in 1814) in the spectrum of the sun was unambiguously due to the element sodium. Further experiments convinced Kirchoff that a substance capable of emitting a certain spectral line has a strong absorptive power for the same line. He concluded that the dark D line in the solar spectrum was produced by absorption by sodium in the atmosphere of the sun. The sun, like the earth, contained sodium! By 1862 Kirchhoff had detected calcium, magnesium, iron, chromium, nickel, barium, copper, and zinc in the sun, in addition to sodium. The analysis of the composition of celestial bodies had begun.

The excitement felt by Bunsen and Kirchhoff during this period may be best conveyed by quoting from a letter written by Bunsen to an English colleague on 15 November 1859:

At present, Kirchhoff and I are engaged in an investigation that doesn't let us sleep. Kirchhoff has made a wonderful, entirely unexpected discovery in finding the cause of the dark lines

10.4   THE BOHR MODEL OF THE ATOM

At the beginning of this century little was known about the atom. (In fact, many doubted that atoms existed at all.) This made the origin of atomic spectra a total mystery, in spite of the fact that spectroscopy was a booming field. In 1911 an important experiment performed by Ernest Rutherford in England revealed that the positive charge in an atom is concentrated in a tiny core—the nucleus (more on this in Historical Notes). This result was quickly followed in 1913 by a model of the atom put forth by the great Danish physicist Niels Bohr (Figure 10.15). Bohr's model of the atom, later modified, as we will see in the next section, successfully explained the nature of atomic spectra. But like Planck's explanation of blackbody radiation, Bohr's model of the atom was based on assumptions that did not seem sensible. The basic features of Bohr's model are the following:

1. The atom forms a miniature "solar system," with the nucleus at the center and the electrons moving about the nucleus in well-defined orbits. The nucleus plays the role of the sun and the electrons are like planets in orbit.

*in the solar spectrum, and he can in-
crease them artificially in the sun's
spectrum or produce them in a continuous spectrum and in exactly the
same position as the corresponding
Fraunhofer lines. Thus, a means has
been found to determine the composition of the sun and the fixed stars with
the same accuracy as we determine
strontium chloride, etc. with our chemical reagents.*

No sooner had Kirchhoff and Bunsen unlocked
the door to the chemical secrets of the universe
beyond the earth than they announced (in 1860) a
new terrestrial metal, cesium (from the Latin *cae-
sius,* "sky blue"), so named because of its brilliant
blue spectral lines. The following year they published the discovery of yet another element, named
rubidium (from the Latin *rubidus,* "dark red") for
its strong, red emission lines. In the years that followed, several other elements were identified spec-
troscopically; thallium (1861), indium (1863), gallium (1875), scandium (1879), and germanium

(1886). In addition, studies of the spectra of the sun
and stars led to the discovery of helium and to the
recognition that hydrogen is the most abundant element in the universe.

It is important to bear in mind that although
Bunsen and Kirchhoff did not understand *how* the
spectra of the different elements were produced,
they were highly successful at exploiting the
uniqueness of such spectra for the purpose of identifying elements in various sources. The ultimate
explanation for atomic spectra did not come until
the work of Max Planck and Niels Bohr at the beginning of the 20th century. Nonetheless it is hardly
an understatement to say that the birth of what
might be called "cosmic chemistry" occurred in
Heidelberg in 1859 as a result of the blending of the
talents of a skilled chemist and a mathematical
physicist who dared to dream.

2. The electron orbits are quantized, that is, electrons can only be in certain
   orbits about a given atomic nucleus. Each allowed orbit has a particular
   energy associated with it, such that the larger the orbit the greater the
   energy. Electrons do not radiate energy (emit light) while in one of these
   stable orbits.

3. Electrons may "jump" from one allowed orbit to another. In going from a
   low-energy orbit to one of higher energy, the electron must *gain* an
   amount of energy equal to the difference in energy it has in the two orbits.
   When passing from a high-energy orbit to a lower energy one, the electron must *lose* the corresponding amount of energy.

Figure 10.16 shows the Bohr model for the simplest atom, that of the element
hydrogen (atomic number = 1). A lone electron orbits the nucleus, in this case
a single proton. The electron can be in any one of a large number of orbits (four
are shown). In each orbit the electrical force of attraction between the oppositely
charged electron and proton supplies the centripetal force needed to keep the
electron in orbit.

When in orbit 1, the electron has the lowest possible energy. The electron has
more energy in each successively larger orbit (Figure 10.17). The electron's

FIGURE 10.15
*Niels Bohr (1885–1962), one of
the giants in the field of atomic
physics.*

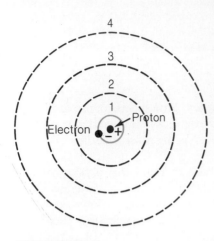

FIGURE 10.16
The Bohr model of the hydrogen atom. The electron can have only certain orbits. Four of these are shown. (The figure is not drawn to scale: the fourth orbit is actually 16 times the size of the first orbit.)

energy while in any of the orbits is negative because it is bound to the nucleus (see the end of Section 3.5). To go to a larger orbit, the electron must gain energy. The maximum energy the electron can have and still remain bound to the proton is called the *ionization energy.* If the electron acquires more than this energy, it breaks free from the nucleus and the atom is ionized. The resulting positive ion in this case is just a bare proton.

How does the Bohr model account for the characteristic spectra emitted by luminous hydrogen gas? When an electron in a larger (higher energy) orbit "jumps" to a smaller orbit, it loses energy. One of the ways it can lose this energy is by emitting light. This process is the origin of atomic spectra.

As an example, imagine that the electron in our hydrogen atom is in the sixth allowed orbit, with its energy represented by E_6. Suppose the electron makes a transition (a jump) to an inner orbit, say the second one, where its energy is E_2. To do so the electron must lose an amount of energy, ΔE, equal to:

$$\Delta E = E_6 - E_2$$

In what is called a *radiative transition,* the electron loses this energy through the *emission of a photon* whose energy is just equal to ΔE (Figure 10.18). The photon, spontaneously created during the transition of the electron from orbit 6 to orbit 2, carries off the excess energy into space in the form of EM radiation. (This is much like Planck's atomic oscillators emitting photons.) Since the energy of a photon equals h times the frequency, we have:

$$\text{Photon energy} = \Delta E = E_6 - E_2$$

And:

$$\text{Photon energy} = hf$$

Therefore:

$$hf = E_6 - E_2$$

The frequency of the emitted light is directly proportional to the difference in the energy of the orbits between which the electron jumped. The larger the energy difference, the higher the frequency of the light given off in the process. A downward transition from orbit 3 to orbit 2 will produce a photon with lower energy and lower frequency, since the energy difference between orbit 3 and orbit 2 is smaller.

$$\Delta E = E_3 - E_2 \quad \text{is smaller than} \quad \Delta E = E_6 - E_2$$

FIGURE 10.17
A hydrogen atom with its electron in (a) orbit 2 and in (b) orbit 3. The electron has more energy when it is in orbit 3.

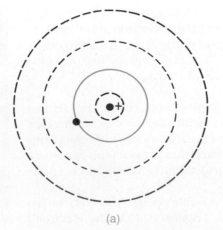

(a)

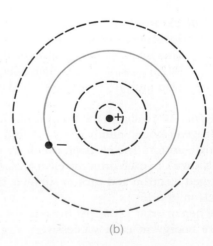

(b)

For the 6 to 2 transition in hydrogen, a violet light photon is emitted. For the 3 to 2 transition, it is a red light photon.

Each possible downward transition from an outer orbit to an inner orbit results in the emission of a photon with a particular frequency. An appropriately heated sample of hydrogen gas will emit light with these different frequencies but no other radiation. This is the line spectrum of hydrogen. (We will discuss this more in Section 10.6.)

A downward electron transition can also occur without the emission of light through what is called a *collisional transition*. In this case a collision between a hydrogen atom and another particle (perhaps another hydrogen atom) can induce the electron in the outer orbit to spontaneously jump to an inner orbit. The energy that the electron loses can be transferred to the other particle, or it can be converted into increased kinetic energy of both colliding particles. (This is much like the collision described in Figure 3.35.) Collision-induced transitions are generally important in dense gases where the numbers of atoms or molecules per volume of gas are large and the likelihood of two or more gas particles colliding is therefore quite high.

So far we have focused on how an electron in an outer orbit may lose energy by jumping to an inner orbit. But the reverse of this process also occurs: an electron in an inner orbit can gain just the right amount of energy and jump to an outer orbit. For example, a hydrogen atom in the lowest energy state might gain the amount of energy needed for its electron to jump from orbit 1 to, say, orbit 5. To do this the electron would have to acquire energy:

$$\Delta E = E_5 - E_1$$

One way to do this would be via a collision: the hydrogen atom could collide with another atom with more energy.

Another way an electron can jump to an outer orbit is by absorbing a photon with the proper energy (Figure 10.19). For example, an electron in orbit 1 can jump to orbit 5 if it absorbs a photon with energy.

$$\text{Photon energy} = \Delta E = E_5 - E_1$$

If a sample of hydrogen gas is irradiated with a broad band of EM waves (like blackbody radiation), many such transitions to outer orbits will occur. Some of the photons in the incident radiation will have just the right energies to induce transitions from inner orbits to outer orbits. In the process, the number of photons with these particular energies will be reduced, and the intensity of the EM radiation at the corresponding frequencies will decrease. This reduction of intensity of light at only certain frequencies after it passes through a gas results in an *absorption spectrum* (Figure 10.20). For example, if white light is passed

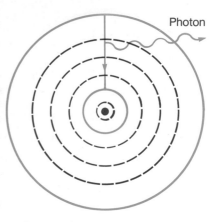

FIGURE 10.18
Photon emission. The electron makes a transition from orbit 6 to orbit 2 and emits a photon. The photon's energy equals the energy lost by the electron.

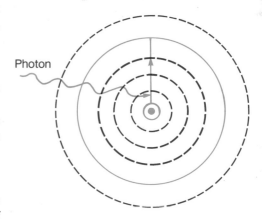

FIGURE 10.19
Photon absorption. The electron makes a transition from orbit 1 to orbit 5 by absorbing a photon.

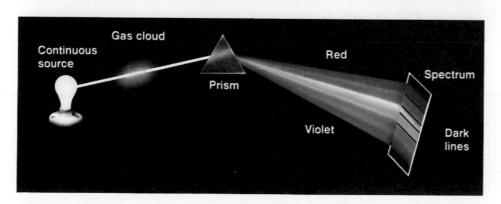

FIGURE 10.20
Absorption spectrum of a gas. The gas absorbs photons in the light passing through it, so at those frequencies (colors) the spectrum is darker.

through hydrogen gas and then dispersed with a prism, dark bands appear at certain frequencies. They are exactly the same frequencies that are in the emission spectrum of luminous hydrogen.

This is how the element helium was discovered. In 1868 some of the absorption lines in the spectrum of sunlight were found not to correspond with any elements known at that time. The existence of a new element in the sun's atmosphere was suggested to account for these lines. This new element was named after the Greek word (*helios*) for sun (see *Physics Potpourri* "What's in a Name?" on page 158).

The Bohr model was very successful at explaining the origin of atomic spectra. But, as we suggested at the beginning of this section, Bohr's model of electron orbits rested on two unexplained assumptions. First, there are only certain allowed orbits. While one can place a satellite in orbit about the earth with any radius, the electron orbits were restricted to specific radii. These radii were determined by a seemingly arbitrary but nonetheless effective condition: that the angular momentum of the electron in its orbit be quantized. Much like the energy of Planck's quantized atomic oscillators, the orbital angular momentum of Bohr's electrons could only have the following values:

Refer back to angular momentum in Section 3.8.

$$\text{Allowed angular momentum} = \frac{h}{2\pi}, \text{ or } 2\frac{h}{2\pi}, \text{ or } 3\frac{h}{2\pi}, \text{ etc.}$$

So quantization was popping up everywhere: in the energy of oscillating atoms, the energy of EM waves, and now the orbital angular momentum of atomic electrons.

The second assumption that physicists of the time found objectionable was that as long as an electron remained in one of its allowed orbits, it did not emit EM radiation. This was a problem because Maxwell's work indicated that whenever a charged object undergoes acceleration, including centripetal acceleration, it will radiate. Put another way, an electron in a periodic orbit is much like a charge oscillating back and forth, and the latter results in the production of an EM wave. An orbiting electron should be continually radiating, and in the process it should lose energy and spiral into the nucleus. According to the physical laws known at the time, atoms shouldn't exist longer than a fraction of a second!

Even though Bohr had a model that worked, clearly the physics behind it was not understood. What was needed was a revolution.

10.5   QUANTUM MECHANICS

The success of Bohr's model of the atom, even though it was at odds with accepted principles of physics, indicated that perhaps a new physics was needed to describe what goes on at the atomic level. The first step in this direction came during the summer of 1923. While working on his PhD in physics, a French aristocrat named Louis Victor de Broglie (rhymes with "Troy") proposed that electrons and other particles possess wavelike properties (Figure 10.21). Einstein had shown that light has both wavelike and particlelike properties, so why not electrons too? The wave associated with any moving particle has a specific wavelength (called the de Broglie wavelength) that depends on the particle's momentum:

$$\lambda = \frac{h}{mv} \quad \text{(de Broglie wavelength)}$$

The higher the momentum of a particle, the shorter its wavelength. High-speed electrons have shorter wavelengths than low-speed electrons.

Once again, Planck's constant shows up. Since h is such a tiny number, de Broglie wavelengths are extremely small. This means that the wave properties of particles are only manifested at the atomic and subatomic level.

What is the de Broglie wavelength of an electron with speed 2.19×10^6 meters/ second? (This is the approximate speed of an electron in the smallest orbit in hydrogen.)

EXAMPLE 10.2

Using the mass given in Table 11.1, the electron's momentum is:

$$mv = 9.11 \times 10^{-31} \text{ kg} \times 2.19 \times 10^6 \text{ m/s}$$

$$mv = 1.995 \times 10^{-24} \text{ kg-m/s}$$

Using the value of h in SI units (Section 10.1):

$$\lambda = \frac{h}{mv} = \frac{6.63 \times 10^{-34} \text{ J-s}}{1.995 \times 10^{-24} \text{ kg-m/s}}$$

$$\lambda = 3.32 \times 10^{-10} \text{ m} = 0.332 \text{ nm}$$

This distance is in the same range as the diameters of atoms.

So yet another radical theory entered the arena of physics. Although submitted by a newcomer, de Broglie's hypothesis was not completely rejected by the physics community because Einstein himself found it plausible. Then, in 1925, the puzzling results of some experiments were interpreted as proof of the existence of de Broglie's waves. In a series of experiments, the American physicist Clinton Davisson (with various collaborators) showed that a beam of high-speed electrons underwent diffraction when sent into a nickel target. The electrons were behaving just like waves. In fact, one gets the same kind of scattering pattern using x-rays or electrons (see Figure 10.22).

FIGURE 10.21
Louis de Broglie, (1892–1987), was the first to suggest that particles can behave like waves.

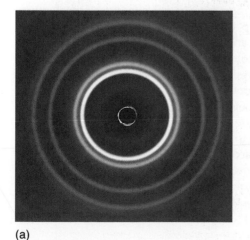

(a)

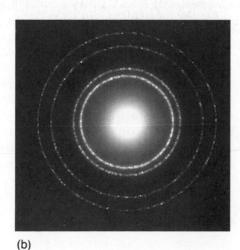

(b)

FIGURE 10.22
These patterns were produced by sending (a) a beam of x-rays and (b) a beam of electrons through an aluminum target. The similarity exists because the electrons act like waves as they interact with the aluminum.

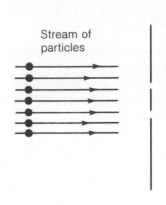

Stream of particles

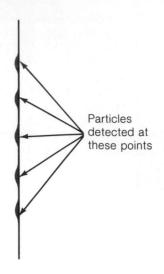

Particles detected at these points

FIGURE 10.23
Particles, such as electrons, undergo interference when passed through narrow slits. They exhibit a property of waves (compare to Figure 9.6)

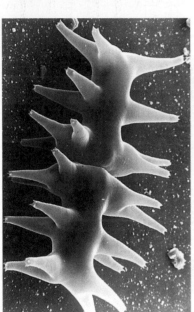

FIGURE 10.24
An image of an algae cell taken with an electron microscope.

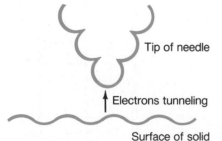

Tip of needle

Electrons tunneling

Surface of solid

FIGURE 10.25
Simplified sketch of a scanning tunneling microscope (STM). The wavelike nature of the electrons allows them to cross the gap from the surface to the needle.

Other experiments have verified the wave properties of electrons, protons, and other particles. Young's classic two-slit experiment, by which he proved that light is a wave, can be used to make particles undergo interference (Figure 10.23). The *electron microscope* exploits the wave properties of electrons. Instead of using ordinary light like a conventional microscope does, an electron microscope uses a beam of electrons that acts like a beam of electron waves. The magnification can be much higher with an electron microscope because the de Broglie wavelength of the electrons is much shorter than the wavelength of visible light (Figure 10.24). Diffraction and interference of waves passing around and between small objects, such as cells, strongly affect image clarity. Shorter wavelength waves (electrons) diffract and interfere much less than do longer wavelength waves (light).

A more recently developed microscope uses the wavelike nature of electrons to form tantalizing images of individual atoms at the surfaces of solids. The *scanning tunneling microscope (STM)* uses an extremely fine-pointed needle that scans back and forth over the surface. A positive voltage maintained on the needle attracts the electrons at the surface of the sample. If electrons were simply particles, they could not traverse the gap—less than 1 nanometer wide—between the surface and the needle. But their wave nature allows the electrons to "tunnel" through the gap and to reach the needle (Figure 10.25). The tiny current of electrons that flows decreases rapidly if the gap widens. As the needle scans back and forth over the surface (like someone mowing a lawn), a feedback mechanism moves the needle up and down to keep this tunneling current constant. The varying height of the needle is recorded and used to draw a contour map of the surface that clearly shows individual molecules and atoms (Figure 10.26 and Figure 4.8). The 1986 Nobel Prize in physics was shared by the inventor of the electron microscope and the codevelopers of the STM.

Another major triumph for de Broglie's wave hypothesis was its ability to explain Bohr's quantized orbits. De Broglie reasoned that since an orbiting electron acts like a wave, its wavelength has to affect the circumference of its orbit. In particular, the electron's wave wraps around on itself, and in doing so it must interfere with itself. Only if the wave interferes constructively (peak matches peak) can the electron's orbit remain stable (Figure 10.27). This means that the circumference of its orbit must equal exactly one de Broglie wavelength, or two, or three, etc.

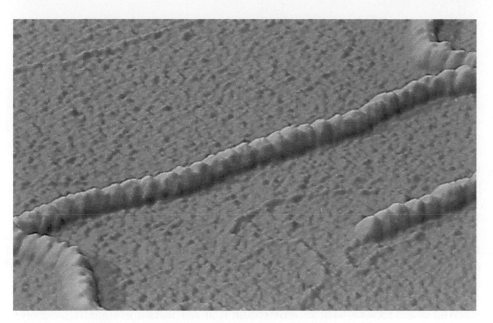

FIGURE 10.26
An STM image (large central object) of DNA-protein complexes (strands of DNA and their protein envelopes). The image was obtained from metal-coated specimens. The structure of the strands is clearly resolved. A DNA strand without its protein envelope, which has a diameter of approximately 1 nanometer, is visible in the lower middle of the image. The images were produced by IBM Zurich Research Center Scientists.

Circumference of orbit = λ, or 2 λ, or 3 λ, etc.

If r is the radius of a circular orbit, then the circumference is $2\pi r$.

$$2\pi r = \lambda, \text{ or } 2\lambda, \text{ or } 3\lambda, \text{ etc.}$$

EXAMPLE 10.3

Using the results of Example 10.2, find the radius of the smallest orbit in the hydrogen atom.

In de Broglie's model, the circumference of the smallest orbit must equal the de Broglie wavelength of the electron. We calculated this wavelength to be 0.332 nm for the smallest orbit. So:

$$2\pi r = \lambda = 0.332 \text{ nm}$$

$$r = \frac{0.332 \text{ nm}}{2\pi} = \frac{0.332 \text{ nm}}{2 \times 3.14} = \frac{0.332 \text{ nm}}{6.28}$$

$$r = 0.0529 \text{ nm}$$

The allowed values for the circumference of an electron's orbit leads to Bohr's allowed values for the electron's angular momentum. Since the de Broglie wavelength is given by

$$\lambda = \frac{h}{mv}$$

the circumference can have the following values.

$$2\pi r = \frac{h}{mv}, \text{ or } 2\frac{h}{mv}, \text{ or } 3\frac{h}{mv}, \text{ etc.}$$

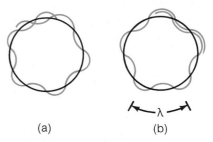

(a)   (b)

FIGURE 10.27
Simplified wave representation of an electron orbiting a nucleus. The orbit in (a) is not possible because the wave interferes destructively with itself. The orbit in (b) is allowed. This corresponds to orbit 5, since it has 5 wavelengths fitted into the orbit.

FIGURE 10.28
German physicist Werner Heisenberg (1901–1976), codeveloper of the science of quantum mechanics.

When we divide both sides by 2π and multiply both sides by mv, we get:

$$mvr = \frac{h}{2\pi}, \text{ or } 2\,\frac{h}{2\pi}, \text{ or } 3\,\frac{h}{2\pi}, \text{ etc.}$$

But mvr is just the angular momentum of the electron. We saw this same relationship at the end of Section 10.4. De Broglie's condition on the circumference of the electron's orbit turns out to be identical to Bohr's condition on the angular momentum of the electron.

The success of quantization and wave-particle duality at the atomic level could not be overlooked. In the later half of the 1920s a flurry of activity resulted in a formal mathematical model that incorporated these ideas—*quantum mechanics*. The two principle founders were Werner Heisenberg (Figure 10.28) and Erwin Schroedinger.

One of the main contributions of Heisenberg was the *uncertainty principle*. Because electrons and other particles on the atomic scale have wavelike properties, we can no longer think of them as being like tiny, shrunken marbles. They are a bit spread out in space, more like tiny, fuzzy cotton balls. In the old particle model of the electron it was possible, at least in theory, to state exactly where an electron is and exactly what its momentum is at any instant in time. Heisenberg stated that the wave nature of particles makes this impossible. One cannot specify both the position and the momentum of an electron to arbitrarily high precision. The more precisely you know the position of the electron, the less precisely you can determine its momentum. If we let Δx represent the uncertainty in the position of a particle, and Δmv represent the uncertainty in the momentum of the particle, then:

$$\Delta x\,\Delta mv = h \text{ (at best)} \qquad \text{(uncertainty principle)}$$

No matter how good the experimental apparatus, the *best* we can do is limited by this equation. On the atomic scale, particles cannot be localized in the same way that they can be on a large scale.

Schroedinger established a mathematical model for the waves associated with particles. In his model a simple system like a hydrogen atom can be described by a *wave function*. The wave function gives the wavelength of the particle and carries information about the likelihood of finding the electron at a given location at a specified time. Knowledge of the wave function allows one to calculate the most probable position of the electron with respect to the nucleus, as well as the electron's average energy. Although often difficult to determine in practice, the wave function of a system contains all the information ever needed about the system.

During the decades beginning the 20th century, the view of matter at the submicroscopic level changed dramatically. EM waves have particlelike properties, and electrons and other particles have wavelike properties. Nature is clearly very different at this level.

10.6   ATOMIC STRUCTURE

The findings of de Broglie, Heisenberg, and Schroedinger force us to revise the simple Bohr model of the atom with its planetary structure. We can no longer represent electrons as little particles moving in nice circular orbits. In reality, each electron is spread out into a wave that represents the probability of finding it at different locations. The atom is pictured as a tiny nucleus surrounded by an "electron cloud" (Figure 10.29). The density of the cloud at each point in space indicates the likelihood of finding the electron there. The different allowed orbits of the electrons appear as clouds with different sizes and shapes.

So the simple drawings of atoms that we used earlier in this text (such as Figure 7.2) *are not strictly correct.* But they served the purpose of presenting the

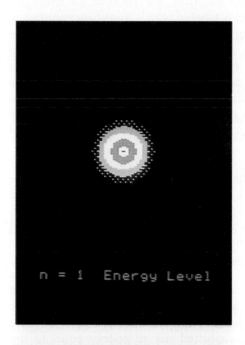

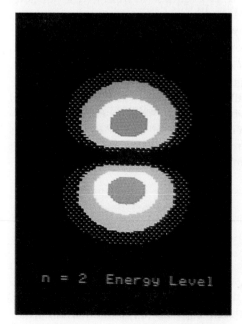

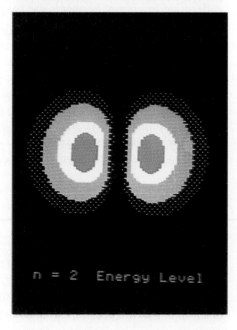

FIGURE 10.29
Computer representations of "electron clouds" for four atomic orbits. The color indicates the probability of finding the electron at that location. Green represents the highest probability, followed by white, maroon, and black. The green areas indicate that the electron is most likely found farther from the nucleus in the n = 2 orbits than in the n = 1 orbit. This corresponds to Bohr's model, in which the n = 2 orbit is physically larger than the n = 1 orbit. (The scale is magnified about 50 million.)

basic structure of the atom without complicating the picture with the wave nature of the electrons.

Since we can't say exactly where the electron is in each of its orbits, it is more useful to concentrate on the electron's energy. After all, when determining the frequency of radiation emitted or absorbed by an atom, the important quantity is the difference between the electron's initial and final energy. In the new model we describe the electrons as being in certain allowed *energy states* or *energy levels*. For the hydrogen atom, the lowest energy level (corresponding to the innermost orbit in the Bohr theory) is called the *ground state*. The higher energy states are referred to as *excited states*. We now represent the structure of the atom schematically using an *energy level diagram,* like the one for hydrogen shown in Figure 10.30. Each energy level is labeled with a *quantum number, n,* beginning with the ground state having $n = 1$ and continuing on up. As the quantum number increases, so does the energy associated with the state. Moreover, as n gets larger, the difference in energy between the adjacent states becomes smaller. The difference in energy between the $n = 4$ and the $n = 3$ states is smaller than the difference in energy between the $n = 3$ and the $n = 2$ states. Finally, we again note the existence of a maximum allowed energy above which the electron is no longer bound to the nucleus. The state is designated $n = \infty$, and its energy is the ionization energy.

The numbers on the left in the energy level diagram are the electron energies for each state. (The negative values indicate that the electron is bound to the nucleus, as we mentioned in Section 10.4.) The transition of an electron from one orbit to another corresponds to the atom going from one energy level to another. The change in energy of the electron as a result of the "energy level transition" is found by comparing the energies of the two states.

Using the energy level diagram, we can give a more complete picture of the emission spectrum of hydrogen. Suppose an atom in the $n = 2$ state undergoes a transition to the $n = 1$ state by emitting a photon. (Typically an atom will remain in an excited state for only about a billionth of a second.) The transition is represented by an arrow drawn from the initial level to the final level (Figure 10.31). The photon that is emitted has an energy equal to the difference in energy between the two levels. In this case:

$$\text{Photon energy} = \Delta E = E_2 - E_1 = -3.4\,\text{eV} - (-13.6\,\text{eV})$$

$$\text{Photon energy} = 10.2\,\text{eV}$$

This is a photon of UV light (refer to Figure 10.7).

In a similar way we can represent all possible downward transitions from higher energy levels to lower energy levels. Figure 10.32 is an enlarged energy-level diagram for hydrogen showing various possible energy-level transitions from higher levels to lower ones. The number with each arrow is the wavelength (in nanometers) of the photon that is emitted.

EXAMPLE 10.4   Find the frequency and wavelength of the photon emitted when a hydrogen atom goes from the $n = 3$ state to the $n = 2$ state.

We first find the energy of the photon, then use that to determine its frequency. The wavelength we get from the equation $c = f\lambda$.

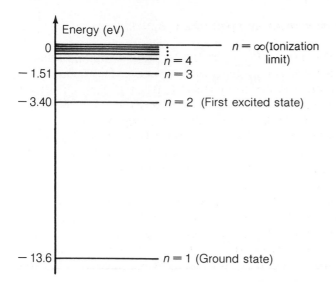

FIGURE 10.30
Energy level diagram for hydrogen. Each level corresponds to one of the allowed electron orbits.

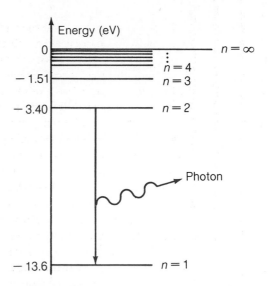

FIGURE 10.31
Downward electronic transition from level n = *2 to level* n = *1 in hydrogen. The transition is accompanied by the emission of a photon with energy 10.2 eV.*

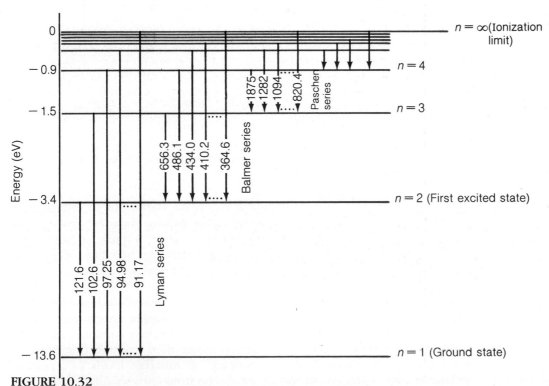

FIGURE 10.32
Energy level diagram for hydrogen showing different possible energy level transitions. The number with each arrow is the wavelength (in nanometers) of the photon that is emitted.

$$\text{Photon energy} = hf = E_3 - E_2$$

$$hf = -1.51 \text{ eV} - (-3.4 \text{ eV})$$

$$hf = 1.89 \text{ eV}$$

Therefore:

$$f = \frac{1.89 \text{ eV}}{h} = \frac{1.89 \text{ eV}}{4.136 \times 10^{-15} \text{ eV/Hz}}$$

$$f = 4.57 \times 10^{14} \text{ Hz}$$

For the wavelength:

$$\lambda = \frac{c}{f} = \frac{3 \times 10^8 \text{ m/s}}{4.57 \times 10^{14} \text{ Hz}}$$

$$\lambda = 6.56 \times 10^{-7} \text{ m} = 656 \text{ nm}$$

From Figure 10.7 we see that this is a photon of visible light. (To be more precise, Table 8.1 shows that it is red light.)

By doing similar calculations for the other transitions, we can draw the following conclusions about the light that can be emitted by excited hydrogen atoms.

1. Transitions from higher energy levels *to the ground state* ($n = 1$) result in the emission of *ultraviolet* photons. This series of emission lines is referred to as the Lyman series.

2. Transitions from higher energy levels to the $n = 2$ state result in the emission of *visible* photons. This series of emission lines is referred to as the Balmer series.

3. Transitions from higher energy levels to the n = 3 state result in the emission of infrared photons. This series of emission lines is referred to as the Paschen series.

4. Downward transitions to other states result in the emission of infrared or other lower energy photons.

An atom in the $n = 3$ state or higher can make transitions to intermediate energy levels instead of jumping directly to the ground state. For example, an atom in, say, the $n = 4$ state may jump to the $n = 1$ state, or it may go from $n = 4$ to $n = 2$ and then from $n = 2$ to $n = 1$. In the latter case, two different photons would be emitted (Figure 10.33). Atoms in the higher energy levels can undergo different "cascades' in returning to the ground state.

We can envision what happens when hydrogen gas in a tube is heated to a high temperature, as in Figure 10.11, or is excited by passing an electric current through it. The billions and billions of atoms will be excited, some to each of the possible higher energy levels. The excited atoms then undergo transitions to lower levels, with different atoms "stopping" at different levels in a random fashion. The hydrogen gas continuously emits photons corresponding to all of the possible energy level transitions. This is hydrogen's emission spectrum. (As in the Bohr model, excited atoms can also lose energy via collisions. No photons are emitted in this case.)

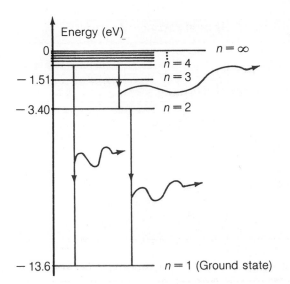

FIGURE 10.33
Two different ways for a hydrogen atom in the n = 4 *state to return to the ground state. The left arrow represents a direct transition with the emission of one photon. The two arrows on the right show a transition to the* n = 2 *level, followed by a transition to the ground state. Two lower energy photons are emitted.*

Upward energy level transitions occur when the atom gains energy, by absorption of a photon with the proper energy, or perhaps by collision. For example, a hydrogen atom in the ground state can jump to the $n = 2$ state by absorbing a photon with energy 10.2 eV. When white light—containing photons with many different energies—passes through hydrogen, those photons with just the right energy will be absorbed, leading to the observed absorption spectrum (see Figure 10.20).

An atom is ionized if it absorbs enough energy to make the electron energy greater than zero. For example, a hydrogen atom in the ground state is ionized if it absorbs a photon with energy greater than 13.6 eV. This process, referred to as "photoionization", is essentially the same thing as the photoelectric effect. Any excess energy the electron has appears as kinetic energy.

We've presented a fairly complete picture of the hydrogen atom, but what of the other elements? The presence of more than one electron complicates things, but the general structure is much the same. Each element has its own atomic energy level diagram, with correspondingly different energy values for each level. The various downward transitions between these levels produce the element's characteristic emission spectrum. Atoms with larger atomic numbers have more protons in the nucleus, so the force on the inner electrons is stronger. This means that the electrons are more tightly bound and that their energies have larger magnitudes but are still negative.

One very common device that uses the emission spectra of elements is the neon sign (Figure 10.34)—so named because neon is one of the most commonly used elements in them. These signs are made by placing low-pressure gas in a sealed glass tube. A high-voltage alternating current power supply connected across the ends of the tube causes electrons to move back and forth through the tube. The electrons excite the atoms of the gas by collision, and the atoms emit photons as they return to their ground states. Since different elements have different emission spectra, signs can be made to emit different colors by being filled with different gases. Neon-filled signs are red because there are several bright red lines in neon's emission spectrum. Fluorescent lights (Section 8.5) make use of the ultraviolet emission lines of mercury.

The structure of atoms with more than one electron is governed by a principle formulated by Wolfgang Pauli in 1925. Pauli, who was awarded the 1945 Nobel

FIGURE 10.34
The light from a neon sign is the emission spectrum of the gas inside the glass tube.

TABLE 10.1	GROUND STATE CONFIGURATIONS OF SOME ATOMS			
		Number of Electrons in Level		
Element	Atomic Number	1	2	3
Hydrogen	1	1	0	0
Helium	2	2	0	0
Lithium	3	2	1	0
Carbon	6	2	4	0
Oxygen	8	2	6	0
Neon	10	2	8	0
Sodium	11	2	8	1

Prize in physics for his work, carefully analyzed the observed emission spectra of different elements as well as the results of other experiments and concluded that only a certain number of electrons can occupy each energy level in an atom. In the case of emission spectra, he noticed that some expected transitions did not occur when the lower energy level was already occupied by its limit of electrons. (We can't sit on a sofa if it is already full of people.) Pauli stated his conclusion as the *exclusion principle:*

> *Two electrons cannot occupy the same quantum state at the same time.*[1]

(Pauli exclusion principle)

There is a set number of quantum states available to electrons in each energy level. Once all of the quantum states in a given energy level are filled, the remaining electrons in the atom must occupy other energy levels that have vacancies. The number of quantum states in the $n = 1$ level is 2, so the maximum number of electrons that can occupy this level is 2. For the $n = 2$ level, the maximum number is 8; for the $n = 3$ level it is 18. The general rule is *for level n, the occupation limit of electrons is $2n^2$.* The ground state of a multielectron atom is one in which all electrons are in the lowest energy levels consistent with the Pauli exclusion principle. If any one electron is in a higher energy level, the atom is in an excited state. Table 10.1 shows the ground-state energy-level populations for several atoms.

The properties of each element are determined to a great extent by the ground-state configuration of its atoms, particularly the number of electrons in the highest energy level that is occupied. Table 10.1 shows that helium in the ground state has the $n = 1$ energy level filled and that neon has the $n = 2$ energy level filled. Because of this, helium and neon have similar properties: they are both gases at normal room temperature and pressure and they are very stable. They don't burn or react chemically in other ways except under special circumstances. Hydrogen, lithium, and sodium have similar properties because they all have one electron in the highest occupied energy level (when in their respective ground states).

---

[1]The exclusion principle applies to an entire class of elementary particles that includes electrons. More on this in Chapter 12.

The periodic table of the elements (Appendix B) was developed by Dmitri Mendeleev in 1869. He arranged the elements known at the time according to their properties: elements with similar properties were placed in the same column. After Pauli's discovery it was determined that the elements in each column have similar ground-state configurations, and that is why they are alike. (For the above example, hydrogen, lithium, and sodium are in the first column, called Group 1A, because each has one electron in its highest occupied energy level.) This is one indication of how quantum mechanics plays a crucial role in the science of chemistry.

LEARNING CHECK

1. A hot, luminous solid emits a _____ spectrum, while a hot, luminous gas produces a _____ spectrum.

2. Is the energy associated with a photon of blue light (a) greater than, (b) less than, or (c) equal to the energy associated with a photon of red light? Why?

3. In the Bohr model of the atom, a photon is _____ when an electron jumps from a large, high-energy orbit to a smaller, low-energy one.

4. An atom is said to have been _____ when it absorbs a photon with sufficient energy to free an electron.

5. (a) Indicate whether the following transitions within the energy level diagram of hydrogen are accompanied by the *emission* or *absorption* of photons.
   (b) Assuming only one photon is emitted or absorbed in the course of each transition, order (from highest to lowest) the frequencies of the photons involved in each case:
   i) $n = 5$ to $n = 3$
   ii) $n = 1$ to $n = 8$
   iii) $n = 8$ to $n = 9$
   iv) $n = 3$ to $n = 5$
   v) $n = 6$ to $n = 5$

10.7   X-RAY SPECTRA

We have seen that the Bohr model of the atom (as refined by quantum mechanics) had great success in explaining the features of hydrogen's spectrum. Another early triumph of the Bohr atom came in connection with the study of x-rays. Soon after Bohr published his model, a young English physicist named H. G. J. Moseley used it to explain the *characteristic spectra* of x-rays.

In Section 8.5 we described how x-rays are produced. Electrons are accelerated to a high speed and are directed into a metal target (Figure 8.34). A band of x-rays is emitted, with different wavelengths having different intensities. Figure 10.35 shows the x-ray spectra when the elements tungsten (W) and molybdenum (Mo)

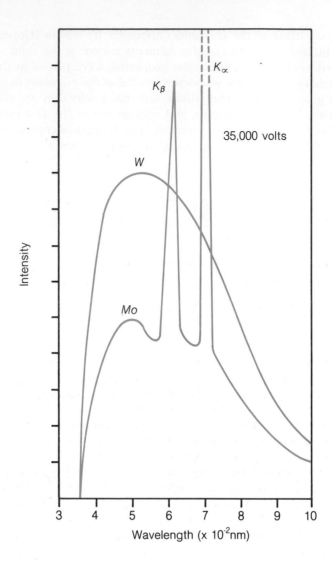

FIGURE 10.35
X-ray spectra of tungsten (W) and molybdenum (Mo) resulting from the bombardment of targets made of each element by electrons having energies of 35,000 eV. Over the wavelength range shown, the spectrum of tungsten is a continuous one produced by bremsstrahlung processes. The spectrum of molybdenum includes two very strong characteristic peaks.

are used as targets. These graphs are like blackbody radiation curves in that they are plots of intensity versus wavelength. Most of the x-rays are produced as the electrons are rapidly decelerated upon entering the target. (This is one example of Maxwell's finding that accelerated charges emit EM waves.) This *bremsstrahlung* (German for "braking radiation") appears as the smooth part of the spectra covering the full range of wavelengths. Bremsstrahlung spectra are much the same for different elements. But the two sharp peaks for the molybdenum target are unique to this element and constitute its characteristic spectrum. Other elements exhibit characteristic spectra but at different wavelengths.

Moseley compared the characteristic x-ray spectra of different elements and found a simple relationship between the wavelengths of the peaks and the atomic number of the element. For each peak a graph of the square root of the frequency versus the atomic number of the element was a straight line (Figure 10.36). This was a very practical discovery because it allowed him to determine the atomic number, and therefore the identity, of unknown elements. All he had to do was determine the wavelengths of the characteristic x-ray peaks and use his graph. This method was a key factor in the discovery of several of the elements (including promethium (atomic number 61) and hafnium (atomic number 72).

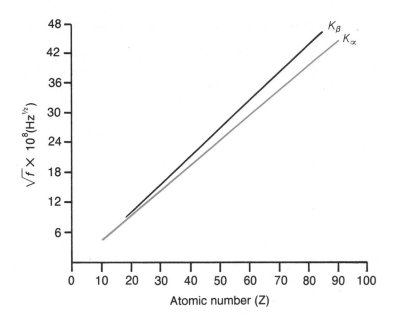

FIGURE 10.36
Moseley's diagram showing the relationship between the square root of the frequency for two characteristic x-ray peaks and the atomic number (Z) of the emitting element.

After Bohr developed his model of the atom, Moseley (Figure 10.37) quickly realized that the characteristic x-rays are much like the emission lines of hydrogen. He showed that they are emitted when one of the innermost electrons jumps from one orbit (energy level) to another. Different elements have x-ray peaks with different wavelengths because the electron energy levels have different energy values. In particular, atoms with higher atomic number have more protons in the nucleus and bind the inner electrons more tightly.

Moseley proposed the following scenario to account for the x-ray spectra of the elements. If a bombarding electron collides with, say, a molybdenum atom and knocks out an electron from the $n = 1$ level, electrons in the upper levels will cascade down to fill the vacancy left by the ejected electron and photons will be emitted. The lowest frequency (longest wavelength) x-ray photon corresponds to the transition $n = 2$ to $n = 1$. Because the lowest energy level was called the "K shell" by x-ray experimenters, this is called the K_α (K-alpha) peak. Electrons jumping from $n = 3$ to $n = 1$ emit photons that form the K_β (K-beta) peak. Notice that the K_α and K_β peaks correspond to the two lowest-frequency lines in hydrogen's Lyman series.

With this understanding we can see why high-speed electrons are needed to produce the characteristic x-rays of heavy elements. The inner electrons are so tightly bound that it takes very high energy electrons to knock them out.

FIGURE 10.37
H. G. J. Moseley (1887–1915), whose work with x-ray spectra confirmed Bohr's model of the atom. Moseley's brilliant career was cut short when he enlisted during World War I. He was killed in action.

10.8   LASERS

The word "laser" is an acronym derived from the phrase light amplification by stimulated emission of radiation. By examining this phrase one piece at a time, we will describe how the laser operates. In so doing we will find yet another example of how our theory of atomic structure provides an explanation for a phenomenon

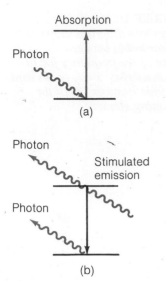

FIGURE 10.38
Stimulated emission. (a) Photon absorption places an atom in an excited state. (b) A photon with the same energy stimulates the excited atom to return to the lower state and emit an identical photon.

FIGURE 10.39
Coherent versus incoherent light. (a) Coherent sources of light, like lasers, emit relatively long wave trains so that points along the wave separated by distances of the order of several wavelengths are in phase with one another. (b) Incoherent sources, like light bulbs, emit very short, randomly oriented wave trains. Over distances of even a few wavelengths, the phase of the beam changes in abrupt and unpredictable ways.

that is found in systems as remote as the Martian atmosphere and as near as the local supermarket.

Suppose we now consider an electron that has been excited to some higher energy level by either a collision or the absorption of a photon with the proper energy, E. Generally such an electron will remain in this excited state for only a short time (around a billionth of a second) before returning to a lower energy level. If this decay to a lower state occurs spontaneously by radiation (which is usually the case in low-density gases when collisions are rare), a photon will be emitted in some random direction.

It turns out, however, that an electron can be *stimulated* to return to its original energy level through the intercession of a second photon with the same energy, E (Figure 10.38). If a group of atoms all having their electrons in this same excited state is "bathed" in light consisting of photons with energy E, the atoms will be stimulated to decay by emitting additional photons with the same energy E. By this process the intensity of a beam of light is increased or amplified as a result of stimulated emission by atoms in the region through which the radiation passes. In other words, we have a laser.

To achieve this amplification, we must first arrange for the majority of the atoms through which the stimulating radiation travels to have their electrons in the same excited state. Otherwise the unexcited atoms will just absorb the radiation. This is not an easy situation to arrange because the excited atoms normally don't stay that way for long.

The problem can be overcome because many atoms possess excited states referred to as *metastable*. Once in such a state, the electrons tend to remain there for a relatively long time (perhaps a thousandth of a second instead of a billionth) before spontaneously decaying to a lower state. During the time it takes to excite more than half of the atoms, the ones already excited don't jump back down. This process, called "pumping," can result in the majority of the atoms being in the same metastable, excited state. This condition is called a *population inversion* because there are more atoms with electrons populating an upper energy level than a lower one, the reverse of the usual situation. This condition is necessary for the generation of laser light.

If we now irradiate the population-inverted atoms with photons of the correct energy, a chain reaction can be established that greatly amplifies the incident light beam. One photon stimulates the emission of another identical one, these two stimulate the emission of two more, these four stimulate the emission of four more, and so on. In the end, an avalanche of photons is produced, all having the same frequency (color). The beam is said to be *monochromatic*. The light amplification can produce a very high-intensity beam of light.

Besides being intense and monochromatic, laser light has an additional property that makes it extremely useful: *coherence*. This means that the stimulating radiation and the additional emitted laser radiation are in phase: the crests and troughs of the EM waves at a given point all match up (Figure 10.39a). An

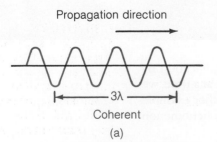

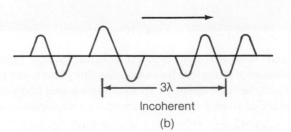

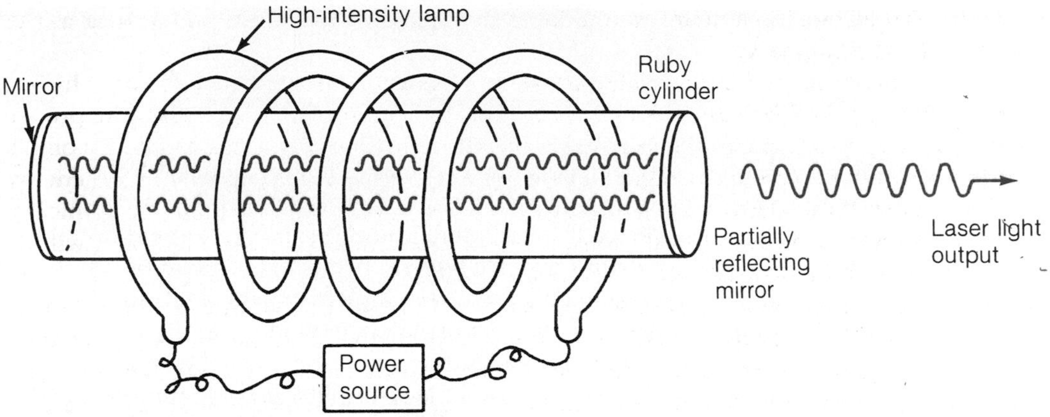

High-intensity lamp

Mirror

Ruby cylinder

Partially reflecting mirror

Laser light output

Power source

ordinary light source such as an incandescent bulb emits incoherent light (different parts of the beam are not in phase with one another) because the excited atoms giving off the radiation do so independently of each other. The emitted photons in this case may be considered to be individual, short "wave trains" bearing no constant phase relation to one another (Figure 10.39b). By contrast, stimulated emission of radiation by excited atoms produces photons that are in phase with the stimulating radiation. The coherence of laser light contributes to its high intensity. Coherent sources like lasers are important for producing interference patterns and *holograms* (described at the end of this section).

One possible type of laser, *the ruby laser,* is shown schematically in Figure 10.40. It consists of a ruby rod whose ends are polished and silvered to become mirrors, one of which is partially (1% to 2%) transparent. The ruby rod is composed of aluminum oxide (Al_2O_3) in which some of the aluminum atoms have been replaced by chromium atoms. It is the chromium atoms that produce the lasing effects. The rod is surrounded by a flash tube capable of producing a rapid sequence of short, intense bursts containing green light with a wavelength of 550 nanometers. The chromium atoms are excited by these flashes from their lower state E_0 to state E in a process referred to as *optical pumping.* (The high-intensity flash tube "pumps" energy into the chromium atoms.) The chromium atoms quickly spontaneously decay back to level E_0, or, in some cases, to the metastable level E_1 (Figure 10.41). With very strong pumping, more atoms can be forced into state E_1, and a population inversion is produced. Eventually a few of the chromium atoms in state E_1 decay to the state E_0, thus emitting photons that stimulate other excited chromium atoms to execute the same transition. When these photons strike the end mirrors, most of them are reflected back into the tube. As they move back in the opposite direction, they cause more stimulated emission and increased amplification. A small fraction of the photons oscillating back and forth through the rod are transmitted through the partially silvered end and make up the narrow, intense, coherent laser beam. The beam produced by a ruby laser has a wavelength of 694.3 nanometers, a deep red color.

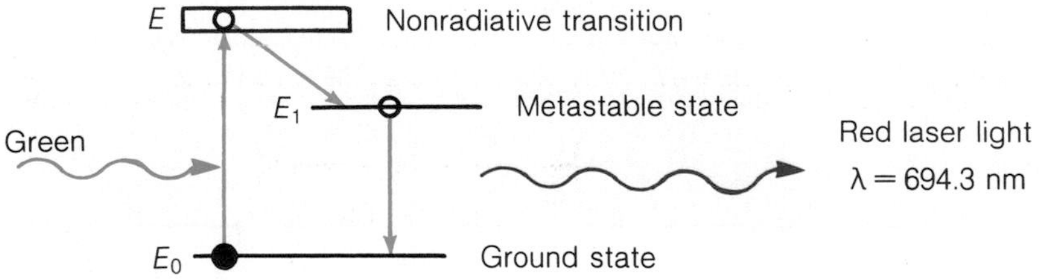

FIGURE 10.40

Schematic of a ruby laser system. The pumping radiation is provided by a high-intensity flash tube. The stimulating photons are reflected back and forth between the parallel end mirrors to build up a beam of high intensity. The laser beam consists of photons that escape through the partially transmitting end reflector at the right. Lasers of this type must be pulsed to avoid overheating the ruby rod and possibly cracking it.

E Nonradiative transition

E_1 Metastable state

Green

Red laser light
$\lambda = 694.3$ nm

E_0 Ground state

FIGURE 10.41

Energy level transitions used by a ruby laser. The chromium atoms are excited by green pumping radiation. A transition to the metastable state follows. Stimulated emission from this level produces the red laser light.

For persons growing up during the 1950s, it was possible to go to see movies "in 3-D." As you entered the theater the usher would hand you a pair of glasses with what appeared to be cellophane lenses, and while wearing them you did get a sense of depth—a third dimension—in the cinematic image.[1] True 3-D objects, however, generally change appearance as you move around them. You see different parts of the object depending upon where you stand or sit. Such was not the case with the old 3-D movies. The same aspects of the actors and actresses were seen from the right side of the theater as from the left—glasses and all. The images were basically two-dimensional because ordinary photographic film records only the intensity of the light striking it and nothing about its phase.

The production of true 3-D images that faithfully reproduce all aspects of the original objects was first accomplished in 1947 by Dennis Gabor. Gabor developed a way to preserve information about the relative phases of light beams emitted by or reflected from different points on an object in a complex *interference pattern*. A modern version of his experimental setup is shown in Figure 10.42. A beam of coherent radiation, in modern applications usually from a laser, is split in two, one beam traveling directly to a piece of photographic film, the second reaching the film after being reflected off some object. At the position of the film, the two beams recombine, interfering with one another to produce a complex pattern of bright and dark areas (Figure 10.43). (See also "Interference" in

FIGURE 10.42
Hologram production. Laser beams reflected from the mirror and from the object combine at the photographic plate to produce an interference pattern. The developed plate is the hologram.

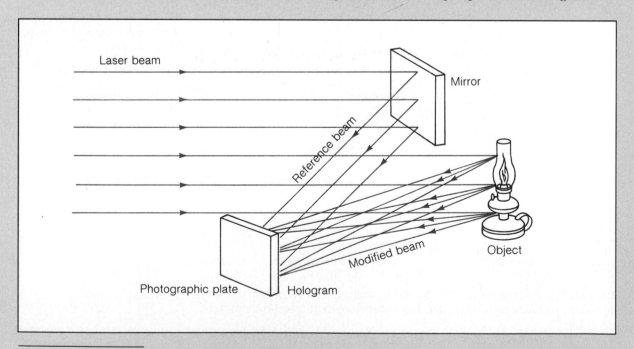

[1]We have a sense of depth or three dimensions largely because each eye sees slightly different views of the same scene. 3-D movies use two projectors with two films shot with slightly different perspectives. To keep the eyes from seeing both scenes at once, each projector is fitted with a Polaroid filter so the two images consist of light with two different planes of polarization. The 3-D viewing glasses have Polaroid filter lenses with the transmission axis of each lens matching the polarization of one of the projectors. Thus each eye sees the image from only one projector.

Section 9.1.) This pattern is recorded on the film and contains information not only about the relative intensities of the object and reference beams but also about their phase differences.

To see the image (ie, to reconstruct the object), coherent light of the same type as that used originally is shone on the film. Depending on the details of the original exposure process, the transmitted or reflected light is viewed obliquely to reveal the image (Figure 10.44). The resulting *hologram* (from the Greek word *holos,* meaning "whole") is a true 3-D image of the original object. Its aspect changes as you move your head; some parts of the object become visible while others disappear. The apparent shape, size, and brightness of the image change just as they would if you were examining the genuine article. Gabor's research earned him the Nobel Prize in physics in 1971.

When first perfected during the 1960s, holography—the science of producing holograms—was viewed as little more than a curiosity, a clever way to generate interesting 3-D images and to illustrate simultaneously some basic physical concepts. In recent years the ability of holograms to store huge quantities of information has been recognized by those in the computer industry involved in data storage and retrieval. (Dennis Gabor once estimated that the contents of an entire encyclopedia could be stored on a hologram no larger than 8 by 11 inches!) Other applications involve component testing and analysis in which two holograms of an object taken at different times are superimposed. Any changes in the object, like deformations produced by, say, forces exerted during industrial stress tests or by tumor growth beneath the surface of the object, show up immediately as interference patterns. Finally, researchers at the University of Michigan under the leadership of Emmett Leith have developed holographic filmstrips that can display art objects, rare artifacts, and old musical instruments in full three-dimensionality for interested students, something ordinary photographs simply cannot do, 3-D glasses notwithstanding (Figure 10.45).

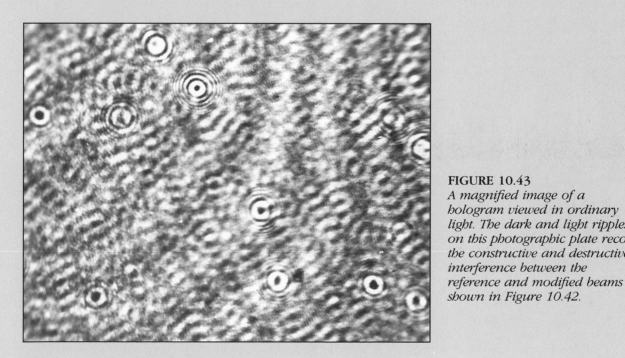

FIGURE 10.43
A magnified image of a hologram viewed in ordinary light. The dark and light ripples on this photographic plate record the constructive and destructive interference between the reference and modified beams shown in Figure 10.42.

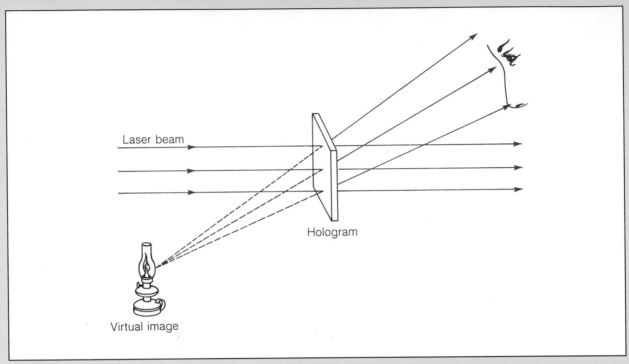

Laser beam

Hologram

Virtual image

FIGURE 10.44
Typical geometry used to view a hologram. It is a virtual image that appears on the other side of the plate.

FIGURE 10.45
Two different views of the same holographic image of chess pieces on a chess board. The three-dimensional quality of the image may be seen easily as a change in viewer perspective on the scene which accompanies a change in viewer position. Note also the change in image focus from one to the other which results from the "depth" of the holograph; different chessmen are at different distances from the camera, which has a fixed focal length.

TABLE 10.2 CHARACTERISTICS OF SOME COMMON LASERS		
Active Medium	Wavelength (nm)	Type
Helium-cadmium	441.6	Continuous
Argon	476.5, 488.0, 514.5	Continuous
Krypton	476.2, 520.8, 568.2, 647.1	Continuous
Helium-neon	632.8	Continuous
Ruby	694.3	Pulsed
Gallium arsenide	780–904[a] (IR)	Continuous
Neodymium	1060 (IR)	Pulsed
Carbon dioxide	10,600 (IR)	Continuous

[a]Depends on temperature.

Another common type of laser is the *helium-neon laser,* in which the lasing material is a mixture of about 15% helium gas and 85% neon gas. In this case the neon gas produces coherent radiation of wavelength 632.8 nanometers (red) as a result of stimulated emission from a metastable level to which it has been initially excited by collisions with the helium atoms. The helium-neon laser is a *continuous* laser in that it produces a steady beam. The ruby laser is a *pulsed* laser: it produces a single, short pulse each time the tube flashes.

Lasers have assumed increasingly large and important roles in a variety of areas since their invention in the late 1950s. Lasers emitting ultraviolet and infrared radiation, in addition to visible light, have been developed (Table 10.2). Today lasers are used to perform surgery to correct certain medical conditions (such as detached retinas) and to remove cancerous tumors of the skin, to precisely cut metals and other materials, to induce nuclear fusion reactions, to measure accurately the distances separating objects (the earth and the moon or two mirrors on opposite sides of a geological fault), to transmit telephonic information along optical fibers (see *Physics Potpourri* "Reflections: Fiber Optics" on page 405), to replay faithfully recordings of music (compact disc players), and to determine the price of goods at the supermarket checkout counter (Figure 10.46). Beyond our own earth, laserlike processes involving CO_2 have been discovered by astronomers in giant molecular cloud complexes sprinkled throughout our galaxy. Regardless of the location or application, lasers all operate according to the same basic physics—physics entirely accessible and comprehensible once the structure of the atom is understood.

HISTORICAL NOTES

It is extremely difficult to describe in a few paragraphs very much of the history associated with the rapid and revolutionary developments that occurred in physics during the 40-year period from 1895 to 1935. It is equally difficult, if not impossible, in the same space to do justice to the many brilliant men and women

(a)

(b)

(c)
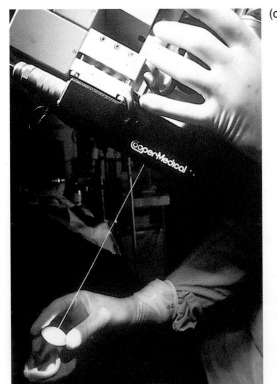

(d)

18

FIGURE 10.46
Applications of laser technology. (a) Laser welding. Dual lasers delivered by fiber optic cables are used to weld wires in auto headlamps. (b) Laser drilling. A 1000-watt laser shown drilling a hole through a 1/2-inch super alloy plate. (c) Laser surgery. An argon laser, with the aid of fiber optics, is used to operate inside the human ear. (d) A laser price scanner.

physicists and chemists who contributed their talents to bring about these developments. Some of the Suggested Readings on page 476 may help to convey the frequency with which fundamental advances in theory and experiment were made during this interval. The list of Nobel Prize recipients in physics and their citations for this same period (Appendix C) may help to establish the significance of the work done by some of the "giants" of this era. But such chronologies only begin to capture the excitement and spirit of adventure that existed within the scientific community at this time. Nowhere was this spirit more in evidence than at the Solvay Conferences held between 1911 and 1933.

Ernest Solvay (1838–1922) was a Belgian industrial chemist who perfected a process for producing sodium carbonate from sea salt, ammonia, and carbonic acid and who subsequently became extremely wealthy as the use of the Solvay process spread across the world. Solvay had long been interested in the fundamental structure of matter and by 1910 had become especially intrigued with the developing crisis between classical physics and the recently introduced quantum theories. Consequently, in an effort to promote discussion and to attempt to resolve the crisis, Solvay held an international conference in October 1911, to which he invited many of the most prominent physicists of the period, including Planck, Einstein, and Madame Curie.

The first Solvay Conference proved so successful that Solvay established a foundation in May 1912, whose objective was "to encourage the researches which would extend and deepen the knowledge of natural phenomena." One of the articles of the statutes of the foundation required that from time to time a conference similar to that held in 1911 be convened, "having as its goal the examination of significant problems in physics or physical chemistry." Over the years the Solvay Conferences (which continue to be held even today) have admirably met their goals. They have served as a living monument to a man who devoted his life and fortune to establishing programs and organizations designed to solve both social and scientific problems facing humanity.

The first Solvay Conference had as its topic radiation theory and the quanta; the second conference, in 1913, dealt with the structure of matter; while the third (1921), fourth (1924), fifth (1927), sixth (1930), and seventh (1933) meetings discussed atoms and electrons, the electrical conductivity of metals, electrons and photons, magnetism, and the structure and properties of atomic nuclei, respectively. Figure 10.47 is a group portrait of the participants of the 1927 Solvay Conference. The companies assembled for the other conferences held during this period were no less luminous than that shown in this photograph.

The progress made in the understanding and interpretation of natural phenomena during these early Solvay Conferences was considerable. In what follows, we will mention some of the highlights of these sessions in an attempt to convey the importance of the work done during these meetings and the intense, spirited, and ofttimes humorous manner in which it was carried out.

During the third conference the two major subjects discussed were the nuclear model of the atom, as proposed and elaborated upon by E. Rutherford, and Bohr's theory. During the discussion following Rutherford's report, three rather remarkable notions emerged—all of which were later confirmed as being substantially correct. (1) Jean Perrin suggested that the mechanism by which the sun derives its energy is the transmutation of hydrogen into helium, whereby radiation equivalent to the mass difference between the reactants and the product is produced according to Einstein's equation $E = mc^2$ (Section 11.7). (2) Rutherford, after agreeing with Perrin that huge quantities of energy could be released in the combination of hydrogen nuclei to form helium, then commented:

FIGURE 10.47

Participants of the 1927 Solvay Conference. Front row: I. Langmuir, M. Planck, M. Curie, H. A. Lorentz, A. Einstein, P. Langevin, Ch. E. Guye, C. T. R. Wilson, O. W. Richardson. Middle row: P. Debye, M. Knudsen, W. L. Bragg, H. A. Kramers, P. A. M. Dirac, A. H. Compton, L. de Broglie, M. Born, N. Bohr. Back row: A. Piccard, E. Henriot, P. Ehrenfest, Ed. Herzen, Th. de Donder, E. Schroedinger, E. Verschaffelt, W. Pauli, W. Heisenberg, R. H. Fowler, L. Brillouin.

It has occurred to me that the hydrogen atoms of the (solar) nebula might consist of particles which one might call the "neutrons," which would be formed by a positive nucleus with an electron at a very short distance. These neutrons would hardly exercise any force in penetrating into matter. They will serve as intermediaries in the assemblage of nuclei of elements of higher atomic weights.

The neutron was later discovered in 1932 by J. Chadwick. (3) Madame Curie argued that the stability of the nucleus could not be accounted for on the basis of electrostatic forces and that such stability required strong but very short-range, attractive forces of another type. These forces are now referred to as "nuclear" forces (Section 11.1).

The 1927 Solvay Conference focused on the new quantum mechanics and initiated what has been called the Einstein-Bohr dialogues, discussions (which were to be continued at the conference of 1930) concerning the implications of the probabilistic nature of this theory. Briefly, Bohr believed that at the subatomic level, the best one could do was to calculate the probabilities of finding a system in a given set of allowed states and that the very act of measurement involved interactions with the system that forced it into one of these allowed states where it is "observed." Numerous repeated measurements of the same type would reveal the system, on the average, to be in that state having the highest quantum mechanical probability.

Einstein was not at all happy with the lack of determinism in the quantum theory and argued that the quantum mechanical description of nature may not have exhausted all the possibilities of accounting for observable phenomena. He believed that, if carried further, the analysis would reveal a means of knowing precisely, with 100% surety, the outcome of an experiment of a particular type on a given quantum system. Einstein's attitude toward quantum mechanics, as regards its being a complete description of nature on a microscopic level, may be summarized in his own words: "To believe this is logically possible without contradiction: but it is so very contrary to my scientific instinct that I cannot forgo the search for a more complete conception."

The Einstein-Bohr dialogues produced some lively debates during the 1927 Solvay meeting. The chairman of the session, H. A. Lorentz, himself an eminent physicist and master of three languages (English, German, and French), tried to maintain order but found it exceedingly difficult to do so, as one speaker after another joined in the fray, each in his own language. At one point things became so confused that one of the participants, Paul Ehrenfest, a friend and colleague of both Einstein and Bohr, went to the blackboard and wrote: "The Lord did there confound the language of all the Earth." And, at a later time, during one of the lectures, Ehrenfest passed a note to Einstein which read: "Don't laugh! There is a special section in purgatory for professors of quantum theory, where they will be obliged to listen to lectures on classical physics ten hours every day." Present-day students of quantum theory might still find solace in Ehrenfest's remark!

It is interesting that what sparked the Einstein-Bohr discussions at the fifth Solvay Conference was the report by A. H. Compton summarizing a number of experimental results that could not be easily understood within the framework of classical physics, including an experiment he had performed that demonstrated a change in the frequency of x-rays scattered off free electrons (the so-called Compton effect). At the end of this report, during its discussion, Madame Curie, as she had done at an earlier Conference, made the prophetic remark that she thought that the Compton effect might have important applications in biology and that the high-voltage techniques used in generating the x-rays would find important uses for therapeutic purposes. As early as 1927, at an international congress of renowned physicists, we find the roots of therapeutic radiology being established side by side with those of the new theory of the quantum.

The foregoing discussion has attempted to present a little of the flavor of some of the early Solvay Conferences dedicated to the elucidation of nature in its smallest dimensions. The significance of the events at these meetings for later generations of physicists has been great indeed. The traditions of excellence established during these early conferences continue even to the present. As Werner Heisenberg, one of the founders of modern quantum theory and a Solvay participant in 1927, 1930, and 1933, wrote in 1974: "There can be no doubt that in those years (1911–1933) the Solvay Conferences played an essential role in the history of physics . . . the Solvay Meetings have stood as an example of how much well-planned and well-organized conferences can contribute to the progress of science. . . ."

SUMMARY

At the turn of the last century, developments in experimental physics, particularly those related to blackbody radiation, the photoelectric effect, and atomic spectra, forced theoretical physicists to introduce revolutionary new concepts that defied the classical physics as articulated by Newton and Maxwell. Key characteristics of these three processes were explained by assuming that things on the atomic scale are quantized. Light quanta (photons) have energy that is proportional to their frequency. The Bohr model of the atom has electrons restricted to certain orbits in which the angular momentum is quantized. Photons are emitted and absorbed when electrons jump from outer orbits to inner orbits, or vice versa.

During the 1920s and 1930s, the field of quantum mechanics emerged as it became clear that particles possess wavelike properties. This allowed a refinement of Bohr's model of the atom and led to a new probabilistic interpretation of physics on the atomic scale. No longer simply tiny particles, electrons are now represented as "clouds" surrounding nuclei.

The atomic theory put forth by Bohr and his colleagues has been widely applied since its inception in 1913. For example, in the years just before World War I, Moseley used Bohr's model to account for the characteristic x-ray spectra from heavy elements. More recently the establishment of the energy level structure of such elements as neon and chromium, including the identification of relatively long-lived metastable states, has been instrumental in the development of laser systems. Today quantum mechanics forms the foundation of our understanding of physics on the microscopic scale and stands as a monument to the combined genius of the men and women who participated in the early Solvay Conferences where the cornerstones of this branch of physics were laid.

SUMMARY OF IMPORTANT EQUATIONS

EQUATION	COMMENTS
$E = hf$	Energy quantum of atomic oscillator. Energy of a photon.
$\Delta E = hf$	Energy of photon emitted or absorbed in energy level transition.
$\lambda = \dfrac{h}{mv}$	de Broglie wavelength
$\Delta x \, \Delta mv = h$	Heisenberg's uncertainty principle

QUESTIONS

1. What does it mean when we say that the energy of something is "quantized?"

2. Of the things that a car owner has to purchase routinely—gasoline, oil, antifreeze, tires, and so on—which are normally sold quantized and which are not?

3. What assumption allowed Planck to account for the observed features of blackbody radiation?

4. What is a photon? How is its energy related to its frequency? its wavelength?

5. Describe the photoelectric effect. Name some devices that make use of this process.

6. If nature suddenly changed and Planck's constant became a much larger number, what effect would this have on things like solar cells, atomic emission and absorption spectra, lasers, and so on?

7. Based on what you learned about image formation in Chapter 9, describe how you might design a photocopying machine that could make a copy that is enlarged or reduced compared to the size of the original.

8. A mixture of hydrogen and neon is heated until it is luminous. Describe what is seen when this light passes through a prism and projects onto a screen.

9. What are the basic assumptions of the Bohr model? Describe how the Bohr model accounts for the production of emission-line spectra from elements like hydrogen.

10. Discuss what is meant by the term *ionization*. Give two ways by which an atom might acquire enough energy to become ionized.

11. Compare the emission spectra of the elements hydrogen and helium (see Figure 10.12). Which element emits photons of red light that have the higher energy?

12. If an astronomer examines the emission spectrum from luminous hydrogen gas that is moving away from the earth at a high speed and compares it to a spectrum of hydrogen seen in a laboratory on earth, what would be different about the frequencies of light?

13. A high-energy photon can collide with an electron and give it some energy. (This is called the Compton effect.) How are the photon's energy, frequency, and wavelength affected by the collision?

14. What is the de Broglie wavelength? What happens to the de Broglie wavelength of an electron when its speed is increased?

15. An electron and a proton are moving with the same speed. Which has the longer de Broglie wavelength? (You may want to look ahead at some useful information in Table 11.1.)

16. Name a device that directly uses the wave nature of electrons.

17. What is the uncertainty principle? What is its implication for the Bohr model of the atom?

18. What does it mean when a hydrogen atom is in its "ground state"?

19. Explain why a hydrogen atom with its electron in the ground state cannot absorb a photon of just any energy when making a transition to the second excited state ($n = 3$).

20. Will the energy of a photon that ionizes a hydrogen atom from the ground state be larger than, smaller than, or equal to the energy of a photon that ionizes another hydrogen atom from the first excited state ($n = 2$)? Why?

21. What would an energy level diagram look like for the quantized cat in Figure 10.3?

22. In what part of the EM wave spectrum does the Lyman series of emission lines from hydrogen lie? the Balmer series? the Paschen series? Describe how each of these series is produced. In what final state do the electron transitions end in each case?

23. The x-ray spectrum of a typical heavy element consists of two parts. What are they? How is each produced?

24. Will the frequency of the K_α-peak in the x-ray spectrum of copper (Cu) be higher or lower than the frequency of the K_α-peak of tungsten (W)? Explain how you arrived at your answer.

25. Describe how the Bohr model may be used to account for characteristics x-ray spectra in heavy atoms.

26. What is the origin of the word *laser*?

27. Distinguish between a metastable state and a normally allowed energy state within an atom. Discuss the role of metastable states in the operation of laser systems.

28. Define what is meant by the term *population inversion*. Why must this condition be achieved before a system can successfully function as a laser?

29. Describe the operation of a pulsed ruby laser.

30. What are the Solvay Conferences? In what ways did these meetings contribute to the development of quantum mechanics during the years 1911 to 1933?

PROBLEMS

1. Find the energy of a photon whose frequency is 1×10^{16} Hz.

2. If your body is emitting infrared radiation of wavelength 9.4×10^{-6} m, what is the energy of the released photons?

3. In what part of the EM spectrum would a photon of energy 9.5×10^{-25} J be found? What is its energy in eV?

4. Gamma rays (γ-rays) are high-energy photons. In a certain nuclear reaction, a γ-ray of energy 0.511 MeV (million electron volts) is produced. Compute the frequency of such a photon.

5. Electrons striking the back of a TV screen travel at a speed of about 8×10^7 m/s. What is their de Broglie wavelength?

6. In a typical electron microscope the momentum of each electron is about 1.6×10^{-22} kg-m/s. What is the de Broglie wavelength of the electrons?

7. During a certain experiment the de Broglie wavelength of an electron is 670 nm = 6.7×10^{-7} m, which is the same as the wavelength of red light. How fast is the electron moving?

8. If a proton were traveling the same speed as electrons in a TV picture tube (see Problem 5), what would its de Broglie wavelength be? The mass of a proton is 1.67×10^{-27} kg.

9. A hydrogen atom has its electron in the $n = 2$ state.
   a) How much energy would have to be absorbed by the atom for it to become ionized from this level?
   b) What is the frequency of the photon that could produce this result?

10. A hydrogen atom initially in the $n = 3$ level emits a photon and ends up in the ground state.
   a) What is the energy of the emitted photon?
   b) If this atom then absorbs a second photon and returns to the $n = 3$ state, what must the energy of this photon be?

11. The following is the energy-level diagram for a particularly simple, fictitious element Kansasium (Ks). Indicate by the use of arrows all allowed transitions leading to the emission of photons from this atom and order the frequencies of these photons from highest (largest) to lowest (smallest) (see Figure 10.48).

E_3 _____ $n = 3$

E_2 _____ $n = 2$

Energy

E_1 _____ $n = 1$

12. Characteristic x-rays emitted by molybdenum have a wavelength of 0.072 nm. What is the energy of one of these x-ray photons?

13. In a helium-neon laser, find the energy difference between the two levels involved in the production of the red light of wavelength 632.8 nm by this system.

14. The carbon dioxide laser is one of the most powerful lasers developed. The energy difference between the two laser levels is 0.117 eV.
    a) What is the frequency of the radiation emitted by this laser?
    b) In what part of the EM spectrum is such radiation found?

15. If you bombard hydrogen atoms in the ground state with a beam of particles, the collisions will sometimes excite the atoms into one of their upper states. What is the *minimum* kinetic energy the incoming particles must have if they are to produce such an excitation?

16. a) Which of the following elements emits a *K*-shell x-ray photon with the highest frequency?
    b) Which emits a *K*-shell photon with the lowest frequency?
       i) Silver (Ag)
       ii) Calcium (Ca)
       iii) Iridium (Ir)
       iv) Tin (Sn)

CHALLENGES

1. Can you think of a reason why metals exhibit the photoelectric effect most easily? Is there a connection between this phenomenon and properties of a good electrical conductor?

2. One serious problem with sending intense laser beams long distances through the atmosphere to receiving targets is that the beams spread out. This effect, called *thermal blooming,* is caused by the fact that the beam heats the air, thereby changing the speed that the light travels. Explain how such a change in the speed of light in the air could produce the observed effect. Will the speed increase or decrease upon heating? Why?

3. How might you explain the concept of quantization to a younger brother or sister, using money as the quantized entity?

4. The muon is a negatively charge particle that is much like an electron but with a mass about 200 times larger. A "muonic" hydrogen atom forms when a muon orbits a proton. The muonic atom's orbits and energy levels follow the basic rules of the Bohr model and the quantum mechanical model of the atom. Using the analysis in Section 10.5, explain why the size of the each Bohr orbit in the muonic atom is much smaller than the corresponding orbit in an ordinary hydrogen atom. How would the energies of each of the energy levels in the muonic atom be different than those in the regular hydrogen atom? What would be different about the emission spectrum?

5. The *aurora borealis* (or "northern lights") is produced when high-speed charged particles emitted by the sun (solar wind particles) enter the earth's upper, rarefied atmosphere and collide with air molecules there. Given this information, do you expect the emitted light to exhibit a continuous spectrum or a line spectrum?

   The rate at which solar wind particles enter the atmosphere is higher during the day than at night, yet the intensity of the auroral emissions remains high well after the sun has set. Can you suggest a means by which the atmospheric molecules might be able to radiate long after the period of collisions with charged particles has ended? (Hint: How long does it take a typical atom to radiate from a normal allowed energy state? How could that time be lengthened?)

SUGGESTED READINGS

Boraiko, Allen A. "A Splendid Light." *National Geographic* 165, no. 3 (March 1984): 335–377. Describes the diverse uses of lasers and holograms. Includes dozens of color photographs and a hologram on the cover.

Burl, Donald M., and Schein, Lawrence B. "Physics of Electrophotography." *Physics Today* 39, no. 5 (May 1986): 46–53. An in-depth look at the photocopying process.

Clark, Ronald W. *Einstein: The Life and Times.* New York: Avon, 1971. Perhaps the best biography of Einstein.

Gamow, George. *Biography of Physics.* New York: Harper and Row, 1961. Chapter 7 deals with the development of atomic physics and quantum mechanics.

Keller, Alex. *The Infancy of Atomic Physics. Hercules in His Cradle.* Oxford, England: Clarendon Press, 1983. A detailed account of the work of atomic scientists in the decades preceding World War I.

Segrè, Emilio. *From X-rays to Quarks.* San Francisco: W. H. Freeman and Co., 1980. An excellent book on the development of modern physics. Chapter 4 is about Planck, Chapter 7 is about Bohr, and Chapter 8 is about quantum mechanics.

"Special Issue: Lasers." *Physics Today* 41, no. 10 (October 1988): 24–63. Contains articles on the history of the laser, military laser research, and medical applications of lasers.

Wickramasinghe, H. Kumar "Scanned-Probe Microscopes." *Scientific American* 261, no. 4 (October 1989): 98–105. Describes the STM and other scanning microscopes that have evolved from it.

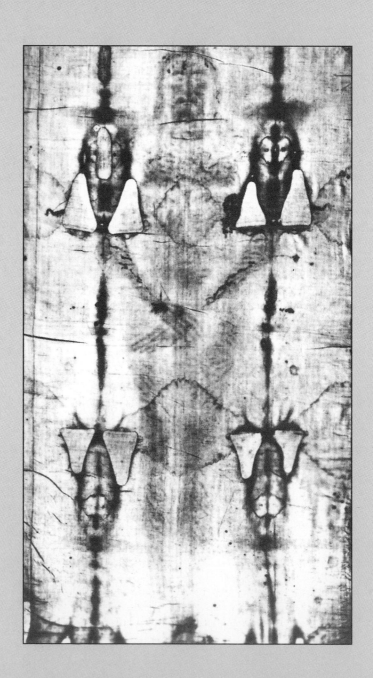

NUCLEAR PHYSICS

PROLOGUE: NUCLEAR SLEUTHS

In the middle of the 14th century a French nobleman revealed an extraordinary 14-foot long cloth that he claimed was the burial shroud of Jesus Christ. Known today as the Shroud of Turin, the cloth is imprinted with a faint image of a man with wounds like those described in the accounts of Christ's death. Since 1578 the cloth has been housed in a cathedral in Turin, Italy. During the intervening centuries, countless thousands of devout followers of Christianity came to regard it as a true religious relic. However, there were many skeptics, and the Catholic church itself never claimed that the shroud was authentic. In 1978 a team of 40 scientists were allowed to test the shroud in a variety of nondestructive ways in an attempt to determine the age of the cloth and the nature of the image. The test results were inconclusive and a source of controversy among the investigators. Then in 1988, a sophisticated dating procedure developed from the study of nuclear physics apparently laid the issue to rest.

For decades the *carbon-14 dating* process has been an accurate way to measure the ages of artifacts that were once alive—like wooden timbers, bones, and cloth. About one out of every trillion carbon atoms in living tissue is carbon-14 (the others are carbon-12 and carbon-13). The carbon-14 atoms "disappear" from the remains of plants and animals at the rate of about 1.2% each century, or 50% each 5,700 years. By measuring how much carbon-14 is still present in an artifact compared to the original amount, its age can be determined. Cloth made from flax grown at the time of Christ would have about 21% fewer carbon-14 atoms than similar cloth today.

The caretakers of the shroud understandably did not allow the standard carbon-14 dating process to be applied because it would have required cutting off a portion the size of a handkerchief. But by 1988 a new method of carbon-14 dating, called the *accelerator mass spectrometer method,* had been perfected to the point that a sample the size of a postage stamp was sufficient. The carbon atoms in the sample are extracted and placed in a special particle accelerator. The carbon-14 atoms, more massive than the carbon-12 and -13 atoms, accelerate more slowly and become separated from them. The machine isolates the carbon-14 atoms and counts how many are present.

Samples were cut from the shroud and were sent to university laboratories in Arizona, Oxford, and Zurich that specialize in this type of archaeological dating. The laboratories separately found that the carbon-14 content corresponded to cloth woven between the years 1260 and 1390. The Shroud of Turin is apparently a medieval fabrication.

Such scientific detective work is often used to determine if works of art are authentic. One of these procedures, called *neutron activation analysis,* is also an offshoot of nuclear physics. A suspect painting is irradiated with neutrons from a nuclear reactor. Different elements in the pigments in the painting absorb the neutrons and then emit nuclear radiation. Each element emits its own characteristic radiation that can be used like a fingerprint to indicate its presence. Prior knowledge of the chemical makeup of paints used by different artists in different time periods makes it possible to tell if a painting is real or a forgery.

The basic properties of nuclei and the many applications of nuclear processes, including carbon-14 dating and neutron activation analysis, are the main topics of this chapter. The first two sections describe the composition of the nucleus and the process of radioactive decay. This leads to the concept of half-life and the use of radioisotopes as natural clocks. Most of the remainder of the chapter deals with the role of energy in nuclear physics, particularly the two principal means of tapping that energy: nuclear fission and nuclear fusion.

11.1   THE NUCLEUS

The nucleus occupies the very center of the atom. It is tiny yet incredibly dense: more than 99.9% of the atom's mass is compressed into roughly *one trillionth* of its total volume. If an atom could be enlarged until it were 2,000 feet across, its nucleus would only be about the size of a pea. The nucleus is impervious to the chemical and thermal processes that affect its electrons. But it is seething with energy—energy that makes the sun and the other stars shine.

The nucleus contains two kinds of particles: protons and neutrons. The particles have nearly the same mass, about 1,840 times that of an electron. The masses are extremely small, and it is convenient to introduce an appropriate unit of mass, the *atomic mass unit, u.*

$$1\,u = 1.66 \times 10^{-27}\,kg$$

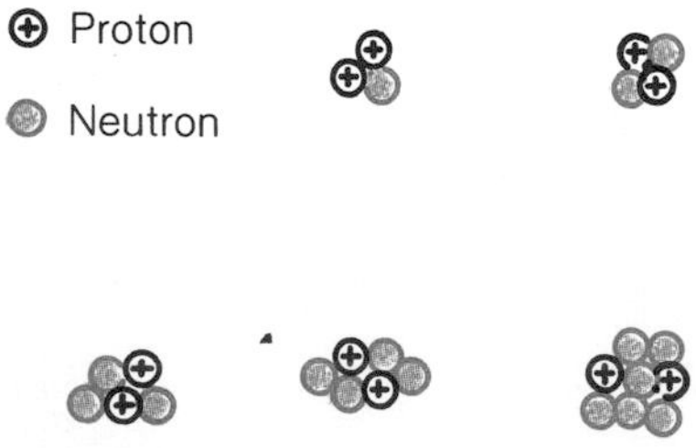

The proton carries a positive charge, and the neutron is uncharged (Table 11.1). The number of protons in the nucleus is the *atomic number Z* of the atom. There is one proton in the nucleus of each hydrogen atom ($Z = 1$), two in that of helium ($Z = 2$), eight in oxygen ($Z = 8$), and so on. This number determines the identity of the atom.

The neutral neutrons have much less influence on the properties of the atom: their main effect is on the atom's mass. In fact, the number of neutrons in the nuclei of a particular element can vary. Most helium atoms have two neutrons in the nucleus, but some have one, three, four, or even six (Figure 11.1). Each is still a helium atom; the only difference is the mass. This allows us to associate another number with each nucleus, the neutron number.

⊕ Proton

◯ Neutron

FIGURE 11.1
The different possible nuclei of helium atoms. All have two protons, but the number of neutrons varies. The electrons in orbit about the nuclei are not shown. With this scale the radius of such an orbit would be about 1,000 feet.

NEUTRON NUMBER   The number of neutrons contained in a nucleus.

TABLE 11.1 PROPERTIES OF THE PARTICLES IN THE ATOM		
Particle	Mass	Charge
Electron	9.110×10^{-31} kg $= 0.00055$ u	-1.602×10^{-19} C
Proton	1.67265×10^{-27} kg $= 1.00728$ u	1.602×10^{-19} C
Neutron	1.67495×10^{-27} kg $= 1.00866$ u	0

N is used to represent the neutron number just as Z is used to represent the atomic number. For all helium atoms $Z = 2$, but N can be 1, 2, 3, 4, or 6.

The mass of an atom is determined primarily by how many protons and neutrons there are in the nucleus. (The electrons are so light that they contribute only a negligible amount to the total mass.) Therefore it is useful to define yet a third number for each nucleus.

> ATOMIC MASS NUMBER   The total number of protons and neutrons in a nucleus.

The atomic mass number is represented by the letter A. Note that

$$A = Z + N$$

The atomic mass number indicates what the mass of a nucleus is, just as the atomic number indicates the amount of electric charge in the nucleus. Protons and neutrons are collectively referred to as *nucleons*. The atomic mass number is just the total number of nucleons in the nucleus. The possible atomic mass numbers for helium are $A = 3$, 4, 5, 6, or 8. Each different possible "type" of helium is called an *isotope*.

> ISOTOPES   Isotopes of a given element have the same number of protons in the nucleus but different numbers of neutrons.

The different isotopes of an element have essentially the same atomic properties: these are determined by the atomic number Z. For example, most of the carbon atoms in your body have six neutrons in their nuclei, but small numbers of the isotopes with seven and eight are also present. The three different carbon isotopes are indistinguishable as far as chemical processes (such as burning) are concerned. The different isotopes of an element often *do* have vastly different nuclear properties. For example, nuclear power plants use the splitting of uranium-235 nuclei; that is, uranium atoms with atomic mass number $A = 235$. Uranium-238 will not work.

Most of the 109 different elements have several isotopes. Some have only a few (hydrogen has 3), and others have more than 20 (iodine, silver, and mercury, to name a few). More than 2,200 different isotopes have been identified and studied. Of these, only about 300 occur naturally. The rest are produced in certain nuclear processes such as nuclear explosions. The vast majority of isotopes are unstable; the nuclei eventually transform into different nuclei through a process known as *radioactive decay*. More on this later.

Each different isotope of an element is designated by its atomic mass number. The most common isotope of helium has $A = 4$ ($Z = 2$ and $N = 2$) and is called helium-4. The other helium isotopes are helium-3, helium-5, helium-6, and helium-8. The three isotopes of carbon in your body are carbon-12, carbon-13, and carbon-14. Two of the isotopes of hydrogen have been given special names: hydrogen-2 is called deuterium, and hydrogen-3 is called tritium. (The prefixes come from the Greek words for "second" and "third".)

Freezing, boiling, burning, crushing, and other chemical and physical processes do not affect the nuclei of atoms. These processes are influenced by the forces between different atoms, forces that involve only the outer electrons. But nuclei are not indestructible: a number of nuclear processes do affect them. Nuclei can lose or gain neutrons and protons, absorb or emit gamma rays, split into smaller nuclei, or combine with other nuclei to form larger ones. These processes are called *nuclear reactions.* Some occur around us naturally, but most are produced artificially in laboratories. Most of the remainder of the chapter deals with the nature of nuclear reactions and their applications.

To diagram nuclear reactions, we use a special notation to represent each isotope. It consists of the atom's chemical symbol with a subscript and a superscript on the left side. The *subscript* is the atom's *atomic number Z,* and the *superscript* is the atom's *atomic mass number A.* Some examples:

$$
\begin{array}{ll}
\text{Helium-4} & {}^{4}_{2}\text{He} \\
\text{Carbon-12} & {}^{12}_{6}\text{C} \\
\text{Carbon-14} & {}^{14}_{6}\text{C} \\
\text{Uranium-235} & {}^{235}_{92}\text{U}
\end{array}
$$

The two numbers indicate the relative mass and charge that the nucleus possesses. The neutron number N can be found by subtracting the lower number (Z) from the upper number (A). For example, each uranium-235 nucleus contains 143 neutrons ($235 - 92 = 143$). Note that both the chemical symbol and the subscript indicate what the element is, so they must always agree. Regardless of what the superscript is, if the subscript is 6, the element is carbon and the symbol must be C.

This notation can be extended to represent individual particles as well. The designations for neutrons, protons, and electrons are

$$
\begin{array}{ll}
\text{Neutron} & {}^{1}_{0}\text{n} \\
\text{Proton} & {}^{1}_{1}\text{p} \\
\text{Electron} & {}^{0}_{-1}\text{e}
\end{array}
$$

For the electron, 0 is used for the mass number because its mass is so small. The -1 indicates that it has the same size charge as a single proton, except it is negative.

Perhaps you've wondered how protons can be bound together in a nucleus. After all, like charges do repel each other. There is indeed a strong electrostatic force acting to push the protons apart, but there is also another force many times stronger that acts to hold them together inside a nucleus. This is called the *strong nuclear force,* one of the four fundamental forces in nature. (The *weak nuclear force* is involved in certain nuclear processes, but it is not important in holding the nucleus together.) Compared to the gravitational and electromagnetic forces, the strong nuclear force is rather strange. It is much stronger than the others, but it has an extremely short range: the attractive nuclear force between particles in the nucleus effectively disappears if the particles become more than about

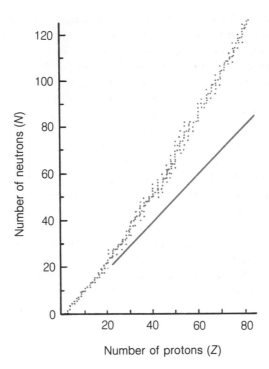

FIGURE 11.2
Plot showing the number of neutrons and the number of protons in the isotopes that are stable. Each dot represents one stable isotope. Its coordinates are the numbers N and Z. The solid line indicates the nuclei that would have the same number of protons and neutrons. The dots curve away from this line, indicating that larger nuclei have successively more neutrons than protons.

3×10^{-15} meters apart. This puts an upper limit on the size that a nucleus can have and still be stable. In a large nucleus protons on opposite sides are far enough apart that the repulsive electric force becomes important. No known stable isotope has an atomic number larger than 83.

The effectiveness of the nuclear force at holding a nucleus together also depends on the relative numbers of neutrons and protons. If there are too many or too few neutrons compared to the number of protons, the nucleus will not be stable. For example, carbon-12 and carbon-13 are stable, but carbon-11 and carbon-14 are not. The ratio of N to Z for stable nuclei is about 1 for small atomic numbers and increases to about 1.5 for large nuclei (Figure 11.2). The stable isotope lead-208 has 126 neutrons and 82 protons in each nucleus.

In summary, the nucleus is a collection of neutrons and protons held together by the nuclear force. This force, while strong, is limited to its ability to hold nucleons together. Nuclei that are too large or that do not have the correct ratio of neutrons to protons are unstable; they eject particles and release energy. There are several different mechanisms used by various unstable nuclei to accomplish this. The principal ones are discussed in Section 11.2.

11.2   RADIOACTIVITY

Radioactivity, also called *radioactive decay,* occurs when an unstable nucleus emits radiation. Isotopes with unstable nuclei are called *radioisotopes.* The majority of all isotopes are radioactive. For the moment we will put aside the question of where radioisotopes come from and concentrate on the processes of radioactive decay.

FIGURE 11.3
The three common types of nuclear radiation are affected differently by a magnetic field. Alpha and beta rays are deflected as they go through the field because they are charged particles. Gamma rays are not deflected.

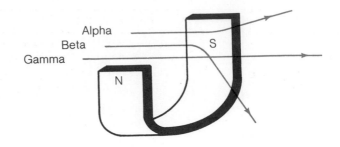

When it was first investigated, nuclear radiation was found to be similar to x-rays. For example, it exposes photographic film. Soon it was determined that there were actually three different types of nuclear radiation, named *alpha* (α), *beta* (β), and *gamma* (γ) radiation. Alpha rays and beta rays are actually high-speed charged particles. They can be deflected with magnetic and electric fields (Figure 11.3). Gamma rays are extremely high-frequency electromagnetic waves (high-energy photons).[1]

There are several different devices used to detect radiation, the most common being the *Geiger counter*. It exploits the fact that nuclear radiation is ionizing radiation. A gas-filled cylinder is equipped with a fine wire running along its axis (Figure 11.4). A high voltage is maintained between the outer wall and the wire, causing a strong electric field. When an alpha, beta, or gamma ray enters the cylinder and ionizes some of the atoms, the freed electrons are accelerated by the electric field. These in turn ionize other atoms, and an avalanche of electrons reaches the wire and causes a current pulse. Most Geiger counters emit an audible "click" each time a ray is detected and also keep track of how many are detected each second. In other words, they indicate the count rate—the number of alpha, beta, or gamma rays detected each second.

The emission of each type of radiation has a different effect on the nucleus. Both alpha decay and beta decay alter the identity of the nucleus (the atomic number is changed) as well as release energy. Gamma-ray emission does not in itself change the nucleus: it simply carries away excess energy in much the same way that photon emission carries away excess energy from excited atoms. Gamma-ray emission often accompanies alpha decay and beta decay; these two processes often leave the nucleus with excess energy. Table 11.2 lists several radioisotopes and their modes of decay.

---

[1]It is now known that there are many more ways that unstable nuclei can undergo radioactive decay. (There are several types of beta decay alone.) For the sake of simplicity, only the three most common forms of radioactive decay, alpha, beta (one type), and gamma decay, will be considered.

FIGURE 11.4
Simplified sketch of a Geiger counter. Nuclear radiation ionizes the gas in the cylinder. The freed electrons are accelerated to the wire and produce a current pulse.

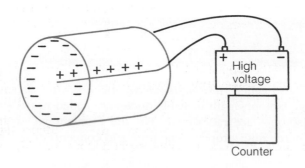

TABLE 11.2	PROPERTIES OF SELECTED ISOTOPES		
Element	Isotope	Decay Mode(s)	Relative Abundance
Hydrogen	^1_1H	. . . (Stable)	99.985%
	^2_1H deuterium	. . .	0.015%
	^3_1H tritium	beta	. . .
Helium	^3_2He	. . .	0.00014%
	^4_2He	. . .	99.9999%
	^5_2He	alpha	. . .
	^6_2He	beta	. . .
	^8_2He	beta, gamma	. . .
Carbon[a]	$^{12}_6C$	. . .	98.90%
	$^{13}_6C$	. . .	1.10%
	$^{14}_6C$	beta	trace
	$^{15}_6C$	beta	. . .
Silver[a]	$^{107}_{47}Ag^{*}$[b]	gamma, then stable	51.84%
	$^{108}_{47}Ag$	beta, gamma	. . .
	$^{109}_{47}Ag^{*}$[b]	gamma, then stable	48.16%
	$^{110}_{47}Ag$	beta, gamma	. . .
Uranium[a]	$^{232}_{92}U$	alpha, gamma	. . .
	$^{233}_{92}U$	alpha, gamma	. . .
	$^{234}_{92}U$	alpha, gamma	0.0055%
	$^{235}_{92}U$	alpha, gamma	0.72%
	$^{236}_{92}U$	alpha, gamma	. . .
	$^{237}_{92}U$	beta, gamma	. . .
	$^{238}_{92}U$	alpha, gamma	99.27%
	$^{239}_{92}U$	beta, gamma	. . .

[a]Not every isotope of the element is given.
[b]Asterisks (*) indicate that the nuclei are in an excited state.

Alpha Decay

An alpha particle is really four particles tightly bound together: two protons and two neutrons. It is identical to a nucleus of helium-4. For this reason an alpha particle can be represented as

$$\text{Alpha particle:} \quad \alpha \quad \text{or} \quad ^4_2He$$

A nucleus that undergoes alpha decay *does not contain a helium-4 nucleus;* this just happens to be a particularly stable combination of nuclear particles that can be ejected *en masse* from the nucleus.

The emission of an alpha particle reduces both the atomic number Z and the neutron number N by 2. The atomic mass number A is reduced by 4. Figure 11.5 is a diagram of the alpha decay of a plutonium-242 nucleus. The atomic number

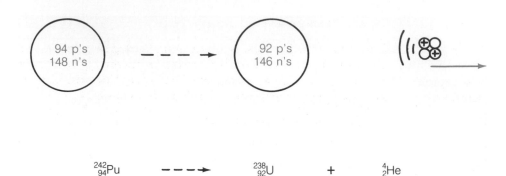

FIGURE 11.5
A nucleus of plutonium-242 undergoing alpha decay.

$$^{242}_{94}\text{Pu} \quad \dashrightarrow \quad ^{238}_{92}\text{U} \quad + \quad ^{4}_{2}\text{He}$$

of the nucleus decreases from 94 to 92: the nucleus is transformed from plutonium to uranium. The original plutonium-242 nucleus is called the "parent", and the resulting uranium-238 nucleus is called the "daughter".

Figure 11.5 shows why the isotopic notation introduced earlier is so convenient. When used to represent a nuclear process, like alpha decay, it clearly shows how the mass and the charge of the nucleus are affected. Both the total electric charge and the total number of protons and neutrons must be the same before and after the process. This means that the sum of the subscripts on the right side must equal the sum of those on the left. The same is true for the superscripts. This allows one to determine what the decay product (the daughter) is when a given nucleus undergoes alpha decay.

EXAMPLE 11.1   The isotope radium-226 undergoes alpha decay. Write the reaction equation, and determine the identity of the daughter nucleus.

From the periodic table of the elements in Appendix B, we find that the atomic number of radium is 88 and its chemical symbol is Ra. So the reaction will appear as follows:

$$^{226}_{88}\text{Ra} \rightarrow \, ? \, + \, ^{4}_{2}\text{He}$$

The atomic number A of the daughter nucleus must be 222 for the superscripts to agree on both sides of the arrow. By the same reasoning, the atomic number Z must be 86. From the periodic table, we find the element *radon* (*Rn*) has $Z = 86$. The daughter nucleus is radon-222.

$$^{226}_{88}\text{Ra} \rightarrow \, ^{222}_{86}\text{Rn} \, + \, ^{4}_{2}\text{He}$$

Because alpha decay causes such a drastic change in the mass of the nucleus, it generally occurs only in radioisotopes with high atomic numbers. The alpha particle is ejected with very high speed (typically around one twentieth the speed of light). Alpha particles are quickly absorbed when they enter matter: even a sheet of paper can stop them.

Beta Decay

Beta decay is easily the oddest of the three kinds of radioactivity. A beta particle is simply an electron ejected from a nucleus. This means that a beta particle has the same symbol as an electron.

Beta particle: β or $_{-1}^{0}e$

But wait a minute: there aren't any electrons in a nucleus. During beta decay, one of the neutrons is spontaneously converted into an electron (the beta particle) and a proton. The electron is ejected with very high speed, and the proton remains in the nucleus.[2] We can represent this process as follows:

$$\text{Neutron} \rightarrow \text{proton} + \text{electron}$$
$$_{0}^{1}n \rightarrow _{1}^{1}p + _{-1}^{0}e$$

The total electric charge remains the same—zero.

A nucleus that undergoes beta decay loses one neutron and gains one proton. Figure 11.6 shows the beta decay of a carbon-14 nucleus. Note that the atomic mass number A of the nucleus is unchanged.

The isotope iodine-131 undergoes beta decay. Write the reaction equation, and determine the identity of the daughter nucleus.

From the periodic table we find that iodine's chemical symbol and atomic number are I and 53. Therefore:

$$_{53}^{131}I \rightarrow ? + _{-1}^{0}e$$

The atomic mass number stays the same, and the atomic number is increased by 1 to 54. We find that this is the element *xenon* (*Xe*). The daughter nucleus is xenon-131.

$$_{53}^{131}I \rightarrow _{54}^{131}Xe + _{-1}^{0}e$$

EXAMPLE 11.2

Often the daughter nucleus in both alpha and beta decay is itself radioactive and decays into another isotope. Plutonium-242 undergoes alpha decay to uranium-238. This is a radioisotope that undergoes alpha decay into thorium-234. This process continues until, after a total of nine alpha decays and six beta decays, the stable isotope lead-206 is reached. This is called a *decay chain*. When the earth was formed from the debris of exploded stars, hundreds of different radioisotopes were present in varying amounts. Geological formations that were originally rich in uranium-238 now contain large amounts of lead-206 as well.

---

[2]Another particle, a *neutrino* (from the Italian for "little neutral one"), is also emitted in beta decay. Neutrinos are very strange little beasts; they have no charge, their mass is extremely small, and they rarely interact with matter. Neutrinos routinely pass through the entire earth without being absorbed, deflected, or otherwise affected. For our purposes the neutrino can be regarded as just part of the energy that is released during beta decay. (More on neutrinos in Chapter 12.)

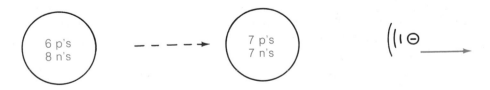

| 6 p's | 7 p's |
| 8 n's | 7 n's |

$_{6}^{14}C$ - - - → $_{7}^{14}N$ + $_{-1}^{0}e$

FIGURE 11.6
A nucleus of carbon-14 undergoing beta decay.

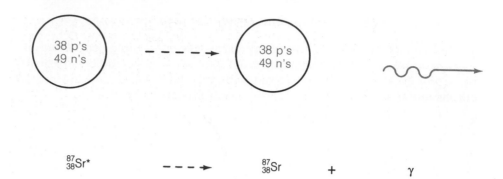

FIGURE 11.7
A nucleus of strontium-87 undergoing gamma decay. The asterisk (*) indicates that the nucleus is in an excited state.

Gamma Decay

Since a gamma ray has no mass or electric charge, gamma-ray emission has no effect on the atomic mass number A or the atomic number Z of a nucleus. Nuclei can exist in excited states in much the same way that orbiting electrons can. Gamma rays are emitted when the nucleus goes to a lower energy state. Figure 11.7 shows the gamma decay of a strontium-87 nucleus. The identity of the nucleus is not changed during the process.

Gamma-ray photons, as indicated in Figure 10.7, have energies from about 100,000 electron volts to more than a billion electron volts. Most gamma-ray photons emitted in gamma decay are around 1 million electron volts. (The unit of energy that is used most often in nuclear physics is the megaelectron volt (MeV), which is 1 million electron volts.)

One way to compare the different decay processes is to focus on three properties of the nucleus: mass, electric charge, and energy. Alpha decay alters all three: an alpha particle carries away considerable mass, two charged protons, and a great deal of kinetic energy. Beta decay has little effect on the mass of the nucleus, but it does increase the positive charge of the nucleus and take away energy in the form of the beta particle's kinetic energy. Gamma decay only takes energy away from the nucleus.

The three types of nuclear radiation differ considerably in their ability to penetrate solid matter. All three are ionizing radiation; they ionize atoms as they pass through matter. Alpha particles are the least penetrating. The positive charge causes them to interact strongly with atomic electrons and nuclei, and they are absorbed after traveling only a short distance. Beta particles are more penetrating than alpha particles. For comparison, a thin sheet of aluminum that will block essentially all alpha particles will stop only a fraction of beta particles. Gamma

FIGURE 11.8
Alpha, beta, and gamma radiation differ a great deal in their ability to penetrate matter. Alpha rays are the least penetrating, while gamma rays are the most.

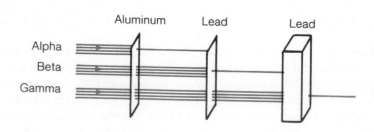

FIGURE 11.9
The electricity used by the Voyager spacecraft is generated from the energy released in radioactive decay.

rays are the most penetrating of the three. Since they have no electric charge and travel at the speed of light, they interact with atoms much less frequently. It requires several centimeters of lead to block gamma rays (see Figure 11.8).

What becomes of the energy of nuclear radiation as it is absorbed? Most of it goes to heat the material. Early experimenters with radioactivity noticed that highly radioactive samples were physically warmer than their surroundings. As long as the substance continued to emit radiation, it stayed warm. Radioactivity has been used as an energy source on some deep-space satellites. The Voyager spacecraft that photographed Saturn, Jupiter, Uranus, and Neptune, use a plutonium isotope in what are called thermoelectric generators (Figure 11.9). The heat from the radioactivity is converted into electricity to operate cameras, radio transmitters, and other on-board equipment.

The interior of the earth is so hot that much of it is molten. Some of this heat reaches the surface in volcanoes, in geysers, and through other geothermal processes. Geologists are not certain why the earth's interior is so hot, but they know what keeps it from cooling off: heat from radioactive decay. In the earth's interior, relatively small amounts of radioisotopes are still present. The energy released as these radioisotopes decay is enough to compensate for the conduction of heat to the earth's surface. Since much of this radioactive material is concentrated in a layer just below the earth's outer crust, this layer is kept in a partially liquid state by heat from the radioactivity. This in turn allows the crustal plates (continental land masses) to slowly slide around over the earth in a process called *continental drift*.

Analysis of radioactive decay is an important tool in nuclear physics. The type of radiation emitted by a particular radioisotope, along with the amount of energy released, provide clues to the structure of the nucleus. Since nuclei are much too small for us to examine with our eyes, we have to use indirect information, such as that carried by radiation when it leaves the nucleus, to learn about them.

11.3   HALF-LIFE

FIGURE 11.10
The exact value of one roll of dice can't be predicted, but the approximate number of times that seven will turn up in 1,000 rolls can be predicted quite closely. The exact time at which a given nucleus will decay can't be predicted, but the fraction of 1 million nuclei that will decay during some time interval can be predicted quite closely.

We now address an aspect of radioactive decay that we have avoided so far—time. If a nucleus is radioactive, when will it decay? There is no way of knowing exactly when a particular nucleus will decay: it may wait a billion years or a millionth of a second. Radioactive decay is a random process, much like throwing dice (Figure 11.10). One can't predict exactly what will come up each throw, but one can analyze the results of hundreds of throws and indicate how likely it is that each possible value will come up. In a similar way, we can predict how much time will elapse, on the average, before a nucleus of a given radioisotope will decay. We find that there is wide variation among the hundreds of different radioisotopes. Nearly all nuclei of some isotopes will decay in less than a second, while only a fraction of the nuclei of other isotopes will decay in 4.5 billion years, the age of the earth. This leads us to the concept of *half-life*.

> **HALF-LIFE**   The time it takes for half the nuclei of a radioisotope to decay. The time interval during which each nucleus has a 50% probability of decaying.

The half-lives of radioisotopes range from a tiny fraction of a second to billions of years (Table 11.3). During the span of one half-life, approximately half of the nuclei in a sample will decay, that is, emit their radiation. Half of the remaining nuclei will decay during the span of a second half-life, leaving only one fourth of the original nuclei. After three half-lives, one eighth remain, and so on. After n half-lives, one half raised to the n power of the original nuclei will remain undecayed.

For example, let's say that we start with 8 million nuclei of a radioisotope with a half-life of 5 minutes. About 4 million of the nuclei will decay in the first 5 minutes. Half of the remaining 4 million will decay during the next 5 minutes, leaving 2 million. After 15 minutes there will be 1 million left undecayed. After 50 minutes (10 half-lives), there will be about 7,800 nuclei left (8 million $\times$ $(\frac{1}{2})^{10}$ = 7,800).

EXAMPLE 11.3   A pure sample of uranium-237 is prepared. As the uranium nuclei decay, the sample becomes "contaminated" with decay products. How much time will elapse before only one fourth of the sample is uranium-237?

One half of the uranium nuclei decay during 6.75 days, the half-life (from Table 11.3). After another 6.75 days, one half of the remaining nuclei decay, leaving one fourth of the original amount. Therefore, after a total of *13.5 days,* only one fourth of the sample will be uranium-237.

In reality it isn't possible to count how many nuclei are left undecayed every 5 minutes. But with a Geiger counter, one can keep track of how the rate of decay, the number of nuclei that decay each minute, decreases. In the previous example, 4 million decay during the first 5 minutes, 2 million during the next 5 minutes, and so on. A Geiger counter placed nearby might show an initial count rate of 10,000 counts per minute. (The count rate would depend on how close the counter is to the sample.) Five minutes later, the count rate would be one half

Element	Isotope	Half-Life
TABLE 11.3 HALF-LIVES OF ISOTOPES IN TABLE 11.2		
Hydrogen	$^{1}_{1}H$	...
	$^{2}_{1}H$ deuterium	...
	$^{3}_{1}H$ tritium	12.3 years
Helium	$^{3}_{2}He$	...
	$^{4}_{2}He$	...
	$^{5}_{2}He$	2×10^{-21} s
	$^{6}_{2}He$	0.805 s
	$^{8}_{2}He$	0.119 s
Carbon[a]	$^{12}_{6}C$	...
	$^{13}_{6}C$	...
	$^{14}_{6}C$	5,730 years
	$^{15}_{6}C$	24 s
Silver[a]	$^{107}_{47}Ag$*[b]	44.2 s
	$^{108}_{47}Ag$	2.42 min
	$^{109}_{47}Ag$*[b]	39.8 s
	$^{110}_{47}Ag$	24.6 s
Uranium[a]	$^{232}_{92}U$	70 years
	$^{233}_{92}U$	159,000 years
	$^{234}_{92}U$	245,000 years
	$^{235}_{92}U$	704,000,000 years
	$^{236}_{92}U$	23,400,000 years
	$^{237}_{92}U$	6.75 days
	$^{238}_{92}U$	4,470,000,000 years
	$^{239}_{92}U$	23.5 min

[a]Not every isotope of the element is given.
[b]Asterisks (*) indicate that the nuclei are in an excited state.

that, about 5,000 counts per minute. In other words, the count rate also is halved during each half-life (Figure 11.11). Thus by monitoring the count rate of the nuclear radiation emitted by a radioisotope, one can determine the isotope's half-life.

DO-IT-YOURSELF PHYSICS

You can simulate the decay of radioactive nuclei with a large number of pennies or similar flat objects. It's best to use 50 or more. Place the pennies in a box or bag and shake them thoroughly. Dump the pennies out onto a flat surface, like a desktop. Treat each penny with "tails" showing as a nucleus that

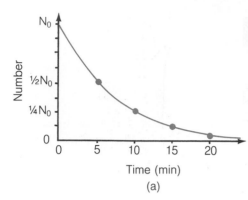

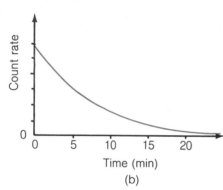

has decayed; push these off to the side. Collect the "undecayed" pennies and repeat the procedure.

Each penny has a 50% chance of turning up "tails" after each throw. Thus each throw represents one half-life for the pennies; on the average, half of them "decay" during each throw.

Notice how many decay with each throw and how the number of pennies left undecayed decreases. If you record these numbers and graph them, you should get graphs like those in Figure 11.11. Notice the statistical nature of the process: after, say, the fifth throw, any surviving penny has just turned up "heads" five times in a row. In spite of this, its chance of turning up "heads" on the next throw is still 50%.

FIGURE 11.11
(a) Graph of the number of remaining nuclei versus time for a radioisotope with a 5-minute half-life. (b) Graph of the count rate versus time for the same radioisotope.

Is it useful to know the half-lives of radioisotopes? Yes, for a number of reasons. Knowing the half-life of a radioisotope allows one to estimate over what period of time a sample will emit appreciable amounts of radiation. Small amounts of certain radioisotopes are sometimes used medically as "tracers" in the human body. The flow of the isotope through the blood stream (for example) can be monitored with a Geiger counter. The radioisotope that is used must have a long enough half-life to remain detectable during the time that it takes to move through the body. Its half-life must not be too long however. To minimize any possible harmful effects, the body should not be subjected to the radiation any longer than necessary.

Dating

The regular rate of decay of radioisotopes can be used as a "clock." *Carbon-14 dating* is a good example of this application. Carbon is a key element in all living things on earth. Carbon enters the food chain as plants take in carbon dioxide from the air. The complex carbon-based molecules formed by plants are ingested by animals and people. About 99% of this carbon is the stable isotope carbon-12. About one out of every trillion carbon atoms is radioactive carbon-14. In the upper atmosphere carbon-14 is constantly being produced as cosmic rays from outer space collide with atoms in air molecules. Some of the fragments are free neutrons that cause the formation of carbon-14 when they collide with nitrogen-14. The reaction goes like this:

$$\,^{1}_{0}n + \,^{14}_{7}N \rightarrow \,^{14}_{6}C + \,^{1}_{1}p$$

(See Figure 11.12.) The carbon-14 atoms can combine with oxygen atoms to form carbon dioxide molecules, then mix in the atmosphere and enter the food chain. A small percentage of all the carbon atoms in plants, animals, and people

FIGURE 11.12
The formation of carbon-14 from nitrogen-14. This process occurs naturally in the upper atmosphere.

are carbon-14. (You are slightly radioactive! The amount of carbon-14 is so small that the radiation—beta particles—is not a hazard.) The half-life of carbon-14 is about 5,700 years, so very little of it decays during the lifetime of most organisms. (An exception would be certain trees that can live thousands of years.)

After an organism dies, no new carbon-14 is added, so the percentage of carbon-14 decreases as the nuclei undergo radioactive decay. This can be used to measure the age of the remains. For example, if a tree is cut down and used to build a shelter, 5,700 years later there will be one half as much carbon-14 in the wood as in a live tree (Figure 11.13). After 11,400 years, there would be one fourth as much, and so on. Thus by measuring the carbon-14 content in ancient logs, charcoal, bones, fabric, or other such artifacts, archaeologists can estimate their age. This process has become an invaluable tool to archaeologists and is quite accurate for material ranging up to 40,000 years old.

Geologists use radioisotopes to estimate the ages of rock formations and the earth itself. The ratio of the amount of the parent radioisotope to the amount of the daughter is used. For example, about 28% of the naturally occurring atoms of the element rubidium are the radioisotope rubidium-87. Its half-life is so long—49 billion years—that only a small percentage of it has decayed since the earth was formed. Rubidium-87 undergoes beta decay into the stable isotope strontium-87. The age of rock formations that contain rubidium-87 can be estimated by measuring the relative amount of strontium-87 that is present. The more strontium-87 present, the older the rock.

Because of radioactive decay, most of the radioisotopes present when the earth formed have long since decayed away. Only radioisotopes with very long half-lives, like rubidium-87 and uranium-238, have survived to this day. A few radio-isotopes with short half-lives occur naturally because they are constantly being produced, carbon-14 being a good example. The human race has added a large number of radioisotopes to the environment in this century through nuclear explosions and nuclear power production. (Concern over the amount of radio-active fallout in the atmosphere produced by nuclear weapons testing led to the limited Test Ban Treaty of 1963.)

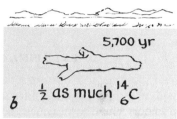

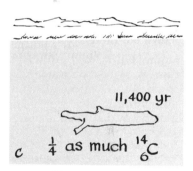

FIGURE 11.13
(a) As soon as a tree dies, the amount of carbon-14 in it begins to decrease. (b) After 5,700 years (one half-life), it contains one half as much. (c) After 11,400 years, it contains one fourth as much.

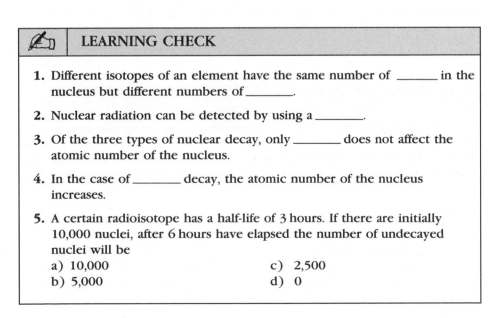

LEARNING CHECK

1. Different isotopes of an element have the same number of _____ in the nucleus but different numbers of _____.

2. Nuclear radiation can be detected by using a _____.

3. Of the three types of nuclear decay, only _____ does not affect the atomic number of the nucleus.

4. In the case of _____ decay, the atomic number of the nucleus increases.

5. A certain radioisotope has a half-life of 3 hours. If there are initially 10,000 nuclei, after 6 hours have elapsed the number of undecayed nuclei will be
   a) 10,000   c) 2,500
   b) 5,000   d) 0

We are all constantly exposed to small amounts of nuclear radiation. This is as much a part of our natural existence on this planet as being exposed to ultraviolet radiation from the sun. In both cases moderate exposure throughout one's lifetime is not considered harmful.

The natural sources of radiation that people have always been exposed to include cosmic radiation from outer space, consisting mainly of high-speed protons and nuclei; radioisotopes in the soil and rocks; and radioisotopes in the human body, like carbon-14 and potassium-40, and in the air we breathe. In the last century, human activity has added to these natural sources of radiation a number of other sources. They include medical and dental x-rays; radioisotopes released by nuclear research activities, the production of nuclear weapons, and the nuclear power industry; and traces of radioactive fallout from nuclear weapons testing (on the decline since the limited Test Ban Treaty of 1963).

The total amount of radiation from these sources is usually too small to be injurious to one's health. But in many cases people have been exposed to much more massive doses. Accidents in research laboratories, nuclear materials processing facilities, and nuclear reactors, such as the Chernobyl power plant accident in 1986; the atomic bombing of Hiroshima and Nagasaki; and some nuclear weapons tests, particularly in the 1950s, have exposed scientists, soldiers, and civilians to high levels of radiation. Such massive doses can be harmful and, if large enough, fatal.

Nuclear radiation damages living tissue because it is ionizing radiation; it ionizes atoms along its path as it travels through matter. A single alpha particle may have enough energy to ionize more than 1,000 atoms. Inside living cells this ionization can break up molecules, thereby killing the cell or affecting its ability to reproduce. The resulting damage to the living organism can be classified as *somatic* or *genetic*.

Somatic effects are direct injury to the organism itself. In the course of normal maintenance, the body replaces millions of dead cells each day. But if radiation kills too many cells, the body cannot keep up, and *radiation sickness* results. This is characterized by nausea, fatigue, loss of hair, reddening of the skin, and other symptoms. Even if a cell is not killed, radiation damage to molecules in a *gene* inside the cell causes a *mutation*. Since genes control the cell's ability to reproduce itself, mutations can lead to abnormal cell growth and cancer.

Genetic effects can occur when the genes of cells specialized for reproduction are damaged. The resulting mutations can cause birth defects in the immediate offspring or defects that do not appear for several generations.

Clearly it is important to be able to measure radiation doses and to know what levels are poten-

11.4 ARTIFICIAL NUCLEAR REACTIONS

Radioactivity and the formation of carbon-14 from nitrogen-14 are examples of natural nuclear reactions. With the exception of gamma decay, each of these results in *transmutation*—the conversion of an atom of one element into an atom of another (the medieval alchemists' dream come true). Many other types of nuclear reactions can be induced artificially in laboratories. A common example of this is the bombardment of nuclei with alpha particles, beta particles, neutrons, protons, or other nuclei. If a nucleus "captures" the bombarding particle, it will become a different element or isotope. (This is called *artificial transmutation.* Note that here something is added to the nucleus, whereas in radioactive decay something leaves the nucleus.)

tially harmful. Radiation exposure can be measured in a couple of ways. The *rad* is a unit used to measure the amount of radiation a substance absorbs. One rad of radiation means that each kilogram of the substance absorbed 0.01 joule of energy from the radiation:

$$1 \text{ rad} = 0.01 \text{ J/kg}$$

The corresponding SI unit is the *gray*. One gray equals 1 J/kg or 100 rads.

The different types of ionizing radiation do not have the same ability to cause harmful effects. For example, 1 rad of alpha radiation tends to do more harm than one rad of gamma radiation. For this reason the *rem* (*rad* equivalent *man*) was introduced. The rem takes into account the relative biological damage that each kind of radiation causes. One rad of gamma radiation is rated at 1 rem, but one rad of alpha radiation is rated at about 20 rems. The corresponding SI unit is the *sievert* (Sv). One sievert equals 100 rems.

Analysis of the effects of radiation exposure has led the government to establish guidelines on maximum safe levels. People who work with sources of radiation are allowed a maximum dose of 5 rems accumulated throughout a year. It is stipulated that the general public's dose from artificial sources should be less than 0.17 rems per person per year. (The *millirem* [mrem] is the unit used for small doses; 0.17 rems equals 170 millirems.)

What levels of radiation are harmful? Since some of the somatic effects like cancer and leukemia may not show up in a victim for decades and since some genetic effects may not show up for a century or more, it is difficult to determine what is the highest dose that will *not* cause *any* harm. Studies of accident and bomb victims do give us a fairly good idea of the effects of massive doses of radiation. Doses below about 25 rems cause no immediate effects. A 100-rem dose causes mild radiation sickness but is not usually fatal. Higher doses lead to more severe radiation sickness and an increasingly larger percentage of fatalities. (A dose of 500 rems is fatal about 50% of the time.) The radiation sickness symptoms can subside for a couple of weeks and then recur. Doses above 800 rems are almost always fatal.

The normal sources of background radiation mentioned earlier usually result in very low doses, with large variations due to locality and other factors. For example, people who live at high altitudes or who often travel by air are exposed to higher amounts of cosmic radiation. The content of radioisotopes in the earth varies a great deal. A major concern in many areas of the country is the accumulation of radioactive radon, a gaseous element, inside buildings, particularly basements. Radon-222, a radioisotope in the decay chain of uranium-238, seeps through the ground, into groundwater, and eventually into basements. Hazardous levels

One example of a useful artificial reaction involves bombardment of uranium-238 nuclei with neutrons. The result is uranium-239.

$$^{238}_{92}\text{U} + ^{1}_{0}\text{n} \rightarrow ^{239}_{92}\text{U}$$

This new nucleus undergoes two beta decays, resulting in plutonium-239.

$$^{239}_{92}\text{U} \rightarrow ^{239}_{93}\text{Np} + ^{0}_{-1}\text{e}$$

$$^{239}_{93}\text{Np} \rightarrow ^{239}_{94}\text{Pu} + ^{0}_{-1}\text{e}$$

The plutonium-239 can be used directly in nuclear reactors, but the original uranium-238 can't. A breeder reactor is a nuclear reactor designed to use neutrons to produce, or "breed" reactor fuel: the plutonium-239. This process also shows how the elements with Z greater than 92 can be produced artificially.

In a similar way, neutron bombardment can be used to produce hundreds of other isotopes, most of which are radioactive. *Neutron activation analysis* is an can accumulate in inadequately ventilated buildings located in regions with relatively high concentrations of uranium in underground rock formations.

The average dose from natural sources of radiation is about 0.13 rems per year. The artificial sources add another 0.07 rems per year on the average. Each medical or dental x-ray gives a dose of around 0.01 to 0.1 rems (10 to 100 millirems), but poor procedures can raise that to more than 1 rem. The radiation from nuclear power plants, weapons testing, and their associated mining and processing operations gives an average annual dose of less than 0.01 rem.

One of the biggest dangers of fallout is that some isotopes tend to enter the food chain and become more concentrated in certain organisms. For example, isotopes of strontium and radium behave chemically much like calcium. The body naturally accumulates them in the bones with the calcium. Once there, the isotopes are a steady source of radiation that is absorbed by the surrounding tissue.

The delayed somatic and genetic effects are now assumed to arise in proportion to the radiation dose; even small doses of radiation increase the probability that a person will develop cancer, leukemia, or genetic defects. This approach is used as the basis for setting maximum recommended doses. Another theory holds that there is a *threshold* level of radiation exposure below which there are no harmful effects.

Radiation is a part of life, and the human race has flourished in spite of it. As long as the exposure is kept low, we still have nothing to fear from it.

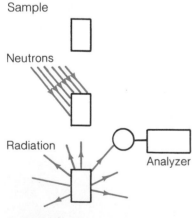

Sample

Neutrons

Radiation

Analyzer

FIGURE 11.14
In neutron activation analysis, the sample is irradiated with neutrons. This produces radioisotopes in the sample. Analysis of the radiation that is emitted makes it possible to determine the original composition of the sample.

accurate method for determining what elements are present in a substance. The material to be tested is bombarded with neutrons. Many of the nuclei in the substance are transformed into radioisotopes. By monitoring the radiation that is then emitted, it is possible to determine which elements were originally present (Figure 11.14). For example, the only naturally occurring isotope of sodium, sodium-23, becomes radioactive sodium-24 by neutron absorption.

$$^{23}_{11}\text{Na} + ^{1}_{0}\text{n} \rightarrow ^{24}_{11}\text{Na}$$

The sodium-24 emits both gamma and beta rays with specific energies. If sodium is present in an unknown substance, it can be detected by bombarding the substance with neutrons and then looking to see whether gamma and beta rays with the proper energies are emitted.

Neutron activation analysis has been used to make some interesting discoveries. An analysis of a lock of Napoleon Bonaparte's hair revealed a high concentration of arsenic. Napoleon may have been poisoned, a possibility that was revealed by neutron activation analysis a century and a half after his death.

Neutron activation analysis is used in some airports to search luggage for explosives. The element nitrogen is abundant in most chemical explosives, including the type of plastic explosive that destroyed a 747 jet over Lockerby, Scotland in 1988, killing 270 people. The bomb detector irradiates luggage with neutrons and then looks for the characteristic radiation emitted by nitrogen that could indicate the presence of an explosive. The device's usefulness is hampered because some legitimate materials are rich in nitrogen, and the detector has difficulty sensing small quantities of explosive.

11.5   NUCLEAR BINDING ENERGY

A very important and useful aspect of a nucleus is its binding energy. Imagine the following experiment being performed: a nucleus is dismantled by removing each proton and neutron one at a time, and the total amount of work done in the process is measured. If these protons and neutrons are reassembled to form the original nucleus, an amount of energy equal to the work done would be released. This energy is called the *binding energy* of the nucleus. It indicates how tightly bound a nucleus is. A useful quantity for comparing different nuclei is the average binding energy for each proton and neutron, called the *binding energy per nucleon.* It is just the total binding energy divided by the total number of protons and neutrons in the nucleus—the atomic mass number. If we measure the binding energy per nucleon for all of the elements, we find that it varies considerably, from about 1 MeV to nearly 9 MeV (see Figure 11.15).

Nuclei with atomic mass numbers around 50 have the highest binding energy per nucleon: the protons and neutrons are more tightly bound to the nucleus than in larger or smaller nuclei, so these nuclei are the most stable. (Iron-56 has the largest binding energy per nucleon.) The high stability of the helium-4 nucleus (alpha particle) is indicated by the small peak in the graph at $A = 4$.

A proton or neutron bound to a nucleus is similar to a ball resting in a hole in the ground. The ball's binding energy is the amount of energy that would have to be provided (the amount of work that would have to be done) to lift it out of the hole. A deeper hole means a higher binding energy.

The graph of binding energy per nucleon indicates how one can tap nuclear energy. Imagine taking a large nucleus with A around 200 and splitting it into two smaller nuclei. The graph shows that each of the smaller nuclei, with A around 100, has higher binding energy per nucleon than the original nucleus. All of the neutrons and protons have become more tightly bound together and have *released energy* in the process. (This is like moving the ball to a deeper hole. Its potential energy is decreased, so it has given up some energy.) The act of splitting a large nucleus, referred to as *nuclear fission,* releases energy.

In a similar manner we can combine two very small nuclei into one larger nucleus and release energy. If a hydrogen-1 nucleus and a hydrogen-2 nucleus are

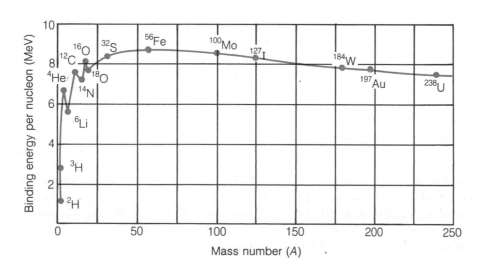

FIGURE 11.15
Graph of the binding energy per nucleon versus atomic mass number. Higher binding energy means the nucleus is more tightly bound.

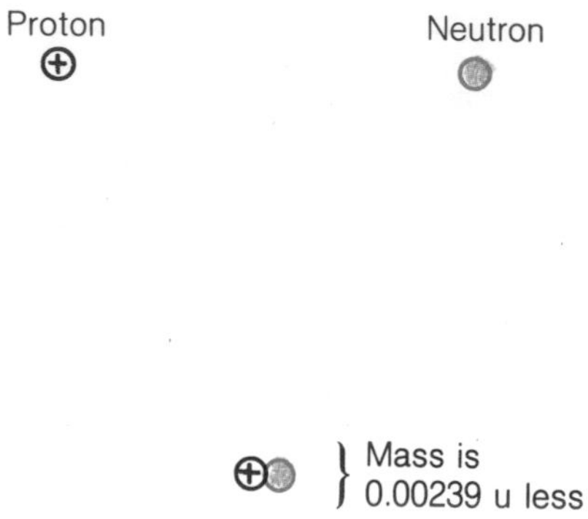

Proton

$\oplus$

Neutron

$\circledcirc$

$\oplus\circledcirc$ } Mass is 0.00239 u less

FIGURE 11.16
A proton and a neutron bound together in a nucleus have less mass *than a proton and a neutron separated. When the two combine, some of their mass is converted into energy—the binding energy of the nucleus. (The electron is not shown.)*

combined to form helium-3, the binding energy of each proton and neutron is increased, and energy is again released. This process is called *nuclear fusion.*

Clearly the graph of binding energy per nucleon versus atomic mass number is very important in nuclear physics because it shows why nuclear fission and nuclear fusion release energy. Fission is exploited in nuclear power plants and atomic bombs, and fusion is the source of energy in the sun, the stars, and in hydrogen bombs. But how does one actually go about measuring the binding energy of a nucleus? It is not practical to actually dismantle a nucleus and to keep track of the amount of work done. (Remember that the largest nuclei contain over 200 nucleons.) The best way to measure binding energy is to use the equivalence of mass and energy.

One of the predictions of Einstein's special theory of relativity, presented in Section 12.1, is that the mass of an object increases when it gains energy. The exact relationship between the energy, E, of a particle and its mass, m, is the famous equation

$$E = mc^2$$

When a particle is given energy, its mass increases; and when it loses energy, its mass decreases. In other words, mass can be converted into energy and vice versa. The amount of mass "lost" (converted into energy) by things around us when we take energy from them is generally much too small to measure. But the quantities of energy involved in nuclear processes are so great that mass-energy conversion can be measured.

For example, the mass of a hydrogen-1 atom (one proton and one electron) is 1.00783 u, and the mass of one neutron is 1.00866 u (see Table 11.1). The total mass of a hydrogen atom and a free neutron is 2.01649 u. But careful measurements of the mass of a hydrogen-2 (deuterium) atom give the value 2.01410 u. When the neutron and the proton are bound together in the hydrogen-2 nucleus, their combined mass is 0.00239 u *less* than when they are apart (Figure 11.16). This much mass is converted into energy when a proton and a neutron combine. This energy is the binding energy of the hydrogen-2 nucleus.

In a similar way we find that the masses of all nuclei are less than the combined masses of the individual protons and neutrons. The energy equivalent of this "mass defect" is the total binding energy of the nucleus. The binding energy divided by the atomic mass number is the binding energy per nucleon of the nucleus. This is a nice way to compute the binding energy per nucleon for the different isotopes.

Incidentally, the atomic mass unit u is defined to be exactly one twelfth of the mass of one atom of carbon-12. In other words, u was defined so that the mass of an atom of carbon-12 is exactly 12 u. Other isotopes could have been used for establishing the size of u, but carbon-12 is quite convenient, since it is very common.

Fission and fusion are processes that convert matter into energy. In both cases the total mass of all the nucleons afterwards is less than the total mass before. Some of the original mass, typically from 0.1% to 0.3%, is converted into energy. In the next two sections we take a closer look at fission and at fusion.

11.6   NUCLEAR FISSION

In the 1930s a discovery was made during neutron bombardment experiments that may be the most fateful in human history: the nuclei of certain isotopes were found to actually split when they absorb neutrons. Uranium-235 and plutonium-239 are the two most important nuclei that do this. Energy is released in the process along with several free neutrons. This process is called *nuclear fission*.

NUCLEAR FISSION   The splitting of a large nucleus into two smaller nuclei. Free neutrons and energy are also released.

The two resulting nuclei are called *fission fragments*. The fissioning nuclei of a particular isotope will not all split the same way. There are dozens of different ways that a nucleus can split, and there are dozens of different possible fission fragments.

Fission is most commonly induced by bombarding nuclei with neutrons. (Protons, alpha particles, and gamma rays also have been used.) Upon absorption of a neutron, the nucleus becomes highly unstable and quickly splits (Figure 11.17). Two of the many possible fission reactions of uranium-235 are:

$$\ _0^1\text{n} + \ _{92}^{235}\text{U} \rightarrow \ _{92}^{236}\text{U}^* \rightarrow \ _{56}^{141}\text{Ba} + \ _{36}^{92}\text{Kr} + 3\ _0^1\text{n}$$

$$\ _0^1\text{n} + \ _{92}^{235}\text{U} \rightarrow \ _{92}^{236}\text{U}^* \rightarrow \ _{54}^{140}\text{Xe} + \ _{38}^{94}\text{Sr} + 2\ _0^1\text{n}$$

barium
krypton
xenon
strontium

The asterisk (*) indicates that the uranium-236 is unstable, so much so that it splits immediately. In both cases energy is released during the fissioning. The total mass of the fission fragments and the neutrons is less than the mass of the original uranium nucleus and the neutron. The missing mass is converted into energy. Most of this energy appears as kinetic energy of the fission fragments; they have high speeds.

The average amount of energy released by the fissioning of a uranium-235 nucleus is 215 million electron volts (3.4×10^{-11} joules). By comparison, the amount of energy released during chemical processes such as burning, metabolism in your body, and chemical explosions is typically about 10 electron volts for each molecule involved. This is why nuclear powered ships and submarines can go years without refuelling, while ships that use diesel or oil must take on tons of fuel for each trip.

In addition to the energy released during fission, two other aspects of this process are extremely important.

1. Almost all of the possible fission fragments are radioactive. The ratio of neutrons to protons in most fission fragments is too high for the nuclei to be stable, and they undergo beta decay. These radioactive fission fragments are important components of the radioactive fallout from nuclear explosions and of the nuclear waste produced in nuclear power plants.

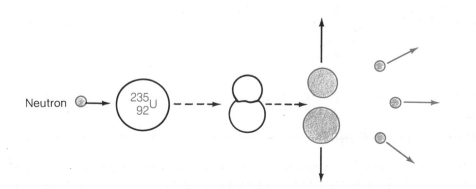

Neutron

$^{235}_{92}$U

FIGURE 11.17
A nucleus of uranium undergoing fission. In this example, three neutrons are released.

2. The neutrons that are released during fission can strike other nuclei and cause them to split. This process is called a *chain reaction*.

For uranium-235, 2.5 neutrons are released on the average by each fission. If just one of the neutrons from each fission induces another fission, a *stable* chain reaction results. The number of nuclei that split each second remains constant; therefore energy is released at a steady rate. This type of reaction is used in nuclear power plants. If, on the average, more than one of the neutrons causes other fissions, an *unstable* chain reaction results—a nuclear explosion. One fission could trigger two, these two could trigger four more, then eight, sixteen, and so on. Energy is released at a rapidly increasing rate. This is the basic process used in atomic bombs.

We can make a comparison between a nuclear chain reaction (fission) and a chemical chain reaction (burning). A stable chain reaction is similar to the burning of natural gas in a furnace or on a cooking stove. The number of gas molecules that burn each second is kept constant, and the energy (heat) is released at a steady rate. An unstable chain reaction is like the explosion of gunpowder in a firecracker. The energy released at the start quickly causes the rest of the powder to burn rapidly until all of it is consumed.

In the remainder of this section we look at some of the details involved in making atomic bombs and nuclear power plants. Keep in mind that there are two different types of nuclear bombs: atomic bombs, which use nuclear fission, and hydrogen bombs (also called thermonuclear bombs), which use both fission and nuclear fusion. The latter will be discussed in the next section.

The key raw material for both atomic bombs and nuclear power plants is uranium. Naturally occurring uranium is approximately 99.3% uranium-238 and 0.7% uranium-235. The uranium-235 fissions readily; the uranium-238 does not (unless irradiated with extremely high-speed neutrons), but it can be "bred" into fissionable plutonium-239.

Atomic Bombs

To produce an uncontrolled fission chain reaction, one must ensure that the fission of each nucleus leads to more than one additional fission. This is accomplished by using a high-density of fissionable nuclei so that each neutron emitted in a fission is likely to encounter another nucleus and cause it to fission (Figure 11.18). In other words, nearly pure uranium-235 or plutonium-239 must be used. Uranium-235 can be extracted from uranium ore through a complicated and costly series of processes referred to as *"enrichment,"* or *"isotope separation."* Since all uranium isotopes have essentially the same chemical properties, the separation processes rely on the difference in the masses of the nuclei.

FIGURE 11.18
Fission chain reaction in pure uranium-235. Since all of the nuclei can fission, each neutron emitted by one fission is likely to strike another nucleus and cause it to fission. This leads to an explosive chain reaction.

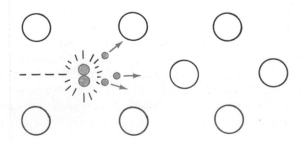

Not only is nearly pure uranium-235 or plutonium-239 necessary, there must be a sufficient amount of it, and it must be put into the proper configuration—a sphere, for example. This is so that each neutron is likely to collide with a nucleus before it escapes through the surface. The minimum amount of fissionable isotope that is needed is called the *critical mass.* With less than this amount, too many of the neutrons escape without causing other fissions. Also, the critical mass must be held together long enough for the chain reaction to cause an explosion. If not, the initial energy and heat from the first fissions will blow the critical mass apart and stop the chain reaction: a "fizzle." (A firecracker is tightly wrapped with paper for a similar reason.)

Another important consideration is timing: a premature explosion is highly undesirable, to say the least. This is prevented by keeping the fissionable material out of the critical mass configuration before the explosion. Two different techniques are used. In a gun-barrel atomic bomb, two subcritical lumps of uranium-235 are placed at opposite ends of a large tube (Figure 11.19). The explosion is triggered by forcing the two lumps together into a critical mass. This type of bomb was dropped on Hiroshima, Japan.

The other type of bomb, the implosion bomb, is used with plutonium-239. A subcritical sphere of the plutonium is surrounded by a shell of specially shaped conventional explosives (Figure 11.20). These explosives squeeze the plutonium-239 so much that it becomes a critical mass, and an explosion occurs. This type of bomb was dropped on Nagasaki, Japan.

A nuclear explosion releases an enormous amount of energy in the form of heat, light, other electromagnetic radiation, and nuclear radiation. The temperature at the center of the blast reaches millions of degrees. Even a mile away, the heat radiation is intense enough to instantly ignite wood and other combustible materials. The air near the explosion is heated and expands rapidly, producing a shock wave followed by hurricane-force winds. In addition to the radioactive fission fragments, large quantities of radioisotopes are produced by the neutrons and other radiation that bombard the air, dust, and debris. As this material settles back to the ground, it is collectively referred to as *fallout.*

The amount of energy released by an atomic bomb explosion is generally expressed in terms of the number of tons of TNT, a conventional high explosive, that would release the same amount of energy. (One ton of TNT releases about 4.5 billion joules when it explodes.) Most atomic bombs are in the 10 to 100 kiloton range: they are equivalent to between 10,000 and 100,000 tons of TNT. (An entire 100-unit freight train can carry only 10,000 tons.) Some of these weapons are small enough to be carried by one person.

Nuclear Power Plants

Nuclear reactors release the energy from nuclear fission in a controlled manner. The key to preventing a fission chain reaction from escalating to an explosion is controlling how the neutrons induce other fissions.

There are dozens of different designs for nuclear reactors. Most of them use only slightly enriched uranium fuel, typically about 3% uranium-235. This keeps the density of fissionable nuclei low so that not all neutrons from each fission are likely to induce other fissions (Figure 11.21). (This is also economical because the enrichment process is very expensive.) It also makes it impossible for a critical mass configuration to arise: a nuclear reactor cannot explode like an atomic bomb.

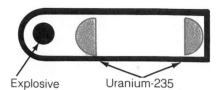

Explosive   Uranium-235

FIGURE 11.19
Gun-barrel atomic bomb. The two lumps of uranium are forced together to form a critical mass.

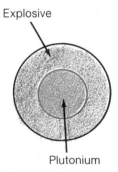

Explosive

Plutonium

FIGURE 11.20
Implosion atomic bomb. The plutonium is subcritical until it implodes under the force of the explosives. This brings it to a critical mass.

FIGURE 11.21
A controlled fission chain reaction in enriched uranium. Since only a small percentage of the nuclei are fissionable, not all of the neutrons induce other fissions. Some are absorbed by uranium-238 nuclei, which do not fission.

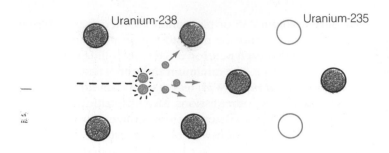

Another important factor in producing and controlling a fission chain reaction in a reactor is the fact that uranium-235 is much better at capturing slow neutrons than fast neutrons. Slowing down the neutrons released in the fissioning of uranium-235 makes it much more likely that they will be captured by other uranium-235 nuclei and cause them to split. This makes it possible to use a low concentration of uranium-235 in a reactor.

How are the neutrons slowed down? This is done through the use of a moderator, a substance that contains a large number of small nuclei. Small nuclei are more effective at taking kinetic energy away from the neutrons in the same way that a golf ball will lose more kinetic energy when it collides with a baseball at rest than with a bowling ball at rest. As neutrons pass through the moderator, they collide with the nuclei and lose some of their energy; they slow down. After many collisions they have the same average kinetic energy as the surrounding nuclei. (For this reason they are called *thermal neutrons* because their energies are determined by the temperature of the moderator.) Almost all nuclear power plant reactors in the United States use water as the moderator because of the hydrogen nuclei in the water molecules. Many reactors in England and the Soviet Union use carbon in the form of graphite as the moderator.

The uranium fuel is shaped into long rods that are separated from each other. (Typical large reactors have tens of thousands of individual fuel rods.) Each rod relies on neutrons from neighboring rods to sustain the chain reaction. This makes it possible to control the chain reaction by lowering *control rods* between the fuel rods (Figure 11.22). Control rods are made of materials that absorb neutrons effectively, such as the elements cadmium and boron. The rate of fissioning in a reactor, and therefore the rate of energy release, is regulated by the number of control rods that are withdrawn.

The energy released in a nuclear reactor is primarily in the form of heat. This heat is used to boil pressurized water into high-temperature steam. The rest of a nuclear power plant is essentially the same as that found in coal-fired power plants (refer to Figure 5.39). The steam turns a turbine, which turns a generator, which produces electricity. In most nuclear reactors, water flows around the fuel rods and carries away the heat in much the same way that the coolant in a car's radiator carries away heat from hot engine parts. If, for some reason, the reactor loses its water (called a "loss of coolant accident"), the reactor can become dangerously overheated.

Nuclear power plants are equipped with a sophisticated array of safety features and emergency backup systems. Millions of gallons of water are poised to flood the reactor if the temperature becomes too high or if the coolant system leaks. The control rods are designed to drop into place at the slightest sign of trouble. The entire nuclear reactor and supporting systems are housed inside a huge steel-

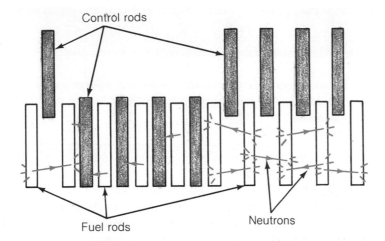

Control rods

Fuel rods

Neutrons

FIGURE 11.22
Simplified diagram of the core of a nuclear reactor. Neutron-absorbing control rods can be inserted between the fuel rods to control the chain reaction. The control rods prevent neutrons released in one fuel rod from inducing fissions in another fuel rod.

reinforced concrete containment building. This building is typically about 200 feet high, dome shaped, with walls that are more than 1-foot thick. It is designed to seal airtight if there is a possibility that radiation has leaked from the reactor. All of this is needed because there is much more radioactivity inside a nuclear reactor than is released by a typical nuclear bomb.

Two notable accidents at nuclear power plants point out the potential for disaster that exists when human operators and designers make mistakes. In 1979 an unlikely series of errors and equipment malfunctions caused the core of a reactor at Three Mile Island, Pennsylvania to lose most of its cooling water for 3 hours. Over half of the core melted. The containment building served its purpose, and very little radiation escaped from the plant. But the costly reactor was ruined, and the core was reduced to a pile of intensely radioactive rubble. The billion dollar cleanup continues after more than a decade.

A major nuclear disaster occurred in April 1986 when a graphite-moderated power plant at Chernobyl, USSR exploded. Operators performing tests on the generator pushed the reactor beyond its limits. Not confined by the kind of containment building at Three Mile Island, tons of radioactive debris were spewed into the atmosphere and were carried by winds literally around the world. The huge mass of graphite was ignited by the heat, adding to the severity of the accident. Thirty persons were reported killed at the site, and over 200 more suffered severe radiation exposure. The surrounding area within 30 kilometers of the plant was evacuated and may remain uninhabitable for decades. Reindeer herds in Sweden and Norway were contaminated by the accident's fallout. Better reactor design and operator training should make a repeat of this disastrous accident unlikely.

Eventually the fissionable isotopes in the fuel rods are consumed, and the spent rods have to be replaced. Spent fuel rods are highly radioactive because they contain large amounts of fission fragments, in the form of more than 200 different radioisotopes. So much radiation is emitted that the rods must be kept under water for months to remain cool. Even after the short–half-life radioisotopes have decayed away and the rods have cooled, they remain dangerously radioactive for thousands of years. Finding a safe way to dispose of spent fuel rods and other nuclear waste is a major concern of government regulatory officials and nuclear power plant authorities.

11.7   NUCLEAR FUSION

As incredible as it may seem, another source of nuclear energy exists that dwarfs nuclear fission: nuclear fusion. It is the source of energy for the sun and the stars and for hydrogen bombs.

> NUCLEAR FUSION   The combining of two nuclei to form a larger nucleus.

Fusion is like the reverse of fission, although it usually involves very small nuclei. One example of a fusion reaction is shown in Figure 11.23. In this case two hydrogen nuclei, one hydrogen-1 and the other hydrogen-2 (deuterium), fuse to form a nucleus of helium-3. There are many other possible fusion reactions that result in a release of energy. Some of these reactions, including the amount of energy released, are:

$$_1^2H + _1^2H \rightarrow _2^3He + _0^1n + 3.3 \text{ MeV}$$

$$_1^2H + _1^3H \rightarrow _2^4He + _0^1n + 17.6 \text{ MeV}$$

$$_1^2H + _2^3He \rightarrow _2^4He + _1^1p + 18.3 \text{ MeV}$$

In each case, energy is released because the total mass of the nucleons after the fusion is less than the total mass before. As with fission, the missing mass is converted into energy. This energy appears as kinetic energy of the fused nucleus and the energy of the proton, neutron, or gamma ray.

Fusion can occur only when the two nuclei are close enough for the short-range nuclear force to pull them together. This turns out to be a major problem in trying to induce a fusion reaction. Nuclei are positively charged; when brought close together, they exert strong repulsive forces on each other and resist fusion.

How can this difficulty be overcome? The most common way is with extremely high temperatures. If nuclei can be given enough average kinetic energy, their momenta when they collide will bring them close enough to fuse (Figure 11.24). This process is called *thermonuclear fusion,* for obvious reasons.

Refer back to Section 5.1.

When we say that the temperatures are high, we really mean it: the hydrogen-1 plus hydrogen-2 fusion reaction requires a temperature of about *50 million* degrees C. Some fusion reactions require several hundred million degrees C. These unearthly temperatures do occur in the interiors of stars and at the centers of nuclear fission explosions.

FIGURE 11.23
Fusion of a hydrogen-1 nucleus and a hydrogen-2 nucleus. A gamma ray is emitted in the process.

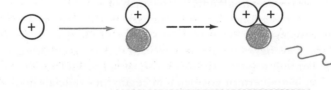

$$_1^1H \quad + \quad _1^2H \quad \longrightarrow \quad _2^3He \quad + \quad \gamma$$

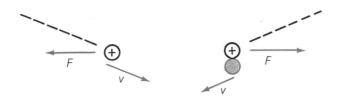

FIGURE 11.24
In a high-speed collision of two nuclei, the repulsive force between them is overcome, and fusion occurs. At extremely high temperatures, nuclei possess such high speeds because of their thermal motions. This is thermonuclear fusion.

Fusion in Stars

The sun glows white hot and emits enormous amounts of energy. This energy originates in a natural fusion reaction in the sun's interior. The sun is composed mostly of hydrogen in a dense, high-temperature plasma. At the sun's core the temperature is around 15 million degrees C, and the pressure is several billion atmospheres. Under these conditions the hydrogen undergoes a series of fusion reactions that results in the formation of helium. Each second, hundreds of millions of tons of hydrogen fuse, and more than 4 million tons of matter are converted into energy.

Stars get their energy from nuclear fusion. Some produce energy by reactions like those in the sun, whereas others that are larger and have hotter cores use other fusion reactions. In large stars with core temperatures in excess of 100 million degrees C, the helium fuses to form larger nuclei such as carbon and oxygen, which in turn may fuse to form silicon and eventually iron. In this way the elements in the periodic table are built up from the basic raw material, hydrogen. The heavier elements are produced during *supernova explosions*—the gigantic explosion of massive stars at the ends of their life cycles (Figure 11.25). The elements contained in earth and everything on it (including yourself) were formed in the cataclysms of supernovas that occurred billions of years ago.

Life as we know it would be impossible on this planet without the sun. Its energy supports the entire web of life. It is rather ironic that solar energy—that tranquil, natural energy source—comes from a violent nuclear process on a scale that is nearly beyond imagination.

Thermonuclear Weapons

The most destructive weapons in the world's arsenals are *thermonuclear warheads,* also called hydrogen bombs. These weapons get most of their explosive energy from the fusion of hydrogen. To produce the high temperatures necessary for the fusion reactions to occur, a nuclear fission explosion is used as a trigger. The fission explosion, an incredibly huge blast in itself, is but a primer for the monstrous fusion explosion. The difference between an atomic bomb and a hydrogen bomb is about the same as that between a firecracker and a stick of dynamite.

Truly enormous amounts of energy are released. Many thermonuclear devices are in the *megaton* range; their energy output is given in terms of *millions of tons* of TNT. The largest weapon ever tested was rated at more than 50 megatons. To put this in perspective, this one blast released more energy than the total of all of the explosions in all of the wars in history, including the two atomic bomb blasts in World War II. (However, it wasn't the most powerful blast to have ever occurred on earth. Some volcanic eruptions, like the one on the island of Krakatoa in 1883, released more energy.)

FIGURE 11.25
The Crab Nebula—debris of a supernova explosion that was observed and recorded by the Chinese in the year 1054.

Currently the world's superpowers possess tens of thousands of nuclear warheads. Indications are that if even a small percentage of these were used in a short "war," much of the life on this planet, including the human race, would be threatened with extinction.

Controlled Fusion

Soon after fusion was discovered, scientists looked for ways to harness it as an energy source. The initial success of nuclear fission reactors gave them hope. But fusion presents technical challenges that have resisted solutions for decades. For a thermonuclear fusion reaction to occur, two conditions must be met:

1. The nuclei must be raised to an extremely high temperature to ignite the fusion reaction.

   This was discussed earlier. It is possible to produce temperatures in the millions of degrees, but the problem is containing (referred to as "confining") a plasma at this high temperature. Containers made of conventional materials would melt long before such temperatures would be reached.

2. There must be a high enough density of nuclei for the probability of collisions, and therefore fusions, to remain high.

   If energy is to be released at a usable rate, a large number of fusions must occur each second. This means that the density of the plasma must be kept sufficiently high for the nuclei to collide often.

Research on controlled fusion is being pursued along a number of avenues. Several of these employ *magnetic confinement* of the plasma; specially shaped magnetic fields are used to keep the plasma confined without letting it come into contact with other matter. This is possible because the nuclei are charged par-

Refer back to Section 8.2.

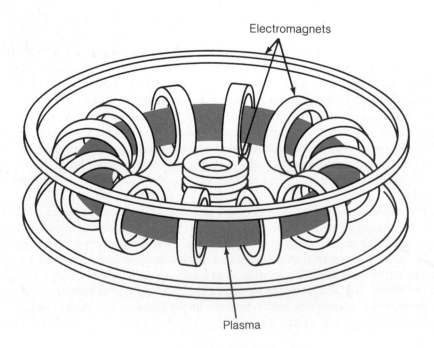

FIGURE 11.26
Schematic of a tokamak fusion device. The plasma containing hydrogen nuclei is trapped by the magnetic field produced by the various electromagnets. (Adapted from Scientific American, *249, no. 4 (October 1983): 63.*

ticles and consequently experience a force when moving in a magnetic field. In essence, a "magnetic bottle" is formed out of magnetic fields, and the plasma is injected into it. One of the most promising designs, called *tokamak*, has the plasma confined inside a toroid—the shape of a donut (Figure 11.26). Fusion reactions have been produced, but so far the amount of energy released has been less than the amount used to produce the reactions. Usually the reaction is maintained for only a fraction of a second because the plasma leaks out of the magnetic bottle.

Another approach is to use extremely intense bursts of laser light to produce miniature fusion explosions. A small pellet of solidified fuel is blasted with laser beams from several directions simultaneously. Figure 11.27 shows the target chamber of the Nova laser fusion device.

There are good reasons why controlled fusion would make a nice energy source. The principal one is that the oceans contain an enormous supply of the fuel: hydrogen nuclei. Compared to fission reactors, much less radioactive waste would be generated. Chief by-products would be helium, which has a number of uses, and tritium, which can be used as fuel. Fusion chain reactions would be easier to control than fission chain reactions. But because of the technical challenges, controlled fusion will not be a viable energy source until well into the next century, if ever.

Cold Fusion

There are ways to induce fusion without using extremely high temperatures. Referred to as *cold fusion*, these processes use other means to bring the fusing nuclei close together. The largest nuclei known, "superheavy" elements with

FIGURE 11.27
Target chamber of the Nova fusion laser.

atomic numbers over 100, are produced in the laboratory using one type of cold fusion. Smaller nuclei are accelerated to high speeds and collide with large nuclei. Under proper conditions the nuclei fuse to form larger nuclei. Between 1958 and 1974, scientists in the United States and in the USSR accelerated very small nuclei to synthesize elements 102 through 106. In the 1980s a team in Germany specialized in creating even larger nuclei by colliding not-so-small nuclei with large nuclei. For example, in 1982 they produced the element 109 by bombarding bismuth-209 with iron-58. The reaction equation is:

$$\,_{26}^{58}\text{Fe} + \,_{83}^{209}\text{Bi} \rightarrow \,_{109}^{266}\text{XX} + \,_{0}^{1}\text{n}$$

(XX—the element has not been named)

The kinetic energy of the incoming nucleus must be just large enough to overcome the electrostatic repulsion between the two nuclei but not so large that the resulting nucleus has enough excess energy to undergo immediate fission. Researchers are optimistic that even large nuclei can be formed using this technique.

Another form of cold fusion seen in the laboratory is produced with the aid of an exotic elementary particle known as the *muon*. (Muons and other elementary particles are discussed in Chapter 12.) Hydrogen atoms normally combine in pairs to form H_2 molecules. If one of the two electrons is removed, the result is two hydrogen nuclei bound together by their attraction to the electron. But the nuclei are too far apart to undergo fusion. The negatively charged muon is basically an overweight electron: its mass is about 200 times larger. Consequently a "muonic atom" can exist that is just a muon in orbit about a hydrogen nucleus. Now if the electron in the molecule described above is replaced by a muon, the two nuclei will be about 200 times closer together, and it is possible for them to fuse. This type of cold fusion has been observed, but it is not likely to be used as a source of energy. Muons are unstable, with a half-life of only 2.2×10^{-6} seconds. Too much energy would be needed to create a constant supply of muons.

The term "cold fusion" became a household word in March 1989 when two scientists reported that they had generated heat using a fusion reaction in a simple apparatus at room temperature. A beaker was filled with *heavy water* (the hydrogen nuclei in the water molecules are the isotope hydrogen-2 instead of the more common hydrogen-1). A piece of the element palladium was immersed in the heavy water and surrounded by wire made of platinum. Then the palladium and platinum were connected to a DC power supply. Supposedly the tiny deuterium nuclei would accumulate in the small gaps between the palladium atoms and get close enough to fuse. Attempts by literally thousands of scientists around the world to recreate the experiment produced mixed results. If the deuterium nuclei fused in the conventional way (see the first reaction equation at the beginning of this section), there would be a large number of neutrons released. Since this was not found to be the case, the energy released was attributed to "unknown nuclear processes."

"Table top fusion" will be remembered as one of the most remarkable stories in the history of science. The discovery was announced at a press conference rather than first being submitted to a journal where it would be subjected to peer review. (This was soon after the sensation created by the new high-T_c superconductor. During that period discoveries came so rapidly that some were announced in the same way.) Attempts to reproduce the findings, an essential way to test the validity of any scientific discovery, were hampered because details about the experimental apparatus were unknown. Confusion and frustration increased when similar experiments performed at reputable laboratories yielded contradictory results. Because table top fusion offered the hope of solving the world's energy problems in a simple way, the world anxiously awaited a simple yes or no verdict. Ordinarily such a reported result can be confirmed or disproved in a few months. But that was not the case. In late 1989 a 20-member panel appointed by the US Department of Energy to review the controversy concluded that there is no convincing proof that a usable fusion reaction is taking place but that there may be something unusual going on.

HISTORICAL NOTES

The history of nuclear physics began less than 100 years ago, at a time when the very existence of atoms was still disputed. It is a remarkable history for many reasons. The pace of discovery, compared to the days of Galileo and Newton, was swift. The way in which physics research was done changed during the period. Instead of lone scientists working in small laboratories, most of the work was done by research groups using increasingly sophisticated equipment. The classification of physicists into experimentalists and theoreticians, starting in the 19th century with Michael Faraday and James Clerk Maxwell, shaped the growing physics community.

The first discoveries in nuclear physics were made before the structure, or even the existence, of the nucleus was known. Henri Becquerel (1852–1908) was the third in a line of four generations of prominent French physicists (see Figure 11.28). Like his father before him, Becquerel studied fluorescence—the emission of visible light from a substance when it is irradiated with ultraviolet light. (This process is exploited in fluorescent lights.)

Becquerel heard of the new x-rays, discovered by Wilhelm Roentgen in 1895, and in early 1896 he began to test fluorescent substances to see if they too emitted x-rays. In one experiment he placed uranium on a photographic film plate that was wrapped to keep out visible light. After exposing the uranium to sunlight, he found that the film had been irradiated with what he thought to be x-rays emitted by the uranium because of the sunlight. He tried to repeat the experiment, but the late winter weather in Paris turned cloudy for several days. Becquerel checked the film and expected very little exposure because of the weak sunlight. To his surprise he found that the radiation from the uranium had been just as strong. During the following weeks he determined that the uranium constantly emitted the radiation and needed no external stimulation. Clearly these were not x-rays. Nuclear radiation, dubbed "Becquerel rays," had been discovered.

At this point another famous family of physicists arrived on the scene, the Curies (see Figure 11.29). Marie Curie (1867–1934), originally Marya Sklodowska, grew up in Russian-occupied Warsaw. Her scientific career began when she joined her sister in Paris in 1891. While working toward her PhD, Marie met Pierre Curie (1859–1906), a French chemist. They were married in July 1985. Two years later, a daughter, Irène, was born. She would follow in her mother's footsteps and become a renowned physicist.

FIGURE 11.28
French physicist Henri Becquerel, the discoverer of radioactivity.

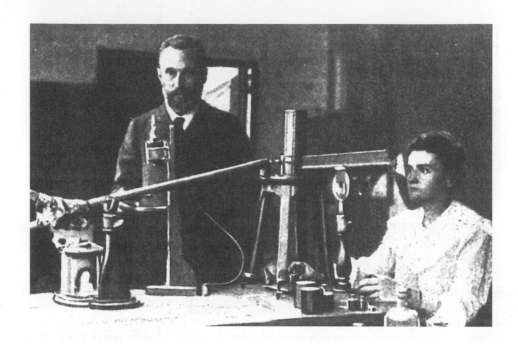

FIGURE 11.29
Pierre and Marie Curie in their laboratory.

At about this time Marie, at the suggestion of Pierre, undertook an investigation of "Becquerel rays" as a thesis project. She tested all of the known elements and found that only uranium and thorium emitted the radiation. Upon testing ore samples from a museum, she found that some minerals were more radioactive than could be accounted for by uranium and thorium. She suspected that some new substance was emitting the radiation. Her husband joined the investigation, and in July of 1898 they announced the discovery of a new radioactive element, which Marie patriotically named polonium, for her native Poland.

In a similar way the Curies discovered another radioactive element, radium, in September of the same year. The task of actually isolating pure samples of the new elements was enormous. To get a tiny sample of radium for analysis, the Curies spent years extracting it from a ton of ore. The conditions in their crude laboratory were primitive, and no one realized the danger of radiation. Marie eventually died from a condition probably caused by radiation poisoning. Even her research notes were found to be radioactive.

Marie and Pierre Curie shared the 1903 Nobel Prize in physics with Henri Becquerel. Three years later Pierre was killed by a runaway carriage. Marie carried on her research and received a second Nobel Prize in 1911.

The most prominent figure in the early years of nuclear physics was Ernest Rutherford (1871–1937; Figure 11.30). Born into a Scottish family in New Zealand, Rutherford developed an interest in physics at an early age. He attended college in New Zealand and won a scholarship to Cambridge. There his brilliance was soon recognized. Rutherford was an excellent experimenter with a reputation for simplicity and elegance. In 1897 he began to study the newly discovered nuclear radiation. The next year he made his first discovery: two different types of radiation are emitted by uranium. These he named alpha and beta. The third type of nuclear radiation, gamma, was discovered later in France.

In 1898 Rutherford took a position at McGill University in Montreal and continued his work. In 1900 he began collaborating with a chemist, Frederick Soddy, and together they established that radioactivity causes transmutation of an ele-

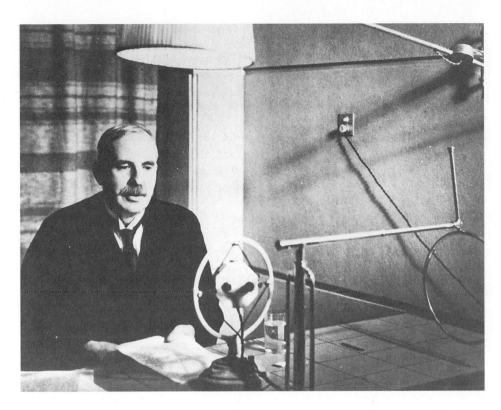

FIGURE 11.30
Ernest Rutherford.

ment. They began to discover other radioactive substances that were later identified as different isotopes of radium. During this time Rutherford reached the controversial conclusion that alpha particles are ionized helium atoms.

In 1907 Rutherford returned to England and took a position at the University of Manchester. The following year he won the Nobel Prize in chemistry. (The fields of chemistry and physics overlap a great deal, particularly in the study of the nucleus.) At this time he experimentally verified the identity of alpha particles. His laboratory flourished, and much of his experimentation was carried out with the help of students and assistants. One such experiment led Rutherford to his most noted discovery, the nuclear model of the atom.

At this time the structure of the atom was a topic of speculation. The accepted model held that the electrons were embedded in some kind of positively charged sphere, somewhat like plums in a pudding. In an experiment performed by a student in Rutherford's laboratory, it was observed that some alpha particles were deflected as they traveled through a thin gold foil. The amount of deflection was much larger than would occur if the positive charge of an atom were spread evenly throughout its volume. Rutherford concluded, and then verified with experiments, that the positive charge is confined to a small region at the center of the atom, which he called the nucleus. This was in 1911, 15 years after the discovery of nuclear radiation.

The work in nuclear physics continued, with an understandable lull during World War I. Rutherford found that alpha particles could be used to disintegrate nuclei of nitrogen. He gave the proton its name and speculated that a similar particle that had no charge might exist in the nucleus. The eventual identification of the neutron was made in 1932 by James Chadwick (1891–1974), a former student of Rutherford's.

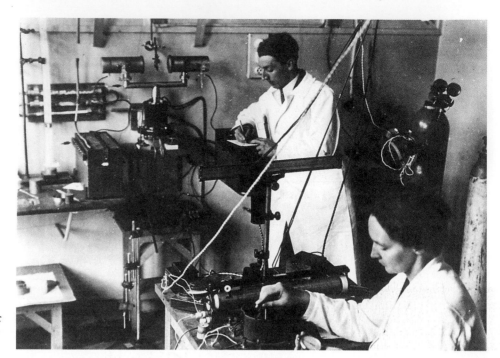

FIGURE 11.31
Husband and wife team Frédéric Joliot and Irène Curie, daughter of Marie and Pierre.

Two of the most prominent experimenters in the 1930s were Irène Curie (1897–1956) and her husband Frédéric Joliot (1900–1958; Figure 11.31). They met when they both worked for Irène's mother, Marie. Early in 1932, Joliot and Curie reported results from experiments with a newly discovered type of radiation that they could not explain. They missed their chance: the radiation was high-speed neutrons. Chadwick quickly performed the necessary experiments and verified their identity. The discovery of the neutron completed Rutherford's nuclear model of the atom and solved some important mysteries in nuclear physics. Chadwick received the 1935 Nobel Prize in physics for his work.

The greatest discovery of Joliot and Curie, which they did interpret correctly, was artificially induced radioactivity. In a one-page paper dated 10 February 1934, they announced that bombarding certain kinds of nuclei with alpha particles resulted in their transmutation into radioactive nuclei. This earned them a Nobel Prize.

One of the first to exploit the discovery of artificial radioactivity was the Italian physicist Enrico Fermi (1901–1954; Figure 11.32). Fermi was a brilliant student and earned his doctorate at the age of 20 years. He became the key figure in a push to restore Italy's greatness in physics. Already famous at the time of Joliot and Curie's discovery, Fermi decided to use neutrons to induce artificial radioactivity. Alpha particles are positively charged and are therefore repelled by nuclei. The uncharged neutrons are much more likely to enter a nucleus. Fermi and his group in Rome started irradiating all of the known elements and soon produced about 40 new radioisotopes (see Section 11.4). This brought Fermi the 1938 Nobel Prize in physics.

Fermi and others missed an opportunity to make a critical discovery when the element uranium was irradiated with neutrons. Although they reported strange results, it was Otto Hahn (1879–1968), once a student of Rutherford in Canada, and Fritz Strassman who discovered barium and other midsized elements in an

FIGURE 11.32
Nuclear physicist Enrico Fermi.

FIGURE 11.33
Lise Meitner and Otto Hahn, two of the discoverers of nuclear fission.

irradiated uranium sample. The correct interpretation of this finding, that some of the uranium had undergone fission, was made by Lise Meitner (1878–1968) and her nephew, Otto Frisch (see Figure 11.33). (Meitner was a colleague of Hahn but had fled to Sweden because of the rise of the Nazis in Germany.)

The news spread to physicists around the world, and many groups quickly confirmed the result. Soon uranium fission was recognized as a potential source of energy, as bombs or in some controlled process. Coming as it did, just as the world was being plunged into the worst war in history, one has to marvel at the timing of this discovery. The incentive to develop nuclear weapons was great.

Volumes have been written about the great atomic bomb project, called the Manhattan Project, during World War II. Some of the greatest physicists of the time, many of them refugees from Hitler's madness, were gathered in top-secret laboratories around the United States. The technical challenges were enormous, but the thought of a German atomic bomb was a powerful motivator. Three bombs were exploded: one was tested near Alamogordo, New Mexico, and the other two were used to destroy the Japanese cities of Hiroshima and Nagasaki. The hundreds of thousands of casualties caused by these explosions closed the final chapter on a war that saw civilians killed by the millions.

The impact of the Manhattan Project on physics was enormous. On the one hand, its result was a triumph in nuclear physics and engineering; but it left a cruel legacy for all involved. More clearly than ever before, it showed how the work of scientists can be used to destroy life. This vivid demonstration of the destructive aspects of applied physics spurred many of the principal scientists to seek careful control of nuclear weapons after the war. But Pandora's box had been opened, and the nuclear arms race was on.

SUMMARY

The nucleus is the tiny, extremely dense core of an atom. It is composed of protons and neutrons held together by the short-range nuclear force. The different isotopes of a given element have the same number of protons in their nuclei but different numbers of neutrons. Isotopes are designated by the name of the element and the atomic mass number, the total number of protons and neutrons in the nucleus, as in uranium-235.

The nuclear force cannot hold a nucleus together if the ratio of protons to neutrons is too high or too low or if the nucleus is too large. In such cases the nuclei are radioactive and emit nuclear radiation. There are three types of nuclear radiation—alpha, beta, and gamma—and each has a different effect on a nucleus. The half-life of a radioisotope is the time it takes for one half of the nuclei in a given sample to decay. There are nearly 2,000 different radioisotopes with half-lives that range from tiny fractions of a second to billions of years. Some radioisotopes, most notably carbon-14, are quite useful for determining the ages of artifacts and of geological formations.

Measurements of the binding energy per nucleon for different elements indicate that large and small nuclei are not as tightly bound as those with A around 50 to 60. This explains why splitting large nuclei and fusing small nuclei release energy. Energy is so concentrated in nuclear processes that Einstein's equivalence of mass and energy can be observed.

Bombarding nuclei with neutrons and other particles can induce a variety of nuclear reactions. Nuclear fission, the splitting of a nucleus into two smaller nuclei, occurs when certain large nuclei are struck by neutrons. The energy that is released is exploited in atomic bombs and in nuclear power plants.

Nuclear fusion is the combining of two small nuclei to form a larger nucleus. The sun, the stars, and thermonuclear warheads use energy released by nuclear fusion. Efforts to harness fusion as a source of energy are hampered by the extreme conditions that are required to induce fusion.

QUESTIONS

1. Why do different isotopes of an element have the same chemical properties?

2. The atomic number of one isotope is equal to its atomic mass number. Which isotope is it?

3. What is the name of the force that holds protons and neutrons together in the nucleus?

4. What aspects of the composition of a nucleus can cause it to be unstable?

5. Describe the common types of radioactive decay. What effect does each have on a nucleus?

6. A concrete wall in a building is found to contain a radioactive isotope that emits alpha radiation. What could be done to protect people from the radiation (short of razing the building)? What if it were gamma radiation being emitted?

7. Explain the concept of half-life.

8. One half of the nuclei of a given radioisotope decays during one half-life. Why doesn't the remaining half decay during the next half-life?

9. How is carbon-14 used to determine the ages of wood, bones, and other artifacts?

10. One cause of uncertainty in carbon-14 dating is that the relative abundance of carbon-14 in atmospheric carbon dioxide is not always constant. If it is discovered that during some era in the past carbon-14 was more abundant than it is know, what effect would this have on the estimated ages of artifacts dated from that period?

11. What is the difference between "somatic" and "genetic" effects of radiation on living things?

12. What are the principal steps in neutron activation analysis?

13. During the normal operation of nuclear power plants and nuclear processing facilities, machinery, building materials, and other things can become radioactive even if they never come into physical contact with radioactive material. What causes this?

14. If the binding energy per nucleon (see the graph in Figure 11.15) increased steadily with atomic number instead of peaking around $A = 56$, would nuclear fission and nuclear fusion reactions work the same way they do now? Explain.

15. How can a nucleus of uranium-235 be induced to fission? Describe what happens to the nucleus.

16. What aspect of nuclear fission makes it possible for a chain reaction to occur? What is the difference between a chain reaction in a bomb and one in a nuclear power plant?

17. Explain how materials that absorb neutrons are used to control nuclear fission chain reactions.

18. What are fission fragments, and why are they so dangerous?

19. Why is a nuclear fusion reaction so difficult to induce?

20. Why are extremely high temperatures effective at causing fusion? What is used to produce such temperatures in a thermonuclear warhead?

21. Why is magnetic confinement being used in fusion research?

22. What is meant by the term "cold fusion"?

PROBLEMS

(Note: In problems 1 through 10 you may need to use the periodic table of the elements in Appendix B.)

1. Determine the composition of the nuclei of the following isotopes.
   a) carbon-14
   b) silver-108
   c) plutonium-242

2. The isotope helium-6 undergoes beta decay. Write the reaction equation, and determine the identity of the daughter nucleus.

3. The isotope silver-110 undergoes beta decay. Write the reaction equation, and determine the identity of the daughter nucleus.

4. The isotope polonium-210 undergoes alpha decay. Write the reaction equation, and determine the identity of the daughter nucleus.

5. The isotope plutonium-239 undergoes alpha decay. Write the reaction equation, and determine the identity of the daughter nucleus.

6. The isotope silver-107* undergoes gamma decay. Write the reaction equation, and determine the identity of the daughter nucleus.

7. The following is a possible fission reaction. Determine the identity of the missing nucleus.

$$_0^1n + {}_{92}^{235}U \rightarrow {}_{92}^{236}U^* \rightarrow {}_{39}^{95}Y + ? + 2_0^1n$$

8. The following is a possible fission reaction. Determine the identity of the missing nucleus.

$$_0^1n + {}_{92}^{235}U \rightarrow {}_{92}^{236}U^* \rightarrow {}_{57}^{143}La + ? + 3_0^1n$$

9. Two deuterium nuclei can undergo two different fusion reactions. One of them is given at the beginning of Section 11.7. In the second possible reaction two deuterium nuclei fuse to form a new nucleus plus a lone proton. Write the reaction equation, and determine the identity of the resulting nucleus.

10. Iron-58 and lead-208 fuse into a large nucleus plus a neutron. Write the reaction equation, and determine the identity of the resulting nucleus.

11. A Geiger counter registers a count rate of 4,000 counts per minute from a sample of a radioisotope. Twelve minutes later the count rate is 1,000 counts per minute. What is the half-life of the radioisotope?

12. Iodine-131, a beta emitter, has a half-life of 8 days. A 2-gram sample of initially pure iodine-131 is stored for 32 days. How much iodine-131 remains in the sample afterwards?

13. An accident in a laboratory results in a room being contaminated by a radioisotope with a half-life of 3 days. If the radiation is measured to be eight times the maximum permissible level, how much time must elapse before the room is safe to enter?

14. The amount of carbon-14 in an ancient wooden bowl is found to be one half that in a new piece of wood. How old is the bowl?

CHALLENGES

1. Geiger counters are not very accurate when the count rates are very high; they indicate a count rate lower than the actual value. Explain why this is so.

2. The deflection of an alpha particle as it passes through a magnetic field is much less than the deflection of a beta particle (see Figure 11.3). Why?

3. One of the types of radioactive decay not discussed in the chapter is *electron capture*. One of the electrons orbiting the nucleus actually enters the nucleus and a "reverse beta decay" takes place. What effect does electron capture have on the parent nucleus? A nucleus of oxygen-15 undergoes electron capture. Write out the reaction equation, and determine the identity of the daughter nucleus.

4. As a rule of thumb, the radioactivity from a particular radioisotope is considered to be reduced to a safe level after 10 half-lives have elapsed. (Obviously the initial quantity is also important.) By how much is the rate of emission of radiation reduced after 10 half-lives? Plutonium-239 is considered to be one of the most dangerous radioisotopes. Its half-life is about 25,000 years. How long would plutonium-239 have to be kept isolated before it is safe?

5. After a fuel rod reaches the end of its life cycle (typically 3 years), most of the energy that it produces comes from the fissioning of plutonium-239. How can this be?

6. When the plutonium bomb was tested in New Mexico in 1945, approximately 1 gram of matter was converted into energy. How many joules of energy were released by the explosion?

7. Using the information given in Section 11.5 and the mass-energy conversion equation, compute the binding energy (in MeV) and the binding energy per nucleon for hydrogen-2 (deuterium).

SUGGESTED READINGS

Armbruster, Peter, and Münzenberg, Gottfried. "Creating Superheavy Elements." *Scientific American* 260 no. 5 (May 1989): 66–72. Describes the synthesis of large nuclei (Z > 100) by fusion of two nuclei with atomic numbers of around 25 and 80.

Cobb, Charles E. "Living with Radiation." *National Geographic* 175, no. 4 (April 1989): 403–437. An account of the effects of radiation on people, from those exposed to radon gas in their homes to survivors of nuclear blasts.

Conn, Robert W. "The Engineering of Magnetic Fusion Reactors." *Scientific American* 249, no. 4 (October 1983): 60–71. Describes the designs of magnetic confinement fusion reactors and how energy would be extracted from fusion reactions.

Craxton, R. Stephen, McCrory, Robert L.; and Soures, John M. "Progress in Laser Fusion." *Scientific American* 255, no. 2 (August 1986): 68–79. Status report on laser fusion, including advances in higher frequency high-power lasers.

Gamow, George. *Biography of Physics*. New York: Harper and Row, 1961. Chapter 8 deals with the development of nuclear physics.

Goodwin, Irwin. "Fusion in a Flask." *Physics Today* 42, no. 12 (December 1989): 43–45. Review of the table-top fusion story and a report on the findings of a panel that disputes the original claims.

Horgan, John. "But Is It Art?" *Scientific American* 257, no. 4 (October 1987): 48–52. Describes several techniques used to test the authenticity of artworks.

Inglis, David R. *Nuclear Energy—Its Physics and Its Social Challenge*. Reading, MA: Addison-Wesley, 1973. A good general source on power plants, weapons, and applications of nuclear radiation.

Lester, Richard K. "Rethinking Nuclear Power." *Scientific American* 254, no. 3 (March 1986): 31–39. Contains information about commercial nuclear power around the world and possible changes in power plant design.

Rafelski, Johann, and Jones, Steven E. "Cold Nuclear Fusion." *Scientific American* 257, no. 1 (July 1987): 84–89. Describes how muons can induce fusion.

Rhodes, Richard. *The Making of the Atomic Bomb*. New York: Simon and Schuster, 1986. Award-winning best seller about the World War II bomb project.

Segrè, Emilio. *From X-rays to Quarks*. San Francisco: W. H. Freeman and Co., 1980. Contains two chapters on Rutherford, one on Fermi and fission (including the Manhattan Project), and more.

Upton, Arthur C. "The Biological Effects of Low-Level Ionizing Radiation." *Scientific American* 246, no. 2 (February 1982): 41–49. Discusses the hazards of radiation to life.

Waldrop, M. Mitchell. "The Shroud of Turin: An Answer Is at Hand." *Science* 30 September 1988, pp. 1750–1751. Describes the events leading up to the final dating of the Shroud.

Weaver, Kenneth F. "The Promise and Peril of Nuclear Energy." *National Geographic* 155, no. 4 (April 1979): 459–493. Contains dozens of color photographs and diagrams relating to many aspects of nuclear energy.

OUTLINE

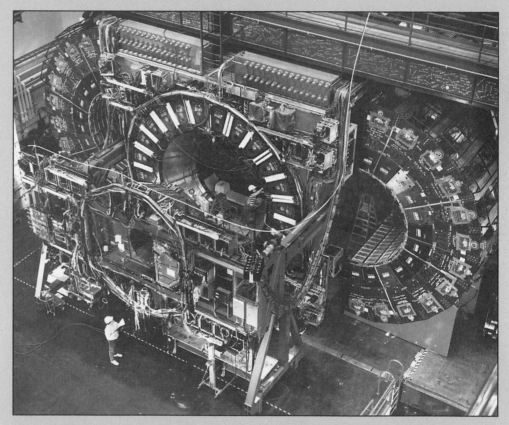

The collider detector, commissioned in October, 1985 at Fermilab. The black arches removed to the sides contain modules of the central calorimeter which were built by a collaboration of scientists from the U.S., Japan, and Italy. The calorimeters contain photomultipliers and electronics used to measure particle energies from the collision. The calorimeters surround the region in the detector where the collisions of protons and antiprotons take place.

SPECIAL RELATIVITY AND ELEMENTARY PARTICLES

PROLOGUE: A BLAST FROM THE PAST

On the evening of 23 February 1987, Ian Shelton, an observatory assistant stationed at Las Campanas in Chile, was one of the first persons to observe the explosive destruction of a star in the Large Magellanic Cloud (LMC), a nearby companion galaxy to our own Milky Way. Shelton detected this supernova event photographically, collecting on a light-sensitive emulsion the photons emitted by highly excited atoms in the rapidly expanding envelope of debris surrounding the once shining star (Figure 12.1). Traveling at the speed of light in the vacuum of intergalactic space, these photons had been in flight for about 170,000 years to cover the distance separating the LMC from earth. What Ian saw was truly a "blast from the past."

Interestingly enough, news of the explosion of Sk − 69°202, the official name of the progenitor star of what has been designated SN (for SuperNova) 1987A, had reached the earth some 18 hours before the optical detection. A short burst of neutrinos was recorded by a collection of 2,000 photomultiplier devices submerged in a tank containing 7,000 tons of ultrapure water buried some 600 meters deep in a salt mine near Cleveland, Ohio. The experiment was being carried out by particle physicists from a consortium of institutions, including the University of Michigan, Brookhaven National Laboratory, and the University of California–Irvine, *not* to detect neutrinos from supernovae but to seek evidence of the spontaneous decay of the proton. Information on the average decay time for protons is of critical importance to theories that attempt to unify the four fundamental forces in nature into a cohesive whole.

As it turned out, the experiment found no evidence of any proton decay events, but it did record the passage of neutrinos emitted during the collapse of the core of Sk − 69°202 to form a neutron star[1] (Figure 12.2). These neutrinos, because of their low interaction probability, escaped directly from the interior regions of the star; they represented the prompt messengers of the conditions in this dense, hot core. The photons produced in this event required more time to work their way outwards through the expanding shell of ejected material and to escape into space. This accounts for the delay in the

FIGURE 12.1
Photographs of a region of the Large Magellanic Cloud (LMC) before, (a), and after, (b), the explosion of Supernova 1987A. The arrow in (a) shows the location of the star that was the progenitor of the supernova.

---

[1]Actually the particles detected were *anti*neutrinos, which interacted with the protons in the water molecules to produce neutrons and positrons. When we investigate particles and their anti's, we will find that they are often created in pairs, so that evidence for the one may constitute evidence for the other.

(a)

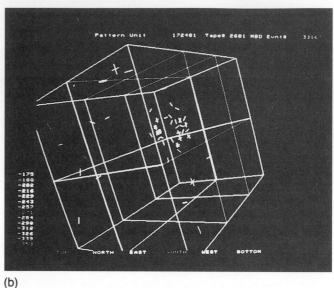

(b)

FIGURE 12.2
*(a) Diver inspecting the array of more than 2,000 light-sensitive devices in the
Irvine-Michigan-Brookhaven (IMB) neutrino detector located outside Cleveland, Ohio.
(b) A computer-generated reconstruction of one of eight neutrino interactions recorded
by the IMB detector due to Supernova 1987A. The grid lines define the faces of the
detector. The crosses in the corner represent photomultiplier tubes that recorded light
signals produced by a high-speed particle—a particle created in the collision between the
(anti) neutrino and a proton in the tank.*

arrival on earth of the photons relative to the neutrinos. Similar experiments
have been undertaken by Japanese and Italian groups; the Japanese also
observed neutrinos from SN 1987A.

By studying the energies and spread in arrival times of the neutrinos, not
only were scientists able to confirm many of their theories about how
supernovae "pop off," but they were also able to set some limits on the mass
of the neutrinos. These data are important to cosmologists because neutrinos
are among the most populous particles in the universe. If their individual
masses are even very small (instead of being strictly zero, as initially
postulated), the accumulated total mass contributed by these particles could
be very large, large enough perhaps to affect the expansion rate of the
universe and its future evolution.

Thus a serendipitous discovery of neutrinos emitted by an exploding star in
another galaxy by an experiment designed to test a theory of elementary
particle physics has led to results that have major implications on the grandest
scale of all! With examples like this, it is small wonder that the physics of the
microscopic and that of the macroscopic are becoming more and more closely
allied.

It may be difficult to imagine how the physics of particles smaller than the
nucleus of an atom could be related to the physics of the largest structure known,
the entire universe, but the links connecting these two domains have been growing stronger with each passing year. We now recognize that the study of elementary particles and that of cosmology are inextricably entwined. Investigations in

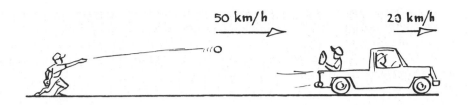

FIGURE 12.3
If you are traveling at 20 km/h, and a friend throws a ball toward you at 50 km/h, you see the ball approaching at 30 km/h.

particle physics yield insights into the formation and evolution of the universe, and observations of the distribution and motion of matter at remote distances provide tests of developing theories in subatomic physics. In this chapter we will explore the connection between the microscopic and macroscopic domains of physics, as well as the attempts of physicists to unify the fundamental forces of nature into one grand theory of everything. We begin our exploration by examining one of the seminal theories in this unification process developed by Albert Einstein, the special theory of relativity.

12.1   SPECIAL RELATIVITY: THE PHYSICS OF HIGH VELOCITY

Imagine the following hypothetical experiment: You are seated in the cargo area of a small pickup truck moving directly away from a companion at a constant speed of 20 kilometers/hour. Your friend tosses a baseball to you with a horizontal speed of 50 kilometers/hour (Figure 12.3). From your point of view, what is the ball's speed? If you answered 30 kilometers/hour, you're right. Clearly, *the speed of the ball with respect to you depends upon your own speed, that is, the speed of the observer.* This is common sense: Galileo and Newton both would have agreed with you completely.

Now consider a second hypothetical experiment: You enter a spacecraft and leave the earth, traveling uniformly at a speed of 200,000 kilometers/second. After a time, a friend sends out a light ray, which moves at the speed of light—300,000 kilometers/second in your direction (Figure 12.4). When the light ray reaches you, what would you measure its speed to be? If you answered 100,000 kilometers/second, you are wrong. Strange as it may seem, you would find the speed of the light to be 300,000 kilometers/second, just what it is for your friend back on

FIGURE 12.4
The light approaches you at 300,000 km/s, <u>even if you are moving away</u> from the source at 200,000 km/s.

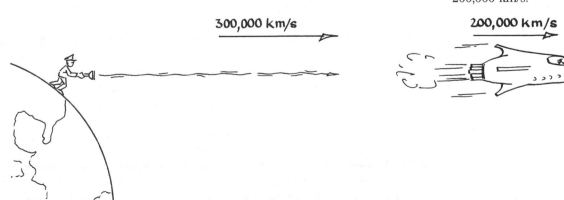

earth. Evidently *the speed of light does not depend upon the speed of the observer.* Light (and all other forms of electromagnetic radiation), behaving as Maxwell predicted, does *not* act as we would expect based on the physics of Newton and Galileo. The theory of Newtonian mechanics and the theory of electromagnetism appear to be in conflict.

Postulates of Special Relativity

Albert Einstein recognized the contradiction between the predictions of classical physics and those of electromagnetism as regards the propagation of light and set about reconciling the two. He began by adopting these two postulates.

1. The speed of light, $c = 300,000$ kilometers/second, is the same for all observers, regardless of their motion.

   Like the gravitational constant, G, in Newton's law of universal gravitation, c is a fundamental constant of nature. The fact that the speed of light is constant was just starting to be accepted at the turn of the century when Einstein began his studies. Precise experiments, like those originally performed in 1887 by A. A. Michelson and E. W. Morley, and others involving astronomical systems have demonstrated beyond doubt that the speed of light is invariant under all circumstances. Thus we are fully justified in accepting Einstein's first postulate.

2. The laws of physics are the same for all observers moving uniformly, that is, at a constant velocity. This is the *principle of relativity.*

   This means that if two observers traveling toward one another at a constant speed perform identical experiments, they will get identical results. Moreover, *no* experiment can be performed by either observer that will indicate whether they are moving or what their speed is. A common example of this principle involves railway travel. If you are aboard a train moving in a straight line at constant speed, any experiments you perform will give the same answers as those obtained in similar experiments done in the train station. The laws of physics are the same in both "laboratories" and cannot be used to demonstrate your uniform motion relative to the train station. But you say you can look out the window and see that you are in motion. True, but how can you *prove* that you are not really at rest and that, by some magic, the trees, fences, and houses are not moving past you in the opposite direction? In point of fact, you can't! Experimentally, the principle of relativity has been well substantiated.

Based on these two postulates, Einstein developed his *special theory of relativity,* which was published in 1905. It describes how two observers, in uniform relative motion, perceive space and time differently. One of the interesting aspects of this theory is that once you have accepted the experimentally verified postulates on which it is based, the fundamental predictions can be understood with only elementary algebra. The equations of special relativity are only a little more difficult than Newton's law of universal gravitation.

Predictions of Special Relativity

Let us consider one of the predictions of special relativity, something called *time dilation.* Imagine that we construct two identical "clocks" consisting of a flash-

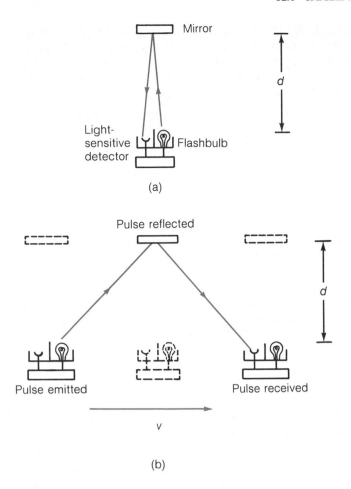

FIGURE 12.5
(a) A "light clock" at rest in a laboratory on earth.
(b) The light clock on board a spacecraft traveling uniformly at speed v, *as it would be seen by an observer on earth.*

bulb, a mirror, and a light-sensitive detector (Figure 12.5). The flashbulb emits flashes of light at some preset rate. Each light pulse is reflected by the mirror into the detector. Each time the detector receives a pulse, a click is emitted like that of a standard ticking clock. Let us now synchronize our two clocks and give one to a friend who is to travel in a spaceship with speed *v* relative to you on earth. The question is: Will the two clocks keep the same time? That is, will they continue to tick at the same rate and indicate the same time? The answer seems obvious: Yes! This is the answer Newton would have provided. Unfortunately, given the postulates of special relativity, it is the wrong answer. Let us see why.

Consider the clock in the spaceship. When your friend took it aboard, you both agreed that it was a properly working, standard clock. Consequently he notes nothing peculiar in its performance as he travels along. Indeed, he *cannot* identify anything different about the clock, since if he did, he could know he was moving. Such a circumstance would violate the principle of relativity, which says that physics is the same for all uniformly moving observers. The clock on the space-craft, *as seen by your friend aboard the craft,* ticks along at the same rate as it did when he first mounted it.

But what about the clock on board the spaceship as seen by you, an external observer? If you track the motion of the clock, say with a powerful telescope, you see that the light, in going from flashbulb to detector, follows a zigzag path, since the clock is moving sideways as the pulse propagates up and down. Evidently the path that the light travels in the moving clock is longer than the path it follows in your laboratory clock. Consequently, since the speed of light is the same in both cases, you conclude that the time it takes the light to reflect back to the detector is longer for the moving clock than for your clock. In other words, the moving clock is running slow: the rate at which it ticks is less than that for your clock.

Of course, if your friend reads your clock from his spaceship, it is *your* clock that appears to be running slow, since from his vantage point, your laboratory appears to be moving uniformly with a speed *v in the opposite direction.* The symmetry between the observations made on earth and in the spaceship is guaranteed by the principle of relativity. But who is *really* right you ask? Whose clock is *really* running slow? Both observers are right, and each clock is really running slow when compared to the other. The observers perceive the rate of flow of time differently because of their relative motion. But because no experiment can determine which observer is *really* in uniform motion, each observer's perception is as good or as true as the other's. *Time, then, is not an absolute, innate quality of nature:* it depends on the observer and his state of *relative* motion.

By how much will the interval between successive ticks differ for the two clocks discussed above? Not very much, unless *v* is very close to the speed of light. If we let Δt be the time between ticks on the clock at rest with an observer and let $\Delta t'$ be the observed time between ticks on the clock moving with speed *v* relative to the observer, then Δt and $\Delta t'$ are related by the following equation:

$$\Delta t' = \frac{\Delta t}{\sqrt{1 - \dfrac{v^2}{c^2}}}$$

Figure 12.6 is a graph of $\Delta t'$ for different speeds *v.* Even for speeds as high as one half the speed of light, 150,000 kilometers/second, $\Delta t'$ is not much larger than Δt; the clocks tick at nearly the same rate. Thus effects of time dilation are virtually unknown in our daily lives, since we do not experience extremely high speeds. However, these effects are as "real" as any other phenomena in physics, and they have been observed in a very interesting manner.

There exists a subatomic particle called the *muon,* which decays spontaneously into an electron plus some other particles in an average time of 0.000002 seconds. These muons are produced in large numbers by collisions between cosmic rays and atmospheric molecules some 10 or more kilometers above the

FIGURE 12.6
Graph of the time between clicks ($\Delta t'$) on a clock moving relative to an observer, versus the clock's speed v. *At speeds even up to 0.5 c, the time $\Delta t'$ is nearly the same as Δt, the time between ticks when the clock is at rest with respect to the observer.*

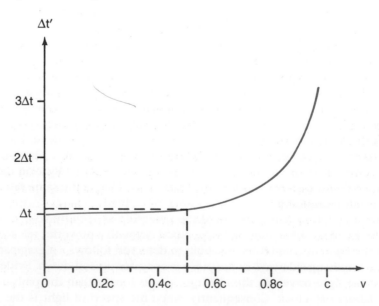

ground. Given their short lifetimes, we should find very few muons reaching the ground, even though they cover the distance between their place of production and the earth's surface at nearly the speed of light. Nevertheless, we detect great numbers of muons at ground level. How can this be possible?

A resolution to this paradox can be had by applying special relativity theory. Because the muons are traveling so rapidly, their internal clocks, which regulate their rate of decay, appear to us to be running some 10 times too slow. Consequently, from our perspective, there is ample time for them to reach the ground and to be detected—which is what happens. Of course, from the point of view of the muons, they do decay in 0.000002 seconds according to their own clocks, and, again from their perspective, it is our clocks that are running 10 times too slow.

What will the mean lifetime of a muon be as measured in the laboratory if it is traveling at 0.90 c with respect to the laboratory? The mean lifetime of a muon at rest is 2.2×10^{-6} seconds.   **EXAMPLE 12.1**

If an observer were moving along with the muon, then the muon would appear to be at rest to such an observer. The muon would decay in an average time $\Delta t = 2.2 \times 10^{-6}$ seconds as seen by this observer. For an observer in the laboratory, the muon lives longer because of time dilation. Applying our equation, we find that the average muon lifetime, $\Delta t'$, as determined in the laboratory is

$$\Delta t' = \frac{\Delta t}{\sqrt{1 - v^2/c^2}} = \frac{2.2 \times 10^{-6}\,\text{s}}{\sqrt{1 - (0.90c)^2/c^2}} = \frac{2.2 \times 10^{-6}\,\text{s.}}{\sqrt{0.19}}$$

$$\Delta t' = 5.0 \times 10^{-6}\,\text{s.}$$

This is about 2.3 times the average lifetime of a muon at rest. To the laboratory observer, the muon's clock appears to be running more than two times too slow. How fast would the muon have to be traveling for its clock to appear to be running 10 times too slow, as mentioned in the text?

Time dilation is one verified prediction of special relativity. There are two others. One is *length contraction,* in which moving rulers are shortened in the direction of motion. A convenient way to measure a distance is to time how long it takes light to traverse it. But if moving clocks run slow, so that the elapsed light-travel-time is smaller, then moving rulers must be too short in the direction of motion. (Remember, distance equals speed × time.) The length of a meter stick moving relative to an observer is decreased by the same factor as the time between ticks in the two light clocks. Thus moving observers disagree on issues involving *both* length and time.

The last consequence of special relativity, and that which is most important in particle physics, is the equivalence of energy and mass. If moving observers disagree on matters involving length and time, then they will also disagree on the velocities of material particles. For example, if a collision between billiard balls occurs in one environment, the initial and final velocities of the balls as determined by an observer in that environment will not agree in general with those determined by another observer moving uniformly relative to the first. Yet both observers *must* agree that the physics of the collision is the same. In particular, both must agree that momentum and energy are conserved during the collision.

It turns out that for the laws of conservation of momentum and energy to be preserved in such cases, the observers must each include in their bookkeeping the *rest energy, E_o,* of the billiard balls, given by Einstein's famous formula

$$E_o = mc^2$$

where m is the ordinary mass of each ball (sometimes called the *rest mass*) and c is the speed of light. Thus, as Einstein wrote in 1921, "Mass and energy are therefore essentially alike; they are only different expressions for the same thing."

In the special theory, then, the total relativistic energy of a particle as measured by an observer is comprised of two parts: the rest energy, mc^2, of the particle plus whatever additional energy the particle has due to its motion; that is, its kinetic energy. The rest energy of a particle is clearly the same for all observers, but the kinetic energy (and hence the total energy) of the particle is not; it depends upon the frame of reference of the observer. Specifically, the energy of a particle of mass m moving with speed v relative to a particular observer from special relativity theory is equal to

$$E_{rel} = KE_{rel} + mc^2 = \frac{mc^2}{\sqrt{1 - v^2/c^2}}.$$

Solving for the relativistic kinetic energy, we find

$$KE_{rel} = \frac{mc^2}{\sqrt{1 - v^2/c^2}} - mc^2.$$

For very low particle speeds, this equation reduces to the familiar Newtonian form *(1/2)mv^2* (see Challenge 1 at the end of this chapter). However, for velocities approaching the speed of light, the energy increases without limit. Thus to accelerate a particle to the speed of light would require an infinite amount of energy. This is the reason why no material particle can ever travel at the speed of light. The energy and power demands for doing so simply cannot be met! The speed of light is not only a constant for all observers, but it is an absolute barrier that no object can cross.

The fact that the (rest) mass of a particle is the same for all observers (that is, an *invariant*) does *not* mean that the mass cannot change. Quite the contrary! In an inelastic collision, mass frequently changes and is transformed into energy. Conversely, in the course of such encounters, energy may be converted into mass. This state of affairs is the direct result of Einstein's recognition of the equivalence of these two things. Thus when two balls of clay collide and stick together, some of the initial energy is converted to heat. Within the framework of special relativity, the energy that has gone into heat (and any other forms of internal excitation present in the final system) is measured exactly by the increase in the rest mass of the final system over that of the initial. Now in most practical situations in everyday life, the changes in mass that accompany interactions of this type are far too small to detect. However, we have discussed some examples in which this type of conversion does lead to very dramatic effects in connection with nuclear reactions in Section 11.5. As we will see shortly, similar conversions taking place in inelastic collisions involving high speed particles are the bread and butter of experimental high-energy physics and lead to the creation of exotic new species seldom seen in nature.

EXAMPLE 12.2   In an x-ray tube (Figure 8.34), an electron with rest mass $m = 9.1 \times 10^{-31}$ kilograms is accelerated to a speed of 1.8×10^8 meters/second. How much en-

ergy does the electron possess? Give the answer in joules and in MeVs (million electron volts).

The total relativistic energy of the electron, E_{rel}, is its relativistic kinetic energy plus its rest energy. From our formula, we see that

$$E_{rel} = KE_{rel} + mc^2 = \frac{mc^2}{\sqrt{1 - v^2/c^2}}.$$

Let us first determine at what fraction of the speed of light the electron is moving.

$$v/c = (1.8 \times 10^8 \text{ m/s})/(3.0 \times 10^8 \text{ m/s}) = 0.60.$$

$$v = 0.60\,c.$$

The electron travels at about 60% the speed of light.

Evaluating the square root in the equation for E_{rel} gives,

$$\sqrt{1 - v^2/c^2} = \sqrt{1 - (0.60)^2} = 0.80.$$

The energy of the electron is then given by

$$E_{rel} = \frac{(9.1 \times 10^{-31} \text{ kg})(3.0 \times 10^8 \text{ m/s})^2}{(0.80)} = 1.02 \times 10^{-13}\text{ J}.$$

But 1 joule $= 6.25 \times 10^{18}$ eV (see Appendix A), so

$$E_{rel} = (1.02 \times 10^{-13})(1 \text{ joule}) = (1.02 \times 10^{-13})(6.25 \times 10^{18} \text{ eV})$$

or

$$E_{rel} = 637{,}500 \text{ eV}.$$

Since 1 MeV $= 1 \times 10^6$ eV, the energy of the electron is approximately 0.638 MeV. (This is actually about four times greater than the typical maximum energies for most x-ray tubes that run near 150,000 eV.)

Notice, with E in MeV, the *equivalent* mass of the electron could be given as 0.638 MeV/c^2. This is a frequently used and very convenient way of representing subatomic particle masses because it eliminates small numbers that necessitate the cumbersome exponential notation.

Let's compare the relativistic kinetic energy of the electron to that given by classical physics. The rest energy of the electron, mc^2, may be easily shown to be 0.511 MeV following the model above. Then

$$KE_{rel} = E_{rel} - mc^2 = 0.638 \text{ MeV} - 0.511 \text{ MeV} = 0.127 \text{ MeV}.$$

According to Newtonian mechanics,

$$KE_{classical} = (1/2)mv^2 = (1/2)(9.1 \times 10^{-31} \text{ kg})(1.8 \times 10^8 \text{ m/s})^2$$

$$= 1.47 \times 10^{-14}\text{ J} = 92{,}100 \text{ eV} = 0.092 \text{ MeV}.$$

The classical result underestimates the electron's kinetic energy by almost 30%.

If we reflect on the special theory of relativity, we see that it accomplishes a profound unification in physics: it reconciles the physics of low speeds with that of high speeds. It is a better, more comprehensive system of kinematics and dynamics than Newtonian mechanics because it works for all particles, regardless of their relative velocities. In the limit of small velocities, we recover the laws of classical mechanics as we specified them in Chapters 1 through 3; for high velocities, we find that Einstein's theory predicts new effects not contained in Newton's physics that are confirmed experimentally. We will return to this theme of unification in Section 12.5 after we consider elementary particles and the forces they mediate, since it has been and remains today one of the overriding goals of physical science.

12.2   FORCES AND PARTICLES

The Four Forces:
Natural Interactions Among Particles

At various points in this book we have mentioned the four fundamental forces of nature. Table 12.1 lists these basic forces and includes some properties of each. It is important to acknowledge that *all* the interactions that occur in our environment are due to these forces. They produce the beauty, variety, and change that we daily witness in the world around us.

Refer to "Physics Potpourri" in Section 2.3.

In Chapter 2 we defined a force as a push or pull acting on a body that usually causes a distortion or a change in velocity (or both). This is a perfectly good description of what we mean by a force in classical physics; but to investigate the realm of particle physics, we must broaden our definition to include every change, reaction, creation, annihilation, disintegration, etc that particles can undergo. Thus when a radioactive nucleus spontaneously decays (see Section 11.2), we will describe this decay in terms of a force that acts between the parent nucleus and its decay products. Similarly, when two particles collide and undergo a nuclear reaction to create new particles (see Section 11.4), we say that there is a force responsible for this transformation.

TABLE 12.1   THE FOUR FUNDAMENTAL FORCES OR INTERACTIONS

Type	Relative Strength	Range (m)	Carrier Particle	Rest Mass (MeV/c^2)	Spin
Strong	1	$\approx 10^{-15}$	Meson*	$> 10^2$	1
Electromagnetic	10^{-2}	Infinite	Photon	0	1
Weak†	10^{-13}	$\approx 10^{-17}$	Z^0, $W^{\pm}$	$\leq 10^5$	1
Gravitational	10^{-38}	Infinite	Graviton	0	2

*At the level of the nucleons, we may regard the messengers of the strong force to be the mesons, although, as discussed in Section 12.4, the true carriers of this interaction are the massless, chargeless *gluons.*

†All the messengers of the basic forces are uncharged, except for two of those of the weak force. The W^+ particles carry one unit of positive charge, while the W^- holds one unit of negative charge. The other carrier of the weak force, the Z^0, is electrically neutral.

Because the roles played by forces in particle physics are somewhat different from those traditionally ascribed to them in classical physics, it is often the case that they are referred to as the *four basic interactions* of nature instead of the four basic forces. In this context we use the word "interaction" to mean the mutual action or influence of one or more particles on another. With this in mind, we return to Table 12.1 and discuss each of the four fundamental interactions briefly, beginning with the most familiar, the gravitational interaction.

Gravity, a very important force in our everyday lives, has been investigated at some length in Chapter 2. Several aspects of this interaction as it pertains to particle physics should be reviewed. First, although gravity affects *all* particles, its importance in particle physics is entirely negligible because its strength is so feeble when compared to the other interactions that can occur. To get a feel for just how inconsequential gravity is on a subatomic level, we can compare the strength of the gravitational attraction between the proton and the electron in a hydrogen atom with the electrical attraction between these oppositely charged particles. A simple calculation (see Challenge 2 at the end of this chapter) shows the electrical interaction to be over 10^{38} times stronger than the gravitational interaction. Comparisons between the strength of gravity and the other interactions of nature are given in Table 12.1; in each case the effects of gravity relative to the remaining forces that may influence the behavior of subatomic particles is too small to be considered seriously.

Before moving to a discussion of the other forces, it is worth remarking upon two aspects of gravity that do have important consequences for *large*-scale interactions: first, gravitational interactions may dominate in circumstances involving bulk matter where charge neutrality prevails. If many particles interact together at once and the number of positive charges balances the number of negative ones, electrical forces may cancel out, leaving gravity the dominant interaction; unlike the electrical interaction, gravity cannot be shielded out or eliminated because there is only one kind of mass. And second, gravity is a long-range interaction. The gravitational force varies inversely as the distance squared, and although it grows ever weaker with separation, it never completely disappears. Thus gravitational effects may reach over vast regions of space, accumulating in such a fashion as to affect the structure and evolution of the entire universe.[2]

Aside from gravity, the next most familiar force or interaction is the **electromagnetic interaction,** discussed in Chapter 8. Unlike the gravitational force that is always attractive, the electromagnetic force can be either attractive or repulsive, depending on the relative signs of the interacting charges. But, like gravity, the electromagnetic force is a long-range force, becoming smaller as the distance separating the charges increases. The electromagnetic force also manifests itself in the magnetic forces associated with moving charges, and it is this interaction that is ultimately responsible for all the various kinds of electromagnetic radiation, from gamma rays to radio waves, which were investigated in Chapter 8.

It is important to note that although the electric and magnetic forces act only between charged particles, the electromagnetic interaction can have an influence on uncharged particles as well. For example, a photon is not a charged particle,

---

[2]To account properly for the interaction between massive objects like stars and galaxies or to develop models for the global structure of the universe that are consistent with observation, use must be made of Einstein's *general theory of relativity.* This is basically a more comprehensive and accurate theory of gravity than that originally proposed by Newton, and it permits scientists to interpret correctly such phenomena as the deviation in the path of starlight passing near the sun and the redshift in the spectral lines of certain very dense stars called white dwarfs.

but the absorption or emission of a photon by an atom is an electromagnetic process.

Next among the cadre of nature's interactions is the **weak nuclear force,** which is responsible for beta decay (the conversion of a neutron to a proton within the nucleus; see Section 11.2). The term "weak" may be interpreted in a variety of ways. For example, this interaction is weak in the sense that it is effective only over very short distances: at least 100 times smaller than the range of the strong nuclear force, and essentially infinitesimal compared to the ranges of gravity and electromagnetism. The "weak" force is also weak because the probability that interactions occur by this force is quite small. Indeed, particle interactions involving the weak force generally happen only as a last resort when all other interaction mechanisms are blocked.

Although it is not apparent from what we have said so far, there exists a very close relationship between the electromagnetic and weak interactions. The similarity between these two was first noted in the late 1950s, when it was pointed out that weak processes could be viewed as involving two currents, not unlike the electromagnetic attraction and repulsion of current-carrying wires (see Section 8.1). Further studies of the connections between the weak force and electromagnetism have led to a unification of these two interactions into one, the *electroweak interaction.* This is much like the unification between electricity and magnetism that occurred in Maxwell's theory. It is now recognized that the source of the electromagnetic and weak forces is the same but that their practical manifestations differ considerably, leading to a separate classification for each. More is said about the issue of unification of forces in Section 12.5.

The last of the forces of nature is the **strong nuclear force.** The strong force is responsible for holding the nuclei of atoms together and is involved in nuclear fusion reactions (see Section 11.7). It is a short-range, attractive interaction that does not depend upon electric charge. Considering the probability that two colliding particles will interact by the strong force as opposed to any one of the other three basic interactions in nature, the "strong" force is indeed quite strong, some 100 times as effective in bringing about a reaction as the electromagnetic force. It is comparisons like these between the relative probability that a reaction will occur via a particular interaction that have been used to establish the measures of the relative strengths of the four forces given in Table 12.1.

Einstein's theory of special relativity has as one of its postulates the finiteness of the speed of light; 300,000 kilometers/second is the maximum speed attainable by particles in the universe. This is also the maximum speed at which information may be propagated through the universe. The fact that a star 150 million kilometers away suddenly explodes *cannot* be known to us until at least 500 seconds (about 8 minutes) later because the particles ejected from the event require that long to make their way to us. Information about this explosion thus comes to us through the intermediary of particles that race out from the interaction site carrying data about the nature of the event to our location.

In the same way that we come to understand the details of a supernova explosion by the particles emitted during the interaction, particle physicists come to know the characteristics of the four forces of nature by the particles involved in these interactions. In fact, current theories associate with each force a *carrier,* or mediator, of the interaction. These carriers are exchanged between the particles experiencing the forces, and they communicate the interaction between the reactants and the products. For example, if we wiggle an electron, the change in its electric field will propagate outwards at the speed of light. The disturbance in the field produces forces on other charged particles in the neigh-

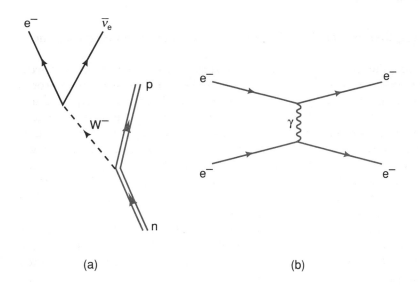

(a)   (b)

FIGURE 12.7
(a) Modern representation of the beta decay of a neutron. The neutron transmutes into a proton after emitting a W^- particle, and the W^- subsequently decays into an electron and an antineutrino. (b) In particle physics, the electric repulsion between two electrons is viewed as being caused by the exchange of photons (γ). This process is shown here schematically in what is called a "Feynman diagram."

borhood of the electron, and communicates to them information about the electron's motion. The role of the field is that of a messenger, and this has led physicists to conceive of the influence of the electric field as being conveyed or carried by a particle messenger—the photon. In this view, all the effects of an electromagnetic field may be explained by the exchange of photons.

Table 12.1 includes the carriers of the four basic interactions as well as their masses (measured in equivalent energy units; see Example 12.2). In the next section we will explore the characteristics of these and other elementary particles in greater detail. Before doing so we show Figure 12.7, which depicts how a particle physicist might represent the weak interaction that converts a neutron inside a nucleus into a proton (a beta decay) and how modern physics views the repulsion between two electrons.

Classification Schemes for Particles

In Chapter 4 we classified matter into solid, liquid, gas, and plasma phases. We also classified matter according to the number and kinds of atoms that are present: elements, compounds, mixtures, etc. We were further able to categorize the properties of matter on the basis of the nature of the forces that acted between the constituents: large forces in solids, smaller forces in liquids, and so on.

Just as we could classify bulk matter in several different ways, it is possible to classify elementary particles on the basis of different schemes. In this section we will consider three ways of doing so: on the basis of spin; on the basis of interaction; and on the basis of mass. Before going much further in this discussion, it is worth defining what we mean by an *elementary particle* and what an *antiparticle* is.

ELEMENTARY PARTICLES   The basic, indivisible building blocks of the universe. The fundamental constituents from which all matter, antimatter, and their interactions derive. They are believed to be true "point" particles, devoid of internal structure or measurable size.

"PARTICLES, PARTICLES, PARTICLES."

FIGURE 12.8

Refer back to the discussion of Dalton's atomic theory in the "Historical Notes" in Chapter 4.

ANTIPARTICLE   A charge-reversed version of an ordinary particle. A particle of the same mass (and spin) but of opposite electric charge (and certain other quantum mechanical "charges").

Every known particle has a corresponding antiparticle. There are antielectrons (positrons), antiprotons, antineutrons, etc. Collections of antiparticles form antimatter, just as collections of ordinary particles form (ordinary) matter. The first antiparticle, the positron, was discovered in 1932 by Carl Anderson in cosmic rays, but particle physicists now routinely create and even store small quantities of antimatter in high-energy accelerators. When matter and antimatter meet, mutual annihilation results, accompanied by a burst of gamma rays. In what follows, we will have several occasions to examine the creation and annihilation of particles and antiparticles. In those reactions we will distinguish antiparticles using the same symbol as that for the corresponding particle but with a "bar" over it. Thus, for example, if n designates a neutron, then $\bar{n}$ (pronounced "en-bar") represents an antineutron.

The kinds of particles that have been termed "elementary" have gradually changed with time. Before 1890 and the discovery of the electron, atoms were regarded as the smallest units of matter, and they were believed to possess no internal structure of their own. In the 1930s it was believed that the basic building blocks of nature consisted of the proton, the neutron, the electron, the positron, the photon, and the neutrino, a massless, chargeless particle that only very weakly interacts with matter and is involved in radioactive decay. Circa 1934, these were the "atoms" (that is, nondivisible particles) sought by the ancient Greeks. In the last 55 years, particle physicists have discovered that not only are there more than six "elementary" particles in nature but that some of the original six are not really elementary at all! They themselves are composed of still more basic and elusive particles. At the time of this writing, there are well over 100 different subatomic particles known (Figure 12.8), and strong evidence exists that the proton and the neutron (among others) are not the ultimate constituents of matter. The majority of the remainder of this chapter will be spent developing the story of our changing perspective on what constitutes a truly "elementary" particle. But, first, a bit more about the properties of subatomic particles in general.

Spin

Spin, like mass and charge, is an intrinsic property of all elementary particles and measures the angular momentum carried by the particle. If we treat an elementary particle as a classical hard sphere—a BB, if you like—then we can picture spin as due to the rotation of the particle about an axis through itself, much like the rotation of the earth about its axis. However, unlike a basketball whirling at the end of one's finger, which can have any amount of spin, the spin of elementary particles is quantized (see Section 10.1) in units of $h/2\pi$, where h is Planck's constant. The spins of all known particles are either integral or half-integral multiples of this basic unit. In other words, spin can take on values of 0, 1/2, 1, 3/2, etc in units of $h/2\pi$. Experiments have shown, for example, that the spin of the electron and the proton is 1/2, while that of the photon is 1.

Stephen Hawking has provided another view of spin that may be helpful in conceptualizing this property of elementary particles. It has the advantage of avoiding any conflicts with quantum mechanics that arise from thinking of par-

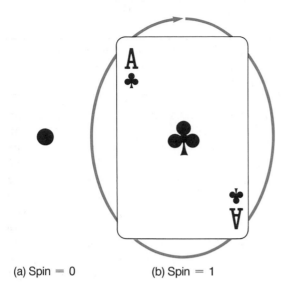

(a) Spin = 0   (b) Spin = 1   (c) Spin = 2

FIGURE 12.9
Illustrations of objects exhibiting symmetry properties analogous to those possessed by elementary particles with certain spins, as described by Stephen Hawking. (a) Dot (spin = 0); (b) Ace of Clubs (spin = 1); (c) Jack of Spades (spin = 2).

ticles as little ball bearings. In Hawking's approach, the spin of a particle tells us what the particle looks like from different directions. Thus a particle of spin zero is like the period at the end of this sentence. It looks the same from all directions (Figure 12.9). A spin 1 particle is like an arrow. It looks different when seen from different directions, but a rotation through a complete circle (360°) restores the original view. A spin 2 particle is similar to a two-headed arrow; rotation by 180° gives the same appearance. In this picture spin 1/2 particles, like electrons, have the remarkable property that *two* complete rotations through 360° are required for the particle to "look" the same. Can you think of a common, everyday object or figure that exhibits this type of symmetry?

Particles possessing half-integral spins are called *fermions,* after Enrico Fermi (see p. 512), who carefully investigated the behavior of collections of such particles. Particles with integer spins are called *bosons,* after S. N. Bose, who, with Einstein, developed the laws describing their behavior. The principal difference between these two types of particles is that the former obey the Pauli exclusion principle (p. 460) while the latter do not. This law, for which Wolfgang Pauli won the Nobel prize in 1945, states that no two interacting fermions of the same type can be in exactly the same state; they must be distinguishable in some manner. Thus in a normal helium atom, when the two electrons are in the lowest atomic energy state (the ground state), the exclusion principle demands that these spin 1/2 particles differ in some way. How can this be achieved? Isn't one electron just like any other? Same mass, same charge, same spin? Yes. But let's return to our rotating BB model for a moment. Relative to the axis of rotation, the BB may spin either clockwise or counterclockwise; for a given total angular momentum, then, there exist two distinct spin states associated with the directions of rotation (Figure 12.10). In the same way one can associate with the electron two different spin configurations, call them "spin-up" and "spin-down," each with the same total amount of spin, $h/4\pi$. With this addition it is now possible to satisfy the Pauli principle for helium by requiring that one of the electrons has spin-up while the other has spin-down. The state of every electron in every atom can be accounted for by this principle.

Bosons, by contrast, do not obey the Pauli principle; they are truly indistinguishable. An unlimited number of bosons can be concentrated in any given volume of space without violating any physical laws. This accounts for the fact

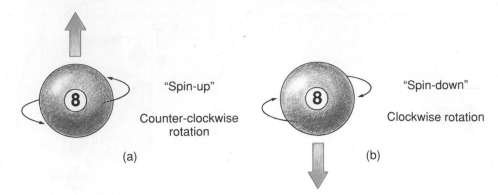

"Spin-up"

Counter-clockwise rotation

(a)

"Spin-down"

Clockwise rotation

(b)

FIGURE 12.10
Spin-up, (a), and spin-down, (b), configurations for a rotating billiard ball.

that there is no restriction on the number of photons (spin 1 particles) that can be packed into a beam of light and hence no (theoretical) limit to the intensity of the beam. There is also no limit to the number of force-carrying particles that can be exchanged in a given reaction, since *all* the mediators of the fundamental interactions are found to have integral spins (Table 12.1). (It may be interesting to mention that the phenomena of superfluidity and superconductivity result from the collective behavior of large numbers of bosons occupying the same state in what is called a *Bose condensate.*)

Elementary Particle Lexicon

Separating particles according to spin divides them into two groups. Establishing additional selection criteria further subdivides these two groups. A very useful way of doing so is to identify which particles participate in strong force interactions and which do not. On this basis we may distinguish four groups of particles, given in Table 12.2: the baryons; the leptons; the mesons; and the intermediate bosons. Examples of several particles of each type are included in the table. An electron is a lepton with spin 1/2 that is unaffected by the strong force; a proton is a baryon with spin 1/2 that does interact via the strong force. Photons, like all the carriers of the basic interactions, are bosons that do not participate in strong interactions.

The names for these groups derive largely from earlier classification schemes based on experimentally determined masses for these particles. The word "baryon" comes from the Greek "baros" meaning heavy, while the word "lepton" means "light one" in Greek. "Mesons" refers to the "middle ones" with intermediate mass. At the time these groups were named, the heaviest particles known

TABLE 12.2 CLASSIFICATION OF PARTICLES BY SPIN AND STRONG FORCE		
	Particles Interacting Via the Strong Force	Particles Not Affected by the Strong Force
Fermions (Half-Integer Spins)	*Baryons* (Protons, neutrons, lambdas, sigmas, . . .)	*Leptons* (Electrons, muons, neutrinos, . . .)
Bosons (Integer Spins)	*Mesons* (Pions, kaons, etas, . . .)	*Intermediate bosons* (Photons, Z^0, $W^\pm$, gravitons)

TABLE 12.3   PROPERTIES OF LONG-LIVED HADRONS

Class	Particle Name	Symbol*	Antiparticle	Rest Mass (MeV/c^2)	B[†]	S[‡]	Lifetime (s)
Baryons	Proton	p	$\bar{p}$	938.3	+1	0	Stable(?)
Spin = 1/2	Neutron	n	$\tilde{n}$	939.6	+1	0	920
	Lambda	Λ^0	$\bar{\Lambda}^0$	1115.6	+1	−1	2.6×10^{-10}
	Sigma	Σ^+	$\bar{\Sigma}^-$	1189.4	+1	−1	0.8×10^{-10}
		Σ^0	$\bar{\Sigma}^0$	1192.5	+1	−1	6×10^{-20}
		Σ^-	$\bar{\Sigma}^+$	1197.3	+1	−1	1.5×10^{-10}
	Xi	Ξ^0	$\bar{\Xi}^0$	1315.	+1	−2	2.9×10^{-10}
		Ξ^-	$\bar{\Xi}^+$	1321.	+1	−2	1.6×10^{-10}
Spin = 3/2	Omega	Ω^-	$\bar{\Omega}^+$	1672.	+1	−3	0.8×10^{-10}
Mesons	Pion	π^+	π^-	139.6	0	0	2.6×10^{-8}
Spin = 0		π^0	Self[11]	135.0	0	0	0.8×10^{-16}
	Kaon[§]	K^+	K^-	493.7	0	+1	1.2×10^{-8}
		K^0	$\bar{K}^0$	497.7	0	+1	0.9×10^{-10}
							5.2×10^{-8}
	Eta	η^0	Self	548.8	0	0	6×10^{-19}

*Superscripts to the right of the particle symbols indicate the charge carried by the particle in units of the proton charge.

[†]Baryon number.

[‡]Strangeness.

[§]There are actually two different types of K^0 particles, one a short-lived particle, K^0_S, and another of somewhat longer lifetime, K^0_L. It is for this reason that two values for the particle lifetime are given in the final column of the table.

[11]Some neutral particles are their own antiparticles. Thus when two π^0's meet, they annihilate one another to form γ-rays.

were among the baryons and the lightest included in the leptons. Recent discoveries, however, have revealed mesons and even a new lepton with masses larger than those of the nucleons (protons and neutrons). Thus it is no longer possible to specify completely the correct class of an elementary particle by its mass alone, although for most species this is still a useful guide.

The force-carrying particles, the intermediate bosons, exhibit a wide variety of mass. Because they are bosons, there is no limit to the number that can be exchanged in any interaction, but there is a close correlation between the range of the force they mediate and their mass. If the carrier particle has a high mass, it will generally be difficult to produce them and to exchange them over long distances. Thus the force carried by massive particles will only have a short range. This is the case for the W and Z particles that mediate the weak interaction. However, if the carrier particles have no mass of their own, like the photon and the graviton, the forces associated with them, in these cases the electromagnetic and gravitational forces, will be of long range.

Before leaving this section, we introduce one additional bit of nomenclature that is used commonly in connection with elementary particles, the word *hadron*. This word is also of Greek origin, coming from "adros", meaning "thick" or "strong." Baryons and mesons are collectively referred to as hadrons (Table 12.3) because they interact by the strong force, a distinction not shared by the leptons or the carrier particles.

In modern physics, governed as it is by probabilities, it is sometimes more important to know what absolutely cannot happen than it is to know anything else. It is in this context that conservation laws play an important role in elementary particle physics. Knowing that no process that contradicts a conservation law can occur, we infer that any process that *does not* contradict a conservation law has some (nonzero) probability of happening. Thus in the world of particle physics, any reaction that is not forbidden may be assumed to take place. Having said this, let us explore an interesting connection between conservation laws and symmetry.

We generally use the word "symmetry" to express an arrangement characterized by some sort of geometrical regularity. For example, the relative positions of atoms in a crystalline solid may repeat over and over in three dimensions so that no matter where you are in the material, things always look the same. Or the locations of objects with respect to a plane may be such that those on one side of the plane appear as mirror images of those on the other (Figure 12.11). Mathematicians discuss symmetry in terms of the operations under which the form of the object remains unchanged. Thus a crystal may exhibit symmetry under the operation of translation, a simple shift in position from one point to another within the crystal. The human body shows symmetry upon reflection about a vertical plane that bisects perpendicularly the shoulder line. Systems like these, which remain unchanged upon translation or reflection (or even rotation), are said to be *invariant* under the application of these operations. Almost every conservation law in physics stems from some fundamental symmetry in nature or, alternatively, from some basic principle of invariance.

The connection between conservation laws, symmetries, and properties of invariance were formulated by one of this century's foremost mathematicians, Emmy Noether (Figure 12.12). A resident of Germany until 1933 when she emigrated to the United States, Noether expressed the relationship between these quantities in a theorem that said that because the laws of physics are unaffected by some operations, physical quantities related to those laws must be conserved, that is, remain constant. This may seem a straightforward enough statement, perhaps even an obvious one. But it requires some very sophisticated mathematics to *prove* that it is true. Rather than dwell on this aspect of the problem, let us now illustrate by example the underlying symmetries associated with some of the better known conservation laws from classical physics.

One of the most important conservation laws is that of mass-energy. In any physical interaction, the total energy, including that bound up as mass, must stay constant: the amount of energy present before the interaction must equal the amount available after the interaction. What fundamental symmetry of nature provides the basis for this conservation law?

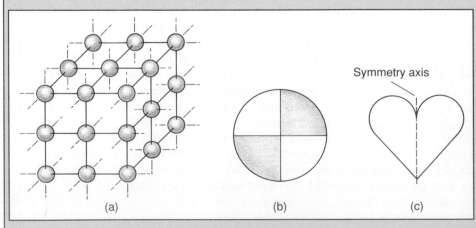

(a)   (b)   (c)

FIGURE 12.11
Examples of objects which show translational, (a), rotational, (b), and reflection, (c), symmetry. From inside the (infinite) crystal lattice, the structure looks the same regardless of where the observer is located. The segmented circle is symmetric with respect to a 180° rotation about its center. The idealized heart shape is symmetric with respect to reflection about the dashed line, called the "symmetry axis."

Symmetry axis

To answer this question, consider an analogy proposed by Heinz Pagels. Suppose the law of gravity were time dependent in such a way that on every Monday the gravitational force was a little bit weaker than it was the rest of the week. Then on Monday you could pump water up to a reservoir high on a hill and on Tuesday let the water run back down to, say, turn a turbine. In the process you would get out more energy than you put in; energy would be created! This is because the amount of work you would have had to do to pump the water upward against a weakened gravitational force on Monday would be smaller than the work done the next day by the strengthened gravitational force in driving the water down hill. This, of course, cannot happen in practice because the law of gravity is time invariant; it doesn't change from day to day or year to year. Conservation of energy, then, derives from the fact that physical laws are invariant with respect to time. Any deviation from this symmetry principle can be shown to lead to the creation of mass or energy where there was none before, in contradiction to observation.

Another of the classical conservation laws that has seen wide use throughout our discussions is the conservation of angular momentum. To get a handle on the symmetry principle on which this conservation law is based, imagine a basketball spinning in empty space. We describe this motion relative to some axis initially fixed in space. But does it matter how we pick this axis? Is one axis to be preferred over another? If you think about it, the answers to these questions should be "No!" In empty space there is no "up" or "down" to provide any guidance when picking an axis about which to characterize the ball's rotation. Any axis will do just as well as another in the absence of such guidance. Indeed, if we choose one axis and then rotate it by some amount, we'd expect to find the equation describing the basketball's motion to remain unchanged. Conservation of angular momentum or spin arises because empty space is isotropic; it looks the same in all directions. It is only when force fields are present that certain "preferred directions" develop and angular momentum fails to remain constant. This was why we emphasized the notion of an "isolated system" when applying this conservation law to the motion of planets and spinning students (see Figure 3.46).

The other classical conservation laws of charge and linear momentum may be similarly interpreted in terms of symmetry principles inherent in nature, and the reader may wish to consult some of the references given at the end of the chapter to find out more about them. (For example, conservation of linear momentum occurs because empty space is homogeneous.) The invariance principles underlying many of the conservation laws used in elementary particle physics involve "internal" symmetries associated with the components of the fields used to describe the particles. The details of these descriptions need not concern us here. Suffice it to say that recognition of these new symmetries has permitted particle physicists to discover the existence of new conserved quantities and new conservation laws that have allowed them to better understand why some things happen in nature and others never do.

FIGURE 12.12
Emmy (Amalie) Noether (1882–1935)

12.3   CONSERVATION LAWS, REVISITED

In Chapter 3 we emphasized the importance of conservation laws in physics, and we used the laws of conservation of mass, energy, linear momentum, and angular momentum to derive information about physical systems on a "before-and-after" basis. Such an approach permitted us to draw some general conclusions about the systems without knowing all the details of the physics of the interaction taking place. A good example of this is the application of the principle of conservation of linear momentum during a collision. Here information about the initial speeds of the colliding objects may be extracted from that concerning the final speeds without knowing the details of the very complicated interactions occurring in crash itself.

In elementary particle physics, the same technique may be applied because the same physical laws are at work. When two particles collide, linear momentum must be conserved: the net momentum present before the collision must equal that after the collision. Likewise, mass-energy must be conserved: the total energy present at the start of an interaction between particles, including that stored as mass, must be the same as that present after the interaction is completed. An important implication of these two laws for particle physics is that they mandate that no single particle can spontaneously decay into particles whose mass exceeds its own. One might imagine, for example, that a high-speed particle could suddenly self-annihilate, converting some of its kinetic energy into mass to form another particle of greater mass moving more slowly. A careful analysis of this situation shows that if this were to happen, either (a) conservation of mass-energy would be violated or (b) linear momentum would not be conserved. (A high-energy photon [γ-ray] may spontaneously break up into an electron and a positron, each of which has more [rest] mass than the photon, but this can only happen in the vicinity of another particle, usually an atomic nucleus, that absorbs the excess momentum to keep the total constant.) Another result of conservation of momentum is that a particle cannot decay into a *single* particle lighter than itself.

In particle physics the law of conservation of angular momentum that we used to analyze spinning ice skaters and orbiting planets in Chapter 3 sees application largely in the preservation of total spin in particle reactions. For example, when a photon turns into an electron-positron pair, spin must be conserved. This means that since the spin of the photon is 1, the spins of the departing electron and positron (each of magnitude 1/2) must both lie along the same direction so that their sum equals 1 (1/2 + 1/2). Similarly, if a particle of spin zero is spontaneously converted into two particles of spin 1/2, their spins must be oppositely aligned so that they add up appropriately (1/2 + [−1/2] = 0).

The three conservation laws discussed above come from classical physics but work equally effectively at the level of subatomic particles. There exists another classical conservation law that also must be obeyed at the submicroscopic level: conservation of charge (see Section 7.1). In any reaction or interaction among particles, the total (or net) charge present before must equal the total (or net) charge present after. Charge can neither be created nor destroyed during an interaction between elementary particles. Continuing with our example of the spontaneous decay of a photon to produce an electron and a positron, we see that the net charge before the decay is zero and that the net charge afterward is also zero, since the electron possesses one unit of negative charge while an antielectron possesses one unit of positive charge; the sum of these is clearly zero. It is interesting that the stability of the electron owes it origin to the classical conservation laws: all the particles lighter than an electron that could be produced in its decay are uncharged. If the electron were to actually undergo a change to produce such particles, a violation of conservation of charge would occur. Consequently, such conversions are forbidden in nature, and the electron is believed to be absolutely stable. As we shall see, such may not be the case for the proton, and the ultimate decay of this baryon may have dire consequences for the fate of the universe as a whole.

New Conservation Laws

The four conservation laws of classical physics were developed from the observed behavior of macroscopic systems, but they are found to apply equally well to submicroscopic systems, in particular to the interaction of elementary particles. However, when experiments involving collisions between particles are carried out, a curious thing is noticed: some reactions that are *not* forbidden by the classical conservation laws are *never* observed. Currently in particle physics it is widely assumed that any reaction that is not strictly forbidden *will* occur, albeit with low probability or frequency. The fact that regardless of the number of times a particular collision happens a certain outcome is never seen has led physicists to suspect that there exist *additional* conservation laws that operate *only* at the submicroscopic scale. These new conservation laws may only be *ad hoc* approximate rules or they may be fundamental statements about nature like that of charge conservation. In either case, the recognition, statement, and use of these additional conservation laws has led to remarkable progress in particle physics, and we take time now to consider several of them. To describe these new regulations on particle reactions, new "charges" or quantum numbers (see p. 456) have had to be invented.

Conservation of Baryon and Lepton Numbers

Consider a collision between two protons. Of the many particles that can be produced in such an interaction, mesons are never seen as the *sole* products, even in cases like the one shown below, which does not violate the four classical conservation laws.

$$p + p \not\rightarrow \pi^+ + \pi^+ + \pi^0$$

(You'll have to take our word for it that this reaction doesn't violate momentum or mass-energy conservation, but you can verify for yourself that it conserves charge and spin using the information in Table 12.3.) Alternatively, mesons *may* be the sole products of the collision between a proton and an antiproton. A possible reaction of this type is the following:

$$p + \bar{p} \rightarrow \pi^+ + \pi^- + \pi^0.$$

Evidently there seems to be a hidden conservation law that prevents the first reaction from happening yet allows the second to go forth.

Investigations of instances similar to the one discussed above have led particle physicists to formulate the *law of conservation of baryon number:* in a particle interaction, the baryon number must remain constant, that is, the number of baryons going into the reaction must equal the number emerging. Like the law of conservation of charge, the law of conservation of baryon number is a simple counting rule. Each baryon is assigned baryon number B = + 1; each antibaryon is assigned baryon number B = − 1. Leptons and mesons have baryon number zero. To apply the conservation of baryon number, simply add up the baryon numbers of the reactants and compare that value to the total baryon number of the products. If the two agree, then baryon number is conserved in the reaction, and the reaction may occur (unless forbidden by other conservation laws). If the values of B before and after the reaction are not equal, baryon conservation is violated, and the reaction cannot occur.

Looking back at the proton-proton reaction given above, we see immediately why it is not observed: the total baryon number at the start is 2, while that at the end is zero—mesons have no baryon number. However, in the interaction between a proton and an antiproton, the initial baryon number is zero. This fact permits the reaction to yield only mesons under certain circumstances.

Conservation of baryon number is obeyed in all interactions, strong, weak, or electromagnetic. It has never been observed to be violated, and it dictates that baryons are created and destroyed in pairs. When a baryon is born in a reaction, an antibaryon must also be created to conserve baryon number. For this reason, the number of baryons in the universe minus the number of antibaryons is constant.

Just as conservation of charge is responsible for the stability of the electron, conservation of baryon number is the root cause of the stability of the proton. The proton is the lightest baryon known; it cannot decay to any other lighter baryons, since none exist. But conservation of baryon number prevents a proton from decaying to any other type of particle, like mesons or leptons, either. Thus it appears that the proton is absolutely stable *if* baryon conservation is a fundamental law and not simply an approximate one. As we will see later in this chapter, some recent theories that attempt to unify the strong force with the electroweak force require that the proton decay spontaneously to lighter particles, in violation of baryon conservation, in an average time somewhere between 10^{30} and 10^{32} years! This result, if true, may not seem to have much relevance for you and me here and now—we are in no immediate danger of spontaneously disappearing in a burst of radioactivity due to the decay of protons in our bodies.[3] But the stability or instability of the proton does have considerable impact on the ultimate fate of the universe, whose age is currently estimated to be around 10^{10} years. But more on this later.

A conservation law similar to that adhered to by baryons exists for leptons. Electrons, muons, neutrinos (typically symbolized by the Greek letter nu, ν), all are created and annihilated in lepton-antilepton pairs. For example, a photon, a nonlepton with lepton number zero, decays into an electron and a positron to conserve lepton number. In beta decay, when a neutron decays into a proton and an electron, an antilepton, in this case an antineutrino, *must* accompany the process to conserve lepton number:

$$n \rightarrow p + e^- + \bar{\nu}_e .$$

(Note: This reaction also conserves charge and baryon number as required.)

There is one additional complication to the application of this lepton counting rule that does not appear when using conservation of baryon number, however. Specifically, leptons belong to *families,* and within each family, lepton number is

---

[3]Physicist M. Goldhaber once said that "we know in our bones that the proton's lifetime is very long if not infinite. If the lifetime were shorter than 10^{16} years, the radioactivity stemming from the decay of protons in our body would imperil our lives."

TABLE 12.4   LEPTON FAMILIES*

Family	Particle Name	Symbol†	Antiparticle	Rest Mass (MeV/c²)	Lepton Number L_e	L_μ	L_τ
Electron	Electron	e^-	Positron (e^+)	0.511	+1	0	0
	Neutrino	ν_e	$\bar{\nu}_e$	0(?)	+1	0	0
Muon	Muon	μ^-	μ^+	105.7	0	+1	0
	Neutrino	ν_μ	$\bar{\nu}_\mu$	0(?)	0	+1	0
Tauon	Tauon	τ^-	τ^+	1784.	0	0	+1
	Neutrino	ν_τ	$\bar{\nu}_\tau$	0(?)	0	0	+1

*The spins of all the leptons, regardless of family, are 1/2. For this reason we have not separately listed this property for each particle.

†Superscripts to the left of the particle symbols indicate the charge carried by the particle in units of the proton charge. Thus the electron possesses one unit of negative charge, while the antimuon carries one unit of positive charge. The neutrinos are chargeless.

conserved separately. Thus the six known leptons (and their anti's) divide into three families: the "electron family," consisting of the electron and its neutrino; the "tau family," containing the tauon and its neutrino; and the "muon family," with the muon and the muon neutrino (Table 12.4). Within each family the particles are assigned lepton number $+1$, and the anti's have lepton number -1. In elementary particle physics reactions, the electron, tauon, and muon lepton numbers before must balance those after.

Conservation of lepton number in this form explains why the following reaction is not observed, even though it does not appear to violate any other conservation laws:

$$\mu^- \not\rightarrow e^- + \gamma.$$

If all leptons had the same kind of lepton number, then this reaction should be possible, since the lepton number at the start is $+1$, and that at the finish is also $+1$. However, experiments have clearly demonstrated that the properties of leptons differ according to family and that because of these differences, a reaction of this type is precluded; it violates lepton number conservation within each lepton family, since the *muon* lepton number on the left is $+1$, while that on the right is zero. Thus the muon cannot disintegrate as shown but may decay to an electron by the following route, which does satisfy all the relevant conservation laws:

$$\mu^- \rightarrow e^- + \bar{\nu}_e + \nu_\mu.$$

Conservation of Strangeness

Beginning in the early 1950s, particle physicists began to detect new particles, which they labeled kaons (K), lambdas (Λ) and sigmas (Σ). These new members of the elementary particle zoo exhibited some very strange properties. First, they were always observed to be formed in pairs. The following two reactions are typical of the way these *strange particles,* as they came to be called, are produced:

$$1.\ \pi^- + p \rightarrow \Lambda^0 + K^0$$

$$2.\ \bar{p} + p \rightarrow K^- + K^0 + \pi^+ + \pi^0.$$

Conversely, the following reaction, which did not seem to violate any then-known [c. 1953] conservation laws, was found not to occur:

$$p + p \not\rightarrow p + \Lambda^0 + \pi^+.$$

Reactions like this that produced single, strange hadrons were never observed.

The second strange aspect about these particles involved their decay rates. The production of these strange species occurred with very high frequency or probability, provided enough energy was available in the collisions. This clearly seemed to indicate that the strong interaction was their source. However, once produced, the lifetimes of these strange particles were much too long for strongly interacting entities: unstable, strongly interacting particles typically decay to other strongly interacting particles on time scales of 10^{-23} seconds. The strange particles decayed to other hadrons only after enormously longer times of the order of 10^{-10} to 10^{-8} seconds. These times turn out to be more characteristic of particles that decay not by the strong interaction but by the weak one!

In 1953, M. Gell-Mann and K. Nishijima independently proposed that certain particles possess another type of "charge" or another quantum number termed "strangeness" (S); this quantity is conserved in strong and electromagnetic interactions but is not conserved in weak interactions. Strangeness is a partially conserved quantity. Table 12.3 lists the strange charges for some of the better known strange particles; their anti's possess strangeness in equal magnitude but of opposite sign. Nonstrange particles, like the nucleons, have zero strangeness.

The impact of the Gell-Mann–Nishijima theory was immediate and revolutionary, since it solved completely all the puzzles presented by the strange particles. When strange particles are produced in strong interactions, the total strangeness of the products must be zero to conserve strangeness; this guarantees that strange particles will be formed in strange-antistrange pairs. (Recall this same behavior is found in strong interactions that produce baryons to conserve baryon number.) Thus in the reaction

$$\pi^- + p \rightarrow \Lambda^0 + K^0,$$

S = 0 initially, since the pion and proton are nonstrange, but S = 0 afterward also, since the strangeness of Λ^0 is -1 while that of K^0 is $+1$. Once created, a strange particle cannot decay via the strong or electromagnetic force to other particles with no strangeness or even to particles of lower strangeness than its own, since this would constitute a violation of the conservation of strangeness. Consequently the only decay channel left to the lonely strange particle is the weak interaction, which, because of its lower probability of occurrence, simply takes longer to happen. In weak processes, strangeness is not conserved, and there can be a net change in the strangeness in such reactions. An example of this is shown below:

$$S = \quad \begin{matrix} \Lambda^0 \\ -1 \end{matrix} \xrightarrow[\text{interaction}]{\text{(weak}} \quad \begin{matrix} p \\ 0 \end{matrix} + \begin{matrix} \pi^- \\ 0 \end{matrix} \quad (\Delta S = +1)$$

Table 12.5 summarizes our knowledge of some of the conservation laws at the present time. Those listed were adequate to explain all the reactions between particles studied up to about 1970. After this date, as we will discover in the next section, several new conservation laws and "charges" had to be invented by particle physicists to explain their observations of new species of elementary particles. The following examples demonstrate how conservation laws may be used to determine whether a given reaction can occur and to predict the identity of "missing" particles in simple reactions.

TABLE 12.5 CONSERVATION LAWS IN PARTICLE PHYSICS	
Physical Quantity or Operation	Interactions in which Quantity is Conserved*
Mass-energy	Strong, electromagnetic, and weak
Linear momentum	Strong, electromagnetic, and weak
Angular momentum	Strong, electromagnetic, and weak
Electric charge	Strong, electromagnetic, and weak
Baryon number	Strong, electromagnetic, and weak
Electronic lepton number	Electromagnetic and weak
Muonic lepton number	Electromagnetic and weak
Strangeness	Strong and electromagnetic only
Parity† (P)	Strong and electromagnetic only
Charge conjugation† (C)	Strong and electromagnetic only
Time reversal† (T or CP)	Strong, electromagnetic, and almost always weak
TCP†	Strong, electromagnetic, and weak

*Lest you get the idea that the weak force is the only major violator of conservation laws on a submi-croscopic scale, we must point out that there exists a quantity called *isospin,* which has nothing to do with ordinary spin but which is conserved only in strong interactions and not in weak or electromagnetic ones.

†These operations are discussed as part of the optional reading in Section 12.3.

Identify the conservation law(s) that would be violated in each of the following reactions: (a) $\pi^+ \rightarrow e^+ + \gamma$; (b) $\pi^- + p \rightarrow K^0 + p + \pi^0$; (c) $\Lambda^0 \rightarrow \pi^- + \pi^+$.

EXAMPLE 12.3

The conservation laws that can be checked easily are those involving charge, spin, baryon number, lepton number, and strangeness. Let us examine each reaction in the light of these regulations.

(a) Since the photon (γ) is neutral, the decay does not violate charge conservation. None of the reactants is a strange particle or a baryon, so conservation laws involving these quantum numbers are irrelevant. But the reaction does violate both conservation of spin angular momentum and conservation of lepton number. Mesons like the π^+ are bosons; they must have integral spins. The positron (e^+), like its mirror twin the electron, is a fermion with spin 1/2; the spin of the photon is 1 (see Table 12.1). The final spin is clearly a half-integer value appropriate for a fermion, not a boson. Conservation of angular momentum precludes the creation of spin where none existed before.

Likewise, since mesons have lepton number 0, as do photons, lepton number is not conserved: $L_e = 0$ initially and $L_e = -1$ finally.

(Remember, an antielectron has lepton number opposite that of an electron.)

(b) Because none of the particles in this reaction are leptons, lepton number conservation need not be considered. All the mesons appearing here (π^-, K^0, π^0) have zero spin; the protons have spin 1/2. Initially the total spin is 1/2. The total spin at the end is also 1/2. This reaction will not violate spin conservation as long as the entering and exiting protons both are spin-up or spin-down.

This interaction clearly violates charge conservation because the total charge at the start of the reaction is zero, but the net charge at the finish is $+1$, the K^0 and π^0 being neutral. (Recall that the proton has one unit of positive charge.)

Although baryon number is conserved in this interaction (mesons have B = 0 while protons have B = +1), strangeness is not. Protons are strongly interacting particles, so this represents a strong interaction that must conserve strangeness. The π's and the p's are nonstrange particles (see Table 12.3); the K^0 has one unit of strangeness. Before the reaction $S = 0$; afterwards, $S = +1$. Strangeness is not constant in this *strong* interaction as it must be.

(c) In this decay, a baryon ($B = +1$ for Λ^0) is converted into two mesons ($B = 0$ for $\pi^\pm$). This is forbidden by conservation of baryon number. Again, because this reaction involves no leptons, conservation of lepton number is moot. However, charge is obviously conserved: $Q_{initial} = 0$ (the Λ^0 is uncharged); $Q_{final} = -1 + 1 = 0$. (In this chapter, we use Q to represent charge, reserving the symbol q for generic reference to quarks beginning in Section 12.4. This notation differs from what we adopted in Chapter 7, and it is not to be confused with the symbol for *heat* used in Chapter 5.)

This reaction also *would not* violate conservation of strangeness because the decay of the strange particle Λ^0 occurs by the *weak* interaction for which S may change by one unit, as it does here. $S_{initial} = -1$; $S_{after} = 0$ (since the mesons are nonstrange). $\Delta S = +1$, which is permitted in weak interactions (see Table 12.5).

Finally, we see that the interaction *might* violate conservation of angular momentum since the Λ^0 is a spin 1/2 fermion, while the π's are both spin zero bosons. Initial spin = 1/2; final spin = 0. Such an interaction is prohibited unless the original spin is converted into orbital angular momentum of the decay products.

EXAMPLE 12.4   Each of the reactions below is missing a *single* particle. Figure out what it must be if these interactions are permitted.

$$(a)\, p + \bar{p} \rightarrow n + \underline{\quad}; \qquad (b)\, \bar{\nu}_\mu + p \rightarrow n + \underline{\quad};$$

$$(c)\, p + p \rightarrow p + \Lambda^0 + \underline{\quad}.$$

(a) Here we have a proton encountering an antiproton and annihilating. The annihilation energy is used to create two new particles. Since the initial charge is zero (the antiproton has one unit of *negative* charge), the missing particle must be neutral like the neutron. But the starting baryon number is also zero ($B = +1$ for p and $B = -1$ for $\bar{p}$). Thus the missing particle must have $B = -1$ to cancel the baryon number ($+1$) of the neutron.

As this is a strong interaction, strangeness must be conserved. Since $S = 0$ at the beginning and n itself is nonstrange (see Table 12.3), the absent particle must also have strangeness zero. To conserve spin, this particle must, in addition, be a spin 1/2 fermion like the p's and n. The only particle satisfying all these criteria is $\bar{n}$, the antineutron.

(b) The key conservation law to use to identify the missing particle is conservation of muonic lepton number. The $\bar{\nu}_\mu$ is a muonic antineutrino with $L_\mu = -1$. The p and the n are clearly baryons with $L = 0$. The absent species must belong to the muon family and have $L_\mu = -1$.

Now to conserve charge, the missing lepton must also possess one unit of positive charge to balance the initial charge of the proton. (Remember, neutrinos have no charge.) The only particle that meets these requirements is the μ^+, an antimuon.

(c) To fill in the missing entity in this reaction, we rely upon conservation of strangeness. Since this is a strong interaction, we know strangeness must remain constant. $S_{initial} = 0$, therefore S_{final} must be zero as well. As it stands, the sum of the strangeness of the products, *excluding* the missing particle is $0 + (-1) = -1$. The particle that is lacking *must* have $S = +1$.

Conservation of charge demands that the missing particle have one unit of positive charge because $Q_{initial} = +2$, while $Q_{final} = +1$ *without* the additional contribution of the missing particle.

The missing particle *cannot* be a baryon, however, because adding another baryon to the products destroys the equality that already exists in baryon number. We are obviously searching for a singly charged, positive meson having one unit of positive strangeness. Examining Table 12.3 shows that we need a K^+, a positive kaon.

The TCP Theorem: Parity *(Optional)*

In the *Physics Potpourri* for this section, we discussed the relationship between conservation laws and symmetry principles and found that underlying each of the classical conservation laws is an invariance under certain mathematical operations, like translation or rotation. In this section we introduce three additional symmetry operations and consider the conservation laws that are associated with each. We begin with the time-reversal operation, T.

Imagine an interaction between two particles, X and Y, taking place through the exchange of a third particle z. The products of the interaction are x and y. The reaction is shown schematically in Figure 12.13(a). According to this diagram, X and Y approach one another from the left, X emits z and decays to x, while Y absorbs z and becomes y. This is the sequence of events with time running forward, the sequence of happenings we would see if we could capture the interaction on film and then run the film forward.

Suppose now that we were to reverse the direction of the film loop and play it backwards instead. This would be like reversing the direction of the flow of time. In this case what we would see is particles x and y merging together from the right; y would then be seen to emit z and become Y. x would collect z to produce X (Figure 12.13[b]). Now obviously this is not the reaction that occurred, but one can imagine such a "time-reversed" interaction and ask: Could it

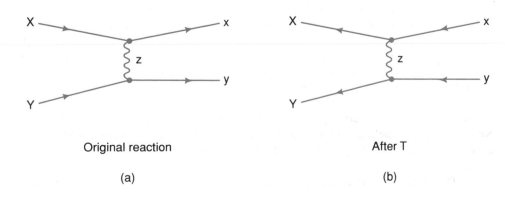

Original reaction

(a)

After T

(b)

FIGURE 12.13
A schematic description (using a Feynman-type diagram) of the reaction $X + Y \rightarrow x + y$ by the exchange of a carrier particle z [wavy line], (a), and its time-reversed (T) analog (b).

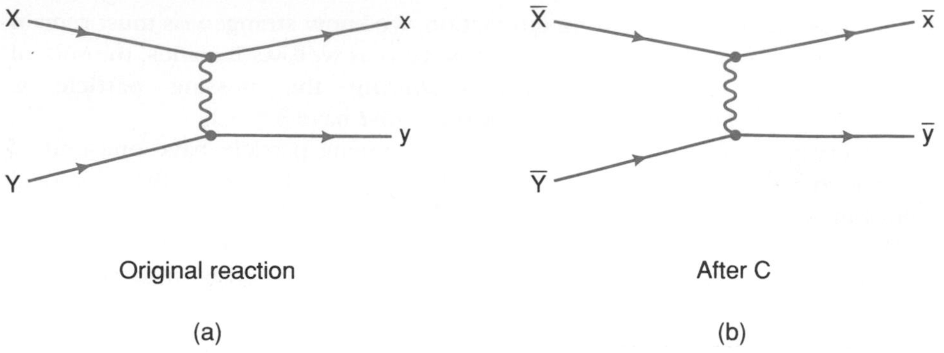

FIGURE 12.14
Feynman diagrams showing the interaction of X and Y as in Figure 12.13(a), and its charge-conjugated (C) equivalent, (b).

Original reaction

(a)

After C

(b)

happen? If so, then we can assert that the original reaction is invariant—unchanged—under the operation T, and we might expect to find a conservation law associated with this symmetry.

Suppose now we visualize the same reaction between X and Y, only now, instead of running the clock backwards, we agree to turn each particle participating in the interaction into its antiparticle (Figure 12.14). Again we ask: Can this so-called "charge-conjugated" reaction occur? If so, we say the original reaction is invariant under the operation of charge conjugation, C, and we anticipate a conservation law embodying this additional symmetry of the reaction.

The last of the three operations to be considered is that of spatial inversion, or *parity* (P). To see the effects of this operation, all you have to do is imagine that you are standing on your head and watching the original reaction going on in a mirror (Figure 12.15). As before, the relevant question is: Will the spatially inverted reaction occur or not? If the answer is affirmative, then the interaction is invariant under the parity operation, and we say parity is conserved in the interaction. Notice that as a result of this operation, both the direction of motion of the particles is reversed *and* so is the spin. (We retain the notion of subatomic particles as tiny rotating spheres for purposes of illustration in this discussion.)

It should be apparent that the operations T, C, and P may be applied separately, in pairs, or altogether to a particular reaction. In the 1950s theorists proved what has become known as the "TCP Theorem." This states that all interactions—strong, weak, and electromagnetic—are invariant under the *combined* operation TCP (*t*ime-reversal, *c*harge-conjugation, *p*arity). This theorem has been tested

FIGURE 12.15
Illustration of the parity (or spatial inversion) operation. The original cue ball moves to the right and spins clockwise (that is, has spin-down). Its spatially inverted partner (or mirror twin) travels to the left and rotates counter-clockwise (spin-up).

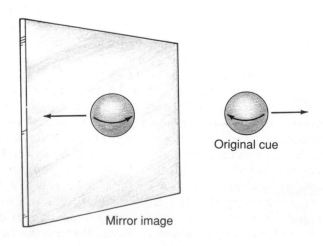

Original cue

Mirror image

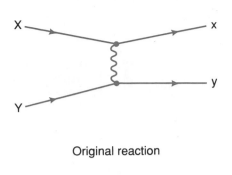

Original reaction

(a)

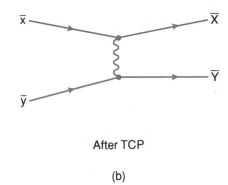

After TCP

(b)

FIGURE 12.16
Feynman diagram showing TCP invariance. According to the TCP Theorem, if the reaction shown in (a) is possible, then the reaction depicted in (b), which is the time-reversed, charge-conjugated, spatially-inverted version of the first, is also possible.

experimentally many times and has always been found to be satisfied. It guarantees that if the reaction shown in Figure 12.16(a) can occur, then the one depicted in Figure 12.16(b) will also happen. But what about reactions in which the operations of T, C, and P are carried out separately or in pair-wise combinations? Are interactions unchanged under TC? or CP? or P?

As an example of the significance of these questions to our understanding of the universe both large and small, let us focus on the issue of conservation of parity, that is, upon the question of whether reactions are invariant under the parity operation. On the surface it would seem that the answer to this inquiry ought to be an obvious "Yes!" After all, part of the parity operation involves what amounts to reflection in a mirror, and this merely interchanges right for left. To deny the conservation of parity is to allege that nature somehow differentiates between left and right. And, up until the 1950s, physicists were unwilling to admit such a possibility. It was believed that every process that took place "to the left" had exactly the same probability of taking place "to the right."

During the middle years of the 1950s, however, evidence began to accumulate that parity was not conserved in certain reactions. In 1956, two young theorists, T. Lee and C. N. Yang, carefully analyzed all the available data and concluded that parity might not be conserved in weak interactions. They even went so far as to suggest some experiments that could be done to test their hypotheses. One such investigation, carried out by C. S. Wu, involved beta decay in radioactive cobalt-60.

A sample of Co-60 was cooled to a very low temperature and placed in a strong magnetic field. In the absence of thermal agitation, the external magnetic field could align the magnetic fields associated with the spinning nucleons, producing an effect much like the alignment found in ferromagnetic materials used to make bar magnets (see Section 8.1). With this "preferred" direction established in the sample, Madam Wu and her collaborators could check the directions in which the electrons were emitted during the beta decay of the cobalt nuclei. What they found surprised everyone. Most of the electrons were emitted in a direction *opposite* to that of the magnetic field. This finding is in clear violation of spatial inversion invariance because the "mirror image" of this reaction is one in which most of the particles are emitted *along* the direction of the magnetic field—and this is simply not what happens in nature! (See Figure 12.17.) For their pioneering work in this area, Lee and Yang received the 1957 Nobel prize in physics.

Since that time other experiments have provided additional evidence that parity is not conserved in weak interactions and that nature truly does make a distinction between left and right. Moreover, it has also been demonstrated that

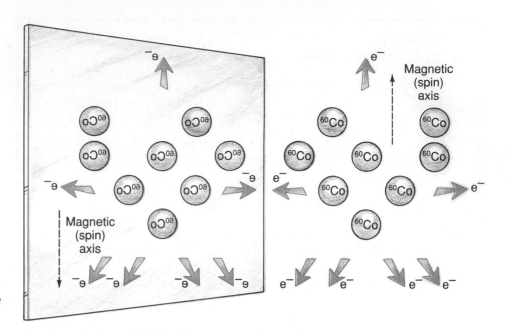

FIGURE 12.17
Illustration of the violation of conservation of parity in the decay of Co–60. Magnetically aligned cobalt nuclei emit electrons predominantly in a direction opposite to that associated with their magnetic field. In the mirror image experiment, most of the electrons are ejected in the direction of the (reversed) magnetic field.

the operation of charge conjugation applied to a weak interaction results in an impossible process. Some weak interaction processes are now known that also violate CP invariance. Referring to Table 12.5, notice that the operations T, C, and P are all conserved in the strong and electromagnetic interactions; the symmetries underlying these invariance principles are all of the "internal" types mentioned in the earlier *Potpourri* in this section.

LEARNING CHECK

1. According to Einstein's special theory of relativity, when compared to an identical clock at rest, a clock moving with a constant velocity will run
   a) slow.   c) at the same rate.
   b) fast.   d) alternately fast and slow.

2. The _____ states that the laws of physics are the same for all observers moving uniformly.

3. Which of the following is(are) NOT fundamental forces in nature?
   a) Friction   d) Strong nuclear
   b) Gravity   e) Electromagnetism
   c) Tension (as in a spring)

4. Particles with integer spins are called _____ , while those with half-integer spins are called _____ .

5. Match each item in column A below with its description from column B. Each entry in column A has only one(1) correct match from column B.

A	B
(i) Baryons	(a) Strongly interacting, spin 0
(ii) Mesons	or 1 particles

continued on next page

(iii) Leptons
(iv) Intermediate bosons

(b) Carrier particles for the fundamental forces
(c) Strongly interacting, spin 1/2 or 3/2 particles
(d) Spin 1/2 particles that do not interact by the strong force

6. In interactions taking place by the strong force, which of the following quantities are conserved? Which are conserved in weak interactions?
   (a) Electric charge
   (b) Baryon number
   (c) Mass-energy
   (d) Strangeness
   (e) Linear momentum
   (f) Angular momentum

7. The positron is the antiparticle corresponding to an ordinary electron. The positron has the same _____ as an electron but the opposite _____ .

12.4 QUARKS: ORDER OUT OF CHAOS

The rapid proliferation of subatomic particles during the 1960s and early 1970s caused many physicists to wonder if what had been thought to be "elementary particles" were really so "elementary" after all. Maybe they themselves were composites of still smaller entities, the *really* elementary" particles. But even before this, in 1956, a Japanese theoretical physicist named Sakata proposed a model in which all hadrons were made out of just six: the proton, the neutron, the lambda, and their anti's. Although there were some peculiar aspects to this model concerning its predictions for the masses and binding energies of some hadrons, it did a pretty good job of accounting for the properties of the then known baryons and mesons, and it violated no then recognized laws of physics.

Shortly after, in 1961, M. Gell-Mann and Y. Ne'eman independently gave a new way to classify hadrons. They grouped them into *supermultiplets,* in which the member particles were treated as different manifestations of a basic state, corresponding to different "orientations" of a new quantum number called "unitary spin." A rotation of the unitary spin in the purely mathematical space in which it is defined changed one particle into another. In the Gell-Mann–Ne'eman model, unitary spin is conserved in strong interactions and has eight components, each of which is a combination of the quantum numbers we have seen before (plus a few others that are too esoteric for us to consider at this level). Because of the eight-part structure of the basic "charge" in this theory and the theory's potential for leading to a deeper understanding of nature, Gell-Mann christened this "the eightfold way," in analogy with "the noble eightfold way" of Buddhism that leads to Nirvana.

Many of the particles listed in Table 12.3 had not yet been discovered at the time these two competing theories for the composition of hadrons were proposed. A test was soon devised to distinguish which of the two, if either, gave the better description of nature. Prior to 1963, the spin of the sigma particles was not known. The Sakata model predicted that these particles should have spin 3/2's;

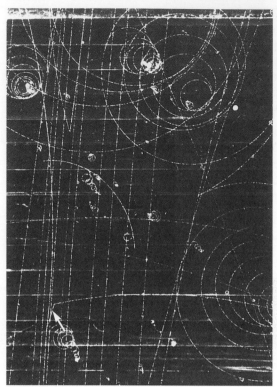

FIGURE 12.18
The first photograph of the Ω^- particle, taken at the Brookhaven National Laboratory in 1964. The path of the Ω^- is marked with an arrow at the lower left of the picture.

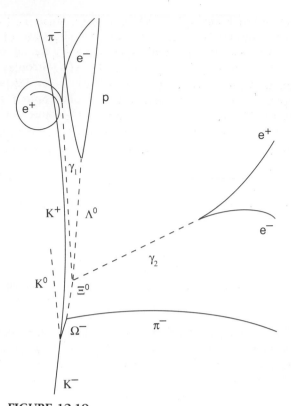

FIGURE 12.19
A schematic reconstruction of some of the particle tracks shown in Figure 12.18. Solid lines indicate the trajectories of charged particles, while dashed lines give the paths of neutral particles (which do not show on the original photograph). The formation and decay of the Ω^- involve the following reactions: (1) $K^- + p \rightarrow \Omega^- + K^+ + K^o$; (2) $\Omega^- \rightarrow \Xi^\circ + \pi^-$.

the eightfold way predicted a spin of only 1/2. When the spins of the Σ^0 and $\Sigma^\pm$ were finally measured, they were found to be 1/2, in agreement with the eightfold way.

Additional support for this model came a little later, with the discovery of the omega minus, Ω^-. Based on the eightfold way, the existence of a single, massive particle with charge -1, strangeness -3, and spin 3/2's, which had never been seen before, was predicted. Particle physicists began to search for this species in new, higher energy collision experiments, and in February 1964, the successful detection of a particle having all the predicted characteristics was announced by scientists at the Brookhaven National Laboratory. Figures 12.18 and 12.19 show the original photograph of the particle tracks in the discovery experiment and the analyses that led to their interpretation in terms of the Ω^-.

Quarks and Their Confinement

Further investigation of the implications of the eightfold way led to a refinement of the theory in 1964. At that time Gell-Mann and G. Zwieg postulated that all hadrons were formed from three fundamental particles, which Gell-Mann called *quarks,* and their anti's. The quarks were designated u (for "up"), d (for "down"), and s (for "strange"). Table 12.6 gives the properties of these three particles and their anti's. Notice that all the quarks have spin 1/2, baryon number 1/3, and charge $\pm 1/3$ or $\pm 2/3$'s (in units of the proton charge). These are fractionally charged particles, unlike anything we have dealt with before!

Surprising as this may appear, the introduction of these noninteger charged particles enabled physicists to describe perfectly all the hadrons discovered prior to about 1970 and, with some extensions, all the heavy particles found since. The two rules governing the formation of hadrons from quarks are simple:

   1. Mesons are composed of quark—antiquark pairs, like $u\bar{u}$ and $d\bar{d}$.

   2. Baryons are constructed out of three-quark combinations; antibaryons are made up of three antiquarks.

For example, a π^+ meson is equivalent to a $u\bar{d}$ pair with their spins oppositely directed. This combination gives a particle of spin 0, baryon number 0 ($=1/3-1/3$), and charge $+1$ ($=2/3+1/3$; see Table 12.6) as required. A K^0 meson may be shown to consist of a combination of a d quark and an $\bar{s}$ quark.

Similarly, a proton is a collection of uud quarks, while the neutron is a udd quark combination. (You should take a few minutes to assure yourself that the addition of quarks as indicated gives the usual properties of the proton and the neutron that you are familiar with. The data in Tables 12.3 and 12.6 will be helpful in this regard.) Other three-quark mixtures give other baryons: $\Sigma^+ = $ uus; $\Lambda^0 = $ uds; and $\Omega^- = $ sss.

To what hadron does the combination of a d and a $\bar{u}$ quark correspond? Assume the spins of the quarks are antiparallel.

EXAMPLE 12.5

A quark-antiquark pair gives a meson, so that is the type of hadron we're looking for. Now a d quark has charge $-1/3$ and a $\bar{u}$ quark has charge $-2/3$. The total charge of our mystery particle is thus -1. Since neither the d or the $\bar{u}$ quark is strange, their combination must yield a nonstrange particle. If the spins of the d and $\bar{u}$ quarks are oppositely aligned, the net spin will be zero for the meson. Evidently we seek a spin zero meson with strangeness 0 and charge -1. Consulting Table 12.3, we find that the particle in question is the π^- meson.

TABLE 12.6	PROPERTIES OF QUARKS*		
Quark	Electric Charge (Q)[†]	Baryon Number (B)	Strangeness (S)
u	$+2/3$	1/3	0
d	$-1/3$	1/3	0
s	$-1/3$	1/3	-1
$\bar{u}$	$-2/3$	$-1/3$	0
$\bar{d}$	$+1/3$	$-1/3$	0
$\bar{s}$	$+1/3$	$-1/3$	$+1$

*The quarks are all spin 1/2 particles.

[†]These values are in units of the proton charge. Thus the charge of an up quark in the international system of units would be $(2/3)(1.6 \times 10^{-19}$ C$) = 1.07 \times 10^{-19}$ C.

EXAMPLE 12.6

Give the quark combination associated with the xi minus, Ξ^-, baryon.

Baryons are composed of three quarks. The Ξ^- has charge -1, strangeness -2, and spin 1/2. Because the quarks are each spin 1/2 particles, we see immediately that the spins of two of the quarks making up the Ξ^- must be paired off (spin-up plus spin-down) so that the net spin is only 1/2. Since the Ξ^- is a strange baryon, it must contain s quarks; they are the only quarks that carry this property. Given that an s quark possesses -1 unit of strangeness, the Ξ^- must contain two of these quarks to have net strangeness -2. If the Ξ^- has two s's, then together they contribute a total charge of $-2/3$ ($= -1/3 - 1/3$). To make up the additional $-1/3$ charge needed to obtain the total charge of -1 for the Ξ^- requires a nonstrange quark of charge $-1/3$. The only candidate is the d quark. Thus a Ξ^- is a dss quark combination.

It is important to emphasize at this point that *the quark model only applies to hadrons.* Leptons, like the electron, the muon, and the neutrinos, are *not* made of quarks. Indeed, the status of the leptons and the quarks is much the same within the realm of particle physics: both are groups consisting of fundamental, irreducible spin 1/2 fermions that together make up the matter in the universe.

The quark model has several distinct advantages over any competing theories of subatomic particles. First, it explains why mesons all have integral spins, that is, why mesons are bosons. Because mesons are two-quark combinations, the mesons, as a class, can only have total spin 0 (when the spins of the two constituent quarks point in opposite directions) or 1 (when the two spins are aligned). Second, the model also accounts for the fact that baryons all are fermions, half-integral spin particles: any arrangement of three quarks will always yield either a spin 1/2 or a spin 3/2 particle, depending on whether the spins of two of the three are paired (spin 1/2) or whether all three spins are parallel (spin 3/2). By the same token, the quark model permits a new interpretation of strangeness and its conservation. Strangeness is just the difference between the number of strange antiquarks and the number of strange quarks making up the particle. Conservation of strangeness in strong interactions may now be seen as the prohibition of the conversion of an s quark to a d or a u quark during a reaction.

Given the abundant success of the quark model, it was not too long before particle physicists began looking for evidence of the existence of quarks. Despite many careful searches among many different types of particle interactions, no free quarks have ever been found, and current wisdom suggests that none will ever be detected, regardless of how high the collision energies are. The argument that has been proposed to account for the absence of free quarks asserts that the interquark force, unlike *any* other force known in nature, *increases* rapidly with the separation between quarks. (Recall from Section 12.2 that the four basic interactions are either very short ranged or *decrease* in strength with increasing distance.) Thus when the quarks are close together, as when bound as a hadron, the force between them is vanishingly small; but when they are as far apart as the diameter of a proton, the restoring forces acting to pull them together become enormous.

One way to envision the quark-quark interaction is to imagine that the quarks are connected to one another by rubber bands (Figure 12.20). When they are near one another, the rubber band is relaxed, and the force between the particles is weak. As the two quarks move apart, the force exerted by the rubber string

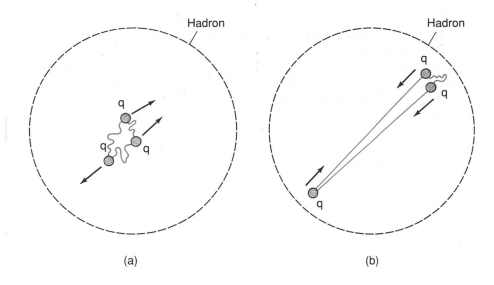

(a)   (b)

FIGURE 12.20
Rubber band model for quark confinement. When the quarks composing hadrons are close together, (a), the force between them is very small; the bands joining them are relaxed. When the distances separating the quarks reach those close to the dimensions of the hadrons themselves, (b), the forces between the quarks become enormous; the bands connecting them are stretched tautly. (Arrows suggest directions of motion of quarks.)

grows larger until finally the two quarks snap back together. The nature of the interquark force thus confines the quarks to remain packed together in hadrons.

This "rubber band" model of quark confinement is useful in explaining yet another interesting aspect of the failure to detect free quarks. Suppose we invested a vast amount of energy in stretching the rubber band holding two quarks to its elastic limit. We might expect that the band would be severed, releasing the quarks. In practice, what happens is that when the energy exceeds that equivalent to the mass of a quark-antiquark pair, the rubber band does break, but instead of freeing the two quarks, two *new* quarks appear at the ends of the broken fragments (Figure 12.21). Again, our attempts to produce free quarks are foiled, and in lieu of unbound quarks, we get mesons! (This is a little like what happens when you cut up a bar magnet; instead of ending up with two fragments, one a north pole and the other a south pole, you always get both poles together in the form of complete, albeit smaller magnets. No matter how finely you divide the magnet, within macroscopic limits, you can never obtain a free magnetic monopole.)

The scenario just described can be carried out in the laboratory by colliding together high energy beams of protons. When one fast moving proton encounters another, they interpenetrate (remember, protons are not *really* like billiard balls!), and some of the available kinetic energy is used to separate the quarks buried in the protons. However, instead of freeing the bound quarks, this energy goes into creating a q-q̄ pair. The newly formed quark may then replace one of the quarks that has been driven out of one of the protons in the course of the reaction, leaving it unchanged. The potentially freed quark then binds to the newly created antiquark to form a meson, and the net result of the interaction is to give back protons plus mesons. No quarks!

To summarize, the rubber band model provides a means of understanding why free quarks *cannot* be produced in particle collisions regardless of how much energy we invest in them. Quarks are forever confined within hadrons, never to be seen as unbound entities. Moreover, the strong force between hadrons is now revealed to be just a shadow of the "*really* strong force" that binds the quarks. But if all of this is true, what evidence *is* there for the existence of quarks? Do we have any reason to believe that the quark model is to be trusted as an accurate view of the world of subatomic physics? Or is it merely a convenient mathematical device with no corresponding physical reality?

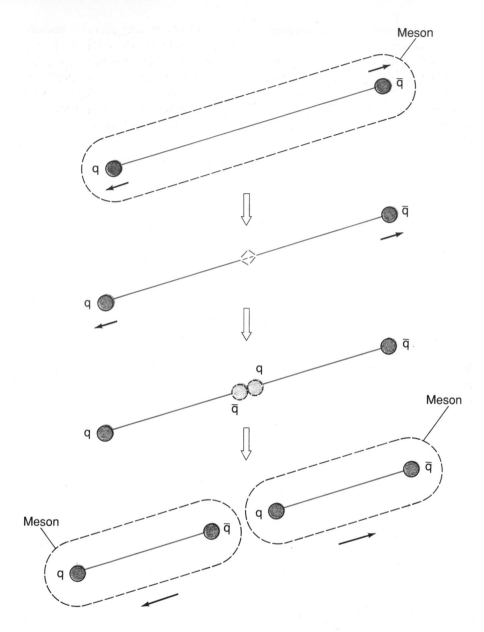

FIGURE 12.21
If the bands linking two quarks are stretched tightly enough, they can break. Instead of producing two free quarks, however, this process leads to the formation of a new quark–antiquark pair at the ends of the severed band. This new pair results from the energy originally invested in stretching the band.

The experimental evidence for quarks comes from two sources: first, the scattering of high-energy electrons off protons and, second, the observation of jets of hadrons coming from collisions between electrons and positrons. In the first instance, a case for the proton's being composed of smaller point particles has been made on much the same basis as that used by Rutherford to argue for the existence of the nucleus (see *Historical Notes* for Chapter 11). If the proton were a uniform spherical distribution of positive charge, then the deflection suffered by a high-speed electron penetrating the interior of a proton would be dependent in large part on how much charge the electron "saw" as it passed through the proton. On the other hand, if the proton were made up of three small, fractionally charged quarks, the deflection of the incoming electron would be slight, unless it happened to hit one of the quarks "head on." Then the electrical force between the electron and the quark would be substantial, and the electron might be scattered through a very large angle. This would occur only relatively rarely of course, since the proton in the quark model is mostly empty space. And what is observed? Just what is expected from the quark picture! In particular, the number of electrons deflected by large amounts is in good agreement with the predictions of the quark model. When these experiments were first reported, they occasioned theoretical physicist Richard P. Feynman to name the pointlike scatterers embedded in the protons "partons"; these later were identified with quarks.

DO-IT-YOURSELF PHYSICS

Probing the interiors of nuclei or subatomic particles to determine if they contain still smaller constituents by bombarding them with high-speed projectiles is a common technique in physics. It provides *indirect* evidence for substructure within larger physical units, and it can even reveal something about the dimensions of the smaller particles. The following exercise is designed to give you some idea of how such indirect measurements may be used to extract information about hidden properties of a system.

Using a photocopier with magnification capabilities, obtain an enlargement (of at least two times) of Figure 12.22, including the rectangular boundaries enclosing the circles. Attach a piece of carbon paper, carbon side down, to the pattern so that it is completely covered, and place the two sheets on a smooth, flat, hard surface. A table top or a tiled floor will do. Drop a marble (or other small hard sphere) onto the papers from a height of a foot or two, *making sure to catch the marble on the rebound so that it only strikes the sheets once*. Repeat this process at least 100 times, making sure you try to cover the entire area of the pattern as completely as possible.

Remove the carbon paper from the target sheet and count the total number

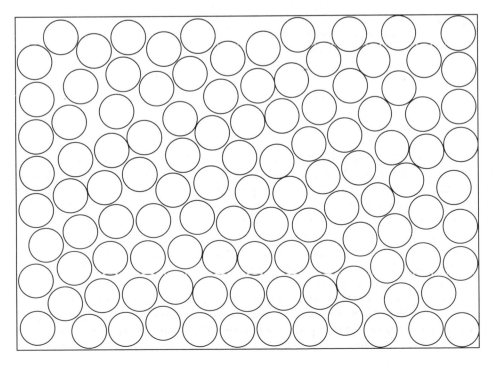

FIGURE 12.22

of marble "hits" (indicated by black carbon dots) that lie within the rectangular boundary defining the target. Next, count only those "hits" that lie wholly within the circles in the target. Using a ruler, measure the length and width of the target; compute the total target area. Then count the number of circles present in the target. If the circles are all uniform and the "hits" randomly distributed, then we expect:

$$\frac{\text{Area of all circles}}{\text{Total area of the target}} = \frac{\text{Number of "hits" within the circles}}{\text{Total number of target "hits"}}.$$

Thus,

$$\text{Area of all circles} = \frac{(\text{Circle "hits"})(\text{Total target area})}{(\text{Total "hits"})}.$$

Or,

$$\text{Area of } one \text{ circle} = \frac{(\text{Circle "hits"})(\text{Total target area})}{(\text{Total "hits"})(\text{Total number of circles})}.$$

From this, the radius of one circle may be found, since the area of a circle is π times its radius squared.

Using your data, compute the area and radius of one of the circular "atoms" in your target. Directly measure the radius of one of these "atoms" using a ruler, and compare your results.

Measurements like this were used by Rutherford to probe the structure of the atom and to estimate the size of the nucleus. Similar types of studies were used by particle physicists to discern the existence of pointlike quarks within hadrons. Demonstrations like this one reveal the power of scattering experiments in physics for defining the ultimate structure of matter.

The second type of evidence for quarks involves the products of high-energy collisions of beams of electrons and positrons. In these reactions the energy derived from the $e^- e^+$ annihilation goes into the creation of q-q̄ pairs. As these particles move off in opposite directions (as required to conserve linear momentum), they produce a stream of new q-q̄ pairs in accordance with the rubber band model. The quarks in this bidirectional jet quickly combine to give various hadrons, so that at the level of the experiment, what is seen is two trains of heavy particles, oriented at 180°, traveling away from one another (Figure 12.23). Similar jets of hadrons are observed in high-energy collisions between neutrinos and protons. Coupled with the data from the electron scattering experiments described in the previous paragraph, these observations provide persuasive evidence for the overall correctness of the quark model and the existence of these elusive particles.

The quark model provides a very nice way of understanding and analyzing decays and reactions in particle physics. Two rules only need be remembered:

1. Quark-antiquark pairs can be created from energy in the form of gamma rays or collisional kinetic energy.

2. The weak interaction alone can change one type of quark into another; the strong and electromagnetic interactions cannot cause such changes. Thus we see quark types are conserved by the strong and electromagnetic forces but not by the weak force.

EXAMPLE 12.7

Analyze the decay $\Lambda^0 \rightarrow p + \pi^-$ in terms of the quark model.

The Λ^0 decay process may be rewritten in terms of the quarks making up the various hadrons present as follows:

$$uds \rightarrow uud + d\bar{u}.$$

Cancelling a u and a d quark on both sides of the arrow yields:

$$s \rightarrow u + d\bar{u}.$$

The fundamental decay is the conversion of an s quark to a d quark by the weak force; the $u\bar{u}$ pair is created in the process from the decay energy.

FIGURE 12.23

Two narrow jets of particles emerge from the collision and annihilation of an electron and a positron (antielectron) in this computer reconstruction of an experiment conducted in Germany. The highly focused nature of these jets suggests each one developed from a single precursor, a quark or an antiquark, instead of directly from the annihilation energy of the $e^-\ e^+$ pair. Solid lines show the paths of electrically charged particles, and dashed lines are used for those of neutral particles. (Adapted from Scientific American, *April 1985).*

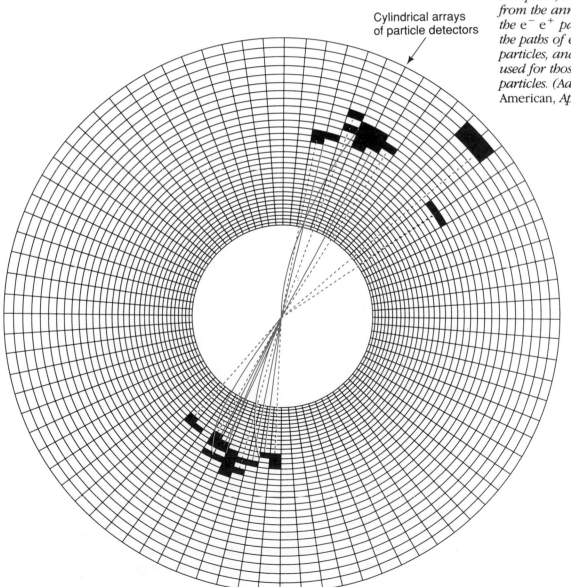

Cylindrical arrays
of particle detectors

EXAMPLE 12.8   Describe the following reaction in terms of the constituent quarks present in the reactants and the products.

$$\pi^- + p \rightarrow \Lambda^0 + K^0$$

This reaction can be written as

$$d\bar{u} + uud \rightarrow uds + d\bar{s}.$$

Each side contains two d's and a u that may be "divided" out of the reaction, leaving

$$\bar{u} + u \rightarrow s + \bar{s}.$$

What we see here is the annihilation of a $u\bar{u}$ pair to produce energy that is then available to create a new $s\bar{s}$ pair.

Quark Colors (*Optional*)

Quarks are fermions; they have half-integral spins, and they should therefore obey the Pauli exclusion principle. But if this is true, how do we explain the existence of the Ω^-? This particle is composed of three strange quarks, all of which share the same mass, spin, charge, etc, that is, the same quantum numbers. It would appear that the Ω^- is made up of three identical, interacting s quarks. Doesn't this violate the exclusion principle? How can we reconcile these circumstances?

One way would be to argue that quarks are somehow different from other fermions and consequently don't have to conform to the Pauli principle. When particles like the Ω^- were discovered, some theoretical physicists did suggest this explanation to resolve the dilemma. However, other scientists were reluctant to make exceptions to the exclusion principle and instead proposed that quarks carry an additional property that makes them distinguishable within hadrons like the Ω^-. For this scheme to work, this new characteristic had to come in three varieties to permit, for example, the three s quarks in the Ω^- to be different from each other. The name that particle theorists gave to this new quantum property or number was "color," although it has absolutely nothing to do with what we commonly refer to as color, that is, our subjective perception of certain wavelengths (or frequencies) in the electromagnetic spectrum. The three quark colors were labeled after the three primary colors of the artist's palette: red, blue, and green. Antiquarks are colored antired, antiblue, and antigreen.

The fact that color is not an *observed* property of hadrons indicates that these particles are "colorless." If they are composed of colored quarks, then the way the colors come together within the hadron must be such as to produce something with no net color, something which is color neutral. For this to be true, the three quarks that make up baryons must each possess a color different from their companions, one red, one blue, and one green. Since the addition of the primary colors produces the result we call "white," baryons containing three different colored quarks are considered to be white or neutral as concerns the color charge. In an analogous manner, for mesons to be color neutral requires them to be made up of a quark of one color and an antiquark possessing the corresponding anticolor. Returning to our previous example of the π^+ meson, we see in the light of this new color physics that if the u quark is red, the $\bar{d}$ quark must be antired.

As you may begin to detect, the language of particle physics is rather whimsical in comparison with the other areas of physics that we have studied. To carry this whimsy one step further, we note that often the different types of quarks, u, d, s (and others that we will shortly introduce), are designated quark *flavors*. Thus quarks will be found to come in six flavors (not counting the antiflavors), and each flavor comes in three colors. Complicated as this may seem to you now, it's still less troublesome than making a choice from among 31 varieties at your local Baskin-Robbins store.

Color is another example of an internal quantum number; it is not directly observed in hadrons, and it cannot be used to classify them. Moreover, it does not influence the interactions among hadrons. So what is the significance of the color charge? What possible role does it play in particle physics?

Current theories now suggest that the color charge is the source of the inter-quark force, much like ordinary electric charge is the source of the electrical force. And just as the electrical force between charged particles is carried by the photon, the color force between quarks is carried by what are called *gluons*. Gluons are electrically neutral, zero-mass, spin 1 particles that are exchanged between the bound quarks in hadrons. There are eight gluons in all, and the emission or absorption of a gluon by a quark can lead to a change in the color of the quark. Such changes occur rapidly and continuously among the quarks forming a given hadron but always in such a way that the overall color of the hadron remains neutral. The color force is sometimes referred to as the "chromodynamic force," from the Greek word *chrome*, meaning "color." The theory of color and its connection to gluons is called *quantum chromodynamics*. As is the case for quarks themselves, evidence for the existence of gluons may be found in high-energy particle collision experiments, particularly ones in which three and four hadron jets are seen.

Charm, Truth, and Beauty

Recall the symmetries mentioned earlier in this section as regards the leptons and the quarks (p. 552). We have repeatedly seen how symmetry principles have guided particle physicists through the maze of possibilities associated with high-energy collision reactions. They also provided guidance to theoreticians attempting to answer the question: How many quarks are there in nature? In the early 1960s, three quarks were enough to account for the known hadrons, but at this time four different leptons were known: the electron and its neutrino and the muon and its neutrino. (The tauon was not discovered until 1975.) Because of this asymmetry in the numbers of quarks and leptons, it was suggested that there should exist a fourth quark to balance out the leptons. Other, "harder" evidence for the existence of a fourth quark came in the form of certain rare, weak inter-actions involving hadrons in which their charges remained constant but their strangeness changed. Together these bits of data convinced particle physicists that another quark, the c quark, was present in nature and that it carried a new quantum "charge" called *charm*. Theory indicated that the charmed quark should have electric charge $+2/3$, strangeness 0, and a mass greater than any of the previously identified quarks.

The search for particles containing the c quark—so-called "charmed" particles—began in earnest in 1974 at Brookhaven and at the Stanford Linear Accelerator Center (SLAC) in California. In November 1974, Samuel Ting, leader of the Brookhaven group, and Burton Richter, head of the SLAC team, jointly

reported the discovery of a short-lived particle produced in $e^- e^+$ annihilation events that seemed to "fit the bill." The particle, called J by Ting's group and ψ (psi) by Richter's, is now believed to be a $c\bar{c}$ combination—one state of several belonging to what has been referred to as "charmonium"—having a mass of 3.1 GeV/c^2. In 1976, Ting and Richter shared the Nobel prize in physics for their work.

The existence of charmonium clearly indicated the presence of a new quark and its anti, but a question lingered about whether these new quarks actually carried the quantum charge called charm. To prove beyond doubt that this special type of quark existed required the identification of a particle, say a meson, composed of one charmed quark (or its anti) plus another, different quark, like a u or a d̄. This was necessary so that the property of charm could be explicitly manifested and not cancel out as it does in the J/ψ. After 2 years of searching, in 1976, a charmed meson, labeled D^0, was found; it appears to be made up of a c$\bar{u}$ quark combination. By the end of the 1970s, other mesons possessing both charm and strangeness were discovered, and even a charmed baryon—the Λ_c = udc! The existence of charm in the universe was secure.

But even more surprises were in store for particle physicists. In 1977, in experiments conducted at Fermilab outside Chicago, a very heavy meson, called the Υ (upsilon), was discovered. It could not be accounted for in terms of the four previously postulated quarks. So to explain the properties of this hadron, researchers proposed that a fifth quark, the b for *beauty* (or bottom), existed. The Υ was taken to be a b$\bar{b}$ pair.

The b quark then carries yet another special quantum charge. It is non-charmed, nonstrange, but beautiful. The discovery of the Υ meson spurred the search for additional mesons composed of a b (or a b̄) quark and a lighter u or d quark, much as the identification of J/ψ led to the search for charmed mesons. Because the masses of these "beautiful" mesons were expected to be very large and their lifetimes extremely short, many, many ultrahigh energy collisions had to be carried out and carefully studied before enough evidence to corroborate the existence of such particles could be gathered. Finally, in 1982, after examining over 140,000 $e^- e^+$ collisions, investigators at Cornell University identified 18 events that they concluded corresponded to decays involving B mesons, hadrons with masses near 10.6 GeV/c^2 composed of b quarks: B^+ = b̄u; B^- = b$\bar{u}$; and B^0 = b$\bar{d}$. Here was strong confirmation of the reality of the fifth quark.

In 1975, the announcement of the discovery of the tauon and its neutrino brought the number of leptons to six. With the identification of the b quark in 1982, the quarks once again lagged the leptons by one. To maintain the symmetry between the two groups, the need to have a sixth quark was pointed out. This newest quark, the t quark, holds a quantum charge called *truth,* and it permits the quarks to be paired off into three families much like the leptons are: (u,d); (s,c); and (b,t). In 1985, in experiments done at Centre Europeén pour la Recherche Nucléaire (CERN) near Geneva, Switzerland, some preliminary evidence for a meson carrying truth or the t quark (sometimes called the "top" quark) was found. Table 12.7 summarizes the properties of the six quarks.

Are there more than six quarks in nature? More than six leptons? Should we expect to see an expansion in the numbers of families of these particles beyond three in the future? And if so, does this provide evidence that these "elementary" particles may themselves be composites of still more fundamental species? The answers to all of these questions cannot be given at this time, but the results of experiments done within the last few years seem to indicate that there cannot be more than three distinct families of quarks and leptons. Assuming future confir-

TABLE 12.7 SUMMARY OF BASIC CHARACTERISTICS OF u, d, s, c, b, AND t QUARKS

| | Quantum Number or Charge* | | | | | |
	Electric Charge	Strangeness	Charm	Beauty	Truth	Mass (MeV/c^2)
u	2/3	0	0	0	0	310
d	−1/3	0	0	0	0	310
c	2/3	0	1	0	0	1500
s	−1/3	−1	0	0	0	500
t	2/3	0	0	0	1	>78000
b	−1/3	0	0	1	0	≈5000

*All the quarks have baryon number 1/3 and spin 1/2.

mation of the existence of the top quark, it appears that we will have found *all* of the fundamental building blocks of nature. But . . .

Table 12.8 summarizes the properties of the three quark and lepton families, as well as those of their anti's, in what is now referred to as the "standard model" of elementary particle physics.

12.5 UNIFICATION OF FORCES: THE ELECTROWEAK INTERACTION AND GUTS (*Optional*)

In Section 12.2 we mentioned that similarities between the electromagnetic and the weak nuclear forces first noted in the 1950s have led scientists to the conclusion that these two interactions are really just two different manifestations of a more basic interaction called the *electroweak interaction.* The theory describing this interaction was developed independently by Steven Weinberg (1967) and Abdus Salam (1968), and is often referred to as the Weinberg-Salam theory. Weinberg and Salam, together with Sheldon Glashow, who had also made contributions to the unification of these two forces, shared the Nobel prize in physics for 1979.

According to the Weinberg-Salam model, at moderate to high energies (of the order of 100 billion electron volts—100 GeVs), the electroweak interaction is carried by four massless bosons and is described by equations that are symmetric (that is, remain unchanged) when we perform mathematically well-defined "rotations" on some specific characteristics of these particles. As the energy of the system is lowered, however, the symmetry is broken spontaneously, with the result that the original family of four massless bosons separates into two subfamilies: the first consisting of the massless photon, which mediates what we call the electromagnetic interaction, and the second including the very massive intermediate bosons ($W^\pm$, Z^0), which carry the weak interaction.

The reasons why some of the originally massless carriers of the electroweak interaction acquire mass as the symmetry of the system is broken are a bit too

Recall our discussion of unitary spin on p. 549.

TABLE 12.8   STANDARD MODEL OF ELEMENTARY PARTICLES

Matter

Quarks

	I	II	III
	Up $\ \ u\ $ $+2/3$ / R,G,B / ~310	Charm $\ \ c\ $ $+2/3$ / R,G,B / 1500	Top/Truth $\ \ t\ $ $+2/3$ / R,G,B / >22500
	Down $\ \ d\ $ $-1/3$ / R,G,B / ~310	Strange $\ \ s\ $ $-1/3$ / R,G,B / 505	Bottom/Beauty $\ \ b\ $ $-1/3$ / R,G,B / ~5000

Leptons

	I	II	III
	Electron $\ \ e^-\ $ -1 / .511	Muon $\ \ \mu^-\ $ -1 / 106.6	Tauon $\ \ \tau^-\ $ -1 / 1784
	Electron Neutrino $\ \ \nu_e\ $ 0 / ~0	Muon Neutrino $\ \ \nu_\mu\ $ 0 / ~0	Tauon Neutrino $\ \ \nu_\tau\ $ 0 / <70

Antimatter

Antiquarks

	I	II	III
	Anti Up $\ \ \bar{u}\ $ $-2/3$ / C,M,Y / ~310	Anti Charm $\ \ \bar{c}\ $ $-2/3$ / C,M,Y / 1500	Anti Top/Truth $\ \ \bar{t}\ $ $-2/3$ / C,M,Y / >22500
	Anti Down $\ \ \bar{d}\ $ $+1/3$ / C,M,Y / ~310	Anti Strange $\ \ \bar{s}\ $ $+1/3$ / C,M,Y / 505	Anti Bottom/Beauty $\ \ \bar{b}\ $ $+1/3$ / C,M,Y / ~5000

Antileptons

	I	II	III
	Positron $\ \ e^+\ $ $+1$ / .511	Antimuon $\ \ \mu^+\ $ $+1$ / 106.6	Antitauon $\ \ \tau^+\ $ $+1$ / 1784
	Electron Antineutrino $\ \ \bar{\nu}_e\ $ 0 / ~0	Muon Antineutrino $\ \ \bar{\nu}_\mu\ $ 0 / ~0	Tauon Antineutrino $\ \ \bar{\nu}_\tau\ $ 0 / <70

Intermediate Bosons

Force	Force Carrier	Rest Mass GeV/c²	Charge	Spin	Relative Strength	Range (m)	Color*
Strong	Gluon	0	0	1	1	$<10^{-15}$	Yes
Weak	W⁺	81	+1	1	10^{-13}	$<10^{-17}$	Color Neutral
	W⁻	81	-1	1			
	Z⁰	93	0	1			
Electro-Magnetism	Photon	0	0	1	10^{-2}	Infinite	Color Neutral
Gravity	Graviton	0	0	2	10^{-38}	Infinite	Color Neutral

Legend

Colors*
R = Red
G = Green
B = Blue
C = Cyan = Antired
M = Magenta = Antigreen
Y = Yellow = Antiblue

I   Generation

Name   Up

Symbol u $+2/3$ — Charge / R,G,B — Color* / ~310 — Rest Mass MeV/c²

* Quark colors are considered in the optional reading in Section 12.4.

complicated for us to explore in any detail, but an example or two of circumstances involving broken symmetry in other contexts might help remove some of the mystery associated with this concept. The first analogy is due to Abdus Salam himself. Imagine being invited to a dinner party at which the table is perfectly circular and the salad plate is placed precisely between one dinner plate and the next (Figure 12.24). The first person to sit down, not being entirely familiar with dining etiquette, would be equally likely to pick as his salad plate the one to his left or the one to his right. Before he makes his choice, the system exhibits perfect symmetry. But once his choice is made, the symmetry is broken, and all the other guests will have to make the same choice as the first if they are all to have salads. Notice that it does not matter which choice is made; either breaks the symmetry. The solution to the first diner's dilemma over the symmetry of the table removes that symmetry.

The second example is given by Stephen Hawking and connects the idea of symmetry within a system to its energy. Consider a roulette ball on a roulette table (Figure 12.25). At high energies when the wheel is spun rapidly, the ball behaves in basically one way: it rolls around and around in one direction in the groove of the wheel. We might say that at high energies there is only one state in which the ball can be found. However, as the wheel slows and the ball's energy decreases, it eventually drops into one of the 37 slots molded into the wheel; at

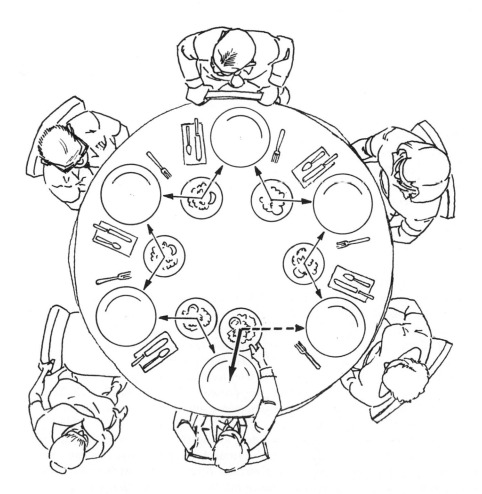

FIGURE 12.24
An illustration of spontaneously broken symmetry. If one of the diners chooses a salad plate, the left-right symmetry which exists for these dishes relative to the place settings is "spontaneously broken." After the selection is made, every other person at the table must make the same choice if they are to have a salad.

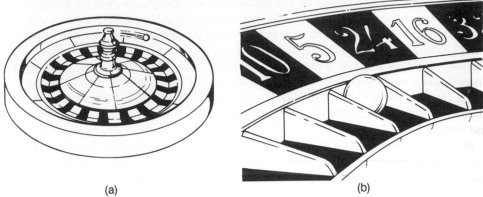

(a)   (b)

FIGURE 12.25
High, (a), and low, (b), energy states at the roulette table. When the wheel spins rapidly, there exists only one state of the system—with the ball whirling around in the groove. When the wheel slows down, there appears to be 37 possible states for the system corresponding to the 37 slots into which the ball can lodge.

low energies it appears that there are 37 different states in which the ball can exist. If we surveyed roulette wheels solely at low energies, we might be led to the conclusion that these 37 possibilities were the only ones allowed to the ball. We would miss the fact that at high energies these different states merge into essentially one. This is analogous to what happens to the carrier family for the electroweak interaction. At high energy these carriers are all massless and behave similarly to one another. But at low energy when the symmetry is broken, this group appears as two families of completely different particles.

At the time Weinberg and Salam first published their theory, particle accelerators did not exist that could produce the energies required to test their predictions, particularly those concerning the existence of the massive $W^{\pm}$ and Z^0 particles that carried the weak nuclear force. By the early 1980s, however, machines were beginning to come on line that could achieve the needed energies, and in January 1983, Carlo Rubbia of CERN reported the detection of the W particles in nine events out of over a million recorded $p\bar{p}$ collisions. In July 1984, the CERN group also announced the discovery of the Z^0 particle together with additional evidence to support their earlier detection of the W's. The 1984 Nobel prize in physics was awarded to Rubbia and his collaborator Simon van der Meer for their search and discovery of the carriers of the weak interaction. These results completely vindicated the Weinberg-Salam model and left no doubts about its correctness.

The success of the unification of the electromagnetic and the weak nuclear forces caused many physicists to wonder if it might not also be possible to unite the strong nuclear force with the electroweak force to produce what has come to be called a *grand unified theory,* or GUT. As Hawking has pointed out, these theories (there are actually several competing variants in the literature) are neither all that grand nor fully unified. They do not include gravity, and they are not complete in the sense that they involve parameters whose values cannot be predicted from the theory but must be selected by comparison with experiment. Nevertheless, they represent the first steps toward a true unification of the interactions of nature, and they do present some interesting implications for the long-term future of the universe.

The basic idea of GUTs is the following: Just as the electromagnetic and weak forces represent different aspects of the same basic force that coalesce at high energies, the electroweak and strong nuclear forces are thought to be different manifestations of yet another even more fundamental force that emerges at still higher energies. The energy at which all three of these forces become fused into a single force is called the *grand unification energy*. Its value is not well known but would probably have to be at least 10^{15} billion electron volts, far above the energies ever likely to be attainable in terrestrial laboratory particle accelerators. The GUTs also predict that at this energy the different spin 1/2 particles, such as quarks and electrons, would behave in essentially the same manner. This is similar to the formation of a single family of electroweak carrier particles from the photons and the intermediate bosons that mediate the separate electromagnetic and weak forces.

Because of the impossibility of reaching the energies needed to test directly the GUTs in the laboratory, scientists have begun searching for indirect, low-energy consequences of these theories. One such search was mentioned in the *Prologue* of this chapter and pertains to the possible decay of the proton. In Section 12.2, we argued that conservation of baryon number in strong interactions demands that the proton be absolutely stable, since there exist no lighter *baryons* to which it can decay. But at grand unification energies, the quarks that make up the proton are basically indistinguishable from leptons, so that the spontaneous decay of a proton (through quark conversion) to lighter particles such as positrons becomes possible. The three quarks bound inside a proton do not normally have sufficient energy to make this transition; that is to say, we live in an environment in which the available energies are typically far below the threshold energy for grand unification. However, very, very occasionally, according to an alternative formulation of the uncertainty principle, for the very briefest of moments, the energy of a quark inside a proton could reach the necessary value, and the proton would then decay. The chances of this happening, as you might guess, are very low. If you could watch a single proton, hoping to catch it in the act of decaying as described, you'd have to wait and watch for something like a million million million million million (or 10^{30}) years! Clearly this is not practical. The entire universe has been in existence for only 15 thousand million years or so.

To increase your chances of witnessing the decay of the proton and hence of validating one version of GUTs, instead of watching only one proton, you could watch lots of them, that is, you could monitor a large quantity of matter containing an enormous number of protons—maybe 10^{31} of them. If this were done for a year, you would expect, on average, to see as many as 10 proton decays. Increasing the number of protons in the sample obviously improves your chances of seeing a decay.

This was precisely the approach taken by the researchers mentioned in the opening *Prologue* when they filled their underground tank with 8 tons of water (each water molecule contributes 10 protons) and used thousands of light-sensitive devices to look for the flashes of radiation emitted by the high-speed positrons believed to accompany the decay of the protons. The results of this experiment were discouraging: no clearly defined proton decay events were recorded over the time frame in which the experiment ran, thus placing the average lifetime of the proton at more than 10^{31} years. This value is greater than that predicted by the simplest GUTs, but there are more elaborate versions of these theories that give longer proton lifetimes. Currently experiments are being devised to test them.

One thing is clear from this discussion. If the proton is found to be unstable, even on such long time frames, then eventually all the matter in the universe will evaporate. This is so because the positrons, as products of the proton decay, will annihilate with electrons to produce gamma rays. Assuming the universe survives for the requisite amount of time, it will ultimately be stripped of all matter and reduced to a cold, dark, expanding space filled with photons (which steadily loose energy) and a few isolated electrons and neutrinos (which escaped destruction)—a grim fate indeed, but one that may well await our now glorious universe. More so than most, this example demonstrates how the physics of the subatomic domain can influence that of the largest scale known. But the reverse is also true.

We have seen that it is not feasible to expect that terrestrial laboratory experiments will ever achieve the energies required to verify directly the predictions of GUTs. Indeed, one might well ask, "Have such energies ever been available in the entire history of the universe?" And the answer is "Yes, in the very first few moments after the birth of the universe in the Big Bang!" Thus we begin to see that the ultimate tests of GUTs (particularly those versions involving "cosmions," unseen particles created in the Big Bang that interact primarily by gravity and hence influence the motions of stars and galaxies) may rest in the field of cosmology, the study of the structure and evolution of the universe on the largest scale possible. The justification of theories involving the smallest of physical entities, quarks and electrons, may depend on our understanding of the largest of physical structures, galaxies and clusters of galaxies.

HISTORICAL NOTES

FIGURE 12.26
Theoretical physicist Murray Gell-Mann, one of the architects of the quark model.

Physics, like almost any other discipline, is full of colorful individuals who, by their powerful personalities and intellects, have exerted far-reaching influence on the progress of the science. In elementary particle physics, few people have done more to shape the development of the field or the language in which it is expressed than Murray Gell-Mann (Figure 12.26).

Murray Gell-Mann was born in New York City and entered Yale University at the age of 15. After graduating in physics, he attended the Massachusetts Institute of Technology (MIT), where he received his PhD in theoretical physics when he was only 21. After a brief stay at the University of Chicago, Gell-Mann moved to Caltech in 1957 where he has remained. A man of incredible intellectual prowess and wide-ranging interests, Gell-Mann is well acquainted with Eastern and Western religions and literature, knows several languages, including Swahili, and, of course, is highly skilled in applying abstract mathematical theory to problems in physics. He is a refined, urbane individual not given to expansive gestures but to calm, orderly discourse peppered with literary references and clever witticisms. Logical progression and mathematical rigor are the hallmarks of his work.

A brief summary of the major contributions to theoretical particle physics made by Murray Gell-Mann is appropriate here: 1953—introduced the idea of "strangeness" to explain the unusual properties of kaons and the lambda; 1961—developed the eightfold way to organize the proliferating "zoo" of elementary particles into a few families having simple, regular symmetries; 1964—introduced quarks as a more fundamental way of explaining the symmetries exhibited in the eightfold way; 1971—introduced "colored" quarks to differentiate three possible varieties of each basic quark type; 1973—recognized "color" as the source of the strong force and coined the name "quantum chromodynamics" to describe the theory of the strong interaction. For his accomplishments up to that time, Gell-Mann received the 1969 Nobel prize in physics, and he justly deserves to be called "one of the heros of our story," as John Polkinghorne refers to him in his chronicle of the evolution of particle physics in the years since the Second World War entitled *Rochester Roundabout*.

As an illustration of the deeply perceptive yet whimsical way in which Gell-Mann approaches physics, the tale of the quarks, as told by Michael Riordan in his book *The Hunting of the Quark,* is perhaps the best. In March 1963, Gell-Mann visited Columbia University to give a series of three talks. At a luncheon during his stay in New York, Gell-Mann was asked why a fundamental group of three particles—a triplet—did not appear in nature, while larger groupings, like those containing eight particles (an octet), did. Gell-Mann's reply was "that would be a funny quirk," and he proceeded to show why: such a triplet, if it existed, would have to be composed of particles with fractional charges equal to $2/3$, $-1/3$, and $-1/3$ the magnitude of the electron charge, respectively. No such fractionally charged entities had ever been observed.

Later that day and into the following morning, Gell-Mann reflected on this situation and recognized that such fractionally charged species were not completely out of the question *if* they remained permanently trapped inside hadrons and never appeared as free particles in nature. In his second lecture at Columbia, Gell-Mann referred to the members of this triplet as "quarks," a nonsense word he had used before and by which he meant "those funny little things." Although relatively little time was devoted to "quarks" in the talk, they were very much the subject of the postlecture coffee hour.

In the fall of that year, Gell-Mann worked out the details of how to construct baryons and mesons out of this fractionally charged triplet of particles and wrote a short, two-page article that was published in a new European journal, *Physics Letters.* (Gell-Mann, when asked why he had not submitted the paper to more prestigious American journals, replied that it "probably would have been rejected.") By this time Gell-Mann had discovered a passage from James Joyce's novel *Finnegan's Wake,* which lent legitimacy to his use of the term "quarks" to designate the fictional triplet:

> "Three quarks for Muster Mark!
> Sure he hasn't got much of a bark,
> And sure any he has it's all beside the mark.
> But O, Wreneagle Almighty, wouldn't un be a sky of a lark
> To see that old buzzard whooping about for uns shirt in the dark
> And he hunting round for uns speckled trousers around by
> Palmerston Park?"

The poem relates the dream of one of Joyce's protagonists as he lies passed out on the floor of a Dublin tavern. The last three lines had, no doubt, a particularly appealing irony in them for Gell-Mann in view of the parallel circumstances of the hapless Muster Mark vainly searching for his clothes in the dark and of the frantic experimentalists equally vainly searching for free, unbound quarks. Indeed, Gell-Mann was one of the last to believe that quarks were anything but mathematical entities. At a conference in Berkeley in 1966, he spoke of "those hypothetical and probably fictitious spin 1/2 quarks," and he frequently referred to Feynman's "partons," which ultimately became identified with quarks and established the latter as bona fide physical objects, as "put-ons." Among his many attributes, caution is one that Gell-Mann also possesses in abundance.

Numerous other stories and anecdotes exist related to the circumstances surrounding Gell-Mann's discoveries. For example, unlike most other major contributions in the field, the eightfold way was *never* published in a physics journal but appears only in a widely circulated Caltech Report, CTSL-20, printed in 1961. In a burst of patriotism, Gell-Mann also originally suggested that the quark colors ought to be signified as red, white, and blue, one of only a few of his ideas that did not take hold in the particle physics community. But this thumbnail sketch may be sufficient to permit some appreciation to be gained of the talents and temperament of this man.

Refer to the "Physics Potpourri" in Section 4.1.

Murray Gell-Mann's achievements have been compared to those of Dmitri Mendeleev. Both men brought order out of chaos, Mendeleev by organizing the elements into the periodic table based on their chemical properties and Gell-Mann by establishing a kind of periodic table of mesons and baryons based on their group symmetry properties. In each case gaps in the patterns led to predictions of missing members that were subsequently discovered experimentally. The direction and impetus provided to chemists and atomic physicists by Mendeleev's work is strongly paralleled by that given to elementary particle physicists by Gell-Mann's.

SUMMARY

Einstein's special theory of relativity extends classical mechanics to the realm of high velocities. It is based on the assumptions that the speed of light is a universal constant for *all* observers, regardless of their motion, and that the laws of physics are the same for *all uniformly moving* observers. Adopting these postulates, Einstein predicted the existence of time dilation, in which moving clocks appear to run slow, and length contraction, in which moving rulers appear shortened in the direction of motion. Both of these effects have been confirmed experimentally in many different circumstances. Special relativity theory also establishes the equivalence between energy and mass, according to which the energy stored in a stationary object of mass m is given by $E_o = mc^2$. The conversion of energy to mass and vice versa are important aspects of the reactions studied in nuclear and particle physics.

There exist four well-established forces or interactions in nature: gravity, electromagnetism, the weak nuclear force, and the strong nuclear force. Only the last three of these are significant in high-energy particle physics. Modern science conceives of these forces as being conveyed through the exchange of carrier particles. These messengers and all other subatomic particles may be classified as either fermions or bosons, according to whether their spins are integral or half-integral multiples of the basic unit of intrinsic angular momentum, $h/2\pi$. Particles may also be segregated by mass and by the types of fundamental interactions in which they participate. The ultimate goal of high-energy physics is to establish the number and characteristics of the truly "elementary" particles, those most basic, structureless, indivisible units of matter and antimatter.

Conservation laws, such as those involving mass-energy, momentum, and charge, play critical roles in the specification of the properties of particles and their anti's. In energetic collisions, anything that is not absolutely forbidden will occur. Thus particle physicists infer the existence of new conservation laws by examining what does *not* happen during an interaction. In this way conservation of baryon number, lepton number, and strangeness have been discovered. Not all of these new conserved quantities (or quantum numbers) remain constant in *all* types of interactions; a few like strangeness are only partially conserved. Moreover, some of these new conservation laws may only be approximately true. Nevertheless, their development has helped scientists to recognize the existence of what may prove to be the most elementary of particles, the quarks.

Quarks are fractionally charged particles that come in six different varieties: up, down, strange, charmed, top (or truth), and bottom (or beauty). Ordinary matter is composed of either 3 quarks or a quark-antiquark pair. The quarks are bound together in such a way that it is impossible for them to exist as free entities outside these combinations. Evidence for quarks comes from scattering experiments such as those in which high-speed electrons are collided with protons; the observed deflections of the electrons substantiate the interpretation that protons are composed of smaller particles with charges $+(2/3)e$, $+(2/3)e$, and $-(1/3)e$, where e is the magnitude of the basic electronic charge. All known species of subatomic particles may be constructed from the 6 quarks and their anti's. Together with the electron, the positron, and their weakly interacting relatives, these particles comprise all the matter and antimatter in the universe.

The unification of the laws of physics is an important theme in science. Maxwell's theory unified the laws of electricity with those of magnetism to form electromagnetism. Einstein unified the physics of low speeds with that of high speeds in his special theory of relativity. In the late 1960s, theoretical physicists Steven Weinberg and Abdus Salam succeeded in unifying electromagnetism with the weak nuclear force to produce the electroweak interaction. At the present time scientists are seeking ways to unite the strong nuclear force with the electroweak force in what are called grand unified theories (GUTs). Although far from complete, this task has led to some startling predictions about such things as the lifetime of the proton. Experimentalists are currently trying to confirm such predictions in elaborate underground laboratories designed to detect elusive entities called neutrinos. Ultimately the determination of whether these grand unified theories are accurate descriptions of nature may have to come from cosmological studies of the universe and its origin in the Big Bang.

SUMMARY OF IMPORTANT EQUATIONS

EQUATION	COMMENTS
$\Delta t' = \dfrac{\Delta t}{\sqrt{1 - v^2/c^2}}$	Time dilation
$E_o = mc^2$	Definition of rest energy; equivalence of mass and energy
$E_{rel} = \dfrac{mc^2}{\sqrt{1 - v^2/c^2}}$	Relativistic total energy
$KE_{rel} = \dfrac{mc^2}{\sqrt{1 - v^2/c^2}} - mc^2$	Relativistic kinetic energy

QUESTIONS

(Asterisks [*] denote questions based on optional reading in the chapter.)

1. Describe the two fundamental postulates underlying Einstein's special theory of relativity.

2. Suppose you were traveling toward the sun at a constant speed of 0.25 c. With what speed does the light streaming out from the sun go past you? Explain your reasoning.

3. Light travels in water at a speed of 2.25×10^8 m/s. Is it possible for particles to travel through water at a speed $v > 2.25 \times 10^8$ m/s? Why or why not? Explain.

4. Galileo used his pulse like a clock to measure time intervals by counting the number of heartbeats. If Galileo were traveling in a spaceship, moving uniformly at a speed near that of light, would he notice any change in his heart rate, assuming the circumstances of his travel produced no significant physiological stress on him? If someone on earth were observing Galileo with a powerful telescope, would he or she detect any change in Galileo's heart rate relative to the resting rate on earth? Explain your answers.

5. Why don't we generally notice the effects of special relativity in our daily lives? Be specific.

6. Does $E_o = mc^2$ apply only to objects traveling at the speed of light? Why or why not?

7. If a horseshoe is heated in a blacksmith's furnace until it glows red hot, does the mass of the horseshoe change? If a spring is stretched to twice its equilibrium length, has its mass been altered in the process? If so, explain how and why in each case.

8. List the four fundamental interactions of nature, and discuss their relative strengths and effective ranges.

9. What is an antiparticle? What may happen when a particle and its anti collide?

10. Distinguish between fermions and bosons in as many different ways as you can.

11. Give some ways by which physicists classify elementary particles.

12. In which of the four basic interactions does an electron participate? A neutrino? A proton? A photon?

13. Who was Emmy Noether, and what was her great contribution to mathematical physics?

14. In your own words, describe what a physicist means when he uses the term "strangeness."

15. Baryon conservation and lepton conservation are laws used frequently by particle physicists to decide whether a reaction involving elementary particles is possible or not. Explain how these laws are applied to make such determinations.

16.* Give an example of time-reversal (or T) invariance in a situation or interaction drawn from everyday experience.

17. What is a quark? How many different types of quarks are now known? What are some of the basic properties that distinguish these quarks?

18. How does the quark model account for the existence of all of the presently known heavy elementary particles (hadrons).

19. Why do physicists believe it is impossible to produce and to detect a free (that is, unbound) quark in any particle reaction, regardless of how much energy is involved?

20.* Quarks are said to possess "color." What does this mean? Are physicists really suggesting that quarks look red like ripe strawberries or green like healthy grass? Explain.

21.* A bumper "snicker" on a car belonging to the chairperson of a physics department reads: "Particle physicists have GUTs!" Explain in your own words the meaning of this little joke or "play on words."

22.* Unification of its basic laws and theories has long been a goal in physics. Describe some ways in which physicists have been successful in unifying certain forces and theories. In what area(s) of physics is the process of unification still ongoing?

23.* Why do scientists care whether the proton is stable or not? Does it really matter if the proton decays in 10^{32} years or "only" 10^{30} years? Discuss.

24. Describe the contributions that physicist Murray Gell-Mann has made in the field of elementary particle physics.

PROBLEMS

(Asterisks [*] denote problems based on optional reading in the chapter. For many of these exercises, reference to Tables 12.3 and 12.8 will be found to be very helpful.)

1. The lifetime of a certain type of elementary particle is 2.6×10^{-8} s. If this particle were traveling at 95% the speed of light relative to a laboratory observer, what would this observer measure the particle's lifetime to be?

2. How fast would a muon have to be traveling relative to an observer for its lifetime as measured by this observer to be 10 times longer than its lifetime when at rest relative to the observer.

3. Calculate the rest energy of a proton in joules and MeVs. What is the mass of a proton in MeV/c²?

4. The tauon is the heaviest of all the known leptons, having a mass of 1784 MeV/c². Find the rest energy of a tauon in MeVs and joules. What is the mass of the tauon in kg? Compare your result with the mass in kg of an electron.

5. In a particular beam of protons, each particle moves with an average speed of 0.8 c. Determine the total relativistic energy of each proton in joules and MeVs.

6. A particle of rest energy 140 MeV moves at a sufficiently high speed that its total relativistic energy is 280 MeV. How fast is it traveling?

7. If the relativistic kinetic energy of a particle is 9 times its rest energy, at what fraction of the speed of light must the particle be traveling?

8. Indicate which of the following decays are possible. For any that are forbidden, state which conservation laws are violated.
    (a) $\Sigma^+ \rightarrow n + \pi^0$
    (b) $n \rightarrow p + \pi^-$
    (c) $v^0 \rightarrow e^+ + e^- + v_e$
    (d) $\pi^- \rightarrow \mu^+ + v_e$

9. Indicate which of the following reactions may occur, assuming sufficient energy is available.
    For any that are forbidden, state which conservation laws are violated.
    (a) $p + p \rightarrow p + n + K^+$
    (b) $\gamma + n \rightarrow \pi^+ p$
    (c) $K^- + p \rightarrow n + \Lambda^0$
    (d) $v^- + p \rightarrow p + e^0 + v_e$

10. Supply the missing particle in each of the following *strong* interactions, assuming sufficient energy is present to cause the reaction:
    (a) $n + p \rightarrow p + p +$ _____ .
    (b) $p + \pi^+ \rightarrow \Sigma^+ +$ _____ .
    (c) $K^- + p \rightarrow \Lambda^0 +$ _____ .
    (d) $\pi^- + p \rightarrow \Xi^0 + K^0 +$ _____ .

11. Identify the missing particle(s) in each of the following *weak* interactions:

(a) $\Omega^- \rightarrow \Xi^o +$ _____ .

(b) $K^o \rightarrow \pi^+ +$ _____ .

(c) $\pi^+ \rightarrow \mu^+ +$ _____ .

(d) $\tau^+ \rightarrow \mu^+ +$ _____ + _____ .

12. A 0.1-kg ball connected to a fixed point by a taut string whirls around a circular path of radius 0.5 m at a speed of 3 m/s. Find the angular momentum of the orbiting ball. Compare this with the angular momentum of a spinning electron. How many times larger is the former than the latter?

13.* List all the capital letters in the English alphabet that exhibit mirror or reflection symmetry with respect to a line in the plane of the page that bisects them vertically.

14.* List all the capital letters in the English alphabet that show rotational symmetry about a line perpendicular to the plane of the page passing through their centers. Give the angle through which these letters must be turned to manifest the symmetry.

15. What is the quark combination corresponding to an antineutron? An antiproton?

16. Give three (3) combinations of quarks (not antiquarks!) that will give baryons with charge: (a) $+1$; (b) -1; and (c) 0.

17. Give all the quark-antiquark pairs that result in mesons that have no charge, no strangeness, no charm, and no beauty. How do you think such particles might be distinguished from one another?

18. Distinguish a particle composed of a combination of $d\bar{u}s$ quarks from one composed of dus quarks in as many ways as you can. Do the same for a $\bar{d}\bar{s}s$ combination and a dss combination.

19. The Δ^- (delta minus) is a short-lived baryon with charge -1, strangeness 0, and spin 3/2. To what quark combination does this subatomic particle correspond?

20. The Ξ^o (xi zero) is a baryon having charge 0, strangeness -2, and spin 1/2. What quarks combine to form this subatomic particle?

21. The F^- meson possesses -1 unit of charge, -1 unit of strangeness, and -1 unit of charm. Identify the quark combination making up this rare subatomic particle.

22. The D^+ meson is a charmed particle with charge $+1$, charm $+1$, and strangeness 0. Work out the quark combination for this species of elementary particle.

23. Analyze the following reactions in terms of their constituent quarks:

(a) $\pi^+ + p \rightarrow \Sigma^+ + K^+$

(b) $\gamma + n \rightarrow \pi^- + p$

(c) $K^- + p \rightarrow \Omega^- + K^+ + K^o$

24. Analyze the following decays in terms of the quark contents of the particles:

(a) $\Sigma^- \rightarrow n + \pi^-$

(b) $n \rightarrow p + e^- + \bar{\nu}_e$

(c) $\Omega^- \rightarrow \Lambda^o + K^-$

CHALLENGES

1. Show that in the limit of low speeds the expression for the relativistic kinetic energy (see p. 526 or the Summary of Important Equations) reduces to the familiar one from classical mechanics, $KE = (1/2)mv^2$. (Hint: For speeds v much smaller than c, the quantity $(1 - v^2/c^2)^{-1/2}$ is approximately equal to $1 + v^2/2c^2$. You should check this for yourself using typical speeds given in the exercises in Chapter 2.)

2. Carry out the calculation suggested in Section 12.2 on p. 529. In particular, find the ratio of the Coulomb force of attraction between a proton and an electron separated by 0.5 nm to their gravitational force of attraction for the same distance. The masses and charges for these two particles are given in Table 11.1.

3. When a proton-antiproton pair at rest annihilates, two photons are created. Find the wavelengths of these two light quanta, given that they share equally in the annihilation energy. In what portion of the electromagnetic spectrum do such wavelengths appear?

4. If the average lifetime of the proton were 10^{16} years, estimate the number of protons per kilogram of human body mass that would decay radioactively each year. Assume that a human being is made entirely of water molecules and that each molecule contributes ten protons. (Hint: The molar weight of water is 18 g, and each mole of any substance contains 6.02×10^{23} particles.)

If each proton decay released 1 MeV of energy, how much energy per kilogram would be produced *in total* by the decaying protons in your body in one year? Give the result in J/kg and in rads (see *Physics Potpourri* on p. 494) If the energy released is mostly in the form of kinetic energy of the positrons produced in the decay, then one rad of this type of ionizing radiation is rated at about 1 rem. Compare the total exposure in rems of a person to the radiation produced by her own body to the maximum safe dose of 0.17 rems for the general public from all other artificial sources.

5. Although doubly charged baryons have been found (that is, baryons with net charge $+2$), no doubly charged mesons have yet been identified. What effect on the quark model would there be if a meson of charge $+2$ were discovered? How could such a meson be interpreted within the quark model?

6. In the electroweak theory, symmetry breaking occurs on a length scale of about 10^{-17}m. Compute the *frequency* of particles whose de Broglie wavelength (see Sec. 10.5) equals this value, and, using this result, determine their energies by the Planck formula (p. 438). Show that the mass of particles having such energies is of the order of the mass of the $W^{\pm}$ particles (see Table 12.8).

SUGGESTED READINGS

Cline, David B. "Beyond Truth and Beauty: A Fourth Family of Particles." *Scientific American* 259, no.2 (August 1988): 60–67. Presentation of experimental evidence supporting the possible existence of a fourth but not more than five families of quarks and leptons.

Cline, David B.; Rubbia, Carlo; and van der Meer, Simon. "The Search for Intermediate Vector Bosons." *Scientific American* 246, no.3 (March 1982): 48–59. Description of particle accelerators and detectors used at CERN to produce and to identify the W and Z particles.

Fritzsch, Harald. *Quarks.* New York: Basic Books, 1983. Nonmathematical but quite detailed introduction of the concept of quarks as the fundamental constituents of matter.

Harari, Haim. "The Structure of Quarks and Leptons." *Scientific American* 248, no.4 (April 1983): 56–68. Discussion of the possibility that quarks and leptons are themselves composed of even more fundamental particles called preons and rishons.

Kaufmann, William J. *Particles and Fields.* San Francisco: W. H. Freeman and Co., 1980. Readings from *Scientific American* spanning the period 1953 to 1979. Includes articles by Gell–Mann, Glashow, Weinberg, and others.

Lederman, Leon M. "The Upsilon Particle." *Scientific American* 239, no.4 (October 1978): 72–80. Description of the discovery of the Υ meson, emphasizing the experimental aspects of the detection.

Mistry, Nariman B.; Poling, Ronald A.; and Thorndike, Edward H. "Particles with Naked Beauty." *Scientific American* 249, no. 1 (July 1983), 106–115. Electron storage ring experiments at Cornell to find the B meson using the CLEO detector are described.

Ne'eman, Yuval, and Kirsh, Yoram. *The Particle Hunters.* Cambridge, England: Cambridge University Press, 1986. Nonmathematical discussion of the development of particle physics from electrons to the W and Z bosons. Highly readable.

Pagels, Heinz R. *Perfect Symmetry: Search for the Beginning of Time.* New York: Simon and Schuster, 1985. A treatment of unification in particle physics and its connection with cosmology. Nonmathematical.

Pais, Abraham. *Subtle is the Lord. . . The Science and Life of Albert Einstein.* New York: Oxford University Press, 1980. Thorough and beautifully written biography of Einstein, the man and his work. The physics discussions become quite technical in places.

Perl, Martin L., and Kirk, William T. "Heavy Leptons." *Scientific American* 238, no. 3 (March 1978): 50–57. Good discussion of the experimental techniques used to identify the tauon.

Polkinghorne, John. *Rochester Roundabout: The Story of High Energy Physics.* New York: W. H. Freeman and Co., 1989. Lively tale about the evolution of high-energy physics since the Second World War told by a former theoretical physicist. The discussion sometimes gets a bit technical, but the anecdotes and personal observations are worth the effort.

Quigg, Chris. "Elementary Particles and Forces." *Scientific American* 252, no. 4 (April 1985): 84–95. A review of the standard model, including quantum chromodynamics and the electroweak theory and its use as guide to grand unification.

Riordan, Michael. *The Hunting of the Quark.* New York: Simon and Schuster, 1987. Engaging account of the 20-year search for the quark written by a research physicist who participated actively in the quest. A very accessible, "insiders'" look at the world of high-energy physics.

Schramm, David N., and Steigman, Gary. "Particle Accelerators Test Cosmological Theory." *Scientific American* 258, no. 6 (June 1988): 66–72. An examination of issues pertaining to the number of families of elementary particles and of the role cosmological theories play in them.

Segrè, Emilio. *From X-rays to Quarks.* San Francisco: W. H. Freeman and Co., 1980. Chapter 11 describes the development of particle accelerators and the pioneering work of E. O. Lawrence in this field. Chapter 12 deals with elementary particles.

Trefil, James S. *From Atoms to Quarks.* New York: Charles Scribner's Sons, 1980. Basic introduction to particle physics, including good discussions of particle classification schemes and search strategies for quarks. Some equations and numerical examples appear.

APPENDIX A   CONVERSION FACTORS AND METRIC PREFIXES

DISTANCE:

1 m = 3.28 ft	1 ft = 0.305 m
1 cm = 0.394 in.	1 in. = 2.54 cm
1 km = 0.621 mi	1 mi = 1.61 km

AREA:

$1 \text{ m}^2 = 10.8 \text{ ft}^2$	$1 \text{ ft}^2 = 0.093 \text{ m}^2$
$1 \text{ cm}^2 = 0.155 \text{ in.}^2$	$1 \text{ in.}^2 = 6.45 \text{ cm}^2$
$1 \text{ km}^2 = 0.386 \text{ mi}^2$	$1 \text{ mi}^2 = 2.59 \text{ km}^2$

VOLUME:

$1 \text{ m}^3 = 35.3 \text{ ft}^3$	$1 \text{ ft}^3 = 0.028 \text{ m}^3$
	$= 7.48 \text{ gallons}$
$1 \text{ cm}^3 = 0.061 \text{ in.}^3$	$1 \text{ in.}^3 = 16.4 \text{ cm}^3$
$1 \text{ m}^3 = 1,000 \text{ liters}$	
1 liter = 1.06 qt	1 qt = 0.95 liter

TIME:

1 min = 60 s
1 h = 60 min = 3,600 s
1 day = 24 h = 86,400 s
1 year = 3.16×10^7 s

SPEED:

1 m/s = 2.24 mph	1 mph = 0.447 m/s
	= 1.47 ft/s
1 m/s = 3.28 ft/s	1 ft/s = 0.305 m/s
	= 0.682 mph
1 km/h = 0.621 mph	1 mph = 1.61 km/h

ACCELERATION:

$1 \text{ m/s}^2 = 3.28 \text{ ft/s}^2$	$1 \text{ ft/s}^2 = 0.305 \text{ m/s}^2$
$1 \text{ m/s}^2 = 2.24 \text{ mph/s}$	$1 \text{ mph/s} = 0.447 \text{ m/s}^2$
$1 \text{ g} = 9.8 \text{ m/s}^2$	$1 \text{ g} = 32 \text{ ft/s}^2 = 22 \text{ mph/s}$
$1 \text{ m/s}^2 = 0.102 \text{ g}$	

MASS:

1 kg = 0.069 slug	1 slug = 14.6 kg
$1 \text{ kg} = 6.02 \times 10^{26} \text{ u}$	$1 \text{ u} = 1.66 \times 10^{-27} \text{ kg}$

FORCE:

1 N = 0.225 lb	1 lb = 4.45 N
	1 ton = 8,896 N
	= 2,000 lb

ENERGY, WORK, AND HEAT:

1 J = 0.738 ft-lb	1 ft-lb = 1.36 J
= 0.239 cal	= 0.0013 Btu
1 cal = 4.186 J	1 Btu = 778 ft-lb
1 cal = 0.004 Btu	1 Btu = 252 cal
1 J = 0.00095 Btu	1 Btu = 1,055 J
1 kW-h = 3,600,000 J	
1 kW-h = 3,413 Btu	1 Btu = 0.00029 kW-h
1 food calorie = 1,000 cal	
$1 \text{ eV} = 1.6 \times 10^{-19} \text{ J}$	
$1 \text{ J} = 6.25 \times 10^{18} \text{ eV}$	

POWER:

1 W = 0.738 ft-lb/s	1 ft-lb/s = 1.36 W
1 W = 0.00134 hp	1 hp = 550 ft-lb/s
	1 hp = 746 W
1 W = 3.41 Btu/h	1 Btu/h = 0.293 W

PRESSURE:

$1 \text{ N/m}^2 = 1.45 \times 10^{-4} \text{ psi}$	$1 \text{ psi} = 6,900 \text{ N/m}^2$
$1 \text{ N/m}^2 = 9.9 \times 10^{-6} \text{ atm}$	1 psi = 0.068 atm
$1 \text{ atm} = 1.01 \times 10^5 \text{ N/m}^2$	1 atm = 14.7 psi
= 76 cm Hg	= 29.92 in. Hg
$= 10.3 \text{ m H}_2\text{O}$	$= 33.9 \text{ ft H}_2\text{O}$

MASS DENSITY:

$1 \text{ kg/m}^3 = 0.0019 \text{ slug/ft}^3$	$1 \text{ slug/ft}^3 = 515 \text{ kg/m}^3$

WEIGHT DENSITY:

$1 \text{ N/m}^3 = 0.0064 \text{ lb/ft}^3$	$1 \text{ lb/ft}^3 = 157 \text{ N/m}^3$

METRIC PREFIXES:

Power of 10	Prefix	Abbreviation
10^{15}	peta	P
10^{12}	tera	T
10^9	giga	G
10^6	mega	M
10^3	kilo	k
10^{-2}	centi	c
10^{-3}	milli	m
10^{-6}	micro	μ
10^{-9}	nano	n
10^{-12}	pico	p
10^{-15}	femto	f

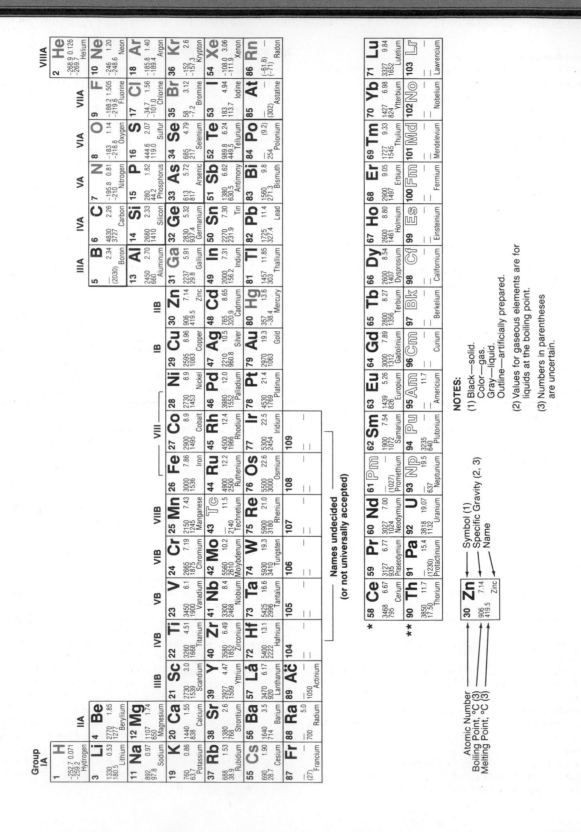

1901 *Wilhelm Roentgen* Discovery of x-rays.

1902 *Hendrik A. Lorentz and Pieter Zeeman* Research on the relationship between magnetism and radiation.

1903 *Antoine H. Becquerel* Discovery of radioactivity.
Pierre Curie and Marie Curie Research in radioactivity.

1904 *Lord Rayleigh* Discovery of argon.

1905 *Philipp E. A. von Lenard* Research on cathode rays.

1906 *Joseph J. Thomson* Studies on conduction of electricity in gases.

1907 *Albert A. Michelson* Optical instruments and investigations using them.

1908 *Gabriel Lippmann* Interference method in color photography.

1909 *Guglielmo Marconi and Carl Ferdinand Braun* Contributions to wireless communications.

1910 *Johannes D. van der Waals* Equation of state for gases and liquids.

1911 *Wilhelm Wien* Studies of blackbody radiation.

1912 *Nils G. Dalén* Automatic regulators for use in coastal lighting.

1913 *Heike Kamerlingh Onnes* Research in low-temperature physics.

1914 *Max von Laue* Diffraction of x-rays by crystals.

1915 *William H. Bragg and William L. Bragg* Studies of crystal structure using x-rays.

1916 No prize

1917 *Charles G. Barkla* Discovery of characteristic x-ray spectra of elements.

1918 *Max Planck* Discovery of quantization of energy.

1919 *Johannes Stark* Effect of electric fields on spectral lines.

1920 *Charles-Édouard Guillaume* Discovery of anomalies in nickel-steel alloys.

1921 *Albert Einstein* Explanation of the photoelectric effect.

1922 *Niels Bohr* Quantum model of the atom.

1923 *Robert A. Millikan* Work on the electron's charge and on the photoelectric effect.

1924 *Karl M. G. Siegbahn* Research in x-ray spectroscopy.

1925 *James Franck and Gustav Hertz* Studies of collisions between electrons and atoms.

1926 *Jean B. Perrin* Studies on the discontinuous structure of matter.

1927 *Arthur H. Compton* Scattering of electrons and photons.
Charles T. R. Wilson Invention of the cloud chamber to study the paths of charged particles.

1928 *Owen W. Richardson* Work on the emission of electrons from hot objects.

1929 *Louis-Victor de Broglie* Discovery of the wave nature of electrons.

1930 *Chandrasekhara Venkata Raman* Studies of the scattering of light by atoms and molecules.

1931 No prize

1932 *Werner Heisenberg* Creation of quantum mechanics.

1933 *Erwin Schroedinger and Paul A. M. Dirac* Contributions to quantum mechanics.

1934 No prize

1935 *James Chadwick* Discovery of the neutron.

1936 *Victor F. Hess* Discovery of cosmic rays.
Carl D. Anderson Discovery of the positron.

1937 *Clinton J. Davisson and George P. Thomson* Discovery of diffraction of electrons by crystals.

1938 *Enrico Fermi* New radioisotopes produced by neutron irradiation.

1939 *Ernest O. Lawrence* Invention of the cyclotron.

1940–1942 No prize

1943 *Otto Stern* Discovery of the magnetic moment of the proton.

1944 *Isador I. Rabi* Nuclear magnetic resonance.

1945 *Wolfgang Pauli* Discovery of the exclusion principle.

1946 *Percy W. Bridgman* Contributions to high-pressure physics.

1947 *Edward V. Appleton* Reflection of radio waves by the ionosphere.

1948 *Patrick M. S. Blackett* Discoveries in nuclear physics.

1949 *Hideki Yukawa* Prediction of mesons.

1950 *Cecil F. Powell* Photographic method of studying nuclear processes.

1951 *John D. Cockroft and Ernest T. S. Walton* Transmutation of atomic nuclei by accelerated particles.

1952 *Felix Bloch and Edward M. Purcell* Research on the magnetic fields of nuclei.

1953 *Frits Zernike* Invention of the phase-contrast microscope.

1954 *Max Born* Research in quantum mechanics.
Walther Bothe Development of the coincidence method.

1955 *Willis E. Lamb* Discoveries concerning hydrogen spectrum.
Polykarp Kusch Determination of the magnetic moment of the electron.

1956 *William Shockley, John Bardeen, and Walter H. Brattain* Discovery of the transistor.

1957 *C. N. Yang and T. D. Lee* Prediction of violation of the conservation of parity.

1958 *Pavel A. Čerenkov, Ilya M. Frank, and Igor Tamm* Discovery and interpretation of the Čerenkov effect.

1959 *Emilio G. Segrè and Owen Chamberlain* Discovery of the antiproton.

1960 *Donald A. Glaser* Invention of the bubble chamber.

1961 *Robert Hofstadter* Studies of electron scattering by nuclei.
Rudolf L. Mössbauer Studies of emission and absorption of gamma rays in crystals.

1962 *Lev D. Landau* Theoretical work on condensed matter, especially liquid helium.

1963 *Eugene P. Wigner* Laws governing interactions between protons and neutrons in the nucleus.
Maria Goeppert-Mayer and J. Hans D. Jensen Discoveries related to nuclear shell structure.

1964 *C. H. Townes, Nikolai G. Basov, and Alexander M. Prochorov* Discovery of the principle behind lasers and masers.

1965 *Sin-Itiro Tomonaga, Julian Schwinger, and Richard P. Feynman* Development of quantum electrodynamics.

1966 *Alfred Kastler* Optical methods for studying atomic structure.

1967 *Hans A. Bethe* Theory of nuclear reactions in stars.

1968 *Luis W. Alvarez* Contributions to elementary particle physics.

1969 *Murray Gell-Mann* Discoveries concerning elementary particles.

1970 *Hannes Alfvén* Work in magnetohydrodynamics.
Louis Néel Discoveries concerning ferrimagnetism and antiferromagnetism.

1971   *Dennis Gabor* Holography.

1972   *John Bardeen, Leon N. Cooper, and J. Robert Schrieffer* Explanation of superconductivity.

1973   *Leo Esaki* Discovery of tunneling in semiconductors.
Ivar Giaever Discovery of tunneling in superconductors.
Brian D. Josephson Prediction of properties of supercurrent through a tunnel-barrier.

1974   *Antony Hewish* Discovery of pulsars.
Martin Ryle Pioneering work in radio astronomy.

1975   *Aage Bohr, Ben Mottelson, and James Rainwater* Discovery of the relationship between collective and particle motion in nuclei.

1976   *Burton Richter and Samuel C. C. Ting* Discovery of the J/ψ particle.

1977   *John N. Van Vleck, Nevill F. Mott, and Philip W. Anderson* Studies of electrons in magnetic solids.

1978   *Arno A. Penzias and Robert W. Wilson* Discovery of the cosmic background radiation.
Peter L. Kapitza Work in low-temperature physics.

1979   *Sheldon L. Glashow, Abdus Salam, and Steven Weinberg* Work on the unification of the weak and electromagnetic interactions.

1980   *Val L. Fitch and James W. Cronin* Experimental proof of charge-parity (CP) violation, which has implications for cosmology.

1981   *Nicolaas Bloembergen and Arthur L. Schawlow* Contributions to the development of laser spectroscopy.

Kai M. Siegbahn Contributions to the development of high-resolution electron spectroscopy.

1982   *Kenneth G. Wilson* Theory of critical phenomena in conection with phase transitions.

1983   *Subrahmanyan Chandrasekhar and William A. Fowler* Work relating to the evolution of stars.

1984   *Carlo Rubbia and Simon van der Meer* Discovery of the W and Z particles.

1985   *Klaus von Klitzing* Discovery of the quantized Hall effect.

1986   *Ernst Ruska* Design of the first electron microscope.
Gird Binning and Heinrich Rohrer Design of the scanning tunneling microscope.

1987   *J. George Bednorz and K. Alex Müller* Discovery of superconductivity in a new class of materials.

1988   *Leon Lederman, Melvin Schwartz, and Jack Steinberger* Experiment that established the existence of a second kind of neutrino and that employed the first beam of neutrinos produced in a laboratory.

1989   *Norman Ramsey, Hans Dehmelt, and Wolgang Paul* Contributions to the development of precision atomic spectroscopy, which laid the basis for cesium atomic clocks and other devices.

1990   *Jerome Friedman, Henry Kendall, Richard Taylor* Experimental verification of the existence of quarks.

APPENDIX D   MATH REVIEW

The following is a review of the math used in this book that goes beyond the simple arithmetic you use nearly every day. Although most of it should be familiar to you, you may have forgotten some of it if you have not used it in some time. Keep in mind that mathematics plays a relatively small role in this book; it is used to express concisely and to apply interrelationships that must first be understood conceptually.

BASIC ALGEBRA

Algebra is essentially arithmetic with letters used to represent numbers. Let's say, for example, that you form an equal partnership with three other people. Your share of any income or expense is then one-fourth of each amount. When an item is purchased for $60, your share of its cost is $15. This situation can be represented by an algebraic relationship—an equation. If we let the letter S stand for your share, and the letter A stand for the amount of the income or expense, then S is equal to one-fourth of A, or 0.25 multiplied by A. The equation is

$$S = 0.25 \times A$$

or

$$S = 0.25A$$

(The multiplication sign, $\times$, is usually omitted in algebra. Whenever a number and a letter, or two letters, appear side by side, they are multiplied by one another.)

S and A are called *variables* because they can take on different values (that is, they can be different numbers). An algebraic equation expresses the mathematical relationship that exists between variables.

To find your share, S, of a $200 payment, you replace A with $200 and perform the multiplication:

$$S = 0.25A$$
$$S = 0.25 \times \$200$$
$$S = \$50$$

The use of algebra in this situation really isn't necessary. All you have to do is to remember to multiply the total payment by 0.25 to compute your share. The algebraic equation is just a shorthand way of expressing the same thing.

This approach can be broadened for use with larger or smaller partnerships. We can let F represent the fraction for the partnership. In the above example, F is 0.25. For a partnership of two people, F is one-half, or 0.5. For 10 people, it is one-tenth or 0.1. Then the share, S, of an amount, A, for each partner is given by

$$S = FA$$

For a partnership of 10 people, the share of a $200 income is

$$S = 0.1 \times \$200$$
$$S = \$20$$

You might be able to compute this mentally, but the equation is a formal statement of the rule that you use to compute the answer.

This type of equation, in which one quantity equals the product of two others, is the one that appears most often in this book. A large number of important relationships in physics take the form of this equation.

The above equation can be used in additional ways. For example, if the share S is known, the equation can be used to find the original amount A. This can be done because any equation can be manipulated in certain ways without altering the precise relationship it expresses. The basic rule is

---

A mathematical operation can be performed on an equation without altering its validity as long as *the exact same thing is done to both sides.*

The following examples illustrate the kinds of operations that can be useful. Numerical values are also used in the examples to show that the relationship expressed by each equation is still correct.

Addition

A number or a variable can be added to both sides of an equation. Let's say we have the equation

$$A = B - 6$$

This means that when B is known to be 8, A is 2:

$$2 = 8 - 6$$

If B is 11, A is 5; if $B = 83$, $A = 77$; and so on.

We can add any quantity to both sides of the equation without affecting its validity. For example, we can add 4 to both sides of the equation $A = B - 6$:

$$A + 4 = B - 6 + 4$$
$$A + 4 = B - 2$$

The relationship has not been changed. This means that any pair of values for A and B that satisfied the original equation will also satisfy the "new" equation. The values $B = 8$ and $A = 2$, which worked in the first equation, also work in the second:

$$2 + 4 = 8 - 2$$
$$6 = 6$$

Why would we want to add something to both sides of an equation? We can do this to change the focus of the equation. The original equation is used to find A when B is the known, or given, quantity. When B is known (or chosen) to be 8, the equation indicates that A must then be 2. If, instead, the value of A is known and we wish to find B, we can change the equation so that B equals something in terms of A. In this case we simply add 6 to both sides:

$$A = B - 6$$
$$A + 6 = B - 6 + 6$$
$$A + 6 = B$$

or

$$B = A + 6$$

The original relationship is preserved because we get 8 for B when A is known to be 2:

$$B = 8 - 6$$
$$B = 2$$

What we have done is *solve the equation for B*, or, equivalently, *express B in terms of* A. This sort of thing is generally the goal when equations are manipulated.

Here's another example, except in this case a variable is added:

$$C = D - E$$

For example, if $D = 13$ and $E = 7$, then C is 6:

$$C = 13 - 7$$
$$C = 6$$

If, instead, we know C and E, then we might want to find the value of D. To do so we add E to both sides:

$$C + E = D - E + E$$
$$C + E = D$$

or

$$D = C + E$$

With C equal to 6 and E equal to 7, we find that D is 13, as before:

$$D = 6 + 7$$
$$D = 13$$

Subtraction

The same variable or number can be subtracted from both sides of an equation.

Subtraction is very similar to addition. We can use it, for example, to reverse the first addition example.

$$B = A + 6$$

We can subtract 6 from both sides.

$$B - 6 = A + 6 - 6$$
$$B - 6 = A$$

or

$$A = B - 6$$

Multiplication

Both sides of an equation can be multiplied by the same variable or number. For example,

$$G = \frac{H}{5}$$

(The right side is the standard way to represent division in algebra. In this case it means "H divided by 5.")

If H is known to be 45, then

$$G = \frac{45}{5}$$
$$G = 9$$

We can multiply both sides of the original equation by 5 if we wish to solve it for H:

$$G \times 5 = \frac{H}{5} \times 5$$
$$5G = \frac{H\,5}{5}$$

The 5's can be canceled, leaving H (5 divided by 5 equals 1).

$$5G = \frac{H\,5}{5} = \frac{H\,\cancel{5}}{\cancel{5}}$$

or

$$5G = H$$

$$H = 5G$$

For $G = 9$, H has to be 45, as before.

$$H = 5 \times 9$$
$$H = 45$$

Choosing to multiply both sides by 5 allows us to solve the equation for H.

This process also works for multiplication by a variable.

$$I = \frac{J}{B}$$

This equation gives a value for I when the values for J and B are known. If $J = 42$ and $B = 7$, then the equation tells us that I must be 6. If we want to express J in terms of I and B, we multiply both sides by B:

$$I \times B = \frac{J}{B} \times B$$

As before, the B's can be canceled.

$$IB = \frac{JB}{B} = \frac{J\cancel{B}}{\cancel{B}}$$
$$IB = J$$
$$J = IB$$

If $I = 6$ and $B = 7$, then J will be 42, as required.

Division

We can divide both sides of an equation by the same variable or number.

Let's go back to the partnership. Suppose one of your partners tells you that your share of a bill is $75. If you want to know what the amount of the bill is, you can solve the "share equation" for A by using division. (Remember that A = amount, F = fraction, and S = your share.)

$$S = FA$$

We can divide both sides by F.

$$\frac{S}{F} = \frac{FA}{F}$$
$$\frac{S}{F} = \frac{\cancel{F}A}{\cancel{F}} = A$$
$$A = \frac{S}{F}$$

So, the original amount A when your share is $75 is

$$A = \frac{\$75}{0.25}$$
$$A = \$300$$

Sometimes it is necessary to use different operations in sequence. For example,

$$T = \frac{U}{C}$$

This tells us that T must be 50 if $U = 400$ and $C = 8$. But what if we are given the values of T and U so that C is unknown? We solve the equation for C by first multiplying both sides by C and then dividing both sides by T:

$$C \times T = \frac{U}{C} \times C$$

$$CT = U$$

$$\frac{CT}{T} = \frac{U}{T}$$

$$C = \frac{U}{T}$$

If we knew that $U = 400$ and $T = 50$, we could find the correct value for C:

$$C = \frac{U}{T}$$

$$C = \frac{400}{50}$$

$$C = 8$$

There is a nice shorthand way of looking at what we've done. The two forms of the equation (before and after) are

$$T = \frac{U}{C}$$

$$C = \frac{U}{T}$$

Note that in the first equation, C is below on the right and T is above on the left. In the second equation, the C and T have been switched: C is above on the left, and T is below on the right. This is a general result for this kind of equation: To move a variable (or number) to the other side of an equation, you must change it from above to below, or from below to above.

Here is another example of using operations in sequence. We will solve the following equation for M. To do this, we use the previous rules to move everything to the other side of the equation except M.

$$P = 3M + 7$$

First, we subtract 7 from both sides:

$$P - 7 = 3M + - 7$$

$$P - 7 = 3M$$

Then we divide both sides by 3:

$$\frac{P - 7}{3} = \frac{3M}{3} = M$$

$$M = \frac{P - 7}{3}$$

Powers

Both sides of an equation can be raised to the same power. That is, we can square or cube both sides, we can take the square root of both sides, and so on.

If we have

$$V = WD$$

Then

$$V^2 = (WD)^2$$

Using numbers, if $W = 3$ and $D = 2$,

$$V = 3 \times 2$$

$$V = 6$$

After the equation is squared:

$$V^2 = (3 \times 2)^2$$

$$V^2 = 6^2$$

$$V^2 = 36$$

So $V = 6$ also fits this equation.

If we start with the equation

$$X^2 = YZ$$

we can take the square root of both sides to find X.

$$\sqrt{X^2} = \sqrt{YZ}$$

$$X = \sqrt{YZ}$$

Or X could be

$$X = -\sqrt{YZ}$$

In most situations in physics, the positive square root is used.

Exercises

Solve the following equations for X.

1. $A = X - 7$
2. $B = X - 25.2$
3. $C = 12 + X$
4. $C = X + 87$
5. $E = X - F$
6. $G = X + 2H$
7. $I = \frac{X}{12}$
8. $J = -6X$
9. $K = \frac{X}{L}$
10. $M = NX$
11. $O = \frac{4}{X}$
12. $P = \frac{20}{X}$
13. $R = 3X - 5$
14. $S = 2X + T$
15. $U = \sqrt{X}$
16. $V = 4X^2$

PROPORTIONALITIES

Much of physics involves studying the relationships between different variables such as speed and distance. It is very useful to know, for example, how the value of one quantity is affected if the value of another is changed. To illustrate this, imagine that you go into a candy store to buy several chocolate bars of the same kind. Since we can choose from different-sized bars with different prices, the number of bars that you buy will depend on the price of each as well as on the amount that you spend. The following equation can be used to tell you the number N of bars costing C dollars each that you will get if you spend D dollars:

$$N = \frac{D}{C}$$

If you spend \$2 on chocolate bars that cost \$0.50 (50 cents) each, the number that you buy is

$$N = \frac{\$2}{\$0.50}$$
$$N = 4$$

Let's look at this relationship in terms of how N changes when one of the other quantities takes on different values. For example, how is N affected when you increase the amount you spend on a particular type of chocolate? Clearly, the more you spend, the more you get. At 50 cents each, \$2 will let you buy four bars, \$3 will let you buy six bars, \$4 will let you buy eight bars, and so on. This relationship is called a *proportionality*: we say that "N is proportional to D," (or "N is directly proportional to D"). If D *is doubled*—as from \$2 to \$4— *then* N *is also doubled,* from 4 to 8 in this case. If D is tripled, N becomes three times as large also.

In a similar way, if the amount you spend, D, is decreased, the number N that you get is also decreased. One dollar will let you buy only two bars at 50 cents each. If D is halved, then N is halved.

$D(\$)$	N
1	2
2	4
3	6
4	8

Now let's say you decide to spend only \$2. How does the cost of the bar you decide to buy affect the number of bars that you get? We know that you get four bars when they cost 50 cents each. If you buy bars that cost \$1 each instead, the number you get is two.

$$N = \frac{\$2}{\$1}$$
$$N = 2$$

If you choose bars that cost \$2 each, you can get only one bar. Thus, when C is increased, N is decreased. In particular, *if C is doubled,* N *becomes one-half as large.* This is called an *inverse proportionality.* We say that "N is inversely proportional to C." If C is decreased, N is increased. At 25 cents each, you can buy eight chocolate bars for \$2.

$C(\$)$	N
0.25	8
0.50	4
1	2
2	1

SCIENTIFIC NOTATION

Scientific notation is a convenient shorthand way to represent very large and very small numbers. It is based on using powers of 10. Since 100 equals 10×10, and 1,000 equals $10 \times 10 \times 10$, we can abbreviate them as follows:

$$100 = 10^2 \quad (\ = 10 \times 10)$$
$$1,000 = 10^3 \quad (\ = 10 \times 10 \times 10)$$

The numbers 100, 1,000, and so on, are powers of 10 because they are equal to 10 multiplied by itself a certain number of times. The number 2 in 10^2 is called the *exponent.* To represent any such power of 10 in this fashion, the exponent equals the number of zeros. So:

$$10 = 10^1$$
$$10,000 = 10^4$$
$$1,000,000 = 10^6$$

In this same fashion, since 1 has no zeros it is represented as

$$1 = 10^0$$

Decimals like 0.1 and 0.001 can also be represented this way by using negative exponents: 10 with a negative exponent equals 1 divided by 10 to the corresponding positive number.

$$0.01 = \frac{1}{100} = \frac{1}{10^2} = 10^{-2}$$

Similarly,

$$0.001 = \frac{1}{1,000} = \frac{1}{10^3} = 10^{-3}$$

To represent decimals by 10 to a negative exponent, the number in the exponent equals the number of places that the decimal point is to the left of the 1. For 0.01, the decimal point is two places to the left of 1, so $0.01 = 10^{-2}$. Other examples are as follows:

$$0.1 = 10^{-1}$$
$$0.0001 = 10^{-4}$$
$$0.000000001 = 10^{-9}$$

To reverse the process, 10 with a positive number exponent equals 1 followed by that number of zeros. Ten with a negative number exponent equals a decimal in which the number of zeros between the decimal point and the 1 is equal to the number in the exponent *minus one.* So, 10^5 equals 1 followed by five zeros—100,000. The number 10^{-5} equals the decimal with four zeros ($5 - 1$) between the decimal point and 1—0.00001.

$$10^4 = 10,000$$
$$10^3 = 1,000$$
$$10^2 = 100$$
$$10^1 = 10$$
$$10^0 = 1$$
$$10^{-1} = 0.1$$
$$10^{-2} = 0.01$$
$$10^{-3} = 0.001$$
$$10^{-4} = 0.0001$$

When two powers of 10 are multiplied, their exponents are added:

$$100 \times 1,000 = 100,000$$
$$10^2 \times 10^3 = 10^5$$

Similarly,

$$10^4 \times 10^8 = 10^{12}$$
$$10^{-3} \times 10^7 = 10^4$$
$$10^{-2} \times 10^{-5} = 10^{-7}$$

When a power of 10 is divided by another power of 10, the second exponent is subtracted from the first:

$$10^6 \div 10^2 = \frac{10^6}{10^2} = 10^4$$
$$10^4 \div 10^9 = 10^{-5}$$
$$10^3 \div 10^{-5} = 10^{3 - (-5)} = 10^8$$
$$10^{-7} \div 10^{-1} = 10^{-7 - (-1)} = 10^{-6}$$

Scientific notation utilizes powers of 10 to represent any number. A number expressed in scientific notation consists of a number (usually between 1 and 10) multiplied by 10 raised to some exponent. For example

$$23,400 = 2.34 \times 10,000$$
$$23,400 = 2.34 \times 10^4$$

When expressing a number in scientific notation, the exponent equals the number of times the decimal point is moved. (For numbers like 23,400 the decimal point is understood to be to the right of the last digit–as in 23,400.–so it was moved four places to the left to get it between the 2 and the 3.) *If the decimal point is moved to the left, the exponent is positive,* as in the above example. Similarly,

$$641 = 6.41 \times 10^2$$
$$497,500,000 = 4.975 \times 10^8$$

If the decimal point is moved to the right, the exponent is negative.

$$0.0053 = 5.3 \times .001 = 5.3 \times 10^{-3}$$
$$0.07134 = 7.134 \times 10^{-2}$$
$$0.000000964 = 9.64 \times 10^{-7}$$

When two numbers expressed in scientific notation are multiplied or divided, the "regular" parts are placed together, and the powers of 10 are placed together.

For multiplication

$$(4.2 \times 10^3) \times (2.3 \times 10^5) = (4.2 \times 2.3) \times (10^3 \times 10^5)$$
$$= 9.66 \times 10^8$$
$$(1.36 \times 10^4) \times (2 \times 10^{-6}) = (1.36 \times 2) \times (10^4 \times 10^{-6})$$
$$= 2.72 \times 10^{-2}$$

For division

$$(6.8 \times 10^7) \div 3.4 \times 10^2) = (6.8 \div 3.4) \times (10^7 \div 10^2)$$
$$= 2 \times 10^5$$
$$(7.3 \times 10^{-5}) \div (2 \times 10^{-8}) = (7.3 \div 2) \times (10^{-5} \div 10^{-8})$$
$$= 3.65 \times 10^3$$

SIGNIFICANT FIGURES

The numbers that are used in physics and other sciences represent measurements, not simply abstract values. A measured distance of, say, 96.4 miles is really some value between 96.35 miles and 96.45 miles. This imprecision must be kept in mind when such numbers are multiplied, divided, and so on. For example, if a car travels 96.4 miles in 3 hours, the speed is found by dividing the distance by the time:

$$\text{Speed} = \frac{\text{distance}}{\text{time}} = \frac{96.4 \text{ miles}}{3 \text{ hours}}$$

The strictly mathematical answer is

$$\text{Speed} = 32.1333333333 \text{ mph}$$

But the uncertainty in the distance measurement means that the exact speed can't be known precisely. Most likely it is some value between 32.1 mph and 32.2 mph. The value for the time is also imprecise. For this reason, the answer should not indicate more precision than the input numbers warrant.

In this case, the first three digits (3, 2, and 1) are said to be *significant figures,* while the remaining digits (all 3's) are *insignificant.* The significant figures represent the limit to the accuracy or precision in the answer. Our proper answer is one that shows only the significant figures—the number 32.1333333333 rounded off to 32.1:

$$\text{Speed} = 32.1 \text{ mph}$$

Mathematical rules indicate how many significant figures there are in an answer, based on the accuracy of the input data. But for the sake of simplicity, you may assume there are three or four significant figures in the answers to the Problems and Challenges. This means that you should round off your answers to three or four digits. Some examples:

$$6.4768 \rightarrow 6.48$$
$$25,934 \rightarrow 25,900$$
$$0.4575 \rightarrow 0.458$$

Exercises

Express the following numbers in scientific notation.

17. 253,000
18. 93,000,000
19. 5,632.5
20. 0.0072
21. 0.00000000338

Use scientific notation to find answers to the following:

22. $(4 \times 10^2) \times (2.1 \times 10^7)$
23. $(6.1 \times 10^{11}) \times (1.4 \times 10^{-5})$
24. $(3.2 \times 10^4) \times (5 \times 10^{-6})$
25. $(7 \times 10^{-3}) \times (8.3 \times 10^{-8})$
26. $(6 \times 10^9) \div (2 \times 10^2)$
27. $(9.3 \times 10^7) \div (4.1 \times 10^{-4})$
28. $(5.23 \times 10^{-8}) \div (3 \times 10^{-6})$
29. $(4.5 \times 10^{-3}) \div (2 \times 10^{-5})$
30. $(2.4 \times 10^4) \div (6 \times 10^9)$

ANSWERS TO ODD-NUMBERED EXERCISES

1. $X = A + 7$
3. $X = C - 12$
5. $X = E + F$
7. $X = 12I$
9. $X = KL$
11. $X = \dfrac{4}{O}$
13. $X = \dfrac{R + 5}{3}$
15. $X = U^2$
17. 2.53×10^5
19. 5.6325×10^3
21. 3.38×10^{-9}
23. 8.54×10^6
25. 58.1×10^{-11} or 5.81×10^{-10}
27. 2.27×10^{11}
29. 2.25×10^2

APPENDIX E   ANSWERS

This appendix contains answers to the odd-numbered Problems, the odd-numbered Challenges that are problems, and the Learning Checks. Worked-out solutions are also given for the odd-numbered Problems, except those that are nearly identical to Examples in the text.

ANSWERS TO ODD-NUMBERED PROBLEMS AND CHALLENGES

CHAPTER 1 PROBLEMS

1. $d = 20$ m

   From conversion table in Appendix A: 1 m = 3.28 ft

   $d = 20$ m $= 20 \times 1$ m $= 20 \times 3.28$ ft

   $d = 65.6$ ft

3. milli- means $^1/_{1000}$ or 0.001; so: 1 second = 1000 milliseconds

   0.0452 s $= 0.0452 \times 1$ s $= 0.0452 \times 1000$ ms

   0.0452 s $= 45.2$ ms

5. $T = 0.8$ s (period)

   $f = \dfrac{1}{T} = \dfrac{1}{0.8 \text{ s}}$

   $f = 1.25$ Hz

7. average speed $= \dfrac{\text{distance}}{\text{time}}$

   $v = \dfrac{d}{t} = \dfrac{1,200 \text{ mi}}{2.5 \text{ h}}$

   $v = 480$ mi/h (mph)

9. In Figure 1.18c, the length of the resultant velocity arrow is 0.66 times the length of the arrow representing 8 m/s (v_1). Therefore the resultant velocity is 0.66×8 m/s $= 5.3$ m/s.

   In Figure 1.18d, the resultant velocity is 12.5 m/s.

11. $d = vt$

    $d = 25$ m/s $\times 5$ s

    $d = 125$ m

    for $v = 250$ m/s: $d = vt$

    $d = 250$ m/s $\times 5$ s

    $d = 1,250$ m

13.

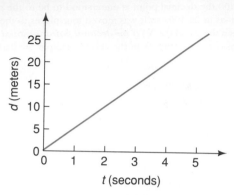

Slope = speed = 5 m/s (See Figure 1.27).

15. (a) $a = 4.06$ m/s^2 (See Example 1.2)

    (b) $a = -8$ m/s^2 (See Example 1.3)

17. $a = 200$ m/s^2 (See Example 1.4)

19. (a) $v = at$ $a = 60$ m/s^2 $t = 40$ s

    $v = 60$ m/s$^2 \times 40$ s

    $v = 2,400$ m/s

    (b) $v = at$

    $7,500$ m/s $= 60$ m/s$^2\ t$

    $\dfrac{7,500 \text{ m/s}}{60 \text{ m/s}^2} = t$ $t = 125$ s

21. (a)

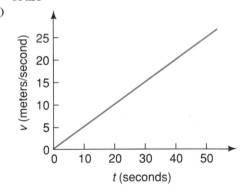

(b)

23. (a) $v = at$ $a = g$ $t = 3$ s
$v = gt = 9.8$ m/s^2 $\times$ 3s
$v = 29.4$ m/s

(b) $d = \frac{1}{2}at^2$ $d = \frac{1}{2}gt^2$
$d = \frac{1}{2}(9.8$ m/s$^2) \times (3$ s$)^2$
$d = 4.9$ m/s^2 $\times$ 9 s^2
$d = 44.1$ m

25. $a = 4.9$ m/s^2
(a) $v = at$ $t = 3$ s
$v = 4.9$ m/s^2 $\times$ 3 s
$v = 14.7$ m/s

(b) $d = \frac{1}{2}at^2$

$d = \frac{1}{2}(4.9$ m/s$^2) \times (3$ s$)^2$
$d = 2.45$ m/s^2 $\times$ 9 s^2
$d = 22.05$ m

27. at "a": $a = $ slope $= \dfrac{500 \text{ m/s}}{0.001 \text{ s}}$
$a = 500{,}000$ m/s^2
at "b": $a = 0$ m/s^2
at "c": $a = -2{,}500{,}000$ m/s^2

CHAPTER 1 CHALLENGES

1. $t = 0.237$ s

3. (a) $v = 1{,}021$ m/s
(b) $a = 0.00271$ m/s^2

5. a $= 7.9$ m/s^2

7. $v = \sqrt{2ad}$

CHAPTER 2 PROBLEMS

1. One example: $W = 150$ lb   1 lb $= 4.45$ N
$W = 150 \times 1$ lb $= 150 \times 4.45$ N
$W = 667.5$ N

$W = mg$
667.5 N $= m\,9.8$ m/s^2
$\dfrac{667.5 \text{ N}}{9.8 \text{ m/s}^2} = m$
$m = 68.1$ kg

3. (a) $W = mg = 30$ kg $\times 9.8$ m/s^2
$W = 294$ N
(b) $W = 294$ N   1 N $= 0.225$ lb
$W = 294 \times 1$ N $= 294 \times 0.225$ lb
$W = 66.15$ lb

5. $F = 600$ N   (See Example 2.1)

7. $F = ma$
10 N $= 2$ kg $\times a$

10 N/2 kg $= a$
$a = 5$ m/s^2

9. $F = ma$
$20{,}000{,}000$ N $= m \times 0.1$ m/s^2

$\dfrac{20{,}000{,}000 \text{ N}}{0.1 \text{ m/s}^2} = m$

$m = 200{,}000{,}000$ kg

11. $m = 4{,}500$ kg $F = 60{,}000\,N$
(a) $F = ma$
$60{,}000$ N $= 4{,}500$ kg $\times a$

$\dfrac{60{,}000 \text{ N}}{4{,}500 \text{ kg}} = a$
$a = 13.3$ m/s^2

(b) $v = at$
$v = 13.3$ m/s^2 $\times$ 8 s
$v = 106.4$ m/s

(c) $d = \frac{1}{2}at^2$
$d = \frac{1}{2}(13.3$ m/s$^2) \times (8$ s$)^2$

$d = 6.65$ m/s^2 $\times$ 64 s^2
$d = 425.6$ m

13. (a) $a = 3$ m/s^2 (See Example 1.2)
(b) $F = 240$ N   (See Example 2.1)

(c) $d = \frac{1}{2}at^2 = \frac{1}{2}(3$ m/s$^2) \times (3$ s$)^2$
$d = 1.5$ m/s^2 $\times$ 9 s^2
$d = 13.5$ m

15. $a = 6\,g$ $g = 9.8$ m/s^2
$= 6 \times 9.8$ m/s^2
$a = 58.8$ m/s^2
$F = ma$ $m = 1{,}200$ kg
$= 1{,}200$ kg $\times 58.8$ m/s^2
$F = 70{,}560$ N

17. $v = 60$ m/s $r = 400$ m $m = 600$ kg

(a) $a = \dfrac{v^2}{r} = \dfrac{(60 \text{ m/s})^2}{400 \text{ m}}$

$a = \dfrac{3{,}600 \text{ m}^2/\text{s}^2}{400 \text{ m}}$

$a = 9$ m/s^2

$g = 9.8$ m/s^2 1 m/s$^2 = \dfrac{g}{9.8}$

$a = 9 \times 1$ m/s$^2 = 9 \times \dfrac{g}{9.8}$

$a = \dfrac{9}{9.8}g$

$a = 0.918\,g$

(b) $F = ma$
$= 600$ kg $\times 9$ m/s^2
$F = 5{,}400$ N

19. $F = m\dfrac{v^2}{r}$

$$60\text{ N} = 0.1\text{ kg} \times \frac{v^2}{1\text{ m}}$$

$$v^2 = \frac{60\text{ N·m}}{0.1\text{ kg}} = 600\text{ m}^2/\text{s}^2$$

$$v = 24.5\text{ m/s}$$

21. a) At *twice* as far away the force is *one fourth* as large. F = 150 lb.

b) F = 66.7 lb

c) F = 6 lb

CHAPTER 2 CHALLENGES

7. (a) F = 44.7 N

(b) v = 3,075 m/s

(c) t = 86,500 s

CHAPTER 3 PROBLEMS

1. $F = \dfrac{\Delta mv}{\Delta t} = \dfrac{(mv)_f - (mv)_I}{\Delta t}$

$mv_I = 1{,}000\text{ kg} \times 0\text{ m/s}$

$mv_I = 0\text{ kg m/s}$

$mv_f = 1000\text{ kg} \times 27\text{ m/s}$

$mv_f = 27{,}000\text{ kg m/s}$

$\Delta t = 10\text{ s}$

$F = \dfrac{(mv)_f - (mv)_I}{\Delta t} = \dfrac{27{,}000\text{ kg m/s} - 0\text{ kg m/s}}{10\text{ s}}$

$F = \dfrac{27{,}000\text{ kg m/s}}{10\text{ s}}$

$F = 2{,}700\text{ N}$

3. v = 39 m/s   (See Example 3.2)

5. $mv_{\text{before}} = 50\text{ kg} \times 5\text{ m/s}$

$mv_{\text{before}} = 250\text{ kg m/s}$

$mv_{\text{after}} = mv = (40\text{ kg} + 50\text{ kg}) \times v$

$mv_{\text{after}} = 90\text{ kg} \times v$

$mv_{\text{before}} = mv_{\text{after}}$

$250\text{ kg m/s} = 90\text{ kg} \times v$

$\dfrac{250\text{ kg m/s}}{90\text{ kg}} = v$

$v = 2.78\text{ m/s}$

7. $mv_{\text{before}} = 0\text{ kg m/s}$

$mv_{\text{after}} = (mv)\text{gun} + (mv)\text{bullet}$

$mv_{\text{after}} = 1.2\text{ kg} \times v + 0.02\text{ kg} \times 300\text{ m/s}$

$mv_{\text{after}} = 1.2\text{ kg} \times v + 6\text{ kg m/s}$

$mv_{\text{before}} = mv_{\text{after}}$

$0\text{ kg m/s} = 1.2\text{ kg} \times v + 6\text{ kg m/s}$

$-6\text{ kg m/s} = 1.2\text{ kg} \times v$

$\dfrac{-6\text{ kg m/s}}{1.2\text{ kg}} = v$

$v = -5\text{ m/s}$ (opposite direction of bullet)

or use: $\dfrac{v\text{ gun}}{v\text{ bullet}} = \dfrac{-m\text{ bullet}}{m\text{ gun}}$

9. Work = 30,000 J   (See Example 3.3)

11. PE = 2,156 J   (See Example 3.7)

13. KE = 4,500 J   (See Example 3.6)

15. $KE = \frac{1}{2}mv^2$

$60{,}000\text{ J} = \frac{1}{2} \times 300\text{ kg} \times v^2$

$\dfrac{60{,}000\text{ J}}{150\text{ kg}} = v^2$

$400\text{ J/kg} = v^2$

$v = 20\text{ m/s}$

17. v = 89.8 m/s   (See Example 3.8)

19. $v^2 = 2\,g\,d$

$(7.7\text{ m/s})^2 = 2 \times 9.8\text{ m/s}^2 \times d$

$59.3\text{ m}^2/\text{s}^2 = 19.6\text{ m/s}^2 \times d$

$\dfrac{59.3\text{ m}^2/\text{s}^2}{19.6\text{ m/s}^2} = d$

$d = 3\text{ m}$ (or use Table 3.2)

21. a) PE = 323,000 J (See Example 3.7)

b) v = 46.4 m/s (104 mph; See Example 3.8)

23. (a) KE = 4,000 J   (See Example 3.6)

(b)

KE at bottom = PE when stopped on hill

$4{,}000\text{ J} = mgd = 80\text{ kg} \times 9.8\text{ m/s}^2 \times d$

$4{,}000\text{ J} = 784\text{ kg m/s}^2 \times d$

$\dfrac{4{,}000\text{ J}}{784\text{ kg m/s}^2} = d$

$d = 5.1\text{ m}$ or use $d = \dfrac{v^2}{2g}$

25. $KE_{\text{before}} = \frac{1}{2}mv^2 = \frac{1}{2} \times 50\text{ kg} \times (5\text{ m/s})^2$

$KE_{\text{before}} = 25\text{ kg} \times 25\text{ m}^2/\text{s}^2$

$KE_{\text{before}} = 625\text{ J}$

$KE_{\text{after}} = \frac{1}{2}mv^2 = \frac{1}{2} \times 90\text{ kg} \times (2.78\text{ m/s})^2$

$KE_{\text{after}} = 45\text{ kg} \times 7.73\text{ m}^2/\text{s}^2$

$KE_{\text{after}} = 348\text{ J}$

$KE_{\text{last}} = 625\text{ J} - 348\text{ J}$

$KE_{\text{last}} = 277\text{ J}$

27. $P = \dfrac{\text{work}}{t}$

$200\text{ W} = \dfrac{10{,}000\text{ J}}{t}$

$$t = \frac{10,000\,J}{200\,W}$$

$$t = 50\,s$$

29. $P = \dfrac{work}{t} = \dfrac{PE}{t}$

$PE = mgd = 1,000\,kg \times 9.8\,m/s^2 \times 30\,m$

$PE = 294,000\,J$

$P = \dfrac{PE}{t} = \dfrac{294,000\,J}{10\,s}$

$P = 29,400\,W$

31. $P = 217\,W$ (See Problem 29)

CHAPTER 3 CHALLENGES

3. $V_b = 281.4\,m/s$

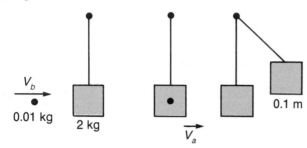

7. (a) $KE = 715.4\,J$

(b) $F = 238.5\,N$

(c) $P = 1,431\,W$

$P = 1.92\,hp$

9. 3.6 revolutions

CHAPTER 4 PROBLEMS

1. $p = \dfrac{F}{A} = \dfrac{2,000,000\,lb}{400\,ft^2}$

$p = 5,000\,lb/ft^2$

$1\,ft^2 = 144\,in.^2$ $p = 5000\,\dfrac{lb}{ft^2} = 5000\,\dfrac{lb}{144\,in.^2}$

$p = \dfrac{5000\,lb}{144\,in.^2}$

$p = 34.7\,psi$

3. $F = pA$

$F = 80\,psi \times 1,200\,in.^2$

$F = 96,000\,lb$

5. $F = 1,176\,lb$ (See Example 4.2)

7. (a) $D = \dfrac{m}{V} = \dfrac{393\,kg}{0.05\,m^3}$

$D = 7,860\,kg/m^3$

$D_w = md \times g = 7,860\,kg/m^3 \times 9.8\,m/s^2$

$D_w = 77,028\,N/m^3$

(b) Iron   (From Table 4.3)

9. (a) $D_w = \dfrac{W}{V}$

water: $D_w = 62.4\,lb/ft^3$ (See Table 4.3)

$62.4\,lb/ft^3 = \dfrac{40,000\,lb}{V}$

$V = \dfrac{40,000\,lb}{62.4\,lb/ft^3}$

$V = 641\,ft^3$

(b) gasoline: $D_w = 42\,lb/ft^3$

$D_w = 42\,lb/ft^3 = \dfrac{40,000\,lb}{V}$

$V = \dfrac{40,000\,lb}{42\,lb/ft^3}$

$V = 952\,ft^3$

11. $m = 162\,kg$ (See Example 4.4 and Table 4.3)

13. $p = 0.433\,d = 0.433 \times 12\,ft$

$p = 5.196\,psi$

15. $p = 3,030,000\,Pa$ (See Example 4.6)

17. About 7 psi

19. $F_b = W_{\text{water displaced}} = (D_w)_{\text{water}} \times V$

$F_b = 62.4\,lb/ft^3 \times 12\,ft^3$

$F_b = 749\,lb$

21. (a) $W = 468\,lb$ (See Example 4.5 and Table 4.3)

(b) $F_b = W_{\text{water}} = (D_w)_{\text{water}} \times V = 62.4\,lb/ft^3 \times 3\,ft^3$

$F_b = 187.2\,lb$

(c)

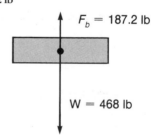

$F_{\text{net}} = 468\,lb - 187.2\,lb$

$F_{\text{net}} = 280.8\,lb$ (downward)

23. (a) $W = 5,720,000\,lb$ (See Example 4.5 and Table 4.3)

(b) $W_{\text{seawater}} = F_b = W_{\text{ice}}$ (since it floats)

$W_{\text{seawater}} = D_w V$ $D_w(\text{seawater}) = 64.3\,lb/ft^3$

$5,720,000\,lb = 64.3\,lb/ft^3 \times V$

$\dfrac{5,720,000\,lb}{64.3\,lb/ft^3} = V$

$V_{\text{seawater}} = 88,960\,ft^3$

(c) $V_{\text{out}} = V_{\text{total}} - V_{\text{underwater}}$

$= 100,000\,ft^3 - 88,960\,ft^3$

$V_{\text{out}} = 11,040\,ft^3 = \dfrac{1}{9} V_{\text{total}}$

25. Scale reading $= 100\,N - F_b$

$F_b = W_{\text{water}} = D_{\text{water}} \times g \times V_{al}$

$W_{al} = D_{al} \times g \times V_{al}$ $V_{al} = 0.00378\,m^3$

$F_b = 37.0\,N$

Scale reading $= 73.0\,N$

CHAPTER 4 CHALLENGES

3. $D = 2,700 \text{ kg/m}^3$

specific gravity = 2.7

$(D_w)_{\text{moon}} = 27.6 \text{ lb/ft}^3$

5. $F = 34.7 \text{ lb}$

CHAPTER 5 PROBLEMS

1. $30°C = 86°F$ (no jacket needed)

3. $\Delta l = -0.336 \text{ ft}$ means it is shorter (See Example 5.1 and Table 5.2)

$l = 699.664 \text{ ft}$

5. Need to shorten diameter from 5 mm to 4.997 mm

$\Delta l = -0.003 \text{ mm}$

$\Delta l = \alpha l \Delta T$ $\alpha = 12 \times 10^{-6}/°C$

$-0.003 \text{ mm} = 12 \times 10^{-6}/°C \times 5 \text{ mm} \times \Delta T$

$\dfrac{-0.003 \text{ mm}}{60 \times 10^{-6} \text{ mm/°C}} = \Delta T$

$\dfrac{-0.003}{60} \times 10^6 \, °C = \Delta T$

$-0.00005 \times 10^6 \, °C = \Delta T$

$\Delta T = -50°C$

7. $\Delta U = \text{work} + Q$ $Q = -2J$ *(heat lost)*

$\text{Work} = Fd = 50 \text{ N} \times 0.1 \text{ m}$

$\text{Work} = 5 \text{ J}$ (Work done *on* gas)

$\Delta U = \text{Work} + Q = 5 \text{ J} + -2 \text{ J}.$

$\Delta U = 3 \text{ J}$

9. $Q = 1,081,000 \text{ J}$ (See Example 5.2 and Table 5.3)

11. (a) $Q = Cm\Delta T$

$C = 2,000 \text{ J/kg-°C}$ for ice

$\Delta T = 25°C$ from Figure 5.36

$Q = Cm\Delta T = 2,000 \text{ J/kg-°C} \times 1 \text{ kg} \times 25°C$

$Q = 50,000 \text{ J}$

(b) $Q = Cm\Delta T$

$C = 4,184 \text{ J/kg-°C}$ for water

$\Delta T = 100°C$ from Figure 5.36

$Q = Cm\Delta T = 4,184 \text{ J/kg-°C} \times 1 \text{ kg} \times 100°C$

$Q = 418,400 \text{ J}$

13. (a) $KE = 375,000 \text{ J}$ (See Example 3.6)

(b) $Q = Cm\Delta T$ $Q = KE$

$KE = Cm\Delta T$ $C = 460 \text{ J/kg-°C}$ for iron

$375,000 \text{ J} = 460 \text{ J/kg-°C} \times 20 \text{kg} \times \Delta T$

$\dfrac{375,000 \text{ J}}{9,200 \text{ J/°C}} = \Delta T$

$\Delta T = 40.8°C$

15. $\Delta T = 31.1°C$ (See Example 5.3 and Table 5.3)

17. (a) Rel. Hum. = 62.5% (See Example 5.6 and Table 5.5)

(b) Rel. Hum. = 5.78% (See Example 5.6 and Table 5.5)

19. The amount of water vapor in the air is given by the water vapor density, which is the humidity.

$\text{Rel. Hum.} = \dfrac{\text{Humidity}}{\text{Sat. Den.}} \times 100\%$

at 20°C, Sat. Den. = 0.0173 kg/m³

$40\% = \dfrac{\text{Humidity}}{0.0173 \text{ kg/m}^3} \times 100\%$

$\dfrac{40\% \times 0.0173 \text{ kg/m}^3}{100\%} = \text{Humidity}$

Humidity = 0.00692 kg/m³

There is 0.00692 kg of water vapor in each m³ of air.

21. $m = DV$ $V = 150 \text{ m}^3$

relative humidity $\dfrac{\text{humidity}}{\text{saturation density}} \times 100\%$

$60\% = \dfrac{D}{0.0228} \times 100\%$ $D = 0.0137 \text{ kg/m}^3$

$m = 0.0137 \text{ kg/m}^3 \times 150 \text{ m}^3$

$m = 2.1 \text{ kg}$

23. Carnot eff. = 34.9% (See Example 5.7)

25. (a) Eff. = $\dfrac{\text{work output}}{\text{energy input}} \times 100\%$

Eff. = $\dfrac{3,000 \text{ J}}{15,000 \text{ J}} \times 100\%$

Eff. = 20%

(b) Carnot Eff. = 88% (See Example 5.7)

CHAPTER 6 PROBLEMS

1. (a) $\rho = 0.167 \text{ kg/m}$ (See Example 6.1)

(b) $v = 15.5 \text{ m/s}$ (See Example 6.1)

3. $v = 20.1 \times \sqrt{T}$ $T = 100°C = 373 \text{ K}$

$v = 20.1 \times \sqrt{373} = 20.1 \times 19.3$

$v = 388 \text{ m/s}$

5. $v = f\lambda$

$v = 4 \text{ Hz} \times 0.5 \text{ m}$

$v = 2 \text{ m/s}$

7. $v = f\lambda$

$80 \text{ m/s} = f \times 3.2 \text{ m}$

$\dfrac{80 \text{ m/s}}{3.2 \text{ m}} = f$

$f = 25 \text{ Hz}$

9. for $f = 20 \text{ Hz}$:

$\lambda = 17.2 \text{ m}$

$\lambda = 56.4 \text{ ft}$ (See Example 6.3)

for $f = 20,000 \text{ Hz}$

$\lambda = 0.0172 \text{ m}$

$\lambda = 0.677 \text{ in.}$ (See Example 6.3)

11. (a) $\lambda = 1.315 \text{ m}$ (See Example 6.3)

(b) $v = f\lambda$ $v = 1,440 \text{ m/s}$ (water, from Table 6.1)

$1,440 \text{ m/s} = 261.6 \text{ Hz} \times \lambda$

$\dfrac{1,440 \text{ m/s}}{261.6 \text{ Hz}} = \lambda$

$\lambda = 5.505 \text{ m}$

13. $v = 347$ m/s   (See Problem 5)

$v = 20.1 \times \sqrt{T} = 347$ m/s

$(20.1)^2 \times T = (347 \text{ m/s})^2$

$T = 298$ K $= 25°C$

15. This is destructive interference, so the difference in the two distances equals one half the wavelength of the sound.

7.2 m $- 7$ m $= 0.2$ m $= \frac{1}{2}\lambda$

$\lambda = 0.4$ m

$f = 860$ Hz   (See Problem 7)

17. (a) $d = vt$ $v = 344$ m/s   (air)

$8{,}000$ m $= 344$ m/s $\times t$

$\dfrac{8{,}000 \text{ m}}{344 \text{ m/s}} = t$

$t = 23.2$ s

(b) $d = vt$ $v = 4{,}000$ m/s   (granite, Table 6.1)

$8{,}000$ m $= 4{,}000$ m/s $\times t$

$\dfrac{8{,}000 \text{ m}}{4{,}000 \text{ m/s}} = t$

$t = 2$ s

19. Distance sound travels underwater in 0.01 second:

$d = vt$ $v = 1{,}440$ m/s

$d = 1{,}440$ m/s $\times 0.01$ s

$d = 14.4$ m

Sound travels to fish, reflects, then returns. Distance to fish is ½ the distance sound traveled.

$d = 7.2$ m

21. For each 10 dB, sound is about 2 times louder.

40 dB $= 4 \times 10$ dB

$2 \times 2 \times 2 \times 2 = 16$

Sound is 16 times louder.

23. **(a)** Harmonics of note have frequencies equal to 2, 3, 4, ... times frequency of note. Harmonics of 4,186 Hz note have frequencies:

8,372 Hz, 12,558 Hz, 16,744 Hz, 20,930 Hz, etc.

We can't hear frequencies over 20,000 Hz. So, highest harmonic we can hear is 16,744 Hz. Including the note itself (4,186 Hz), we can hear 4 harmonics.

(b) The note one octave below has ½ the frequency

$f = 2{,}093$ Hz Harmonics:

4,186 Hz, 6,279 Hz, 8,372 Hz, 10,465 Hz, 12,558 Hz, 14,651 Hz, 16,744 Hz, 18,837 Hz, 20,930 Hz

The highest harmonic we can hear is 18,837 Hz. Including the note itself (2,093 Hz), we can hear 9 harmonics.

CHAPTER 6 CHALLENGES

1. Amplitude is decreased.

3. In air: $t = 0.581$ s

In steel: $t = 0.038$ s

Jack hears the sound in the rail 0.543 seconds before he hears the sound in the air.

7. 1,000,000 times

CHAPTER 7 PROBLEMS

1. $I = \dfrac{q}{t} = \dfrac{5 \text{ C}}{20 \text{ s}}$

$I = 0.25$ A

3. $I = \dfrac{q}{t}$ $t = 1$ min $= 60$ s

0.7 A $= \dfrac{q}{60 \text{ s}}$

0.7 A $\times 60$ s $= q$

$q = 42$ C

5. $V = IR$

120 V $= 12$ A $\times R$

$\dfrac{120 \text{ V}}{12 \text{ A}} = R$

$R = 10\ \Omega$

7. $I = 1.8$ A   (See Example 7.1)

9. $V = 20$ V   (See Example 7.2)

11. $P = 1{,}440$ W   (See Example 7.4)

13. $P = 40$ W $I = 3.33$ A   (See Example 7.5)

$P = 50$ W $I = 4.17$ A   (See Example 7.5)

15. $E = 9{,}600{,}000$ J   (See Example 7.6)

or: $P = 4$ kW $t = 40$ min $= \dfrac{2}{3}$ h

$E = 4$ kW $\times \dfrac{2}{3}$ h

$E = 2.67$ kW-h

17. Hair dryer: $E = 360{,}000$ J   (See Example 7.6)

Lamp: $E = 2{,}160{,}000$ J   (See Example 7.6)

19. (a) $P = 1{,}080$ W   (See Example 7.4)

(b) $E = 64{,}800$ J   (See Example 7.6)

or: $P = 1.08$ kW $t = 1$ min $= \dfrac{1}{60}$ h

$E = Pt = 1.08$ kW $\times \dfrac{1}{60}$ h

$E = 0.018$ kW-h

21. (a) $P = IV$

$1{,}000{,}000{,}000$ W $= I \times 24{,}000$ V

$\dfrac{1{,}000{,}000{,}000 \text{ W}}{24{,}000 \text{ V}} = I$

$I = 41{,}670$ A

(b) $E = Pt$ $t = 24$ h $= 24 \times 3600$ s

$t = 86{,}400$ s

$E = 1{,}000{,}000{,}000$ W $\times 86{,}400$ s

$E = 8.64 \times 10^{13}$ J

in kW-h: $E = 1{,}000{,}000$ kW $\times 24$ h

$E = 24{,}000{,}000$ kw-h

(c) 10¢ $= \$0.1$ Revenue $= \$2{,}400{,}000$

CHAPTER 7 CHALLENGES

1. $F = -8.2 \times 10^{-8}\,\text{N}$

3. Number of electrons $= 1.25 \times 10^{18}$

7. $V = 12\,\text{V}:$ $\quad I = 100\,\text{A}$
$V = 30\,\text{V}:$ $\quad I = 40\,\text{A}$
$V = 60\,\text{V}:$ $\quad I = 20\,\text{A}$
$V = 120\,\text{V}:$ $\quad I = 10\,\text{A}$

The lower voltage dryers would have to have successively larger wires to handle the larger currents.

9. $P = I^2 R$

4 times as much (40 kW·h per day)

CHAPTER 8 PROBLEMS

1. $\dfrac{V_o}{V_i} = \dfrac{N_o}{N_i}$

120 V input, 9 V output

$\dfrac{120\,\text{V}}{9\,\text{V}} = \dfrac{N_o}{N_i}$

ratio $= \dfrac{N_o}{N_i} = 13.3$

3. Example: $\quad f = 92.5\,\text{MHz} = 92{,}500{,}000\,\text{Hz}$

$c = f\lambda$

$3 \times 10^8\,\text{m/s} = 9.25 \times 10^7\,\text{Hz} \times \lambda$

$\dfrac{3 \times 10^8\,\text{m/s}}{9.25 \times 10^7\,\text{Hz}} = \lambda$

$\lambda = 3.24\,\text{m}$

5. $c = f\lambda$

$3 \times 10^8\,\text{m/s} = f \times 0.0254\,\text{m}$

$\dfrac{3 \times 10^8\,\text{m/s}}{0.0254\,\text{m}} = f$

$f = 1.18 \times 10^{10}\,\text{Hz} = 11{,}800\,\text{MHz}$

7. UV band: $\quad f = 7.5 \times 10^{14}\,\text{Hz to } f = 10^{18}\,\text{Hz}$

$\lambda\text{'s:}$

$c = f\lambda$ $\qquad\qquad\qquad$ $c = f\lambda$
$3 \times 10^8\,\text{m/s} = 7.5 \times 10^{14}\,\text{Hz} \qquad 3 \times 10^8\,\text{m/s} = 10^{18}\,\text{Hz} \times \lambda$
$\qquad\qquad \times \lambda$

$\dfrac{3 \times 10^8\,\text{m/s}}{7.5 \times 10^{14}\,\text{Hz}} = \lambda \qquad\qquad \dfrac{3 \times 10^8\,\text{m/s}}{1 \times 10^{18}\,\text{Hz}} = \lambda$

$\lambda = 4 \times 10^{-7}\,\text{m} \qquad\qquad \lambda = 3 \times 10^{-10}\,\text{m}$

9. Energy emitted $\propto T^4$. From 300 K to 3000 K, T increases 10 times. Energy output increases $10^4 = 10{,}000$ times

11. $\lambda_{max} = \dfrac{0.0029\,\text{m·k}}{T}$

$\lambda_{max} = \dfrac{0.0029\,\text{m·k}}{10{,}000{,}000\,\text{k}}$

$\lambda_{max} = 2.9 \times 10^{-10}\,\text{m} \qquad \text{X-rays (See Figure 8.28)}$

13. $\lambda = \dfrac{0.0029}{T}$

$T \times \lambda = T \times 0.0000012\,\text{m} = 0.0029$

$T = 2{,}420\,\text{K}$

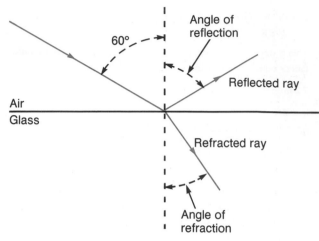

1. By the Law of Reflection, the angle of reflection equals the angle of incidence, 60°. Therefore the reflected ray makes an angle of 60° with respect to the normal.

For the refracted ray, the light slows upon entering the glass, so the ray is bent towards the normal—the angle of refraction is smaller than the angle of incidence. The actual angle is found using Figure 9.34. The refracted ray makes an angle of about 35° with respect to the normal.

3. $p = 35.3\,\text{cm}$ (See Example 9.3)

5. $M = \dfrac{\text{image height}}{\text{object height}} = \dfrac{-0.005\,\text{m}}{2\,\text{m}}$

$M = -0.0025$

7. p = 15 cm (to the right of the lens) $\quad$ (See Example 9.3)
Image is real and inverted.

CHAPTER 9 CHALLENGES

5. (a) $p = 58.3\,\text{mm}$
(b) $M = -0.167$ $\quad$ image height $= -16.7\,\text{mm}$
(c) $p = 50.7\,\text{mm}$ $\quad M = -0.0145$ $\quad$ image height $= -1.45\,\text{mm}$

CHAPTER 10 PROBLEMS

1. $E = 41.4\,\text{eV}$ $\quad$ (See Example 10.1)

3. First find the frequency, then use Figure 10.7.

$E = hf \qquad h = 6.63 \times 10^{-34}\,\text{J/Hz}$

$9.5 \times 10^{-25}\,\text{J} = 6.63 \times 10^{-34}\,\text{J/Hz} \times f$

$\dfrac{9.5 \times 10^{-25}\,\text{J}}{6.63 \times 10^{-34}\,\text{J/Hz}} = f$

$\dfrac{9.5}{6.63} \times \dfrac{10^{-25}}{10^{-34}}\,\text{Hz} = f$

$f = 1.43 \times 10^9\,\text{Hz} \qquad \text{low frequency microwave}$

$1\,\text{eV} = 1.6 \times 10^{-19}\,\text{J} \qquad 1\,\text{J} = \dfrac{1\,\text{eV}}{1.6 \times 10^{-19}}$

$E = 9.5 \times 10^{-25}\,\text{J}$

GLOSSARY OF KEY TERMS AND LAWS

This glossary includes highlighted definitions and laws from the text, along with definitions of selected terms that are found in more than one place in the text. Terms that are narrow in scope or are used in only one section or chapter (such as *optic fiber* and *reverberation*) are not included.

acceleration (a) Rate of change of velocity. The change in velocity divided by the time elapsed.

alternating current (AC) An electric current consisting of charges flowing back and forth.

amplitude Maximum displacement of points on a wave, measured from the equilibrium position.

angular momentum The momentum associated with rotational or circular motion. There are two types: spin and orbital.

antiparticle A charge-reversed version of an ordinary particle. A particle of the same mass (and spin) but of opposite electric charge (and certain other quantum mechanical "charges").

Archimedes' principle The buoyant force acting on a substance in a fluid at rest is equal to the weight of the fluid displaced by the substance.

atom The smallest unit or building block of a chemical element.

atomic mass number (A) The total number of protons and neutrons in a nucleus.

baryon A strongly interacting particle that is composed of 3 quarks. Protons and neutrons are examples. Baryons are fermions.

Bernoulli's principle When the speed of a moving fluid *increases,* the pressure *decreases,* and vice versa.

blackbody A hypothetical object that absorbs all light and other electromagnetic waves that strike it. The radiation it emits is called blackbody radiation (BBR).

boson A particle with integral spin. It does not obey the Pauli Exclusion Principle.

buoyant force The upward force exerted by a fluid on a substance partly or completely immersed in it.

centripetal acceleration Acceleration of an object moving along a circular path. It is directed toward the center of the circle.

centripetal force Name applied to the force acting to keep an object moving along a circular path. It is directed toward the center of the circle.

compound A pure substance consisting of two or more elements combined chemically.

conduction The transfer of heat by direct contact between atoms and molecules.

convection The transfer of heat by buoyant mixing in a fluid.

Coulomb's law The force acting on each of two charged objects is directly proportional to the net charges on the objects and inversely proportional to the square of the distance between them.

current (I) Rate of flow of electric charge. The amount of charge that flows past a given point per second. The unit of measure is the ampere.

diffraction The bending of a wave as it passes around the edge of a barrier. Diffraction causes a wave passing through a gap or a slit to spread out into the shadow regions.

direct current (DC) An electric current that is a steady flow of charge in one direction.

Doppler effect The apparent change in frequency of a wave due to motion of the source of the wave, the receiver, or both.

echolocation Process of using the reflection of a wave to locate objects.

elastic collision Collision in which the total kinetic energy of the colliding objects is the same before and after the collision.

electric charge (q) An inherent physical property of certain subatomic particles that is responsible for electrical and magnetic phenomena. The unit of measure is the coulomb.

electric field The agent of the electric force. It exists in the space around charged objects and causes a force to act on any other charged object.

electromagnetic wave A transverse wave consisting of a traveling combination of oscillating electric and magnetic fields.

electron Negatively charged particle, usually found orbiting the nucleus in an atom.

element One of the 109 different fundamental substances that are the simplest and purest form of matter.

elementary particles The basic, indivisible building blocks of the universe. The fundamental constituents from which all matter, antimatter, and their interactions derive. They are believed to be true "point" particles, devoid of internal structure or measurable size.

energy (E) The measure of a system's capacity to do work. That which is transferred when work is done.

fermion A particle with half-integral spin. It obeys the Pauli Exclusion Principle.

first law of thermodynamics The change in internal energy of a substance equals the work done on it plus the heat added to it.

force (F) A push or a pull acting on a body. Usually causes some distortion of the body, a change in its velocity, or both.

frequency (f) The number of cycles of a periodic process that ocur per unit time. The unit of measure is the hertz (Hz).

frequency (f) (of a wave) The number of cycles of a wave passing a point per unit time. It equals the number of oscillations per second in the wave.

friction A force of resistance to relative motion between two bodies or substances.

gamma rays The highest-frequency electromagnetic waves. One type of nuclear radiation.

gravitational force The attractive force that acts between all pairs of objects.

half-life The time it takes for one-half of the nuclei of a radioisotope to decay. The time interval during which each nucleus has a 50% probability of decaying.

heat (Q) The flow of internal energy. The form of energy that is transferred between two substances at different temperatures.

heat engine A device that converts heat energy into mechanical energy or work. It absorbs heat from a hot source such as burning fuel, converts some of this energy into usable mechanical energy or work, and outputs the remaining energy as heat at some lower temperature.

humidity The mass of water vapor in the air per unit volume. The density of water vapor in the air.

inelasitc collision Collision in which the total kinetic energy of the colliding bodies after the collision is not equal to the total kinetic energy before.

infrared (IR) Electromagnetic waves with frequencies just below that of visible light. Infrared is the main component of the heat radiation.

interference   The consequence of two waves arriving at the same place and combining. Constructive interference occurs wherever the two waves meet in phase (peak matches peak); the waves add together. Destructive interference occurs wherever the two waves meet out of phase (peak matches valley); the waves cancel each other.

intermediate boson   A force-carrying elementary particle that mediates the fundamental interactions of nature. The photon is an example.

internal energy (U)   The sum of the kinetic energies and potential energies of all the atoms and molecules in a substance.

isotopes   Isotopes of a given element have the same number of protons in the nucleus but different numbers of neutrons.

kinetic energy (KE)   Energy due to motion. Energy that an object has when it is moving.

kinetic friction   Friction between two substances that are in contact and moving relative to each other.

laser   A beam of coherent, monochromatic light. Acronym for "light amplification by stimulated emission of radiation."

law of conservation of angular momentum   The total angular momentum of an isolated system is constant.

law of conservation of energy   Energy cannot be created or destroyed, only converted from one form to another. The total energy in an isolated system is constant.

law of conservation of linear momentum   The total linear momentum of an isolated system is constant.

law of fluid pressure   The (gauge) pressure at any depth in a fluid at rest equals the weight of the fluid in a column extending from that depth to the "top" of the fluid divided by the cross-sectional area of the column.

law of reflection   The angle of incidence equals the angle of reflection.

law of refraction   A light ray is bent towards the normal when it enters a transparent medium in which light travels more slowly. It is bent away from the normal when it enters a medium in which light travels faster.

lepton   An elementary particle that does not interact via the strong nuclear force. Electrons and neutrinos are examples. Leptons are fermions.

linear momentum   The mass of an object times its velocity.

longitudinal wave   Wave in which the oscillations are along the direction the wave travels.

magnetic field   The agent of the magnetic force. It exists in the space around magnets and causes a force to act on each pole of any other magnet.

magnetic resonance imaging (MRI)   Medical imaging process that involves stimulating protons in tissue to emit radio waves, then using these waves to form an image of internal organs.

mass (m)   A measure of an object's resistance to acceleration. A measure of the quantity of matter in an object.

mass density (D)   The mass per unit volume of a substance. The mass of a quantity of a substance divided by the volume it occupies.

meson   A particle that is composed of a quark and an antiquark. Mesons are bosons.

microwaves   Electromagnetic waves with frequencies between those of radio waves and infrared light.

molecule   Two or more atoms combined to form the smallest unit of a chemical compound.

moment of inertia   For a particle, it equals the mass multiplied by the square of the distance from the axis of rotation. For an object, it is the sum of the moments of inertia of each component part.

neutrino   An elementary particle with no charge and little or no mass. Many important processes in astronomy release neutrinos, including supernova explosions and fusion reactions in stars. Neutrinos are fermions.

neutron   Electrically neutral particle residing in the nucleus of an atom.

neutron number (N)   The number of neutrons contained in a nucleus.

Newton's first law of motion   An object will remain at rest or in motion with constant velocity unless acted on by a net external force.

Newton's law of universal gravitation   Every object exerts a gravitational pull on every other object. The force is proportional to the masses of both objects and inversely proportional to the square of the distance between them.

Newton's second law of motion   An object is accelerated whenever a net external force acts on it. The net force equals the object's mass times its acceleration.

Newton's third law of motion   Forces always come in pairs: when one object exerts a force on a second object, the second exerts an equal and opposite force on the first.

nuclear fission   The splitting of a large nucleus into two smaller nuclei. Free neutrons and energy are also released.

nuclear fusion   The combining of two small nuclei to form a larger nucleus.

Ohm's law   The current in a conductor is equal to the voltage applied to it divided by its resistance.

orbit   The path of a body as it moves under the influence of a second body. An example is the path of a planet or a comet as it moves around the sun.

orbital angular momentum   The product of a particle's mass, its speed, and the radius of its path.

Pascal's principle   Pressure applied to an enclosed fluid is transmitted undiminished to all parts of the fluid and to the walls of the container.

Pauli Exclusion Principle   Two interacting fermions of the same type, such as electrons in an atom, cannot be in exactly the same state.

period (T)   The time for one complete cycle of a periodic process.

photon   The quantum of electromagnetic radiation. The energy of a photon equals Planck's constant times the frequency of the radiation.

plasma   An ionized gas, often referred to as the fourth state of matter. It consists of positive ions and electrons.

potential energy (PE)   Energy due to an object's position. Energy that a system has because of its configuration.

power (P)   The rate of doing work. The rate at which energy is transferred or transformed. Work done divided by the time. Energy transferred divided by the time.

pressure (p)   The force per unit area that acts on a surface. The force acting on a surface divided by its area.

proton   Positively charged particle residing in the nucleus of an atom.

quark   A fractionally charged elementary particle found only in combination with other quarks in mesons and baryons. Quarks are fermions.

refraction   The bending of a wave as it enters a different medium.

relative humidity   Humidity expressed as a percentage of the saturation density.

resistance (R)   A measure of the opposition that a substance has to current flow. The unit of measure is the ohm.

scanning tunneling microscope (STM)   Device that uses the tunneling of electrons to form images of individual atoms at the surface of a solid.

second law of thermodynamics   No device can be built that will repeatedly extract heat from a heat source and deliver mechanical work or energy without ejecting some heat to a lower-temperature reservoir.

SI   An internally consistent system of units within the metric system.

sound level (SL)   A quantity used to measure sound amplitude. The unit of measure is the decibel (dB).

speed (v)   Rate of movement. Rate of change of distance from a reference point. The distance that something travels divided by the time elapsed.

spherical aberration   Imperfect focusing of light by concave mirrors or converging lenses that arises when their surfaces have spherical shapes.

spin   Intrinsic angular momentum possesed by subatomic particles. It is quantized in integral or half-integral multiples of $h/2\pi$

spin angular momentum   The moment of inertia of a spinning body multiplied by its angular speed.

strong nuclear force   One of the four fundamental forces in nature. It is the force that holds neutrons and protons together in the nucleus.

superconductor   A substance that has zero electrical resistance. Electrical current can flow in it with no loss of energy to heat.

supernova   An explosion that marks the endpoint of the luminous history of a massive star. Supernova events often accompany the formation of neutron stars or black holes.

temperature (T)   The measure of hotness or coldness. The temperature of matter depends on the average kinetic energy of atoms and molecules.

transverse wave   Wave in which the oscillations are perpendicular (transverse) to the direction the wave travels.

ultraviolet radiation (UV)   Electromagnetic waves with frequencies just above that of visible light.

vacuum   Completely empty space devoid of all matter including air.

velocity   Speed with direction.

voltage (V)   The work that a charged particle can do divided by the size of the charge. The energy per unit charge given to charged particles by a power supply. The unit of measure is the volt.

wave   A traveling disturbance consisting of coordinated vibrations that carry energy with no net movement of matter.

vavelength (λ)   The distance between two successive "like" points on a wave. An example is the distance between two adjacent peaks or two adjacent valleys.

weak nuclear force   One of the four fundamental forces in nature. It is responsible for beta decay.

weight (W)   The force of gravity acting on a substance.

weight density (D_w)   The weight per unit volume of a substance. The weight of a quantity of a substance divided by the volume it occupies.

work   The force that acts times the distance moved in the direction of the force.

x-rays   Electromagnetic waves with frequencies in the range from higher-frequency ultraviolet to low-frequency gamma rays.

INDEX

Reference numbers are to pages. Italic page numbers indicate topic discussion in: *t* tables; *f* figures; and *n* notes.